T0215652

Communications in Computer and Information Science 1138

Commenced Publication in 2007
Founding and Former Series Editors:
Phoebe Chen, Alfredo Cuzzocrea, Xiaoyong Du, Orhun Kara, Ting Liu,
Krishna M. Sivalingam, Dominik Ślęzak, Takashi Washio, Xiaokang Yang,
and Junsong Yuan

Huansheng Ning (Ed.)

Cyberspace Data and Intelligence, and Cyber-Living, Syndrome, and Health

International 2019 Cyberspace Congress, CyberDI and CyberLife
Beijing, China, December 16–18, 2019
Proceedings, Part II

 Springer

Editor
Huansheng Ning ⒾⒹ
University of Science and Technology
Beijing, China

ISSN 1865-0929 ISSN 1865-0937 (electronic)
Communications in Computer and Information Science
ISBN 978-981-15-1924-6 ISBN 978-981-15-1925-3 (eBook)
https://doi.org/10.1007/978-981-15-1925-3

This Springer imprint is published by the registered company Springer Nature Singapore Pte Ltd.
The registered company address is: 152 Beach Road, #21-01/04 Gateway East, Singapore 189721, Singapore

Preface

This volume contains the papers from the 2019 Cyberspace Congress which includes the International Conference on Cyberspace Data and Intelligence (CyberDI 2019) and the International Conference on Cyber-Living, Cyber-Syndrome, and Cyber-Health (CyberLife 2019) held in Beijing, China, during December 16–18, 2019.

The explosion of smart IoT and artificial intelligence has driven cyberspace entering into a new prosperous stage, and cyberspace in turn cultivates novel and interesting domains such as cyberspace data and intelligence, cyber-living, and cyber-life. Therefore, CyberDI 2019 and CyberLife 2019 were organized to explore cutting-edge advances and technologies pertaining to the key issues of data and intelligence, cyber syndrome, and health in cyberspace. The aim of the conference was to bring together researchers, scientists, as well as scholars and engineers from all over the world to exchange brilliant ideas, and the invited keynote talks and oral paper presentations contributed greatly to broadening horizons and igniting young minds.

Generally speaking, cyberspace attracts more and more attention serving as an emerging and significantly profound research area. However, some challenges still exist in the comprehensive understanding of basic theories, philosophy, and science. In CyberDI 2019 and CyberLife 2019, there were four main topics which the papers addressed, and the authors focused on the advanced theories, concepts, and technologies relating to data and intelligence, cyber-syndrome, and cyber-living in order to fully mine and dig out the hidden values in cyberspace.

CyberDI 2019 discussed intelligent ways of making full use of data and information, which are regarded as the foundation and core of cyberspace. It focused on data communication and computing, as well as knowledge management with different intelligent algorithms such as deep learning, neural networks, knowledge graphs, etc.

CyberLife 2019 focused on health challenges faced in existing cyberspace, and provided an interactive platform for exploring cyber-syndrome, cyber-diagnosis, as well as the way to healthy living in cyberspace.

In order to ensure the high quality of both conferences, we followed a rigorous review process in CyberDI 2019 and CyberLife 2019. CyberDI 2019 and CyberLife 2019 received 160 qualified submissions, and nearly 64 regular papers and 18 posters were accepted. All manuscripts were reviewed by peer-reviewers in a single-blind review process, and each paper had three reviewers on average chosen by the Program Committee members considering their qualifications and experience.

The proceeding editors wish to thank the dedicated Conference Committee members and all the other reviewers for their contributions. Sincerely, we hope that these proceedings will help a lot for interested readers, and we also thank Springer for their trust and support in publishing the proceeding of CyberDI 2019 and CyberLife 2019.

October 2019

Huansheng Ning

Organization

CyberDI 2019

Scientific Committee

Rongxing Lu	University of New Brunswick, Canada
J. Christopher Westland	University of Illinois – Chicago, USA
Cuneyt Gurcan Akcora	University of Texas at Dallas, USA
Fu Chen	Central University of Finance and Economics, China
Giancarlo Fortino	University of Calabria, Italy
Guangjie Han	Hohai University, China
Richard Hill	University of Huddersfield, UK
Farhan Ahmad	University of Derby, UK
Jin Guo	University of Science and Technology Beijing, China
Octavio Loyola-González	Tecnologico de Monterrey, Mexico
Constantinos Patsakis	University of Piraeus, Greece
Lianyong Qi	Qufu Normal University, China
Tie Qiu	Tianjin University, China
Xiang Wang	Beihang University, China
Qi Zhang	IBM T J Watson, USA
Chunsheng Zhu	Southern University of Science and Technology, China
Asma Adnane	Loughborough University, UK
Humaira Ashraf	International Islamic University Islamabad, Pakistan
James Bailey	The University of Melbourne, Australia
Ari Barrera-Animas	Tecnologico de Monterrey, Mexico
Winnie Bello	University of Huddersfield, UK
Milton Garcia Borroto	Instituto Superior Politécnico José Antonio Echeverría, Cuba
Fatih Kurugollu	University of Derby, UK
Manuel Lazo	Tecnologico de Atizapan, Mexico
Satya Shah	University of Bolton, UK
Yong Xue	University of Derby, UK
Danilo Valdes Ramirez	Tecnologico de Monterrey, Mexico
Ahmad Waqas	Beijing Normal University, China
Reza Montasari	University of Huddersfield, UK
Majdi Mafarja	Birzeit University, Palestine
Abdelkarim Ben Sada	University of Science and Technology Beijing, China
Mohammed Amine Bouras	University of Science and Technology Beijing, China
Lázaro Bustio-Martínez	National Institute of Astrophysics, Optics and Electronics, Mexico
Yaile Caballero	Universidad de Camaguey, Cuba

Barbara Cervantes   Tecnologico de Monterrey, Mexico
Leonardo Chang   Tecnologico de Monterrey, Mexico
Rania El-Gazzar   University South-Eastern, Norway
Hugo Escalante   National Institute of Astrophysics,
                                     Optics and Electronics, Mexico
Fadi Farha   University of Science and Technology Beijing, China
Virginia Franqueira   University of Kent, UK
Luciano García   Tecnologico de Monterrey, Mexico
Mario Graff   CONACYT Researcher – INFOTEC, Mexico
Andres Gutierrez   Tecnologico de Monterrey, Mexico
Raudel Hernández-León   Advanced Technologies Application Center, Cuba
Amin Hosseinian-Far   University of Northampton, UK
Rasheed Hussain   Innopolis University, Russia
Chaker Abdelaziz Kerrache   University of Ghardaia, Algeria
Mehmet Kiraz   De Montfort University, UK
Fatih Kurugollu   University of Derby, UK
Noureddine Lakouari   Instituto Nacional de Astrofísica, Óptica y Electrónica
                                     (INAOE), Mexico
Ismael Lin   Tecnologico de Monterrey, Mexico
Adrian Lopez   CIMAT, Mexico
Arun Malik   Lovely Professional University, UK
Diana Martín   CUJAE, Cuba
Safdar Nawaz Khan Marwat   University of Engineering & Technology, Peshawar,
                                     Pakistan
Arquimides Mendez Molina   Instituto Nacional de Astrofísica, Óptica y Electrónica
                                     (INAOE), Mexico
Senthilkumar Mohan   VIT University, India
Raúl Monroy   Tecnologico de Monterrey, Mexico
Abdenacer Naouri   University of Science Technology Beijing, China
Alberto Oliart   Tecnologico de Monterrey, Mexico
Felipe Orihuela-Espina   INAOE, Mexico
Nisha Panwar   University of California, Irvine, USA
Simon Parkinson   University of Huddersfield, UK
Luis Pellegrin   Universidad Autónoma de Baja California (UABC),
                                     Mexico
Airel Perez   CENATAV, Cuba
Claudia Pérez   Tecnologico de Monterrey, Mexico
Julio César Pérez   INAOE - Cátedra CONACyT, Mexico
  Sansalvador
Laura Pinilla-Buitrago   Instituto Nacional de Astrofísica, Óptica y Electrónica
                                     (INAOE), Mexico
Muthu Ramachandran   Leeds Beckett University, UK
Kelsey Ramírez Gutiérrez   Instituto Nacional de Astrofísica, Óptica y Electrónica
                                     (INAOE), Mexico
Praveen Kumar Reddy   Vellore, Tamil Nadu, Mexico
  Maddikunta

Jorge Rodriguez	Tecnologico de Monterrey, Mexico
Cosme Santiesteban-Toca	Centro de Bioplantas, Cuba
Dharmendra Shadija	Sheffield Hallam University, UK
Nasir Shahzad	University of Leicester, UK

Organizing Committee

Kim-Kwang Raymond Choo	The University of Texas at San Antonio, USA
Ravi Sandhu	University of Texas at San Antonio, USA
Weishan Zhang	China University of Petroleum, China
ZhangBing Zhou	China University of Geosciences (Beijing), China
Huansheng Ning	University of Science and Technology Beijing, China
Qinghua Lu	CSIRO, Australia
Shaohua Wan	Zhongnan University of Economics and Law, China
Bradley Glisson	Sam Houston State University, USA
Haijing Hao	Bentley University, USA
Pengfei Hu	China Mobile Research Institute, China
Miguel Angel Medina Perez	Tecnologico de Monterrey, Mexico
Ata Ullah	National University of Modern Languages, Pakistan
Liming Chen	De Montfort University, UK
Mahmoud Daneshmand	Stevens Institute of Technology, USA
Keping Long	University of Science and Technology Beijing, China
Chunming Rong	University of Stavanger, Norway
Chonggang Wang	InterDigital, USA

CyberLife 2019

Scientific Committee

Mariwan Ahmad	IBM, UK
Karim Mualla	University of Leicester, UK
Yudong Zhang	University of Leicester, UK
Muhammad Ajmal Azad	University of Derby, UK
Hussain Al-Aqrabi	University of Huddersfield, UK
Junaid Arshad	University of West London, UK
Liangxiu Han	Manchester Metropolitan University, UK
Yong Hu	The University of Hong Kong, China
Chenxi Huang	Nanyang Technological University, Singapore
Jianxin Li	Beihang University, China
Liying Wang	Nanjing Normal University, China
Kaijian Xia	China University of Mining and Technology; Changshu No. 1 People' Hospital, China
Likun Xia	Capital Normal University, China
Xiaojun Zhai	University of Essex, UK
Meifang Zhang	University of Science and Technology Beijing, China

Rongbo Zhu	South Central University for Nationalities, China
Nasser Alaraje	Michigan Technological University, USA
Abbes Amira	De Montfort University, UK
Kofi Appiah	Nottingham Trent University, UK
Hamza Djelouat	University of Oulu, Finland
Klaus D. McDonald-Maier	University of Essex, UK
Yanghong Tan	Hunan University, China
Bin Ye	Queens University of Belfast, UK
Wangyang Yu	Shaanxi Normal University, P.R. China
Bo Yuan	University of Derby & Data Science Research Center of University of Derby, UK
Yongjun Zheng	University of West London, UK
Zhifeng Zhong	Hubei University, China
Xingzhen Bai	Tongji University, P.R. China
Kieren Egan	Department of Computer and Information Sciences, University of Strathclyde Glasgow, UK
Preetha Phillips	West Virginia School of Osteopathic Medicine, USA
Pengjiang Qian	Jiangnan University, China
Junding Sun	Henan Polytechnic University, China
Xiaosong Yang	National Center for Computer Animation, Bournemouth University, UK
Maher Assaad	College of Engineering, Ajman University, United Arab Emirates
Yuan Yuan Chen	Beijing Fistar Technology Co., Ltd., China
Ming Ma	School of Medicine, University of Stanford, USA
Wajid Mumtaz	Department of Computer Science, Faculty of Applied Sciences, University of West Bohemia, Czech Republic
Li Tan	School of Computer and Information Engineering, Beijing Technology and Business University, China
Kedi Xu	Qiushi Academy for Advanced Studies, Zhejiang University, China
Susu Yao	Institute for Infocomm Research, A*STAR, Singapore
Da Zhang	College of Information Engineering, Capital Normal University, China
Shaomin Zhang	Qiushi Academy for Advanced Studies, Zhejiang University, China
Yizhang Jiang	Jiangnan University, China
Jin Yong	Changshu Institute of Technology, China
Vishnu Varthanan Govindaraj	Kalasalingam Academy of Research and Education, India
Li Yuexin	Hubei University, China
Shan Zhong	Changshu Institute of Technology, China
Haider Ali	University of Derby, UK
Nyothiri Aung	University of Science and Technology Beijing, China
Ra'ed Bani Abdelrahman	Loughborough University, UK

Luis Castro	Instituto Tecnológico de Sonora, Mexico
Yuanyuan Chen	Beijing Fistar Technology Co., Ltd., China
Federico Cruciani	Ulster University, UK
Shoaib Ehsan	University of Essex, UK
Mark Fox	University of Toronto, Canada
James Hardy	University of Derby, UK
Anju Johnson	University of Huddersfield, UK
Junhua Li	University of Essex, UK
Wenjia Li	New York Institute of Technology, USA
Zhi Li	Guangxi Normal University, China
Feng Mao	Walmart Labs, USA
Fanlin Meng	University of Essex, UK
Kun Niu	University of Posts and Communications, China
Doris Oesterreicher	University of Natural Resources and Life Sciences, Vienna, Austria
John Panneerselvam	University of Derby, UK
Sangeet Saha	University of Essex, UK
Minglai Shao	Beihang University, China
Dhiraj Srivastava	Indian Institute of Technology, India
Jinya Su	Loughborough University, UK
Peipei Sui	Shandong Normal University, China
Shuai Zhang	Ulster University, UK

Organizing Committee

Liming Chen	De Montfort University, UK
Huansheng Ning	University of Science and Technology Beijing, China
Chunming Rong	University of Stavanger, Norway
Chonggang Wang	InterDigital, USA
Changjun Jiang	Donghua University, China
Lu Liu	University of Leicester, UK
Faycal Bensaali	Qatar University, Qatar
Zhengchao Dong	Columbia University, USA
Colin Johnson	University of Kent, UK
Tan Jen Hong	National University of Singapore, Singapore
Rusdi Abd Rashid	University of Malaya Center for Addiction Sciences (UMCAS), Malaysia
Wenbing Zhao	Cleveland State University, USA
John Panneerselvam	University of Derby, UK
Po Yang	University of Sheffield, UK

Local Committee for CyberDI 2019 and CyberLife 2019

Huansheng Ning	University of Science and Technology Beijing, China
Rui Wang	University of Science and Technology Beijing, China
Bing Du	University of Science and Technology Beijing, China
Sahraoui Dhelim	University of Science and Technology Beijing, China
Qingjuan Li	University of Science and Technology Beijing, China
Feifei Shi	University of Science and Technology Beijing, China
Zhong Zhen	University of Science and Technology Beijing, China
Xiaozhen Ye	University of Science and Technology Beijing, China
Dawei Wei	University of Science and Technology Beijing, China
Yang Xu	University of Science and Technology Beijing, China

Contents – Part II

Communication and Computing

CyberLife 2019: Cyber Philosophy, Cyberlogic and Cyber Science

CyberLife 2019: Cyber Health and Smart Healthcare

Contents – Part I

CyberDI 2019: Cyber and Cyber-Enabled Intelligence

Communication and Computing

A Markov Approximation Algorithm for Computation Offloading and Resource Scheduling in Mobile Edge Computing

Haowei Chen[1], Mengran Liu[1], Yunpeng Wang[1], Weiwei Fang[1(✉)], and Yi Ding[2]

[1] School of Computer and Information Technology,
Beijing Jiaotong University, Beijing 100044, China
wwfang@bjtu.edu.cn
[2] School of Information, Beijing Wuzi University, Beijing 101149, China

Abstract. Mobile edge computing has become a key technology in IoT and 5G networks, which provides cloud-computing services in the edge of the mobile access network to realize the flexible use of computing and storage resources. While most existing research focuses on network optimization in small-scale scenes, this paper jointly considers the resources scheduling of servers, channels and powers for mobile users to minimize the system energy consumption. It's an NP-hard problem which can only be solved through the exhaustive search with complexity of exponential level. The lightweight distributed algorithm proposed in this paper based on the Markov approximation framework can make the system converge to an approximate optimal solution with only linear level complexity. The simulation results show that the proposed algorithm is able to generate near-optimal solutions and outperform other benchmark algorithms.

Keywords: Mobile edge computing · Resource scheduling · Markov approximation · Computation offloading

1 Introduction

With the rapid development of the Internet and the continuous innovation of computer science and technology, mobile devices' demands for network communication and computing resources are on a higher level. The commonly used Mobile Cloud Computing (MCC) technology which can extend the capabilities of mobile devices in computing, communication, and caching by exchanging data from remotely centralized clouds over the Internet and the carrier's core network, could overcome the above problems to a certain extent. But at the same time, MCC will impose a huge traffic load on mobile networks and bring high communication delays over long distances between mobile devices and public clouds. As a result, MCC technology cannot meet the user needs of delay-driven and compute-intensive applications such as augmented reality, interactive online gaming and autopilot.

To solve the above problems better, Mobile Edge Computing (MEC) technology emerges: mobile devices offload high-energy, high-complexity computing tasks to MEC

© Springer Nature Singapore Pte Ltd. 2019
H. Ning (Ed.): CyberDI 2019/CyberLife 2019, CCIS 1138, pp. 3–20, 2019.
https://doi.org/10.1007/978-981-15-1925-3_1

servers deployed on nearby base stations to obtain close-range, low latency IT environments and cloud computing services according to [1]. MEC technology efficiently combines cloud computing and mobile networks to "sink" the data and service functions of traditional centralized cloud computing platforms to the edge of the access network. It re-integrates computing, communication and storage resource allocation methods at the edge of mobile networks. The use of the distributed deployment of MEC servers can reduce the latency of transmissions and services, thereby improving the client service experience.

However, with the in-depth development of MEC technology, the following problems have also arisen in the research of mobile edge computing:

(1) Offloading decision design is complicated [2]. Computation offloading is an indispensable part of MEC, and the key is how to design an offloading decision. If the computation task is separable, should it be offloaded to a single MEC server or multiple different servers? If the computation task is inseparable, is the computing task selected to be executed locally or offloaded to the MEC server for execution? Therefore, the design of the offloading decision framework of MEC system is of great significance for the research of MEC resource allocation.

(2) Multi-resource joint optimization is difficult [3]. In the process of the mobile device offloading some or all of the computation tasks to the MEC server, the involved resources include the communication resources and the computing resources. There is a strong correlation between various resources which makes the properly arrangement between various resources is a difficult problem to be solved.

(3) Optimization goal is difficult to determine [2, 4]. The ultimate goal of MEC system optimization is to provide users with low-latency and low-energy edge services. The general optimization goals include reducing latency, reducing energy consumption, and balancing power consumption and latency. The choice of optimization goals depends on the specific MEC system. That is to say, the optimization problem needs to be determined according to different scene requirements and system modeling.

In recent years, the focus of many studies in the MEC system is still at the single server level. Zhao et al. studied the optimization of joint radio resources and computing resources of multiple users. The heuristic strategy based on delay calculation aims to reduce the energy consumption of mobile devices [5]. Although Zhou et al. performed a collaborative optimization of network resources in a multi-user scenario, with exhaustive search and random method as baseline approaches, it was not considered that the mutual interference resulted from preempting the same channel when the task was offloaded [6].

This paper is aimed at new problems and new challenges in mobile edge computing. In multi-mobile devices, multiple MEC servers, and multi-channel large-scale scenarios, the exploration and research are carried out to keep low energy consumption and low latency as optimization goals. The distributed algorithm is designed by Markov approximation framework to realize multi-resource joint optimization, so as to promote the development and practical application of MEC technology.

In summary, the main contributions made in this paper are the following three points, which correspond to the three major issues raised above:

(1) We first establish a model of energy consumption minimization which satisfy users' quality of experience, then design the task offloading decision based on Markov approximation algorithm, and finally achieve convergence to approximate optimal solution in a short time, and obtain more efficient system computing resource scheduling.

(2) For the multi-resource joint optimization problem, the original problem is transformed into the approximation problem of minimum weight configuration by Log-Sum-Exp function. A lightweight distributed algorithm is proposed based on Markov approximation framework. And we analyze the algorithm from three aspects: feasibility, stability and optimality by adjusting the parameters.

(3) We improve the scale complexity of the MEC system, considering the coupling constraints between various resources in the edge computing, and collaboratively optimizing the communication and computing resources in this scenario to ensure the users' quality of experience under the delay-sensitive service.

The rest of this paper is organized as follows. Section 2 presents the system model and the problem definition. Then, the proposed Markov approximation-based algorithm is introduced in Sect. 3. Section 4 demonstrates the simulation results. Finally, Sect. 5 summarizes the conclusions and outlines future work.

2 System Model and Problem Formulation

2.1 System Model

The scene diagram studied in this paper is shown in Fig. 1. Assume within the designated area, there are M MEC servers $\{1, 2, \ldots, M\}$, N mobile devices $\{1, 2, \ldots, N\}$ and S channels $\{1, 2, \ldots, S\}$. The mobile device offloads the task to any of the edge servers M through the channel. This paper is based on the following assumptions: ① The task of any mobile device can only be connected to one edge server through one channel. ② The mobile device must transfer all information about the task being offloaded to the connected server. ③ The task for each mobile device must be processed only by the edge server. ④ For each edge server and wireless channel, it may or may not be occupied. ⑤ The wireless channels are orthogonal to each other, so there is no noise interference between the different channels. ⑥ To prevent the collision, there is at most one device that can connect to the same server through the same channel.

2.2 Problem Formulation

We define the binary variable $a_{nm}^s \in \{0, 1\}$ to indicate whether the mobile device n is connected to the edge server m through the channels:

$$a_{nm}^s = \begin{cases} 1, & \text{the device } n \text{ connects to the edge server } m \text{ through the channel } s \\ 0, & \text{otherwise} \end{cases} \tag{1}$$

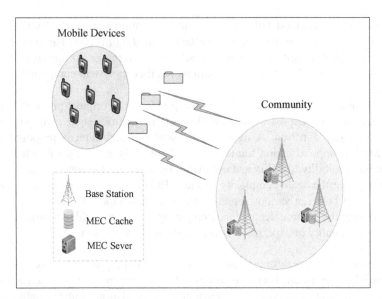

Fig. 1. Multi-channel multi-user device scene graph of multiple MEC servers

According to assumption ①③⑥, we get the constraint:

$$\forall n \in N, s \in S, m \in M \begin{cases} \sum_m \sum_s a_{nm}^s = 1 \\ \sum_n a_{nm}^s \leq 1 \end{cases} \tag{2}$$

In real life, the task is transmitted from the mobile device to the edge server for computing and then returned to the user in three phases: offloading, computing, and return. In order to satisfy the user's Quality of Experience (QoE), the total duration of the MEC cannot exceed the maximum delay for each mobile device according to [7]. According to the existing literature, the delay of the edge server m returning the computation result to the device n is so short that we can ignore it for simplifying the problem:

$$T_n^{offlo} + T_n^{compu} \leq L_n \tag{3}$$

where T_n^{offlo}, T_n^{compu} respectively represent the time of offloading and computing between the mobile device n and the edge server m to which it is connected.

We analyzed the total time overhead of the process in two steps.

(1) Task offloading phase. The task can be offloaded from any mobile device to any edge server via any channel. The distance between them is d_{nm}. We assume the channel passed is Rayleigh fading channel during the uplink transmission where gain is marked as $g_{nm}^s = d_{nm}^{-2}$. For each edge server, the status information of the channel is globally known. Based on the assumption ⑤, we know there is no noise between different channels, but there is mutual interference caused by resource

preemption for the same channel. According to the Shannon capacity formula, the upload rate of a mobile device can be expressed as:

$$R_{nm}^s = B_s \log_2 \left(1 + \frac{P_n^s g_{nm}^s}{\sigma^2 + \sum_{m' \neq m} \sum_{n' \in N_{m'}} a_{n'm'}^s P_n^s g_{nm}^s} \right) \tag{4}$$

$$R_{nm} = \sum_s a_{nm}^s R_{nm}^s \tag{5}$$

Where σ^2 represents the additive white Gaussian noise power. B_s is the bandwidth of channel and $N_{m'}$ is the set of mobile devices which are connected to edge server m'. P_n^s represents the adjustable power which ranges from P^{min} to P^{max} in this paper.

For the mobile device n, we assume the number of its task bits is D_n. Hence, its task offloading delay is:

$$T_n^{offlo} = \sum_s \sum_m \frac{D_n a_{nm}^s}{R_{nm}} \tag{6}$$

(2) Task computing phase. After the device n offloads its task to the edge server through the channel, it is assumed that the number of commands to be executed is $W_n = \mu D_n$, where μ is the command coefficient. With computing task offloaded, each edge server uses the resource sharing policy to execute the instructions. Assuming that the instruction execution rate of the server m is F_m (number of instructions/sec), the allocated computing resource for the device n connected to the server is $f_{nm} = \frac{F_m}{\sum_s \sum_n a_{nm}^s}$, whose task calculation delay is:

$$T_n^{compu} = \sum_s \sum_m \frac{W_n a_{nm}^s}{f_{nm}} \tag{7}$$

We substitute formulas (6) and (7) into formula (3), then get the MEC delay of device n should satisfies the following constraint:

$$\sum_s \sum_m \frac{W_n a_{nm}^s}{f_{nm}} + \sum_s \sum_m \frac{D_n a_{nm}^s}{R_{nm}} \leq L_n \tag{8}$$

Obviously, the energy consumption of all mobile devices in the system is $\sum_m \sum_n \sum_s P_n^s a_{nm}^s$. Our goal is to properly allocate multi-server multi-channel computing resources to achieve minimum system energy consumption while satisfying (8). Our model can be expressed as:

$$\begin{aligned}
\min \quad & \sum_m \sum_n \sum_s P_n^s a_{nm}^s \\
s.t. \quad & \sum_m \sum_s a_{nm}^s = 1, m \in M, n \in M, s \in S \\
& \sum_n a_{nm}^s \leq 1, m \in M, n \in N, s \in S \\
& T_n^{trans} + T_n^{proc} \leq L_n, n \in N \\
& P_n^s \in P, n \in N, s \in S \\
& a_{nm}^s \in \{0, 1\}, m \in M, n \in N, s \in S
\end{aligned} \tag{9}$$

This problem is an NP-hard problem of mixed-integer nonlinear programming. The number of connection schemes between device, channel, and edge server is exponentially related to the number of the three. In order to find a global optimal solution, it is necessary to traverse each connection scheme, and the adjustable power set P for each device in every connection scheme on the basis of satisfying the user's QoE. However, there are cases of mutual interference between devices. To solve this combinatorial optimization problem, the optimal solution can only be obtained by exhaustive search [8], but its complexity is exponential, making it is not suitable in actual situations. Therefore, this paper proposes a distributed algorithm based on the Markov approximation framework, which can obtain an approximate optimal solution of the problem in a short time.

3 Markov Approximation and Algorithm Design

Based on the existing Markov approximation framework [8], we construct a distributed algorithm to solve this combinatorial optimization problem. In this section, we first transform the NP-hard problem into the minimum weight configuration problem by using Log-Sum-Exp function as [8] does. Then we design a specific Markov chain where we approximate the optimal solution by a state transition. Finally, we present the specific steps of the distributed algorithm.

3.1 Log-Sum-Exp Function

We assume that $f = \left\{ a_{nm}^s, P_n^s | m \in M, n \in N, s \in S \right\} \in F$ is one of the feasible solutions to this problem where F is the solution set that satisfies all user's QoE constraints. Next, we converted (9) to its equivalent Minimum Weight Independent Set (MWIS) problem:

$$\min_{\rho \geq 0} \sum_{f \in F} \rho_f \sum_{n \in N} P_n(f)$$
$$s.t. \sum_{f \in F} \rho_f = 1 \tag{10}$$

Where p_f is the percentage of time that system stays in f and $\sum_{n \in N} P_n(f)$ is the weight of f. Hence, our goal is to find the minimum weighted configuration which is the core of the MWIS problem.

This problem is a combinational optimal question which can only be accurately optimized by exhaustive search. However, the size of the feasible solution set F increases exponentially with the increase in the number of devices, which makes it necessary to solve the problem through a large amount of computation. Here, we introduce the Log-Sum-Exp function.

In Log-Sum-Exp function, there is a relationship for positive β and n non-negative real variables:

$$\min(z_1, z_2, \ldots, z_n) - \frac{1}{\beta} \log n \leq -\frac{1}{\beta} \log\left(\sum_{i=1}^{n} \exp(-\beta z_i)\right) \leq \min(z_1, z_2, \cdots, z_n)$$

$$(11)$$

When $\beta \to \infty$, we can get the following equation from (11):

$$-\frac{1}{\beta} \log\left(\sum_{i=1}^{n} \exp(-\beta z_i)\right) = \min(z_1, z_2, \ldots, z_n) \tag{12}$$

Let the n non-negative real variables $z_1, z_2, \ldots, z_n$ correspond to the |F| weights of: $\sum_{i \in N} P_i(f), f \in F$:

$$z_i = \sum_{i \in N} P_i(f), f \in F, i = 1, \cdots, N \tag{13}$$

This gives the following relationship:

$$\min_{f \in F} \sum_{i \in N} P_i(f) \approx -\frac{1}{\beta} \log\left(\sum_{i=1}^{n} \exp(-\beta z_i)\right) \tag{14}$$

We can convert the (8) approximation into:

$$\min_{\rho \geq 0} \sum_{f \in F} \rho_f \sum_{n \in N} P_n(f) + \frac{1}{\beta} \sum_{f \in F} \rho_f \log(\rho_f)$$
$$s.t. \sum_{f \in F} \rho_f = 1 \tag{15}$$

(15) is quadratic, monotonically increasing for all ρ_f, and is a strict convex function whose all constraints are linear. The optimal solution of the above formula can be solved based on the Karush-Kuhn-Tucker (KKT) condition:

$$\rho_f^* = \frac{\exp\left(-\beta \sum_{i \in N} P_i(f)\right)}{\sum_{f \in F} \exp\left(-\beta \sum_{i \in N} P(f')\right)}, f \in F \tag{16}$$

Where ρ_f^* is the stationary distribution of the Markov chain which will be used in the next section. We can find that there is an optimization gap in the approximate solution relative to the original problem (9):

$$r^* = \max \frac{1}{\beta} \sum_{f \in F} \rho_f \log(\rho_f) \tag{17}$$

Absolutely, there is a negative correlation between β and r^* As the value of β increases, we get a relatively more accurate approximate optimal solution. Through the Markov approximation, the approximate optimal solution can be expressed as:

$$\sum_{f \in F} \rho_f^* \sum_{n \in N} P_n(f) \tag{18}$$

3.2 Markov Chain Design

The core idea to solve the above problem is to construct a specific Markov chain by combining the state space F of the feasible solution and the stationary distribution $\rho_f^*(x)$. The states in the Markov chain and the jump probability between each state are unique to this scene. Next, we obtain the approximate optimal solution based on a distributed algorithm. As the Markov chain tends to converge, the percentage of time that the system stays in the approximate optimal configuration will be significantly higher than in other states.

The literature has proved that for any probability distribution satisfying ρ_f^*, there is at least one steady distribution of the Markov chain with continuous time-reversible traversal. To construct this specific Markov chain, we will work with $f, f' \in F$ as the state space, and $q_{f,f'}$ representing the transfer rate from the state f to the state f'. And the state space meets the following two conditions according to [9]:

(1) Any two states are reachable and reversible.
(2) The following balance equation is satisfied.

$$\rho_f q_{f,f'} = \rho_{f'} q_{f',f}$$
$$\exp\left(-\beta \sum_{i \in N} P_i(f)\right) q_{f,f'} = \exp\left(-\beta \sum_{i \in N} P_i(f')\right) q_{f',f} \tag{19}$$

There are a lot of choices for transfer rate design. According to literature [10], we consider the following condition:

$$q_{f,f'} + q_{f',f} = \exp(-\tau) \tag{20}$$

Where τ is a positive constant. We can obtain the following expression, by combining (19) and (20).

$$q_{f,f'} = \frac{\exp(-\tau)}{1 + \exp\left(-\beta\left(\sum_{i \in N} P_i(f') - \sum_{i \in N} P_i(f)\right)\right)} \tag{21}$$

According to the symmetry, the transition rate $q_{f',f}$ of the state f' to the state f can be obtained:

$$q_{f,f'} = \frac{\exp(-\tau)}{1 + \exp\left(-\beta\left(\sum_{i \in N} P_i(f') - \sum_{i \in N} P_i(f)\right)\right)} \tag{22}$$

The above equation is a logical function of system power and has a closed expression represented by $\rho_f^*(x)$ and $\rho_{f'}^*(x)$:

$$q_{f,f'} = \exp(-\tau) \frac{\rho_{f'}^*}{\left(\rho_{f'}^* + \rho_f^*\right)}$$

$$q_{f',f} = \exp(-\tau)\frac{\rho_f^*}{\left(\rho_{f'}^* + \rho_f^*\right)} \tag{23}$$

In the Markov chain, by adjusting the value of β, $f \to f'$ has the following three cases:

(1) $\sum\limits_{i \in N} P_i(f) < \sum\limits_{i \in N} P_i(f')$: the probability that the system stays in configuration f
$q_{f',f} \approx 1$.

(2) $\sum\limits_{i \in N} P_i(f) > \sum\limits_{i \in N} P_i(f')$: the probability that the system jumps to configuration
f' $q_{f,f'} \approx 1$.

(3) $\sum\limits_{i \in N} P_i(f) = \sum\limits_{i \in N} P_i(f')$: the probability that the system stays in configuration f or
jumps to configuration f' are equal $q_{f,f'} = q_{f',f} = \frac{1}{2}$.

3.3 Markov Approximation Algorithm

The Markov approximation algorithm is a distributed algorithm that minimizes the system power by establishing a link $f \leftrightarrow f'$. For the multi-device multi-channel multi-server and multi-adjustable power scheme system in this scenario, we assign an exponentially distributed random number to each mobile device as the countdown time. At the beginning of each iteration, we start counting down. If any device completes the countdown, we change the state of the device (which is counted down first) at a time to realize the jump from configuration f to f' and achieve the traversal of the Markov chain.

In our design, the algorithm is given in Algorithm 1. First, we initialize the mobile device delay constraint, the number of task bits, and computing resources of each edge server. Based on the anti-collision assumption, each mobile device makes a random selection for channel and edge server to establish a connection. In each iteration, we assign each device an exponential distribution random reciprocal time with a mean of $\frac{2\exp(\tau)}{(|MS|-|N|+1)|P|-1}$, where M, S, and P are the number of elements of the edge server, the channel, and the adjustable power set.

Algorithm 1. The Markov approximation algorithm

Algorithm 1 The MA Algorithm

Initialization:

1: **for** each edge server $m \in M$ do

2:     Randomly select geographic coordinates, and there can be at most one server on one coordinate point;

3:     Randomly select the execution rate from F_m;

4: **end for**

5: **for** each device $n \in N$ do

6:     Generate an exponentially distributed random number;

7:     Randomly select geographic coordinates, and there can be at most one server or device on one coordinate point;

8:     Assign task data volume D_n and $W_n \leftarrow \mu D_n$;

9:     Choose a feasible connection scheme from F which satisfies the user's QoE. And update f_{nm} and R_{nm};

10: **end for**

Procedure:

11: **for** each device $n \in N$ do

12: Begin count down by the random number;

13: **If** the random number of any device expires **then**

14:     The device enters choice (i) or enters choice (ii) with a certain probability;

15: **If** the device chooses (i) **then**

16:     The device n chooses to switch the new connection state, and the device will randomly switch to a new channel or a new edge server and re-adjust the transmission power.;

17: **end if**

18: **If** the device chooses (ii) **then**

19:     The device n only adjusts the transmit power and does not change the status;

20: **end if**

21: Each device recalculates the delay.

22: **If** there is a device that is not satisfied, the device n goes back to the configuration f. **Otherwise**, the device stays in the new configuration by the probability of $q_{ff'}$ or goes back by the probability of $1 - q_{f,f'}$;

23: Each user generates a new exponentially distributed random number;

24: **end if**

25: end for

26: **If** the iteration number reaches I, the algorithm terminates. **Otherwise**, return to step 11;

At the meantime, each device calculates its power in the current state f and broadcasts it to make it globally known and starts counting down. The device n, which is terminated first, has two choices with different probabilities to update:

(1) Device n enters here with the probability of $\frac{|MSP - NP|}{(|MS| - |N| + 1)|P| - 1}$. It randomly choosing a new channel or edge server and adjusts the transmission power.

(2) Device n enters here with the probability of $\frac{|P| - 1}{(|MS| - |N| + 1)|P| - 1}$. It only adjusts the transmission power and does not change the connection state.

After that, the device n broadcasts it to other devices to stop counting down. Each device recalculates the delay and determines whether the QoE constraint is met. If there is any device that is not satisfied, the device n jumps back to the configuration f. We denote the new status as configuration f', and calculate the transmit power in this new state. Based on it, we choose to stay in the new state f' with probability $\frac{\exp(-\beta \sum_{i \in N} P_i(f'))}{\exp(-\beta \sum_{i \in N} P_i(f'))+\exp(-\beta \sum_{i \in N} P_i(f))}$ or jump back to the old state f with the probability of $1 - \frac{\exp(-\beta \sum_{i \in N} P_i(f'))}{\exp(-\beta \sum_{i \in N} P_i(f'))+\exp(-\beta \sum_{i \in N} P_i(f))}$. We repeat this process until the number of iterations is reached and the Markov chain converges.

Obviously, the algorithm is a distributed algorithm running on each independent device, and the connection state and transmission power are adjusted to make the system stay in a better configuration longer. In the Markov chain, the device makes decisions in the old and new states, if $\sum_{i \in N} P_i(f) > \sum_{i \in N} P_i(f')$, then the system will have a greater probability of $\frac{\exp(-\beta \sum_{i \in N} P_i(f'))}{\exp(-\beta \sum_{i \in N} P_i(f'))+\exp(-\beta \sum_{i \in N} P_i(f))}$ to jump to the new state f'; otherwise, the system will be more likely to stay in the old state f. Among them, the jump probability is closely related to the value of β. The larger the value of β, the faster the system will converge to the expected steady distribution.

3.4 Algorithm Performance Analysis

Theorem 1. In the Markov chain, the transition rate from configuration f to configuration f' is $\frac{\exp(-\tau)}{1+\exp(-\beta(\sum_{i \in N} P_i(f)-\sum_{i \in N} P_i(f')))}$.

Proof 1. In our Markov approximation algorithm, the system only jumps between two directly connected states. Denote the current state as f, where all devices obtain an exponential distribution random reciprocal time with a mean of $\frac{2\exp(\tau)}{(|MS|-|N|+1)|P|-1}$ and start countdown at the same time. Assuming that the device n counts down first, then the jump $f \to f'$ is divided into two cases: go to choice (i) and go to choice (ii). Hence, the following is divided into two parts:

(i) The system completes the state jump by choice (i):

$$\Pr\left(f \xrightarrow{(i)} f'\right) = \Pr(n \text{ transmits } | n's \text{ timer expires}) \Pr(n's \text{ timer expires})$$

$$= \frac{|MSP - NP|}{(|MS| - |N| + 1)|P| - 1} \cdot \frac{1}{|MSP - NP|}$$

$$\cdot \frac{\exp(-\beta \sum_{i \in N} P_i(f'))}{\exp(-\beta \sum_{i \in N} P_i(f')) + \exp(-\beta \sum_{i \in N} P_i(f))} \cdot \frac{\frac{(|MS|-|N|+1)|P|-1}{\exp(\tau)}}{\sum_{n \in N} \frac{(|MS|-|N|+1)|P|-1}{\exp(\tau)}}$$

$$= \frac{1}{\sum_{n \in N} (|MS| - |N| + 1)|P| - 1} \cdot \frac{\exp(-\beta \sum_{i \in N} P_i(f'))}{\exp(-\beta \sum_{i \in N} P_i(f')) + \exp(-\beta \sum_{i \in N} P_i(f))} \quad (24)$$

(ii) The system completes the state jump by choice (ii):

$$\Pr\left(f \xrightarrow{(ii)} f'\right) = \Pr(n \text{ transimits } | n's \text{ timer expires}) \Pr(n's \text{ timer expires})$$

$$= \frac{|P| - 1}{(|MS| - |N| + 1)|P| - 1} \cdot \frac{1}{|P| - 1}$$

$$\cdot \frac{\exp(-\beta \sum_{i \in N} P_i(f'))}{\exp(-\beta \sum_{i \in N} P_i(f')) + \exp(-\beta \sum_{i \in N} P_i(f))} \cdot \frac{\frac{(|MS| - |N| + 1)|P| - 1}{\exp(\tau)}}{\sum_{n \in N} \frac{(|MS| - |N| + 1)|P| - 1}{\exp(\tau)}}$$

$$= \frac{1}{\sum_{n \in N}(|MS| - |N| + 1)|P| - 1} \cdot \frac{\exp(-\beta \sum_{i \in N} P_i(f'))}{\exp(-\beta \sum_{i \in N} P_i(f')) + \exp(-\beta \sum_{i \in N} P_i(f))} \quad (25)$$

Combine (24) and (25), we get:

$$\Pr(f \to f') = \Pr\left(f \xrightarrow{(i)} f'\right) + \Pr\left(f \xrightarrow{(ii)} f'\right)$$

$$= \frac{2}{\sum_{n \in N}(|MS| - |N| + 1)|P| - 1} \cdot \frac{\exp(-\beta \sum_{i \in N} P_i(f'))}{\exp(-\beta \sum_{i \in N} P_i(f')) + \exp(-\beta \sum_{i \in N} P_i(f))} \tag{26}$$

In our designed Markov chain, device n counts down at the rate of $\frac{(|MS| - |N| + 1)|P| - 1}{2\exp(\tau)}$. Therefore, the rate of leaving configuration f is $\sum_{n \in N} \frac{(|MS| - |N| + 1)|P| - 1}{2\exp(\tau)}$. Considering that the probability of $f \to f'$ is $\Pr(f \to f')$, we get the transition rate as:

$$q_{f,f'} = \frac{2}{\sum_{n \in N}(|MS| - |N| + 1)|P| - 1}$$

$$\cdot \frac{\exp(-\beta \sum_{i \in N} P_i(f'))}{\exp(-\beta \sum_{i \in N} P_i(f')) + \exp(-\beta \sum_{i \in N} P_i(f))} \cdot \sum_{n \in N} \frac{(|MS| - |N| + 1)|P| - 1}{2\exp(\tau)}$$

$$= \exp(-\tau) \frac{\exp(-\beta \sum_{i \in N} P_i(f'))}{\exp(-\beta \sum_{i \in N} P_i(f')) + \exp(-\beta \sum_{i \in N} P_i(f))} \tag{27}$$

(27) is equal with (21) which proves our transition rate design is true.

Theorem 2. Markov approximation algorithm realizes a time-reversed Markov chain whose steady distribution is $\frac{\exp(-\beta \sum_{i \in N} P_i(f))}{\sum_{f \in F} \exp(-\beta \sum_{i \in N} P_i(f'))}$, $f \in F$.

Proof 2. Substituting the above formula into a fine balance Eq. (19), the equation is satisfied which proves our Markov chain is time-reversed and its expression of steady distribution is $\frac{\exp(-\beta \sum_{i \in N} P_i(f))}{\sum_{f \in F} \exp(-\beta \sum_{i \in N} P_i(f'))}$, $f \in F$, according to [11, Theorems 1.3 and 1.14].

4 Simulation Result and Analysis

4.1 Simulation Setup

In this section, we implement the proposed Markov approximation algorithm (denoted by MA) in order to evaluate the performance. The simulation parameters are listed in Table 1. For better evaluation, we compare it with the following two baseline approaches:

(1) Exhaustive Search. It obtains the optimal solution by visiting all the feasible solutions and finding the minimum power consumption, which leads to high computation complexity and only feasible for small-sized problems.
(2) Random Method. The system assigns the server, channel and adjustable power to devices randomly for multiple times based on the above constraints, during which we record the minimum and maximum power consumption.

We examine the convergence and efficiency of MA approach firstly, and further verify the impacts of parameters on the performance of MA algorithm.

Table 1. Simulation parameters

Parameter	Definition	Values
$|M|$	Total number of MEC servers	4
$|N|$	Total number of devices	20
$|S|$	Number of channels	20
D_n	Size of data of device n	$\{0.2 + 0.02 \times i | 0 \le i \le |N|\}$
μ	Linear coefficient of command	4
W_n	Size of commands of device n	$\mu \times D_n$
B_s	Bandwidth of channels	6.4×10^2
β	Convergence control coefficient	0.2
F_m	Server command execution capacity	$[5.5, 6.25, 7, 8.5] \times 10^9$
$|P^{min}, P^{max}|$	Power range of devices	$[10 \times 10^{-5}, 450 + 10 \times 10^{-5}]$
L_n	Maximum delay constraint	100

4.2 Performance Evaluation

Convergence of the MA Algorithm. At first, we examine the convergence of MA approach under different settings. Figure 2 shows the convergence of our MA approach and the results show that MA converges quickly to near optimal only after 260 iterations. Furthermore, we investigate the validity and stability of the MA algorithm, by using the exhaustive method to compare with our MA approach, under the same initialization conditions of computing resources.

Obviously, the complexity of exhaustive is too high. For simplicity, we scale down to 3 tasks, 2 servers, 4 channels and 15 power values (initial value 0.00001, interval 0.06). There are 1 134,000 possibilities. In Fig. 2, the blue curve is the result of the average of the 100 sets of MA algorithms, and the red is the optimal result of the exhaustive method under the same initialization conditions. Compared with the 1 134,000 iterations of the exhaustive method, the MA algorithm has a strong fast convergence. In addition, the convergence result is only higher than the optimal value with a negligible gap which determines the effectiveness and validity of the MA algorithm. At last, the variance (the length of shorter bars inserted in the blue curve) of the 100 sets of MA algorithm gets smaller with iterations, indicating the algorithm has the capacity to deal with the inconstant data, which verifies its stability.

In order to further investigate the efficiency of the MA algorithm, we compare the MA algorithm with the random method. The experimental design is that under the same conditions, the random method is performed 1000 times during which we record its minimum value and maximum value. At the same time, we obtain the average value of MA algorithm by running 100 sets of it 1000 times.

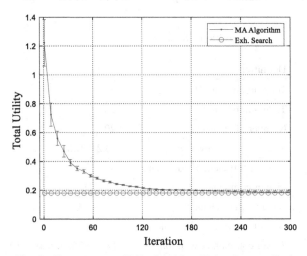

Fig. 2. Convergence of MA algorithm (Color figure online)

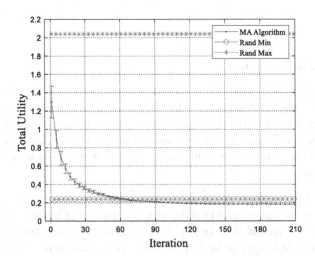

Fig. 3. System utilities of different algorithms (Color figure online)

We extract the maximum and minimum values of the 1000 random methods, as the red line shown. It can be seen from Fig. 3 that the maximum value obtained by the random method is higher than the average initial value of the 100 sets of MA algorithms. Although the system efficiency required by the initial MA algorithm is high, as the number of iterations increases, the system utility of the MA algorithm decreases continuously, reaching convergence at about 180 times, and the convergence result is smaller than the minimum value of the random method, which reflects the high efficiency and practicability of the algorithm. In addition, in the initial stage of MA algorithm, the variance of the 100 sets of experimental data is large, indicating that the initial data is unstable and the volatility is large, but as the result converges, the data variance approaches 0, indicating the volatility is very small and the values are very close. This shows the stability of the MA algorithm.

Impact of Parameters. *Impact of Parameter L_n.* Figure 4 demonstrates the impact of L_n on system utility and convergence time. When the task is transiting, it must satisfy the condition that the task processing time does not exceed the maximum time limit as (8) shown. The smaller the L_n is, the shorter the task processing time is allowed, which means the limit is relatively stricter. It can be clearly seen from Fig. 4 that as the value of L_n decreases, the convergence result of the MA algorithm increases. It's because the smaller the time limit, the higher energy consumption task requires according to (4) and (8).

It can be seen that with only L_n increasing, R_{nm} decreases, that is, P_n^s increases. What's more, in the case where the initial value differs little, when $L_n = 3$, the convergence result is about 100, but when $L_n = 100$, the convergence result is nearly 600. All shows that L_n has a great influence on the degree of convergence.

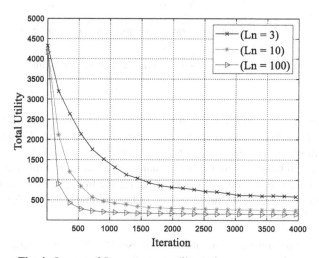

Fig. 4. Impact of L_n on system utility and convergence time

Impact of Parameter β. Figure 5 shows the effect of different parameters β on the convergence speed and system utility under the MA algorithm. From the figure, we can see that when β increases from 0.02 to 0.2, the convergence effect of system utility changes from unstable to stable, and its convergence results show a decreasing trend. When β is equal to 0.02, its convergence value is slightly lower than 1000 and is in a fluctuating state; when β is equal to 0.2, its convergence value is around 100, and the convergence state is stable and closer to the optimal value. In the MA algorithm, when the gap between the new state and the old state is larger, the system will jump to a better configuration with a greater probability. The transition rate is positively correlated with β, and as β increases, the system is more likely to jump to a better configuration. It can be seen from the figure that as β increases, the convergence effect is optimized and the convergence time is prolonged.

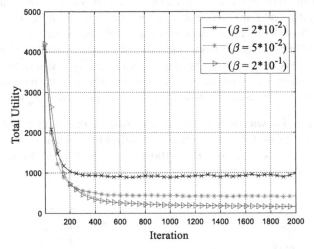

Fig. 5. Impact of β on system utility and convergence time

Impact of Parameter D_n. In Fig. 6, we study that as the amount of input data increases, the convergence result obtained by the MA algorithm becomes larger. Because of within the same maximum processing time limit, the server needs more power to offload processing tasks. It can be seen from the figure that with iterating, at 4000 times, the curve with the largest amount of data still has no convergence trend, and the current value is slightly higher than 1500, while the curve whose data volume is the smallest has already converged and the convergence result is only about 100.

In addition, as the amount of input data increases, the number of iterations required for system convergence increases. As shown in the figure, the convergence rate of the curve with smaller data volume is significantly higher than that of the data volume.

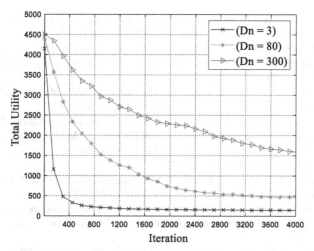

Fig. 6. Impact of D_n on system utility and convergence time

5 Conclusion

The contribution of this paper is to propose a distributed algorithm based on the Markov approximate framework to solve the problem of computational resource combination optimization in complex scenarios. The scene has multi-user, multi-server, multi-channel, multi-adjustable power features, making the solution very difficult. Based on the Markov approximation framework, we design a specific Markov chain and transition rate, so that the system can converge to an approximate optimal solution in a short time by transition between two directly connected states. To illustrate the rationality of our design, theoretical proof of Markov chain design and transition rate design is given in this paper. Finally, we operate a series of simulation experiments comparing with exhaustive search, random method and other benchmark algorithms to analyze the feasibility, robustness, and scalability of the MA algorithm. As a result, the new algorithm shows great advantages in the above aspects.

Acknowledgment. This work was funded by the Beijing Intelligent Logistics System Collaborative Innovation Center under Grant No. BILSCIC-2019KF-10, and the Fundamental Research Funds for the Central Universities under Grant No. 2019JBM027.

References

1. Shi, W., Zhang, X., Wang, Y., Zhang, Q.: Edge computing: state-of-the-art and future directions. J. Comput. Res. Dev. **56**(1), 69–89 (2019)
2. Xie, R., Lian, X., Jia, Q., Huang, T., Liu, Y.: Survey on computation offloading in mobile edge computing. J. Commun. **39**(11), 142–159 (2018)
3. Shirazi, S.N., Gouglidis, A., Farshad, A., et al.: The extended cloud: review and analysis of mobile edge computing and fog from a security and resilience perspective. IEEE J. Sel. Areas Commun. **35**(11), 2586–2595 (2017)

4. Mao, Y., You, C., Zhang, J., et al.: A survey on mobile edge computing: the communication perspective. IEEE Commun. Surv. Tutor. **19**(4), 2322–2358 (2017)
5. Zhao, Y., Zhou, S., Zhao, T., et al.: Energy-efficient task offloading for multiuser mobile cloud computing. In: IEEE/CIC International Conference on Communications in China, Shenzhen, China. IEEE, pp. 789–803 (2015)
6. Zhou, W., Fang, W., Li, Y., et al.: Markov approximation for task offloading and computation scaling in mobile edge computing. Mob. Inf. Syst. **01**(23), 01–12 (2019)
7. Daming, W., Song, C., Weijia, C., Qiang, W.: QoE utility function-based cross-layer resource allocation in multi-user MIMO-OFDM systems. J. Commun. **35**(9), 175–183 (2014)
8. Chen, M., Liew, S.C., Shao, Z., et al.: Markov approximation for combinatorial network optimization. In: INFOCOM. IEEE (2010)
9. Chen, M., Liew, S.C., Shao, Z., Kai, C.: Markov approximation for combinatorial network optimization. IEEE Trans. Inf. Theory **59**(10), 6301–6327 (2013)
10. Moon, S., Oo, T.Z., Kazmi, S.M.A., et al.: SDN-based self-organizing energy efficient downlink/uplink scheduling in heterogeneous cellular networks. IEICE Trans. Inf. Syst. **E100**(D5), 939–947 (2017)
11. Kelly, F.: Reversibility and Stochastic Networks. Wiley, Chichester (1979)

A Geographic Routing Protocol Based on Trunk Line in VANETs

Di Wu[1], Huan Li[2(✉)], Xiang Li[2], and Jianlong Zhang[2,3]

[1] State Grid Xi'an Electric Power Supply Company,
Xi'an 710032, China

[2] Xidian University, Xi'an 710071, China
I.huanli_Yoly@163.com

[3] Shaanxi Key Laboratory of Integrate and Intelligent Navigation,
Xian 710071, China

Abstract. To make full use of historical information and realtime information, a Trunk Road Based Geographic Routing Protocol in Urban VANETs (TRGR) is proposed in this paper. This protocol aims to solve the problem of data acquisition in traditional trunk coordinated control system. Considering the actual physical characteristics of trunk lines, it makes full use of the traffic flow of the trunk lines and the surrounding road network, provides a real-time data transmission routing scheme, and gives a vehicle network routing protocol under this specific condition. At the same time, the TRGR protocol takes into account the data congestion problem caused by the large traffic flow of the main road, which leads to the corresponding increase of the information the flow of the section, and the link partition problem caused by the insufficient traffic flow. It introduces different criteria for judgment and selection, which makes the TRGR protocol more suitable for the application of coordinated control of the main road in the urban environment. Simulation results show that the TRGR protocol has better performance in end-to-end delay, delivery rate and routing cost under the scenario of urban traffic trunk lines comparing with other IOT routing protocols. TRGR protocol can effectively avoid data congestion and local optimum problems, effectively increase the delivery rate of data packets, and is suitable for routing requirements in this application scenario.

Keywords: VANETs · Congestion control · Data transmission routing scheme

1 Introduction

The Vehicular Ad-hoc Network (VANET) is regarded as a mobile network established for moving vehicles on the road. It allows vehicles to play the role of nodes and realize the connection between nodes by applying the routing protocol [1,2]. This feature makes the VANET a significant part in the intelligent transportation system (ITS) [3,4]. At the same time, VANET has huge potential in ensuring

© Springer Nature Singapore Pte Ltd. 2019
H. Ning (Ed.): CyberDI 2019/CyberLife 2019, CCIS 1138, pp. 21–37, 2019.
https://doi.org/10.1007/978-981-15-1925-3_2

road safety and improving traffic efficiency. After obtaining the information on nodes and roads, VANET could provide road warning, route planning and other services for drivers and passengers. Since the dynamic of the network application brings various changes and problems, the routing protocol in VANET has become an active research field. Due to the complex topology of urban roads, different application scenarios generate different routing requirements [5]. Therefore, the thesis of this paper is to design a routing protocol that could meet the requirements of certain application scenarios.

The routing protocols in VANETs can be divided into four categories: topology-based routing, location-based routing, broadcast routing protocols, and cluster-based routing protocols. The GPSR (Greedy Perimeter Stateless Routing protocol) [6,7] protocol is one of the most classic location-based protocols proposed by Brad Karp. GPSR protocol uses greedy forwarding and peripheral forwarding in greedy forwarding failure zone to transmit data packets, but this algorithm still has some problems in urban scenes, such as excessive hops, routing ring. In 2016, Ibtihal Mouhib analyzed the performance of Q-AODV [8] and GPSR protocols in the context of on-board cloud. First, use the SUMO tool to generate the migration scenario. And then network simulation with NS2. It is concluded that Q-AODV is superior to GPSR in throughput and packet loss rate, end-to-end latency, packet delivery ratio, and throughput. AODV (ad-hoc On Demand Vector) [9] is a topology-based routing protocol for Mobile ad-hoc NETworks (MANETs). When there is a communication demand, the sending node broadcasts a RREQ (routing request packet) to the neighbor node, and the neighbor node updates the local routing table immediately after receiving it, and then continues to forward RREQ to its neighbor node, repeating this process until RREQ is sent to the destination node. However, due to the fast moving speed of vehicles on the road, the topology of vehicle-link network changes quickly, and the communication link quality is low and easy to break.

The main idea of TRGR protocol can be summarized as the following three parts:

- Connect Probe Packet (CPP) is used to judge the connectivity of adjacent intersection sections to choose the connected road segment set;
- The optimal road segment is selected under different vehicle densities as the next segment of data packet transmission.
- On the selected optimal section, according to the next hop selection algorithm obtained by the analytic hierarchy process (AHP) [10], the next-hop node is selected.

The rest of the paper is organized as follows. Geographic Routing Protocol Based On Traffic Line is introduced in Sect. 2. Simulation results and the analyses are given in Sect. 3, Sect. 4 concludes the paper.

2   Geographic Routing Protocol Based on Traffic Line

In the application scenario of urban trunk lines carrying large traffic flow, vehicles bear the needs of various on-board communications [11]. To meet the needs of trunk coordinated control, a routing protocol is needed to ensure data packet delivery rate and low routing overhead [12].

2.1 System Model

Assuming that each vehicle generates messages at a rate, according to the results in [13,14], it is possible for the vehicle to send a packet in a randomly selected slot:

$$\tau_s = \frac{2(1-p)^2}{2+pW_s-3p}(T_s\beta_s) \tag{1}$$

Among them, T_s represents the length of a slot and W_s represents the minimum competition window. Suppose there is a car competing for the same channel as the sending vehicle. Then the collision probability of group access information can be expressed as:

$$p_c = 1 - (1-\tau_s)^N \tag{2}$$

2.2 Protocol Content

Link Connectivity Decision Mechanism for Adjacent Intersections. In order to facilitate the trunk terminal roadside unit to effectively collect the relevant information of vehicles entering the trunk line, we use the link between adjacent intersections as the basic unit. The purpose of the Connect Probe Packet (CPP) is to detect the link connectivity between two intersections. The format of the CPP package is shown in Table 1.

Table 1. Connection Probe Packet (CPP)

Originator: Vehicle ID	ToG: Time stamp	From: Intersection ID	To: Intersection ID
Transmission delay			
Other information			

ToG represents the generation time of connection probe packets, and other information is additional routing information, such as average density and average speed. When the connection detection packet arrives at the intersection, the vehicle nearest to the current intersection is responsible for generating the updated weights. Then the information is released through the intersection and sent back to the initiator. When the connection detection data packet is generated, the timer is triggered. Before the timer fails, if the node receives the data packet from the node, the link is determined to be a connection link. Otherwise, it is determined that the section is invalid.

In summary, the link-connected and low-latency segments are selected by using the connection detection packet (CPP) with smaller memory as the alternative route for data packet transmission. and the delivery rate is the primary consideration to select the alternative sections [15,16] because it is very easy to cause data congestion [11,17]when urban trunk line carries a large amount of traffic flow [18].

Bidirectional Multi-lane Multi-hop Packet Delivery Rate Modeling.
In this section, we will give a flow chart for calculating the delivery rate of two-way multi-lane data to solve the problem of data congestion caused by the high the density of vehicles on trunk lines. According to Formula 3, the single-hop transmission rate equals the probability of sending a packet successfully before the specified retry limit:

$$PDR_{one-hop} = 1 - p_c^r \tag{3}$$

It is defined as every hop process of a sending node in the lane. It is assumed that the average density of vehicles in the first lane follows a uniform distribution with parameters λ_i . R represents the transmission radius of a node. Its cumulative distribution function (CDF) can be expressed as:

$$P(d_i) = \frac{e^{-\lambda_i(R-d)}(1 - e^{-\lambda_i d})}{1 - e^{-\lambda_i R}} \tag{4}$$

Then, the cumulative distribution function (CDF) of the expected multi-lane hopping process is as follows:

$$P(d) = P_1(d)P_2(d)...P_n(d)$$
$$= \frac{e^{-(\lambda_1+\lambda_2+...+\lambda_n)(R-d)}(1-e^{-\lambda_1 d})(1-e^{-\lambda_2 d})...(1-e^{-\lambda_n d})}{(1-e^{-\lambda_1 R})(1-e^{-\lambda_2 R})...(1-e^{-\lambda_n R})} \tag{5}$$

Therefore, the expected multi-lane hopping process is as follows:

$$\bar{d} = \int_0^R x dF(x) \tag{6}$$

After knowing the process of each jump in multi-lane, the expected number of hops can be approximated as follows:

$$h = \frac{L}{\bar{d}} \tag{7}$$

Success rate of multi-hop packet transmission: The overall estimated packet transmission rate of the section can be calculated as follows:

$$PDR_{multi-hop} = (1 - p_c^r)^h \tag{8}$$

When the density of vehicles on the trunk line is low, data packet transmission will encounter the problem of link partition, so only vehicle carrying and forwarding can be selected, which will greatly increase the transmission delay [20]. So in this case, the connection probability of vehicle network is considered as the primary index.

Complete Connection Probability Modeling. In the transmission phase of Connection Detection Packet (CPP), when the link connection is interrupted due to low vehicle density, the node can not receive the updated link weight. The link weight is estimated by using historical information through the connectivity of the network, and the reference parameter is the probability that all nodes connect to each other through wireless communication.

Next, a two-way multi-lane network connection analysis model for urban trunk lines is given. As shown in Fig. 1, for urban trunk lines, we take two-way six-lane as an example, and assume that the average density of vehicles in this lane obeys the parameters of $\lambda_1 \lambda_2\ \lambda_n$,. The uniform distribution of Lane1 is λ_1 and Lane2 is λ_2. Lane n is λ_n. The number of vehicles on the first lane is within the interval, which obeys Poisson distribution.

$$f(k_i, l) = \frac{(\lambda_i l)^{k_i}}{k_i!} e^{-\lambda_i l} \tag{9}$$

The distribution function of workshop distance can be expressed as:

$$F_i(x) = 1 - e^{-\lambda_i x} \tag{10}$$

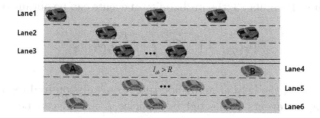

Fig. 1. Overall schematic diagram of network connection

In the same lane, when the distance between two vehicles running continuously exceeds the transmission distance of nodes, it is defined as chain breaking. As shown in Fig. 1, the link between vehicle A and vehicle B is disconnected on Lane 4 lane. For multi-lane link failure, adjacent lanes can be used to improve the connectivity effect.

If the following conditions are met, the links between nodes A and B are determined to be connected: (1) condition η_1: there is at least N_{ab} vehicles in the lane in the gap between vehicles A and B. (2) Conditions η_2: In continuous traffic flow, any workshop distance is shorter than R. (3) Conditions η_3: Length of multiple link coverage Longer than $l_{ab} - R$.

Suppose there is a gap between Lane 3 of opposite lane including N_{ab_left} vehicles, Lane 5 of adjacent lane including N_{ab_right} vehicles and Lane 4 of vehicles. If the lane is located on the roadside and there is only one adjacent lane, the probability of one adjacent lane is calculated. According to Formula 5, the probability of the event is:

$$\Pr(\eta_1) = \frac{(\lambda_3 l_{ab})^{N_{ab\_left}} e^{-\lambda_3 l_{ab}}}{N_{ab\_left}!} + \frac{(\lambda_5 l_{ab})^{N_{ab\_right}} e^{-\lambda_5 l_{ab}}}{N_{ab\_right}!} + \frac{(\lambda_4 l_{ab})^{N_{ab}} e^{-\lambda_4 l_{ab}}}{N_{ab}!} \tag{11}$$

Medium: $N_{ab\_left} \geq N_{ab}, N_{ab\_right} \geq N_{ab}$

According to Formula 10, the probability distribution function of the workshop distance less than the maximum transmission distance R of the vehicle is recorded is defined as $F_i(R) = 1 - e^{-\lambda_i R}$. Represents the number of lanes, so the probability of event η_2 is:

$$F_i(x) = 1 - e^{-\lambda_i x} \tag{12}$$

The probability density function of the corresponding workshop distance is $f_i(x) = \frac{\lambda_i e^{-\lambda_i x}}{1 - e^{-\lambda_i R}}$: the probability density function of the sum of all vehicle workshop distances can be obtained as follows:

$$f_i(S) = \frac{\lambda_i^m S^{m-1}}{(1 - e^{-\lambda_i R})^m (m-1)!} e^{-\lambda_i S} \tag{13}$$

According to Formula 13, the probability of events can be obtained as follows:

$$\Pr(\eta_3) = 1 - \sum_{i-1}^{i+1} \int_0^{l_{ab}-R} f_i(S) dS \tag{14}$$

Therefore, with the help of multi-hop relay transmission in directional lanes, the probability of events connected by vehicles A and B can be expressed as follows:

$$\Pr(\eta) = Pr(\eta_1) Pr(\eta_2) Pr(\eta_3) \tag{15}$$

According to [19], the average number of Lane failure links can be expressed as $N_l = \frac{L}{(e^{\lambda_i R}-1)(\frac{1}{\lambda_i} - \frac{Re^{-\lambda_i R}}{1-e^{-\lambda_i R}})+R+\frac{1}{\lambda_i}}$, So the probability that all vehicles in lane4 will be fully connected by multi-hop transmission is:

$$p_l = [Pr(\eta)]^{N_l} \tag{16}$$

Next Section Selection. Because of the frequent changes of the topological network and the large scale of the network, it is difficult to get a global understanding of the instantaneous information of the network topology. Therefore, an adaptive intersection selection algorithm is adopted, which can select intersections one by one according to the specified requirements, and calculate the path of each intersection partially continuously. Using more updated traffic information, choose the next route to forward the data package. The specific process is as follows:

(1) Determining candidate intersections: When a packet arrives at an intersection, there are different candidate intersections defined as adjacent intersections and current intersections. In order to reduce the cost of traversing all intersections to find the best path, we first select the appropriate intersections according to their shortest geographical distance to the destination.

(2) Choose the best route: In VANETs, the priority is to select the link in all routes, because the data transmission depends on wireless communication technology when network connection, which will bring lower delay. However, in the presence of multiple links, the path with the greatest connectivity is likely the path with the greatest degree of congestion. This may lead to more data collisions, resulting in large latency or low available bandwidth. Therefore, in this case, the transmission rate is considered to be the key parameter to reflect the channel quality. When there are multiple connected routing paths, the aggregation is expressed. We choose the path with the highest packet transmission rate as the optimal path, that is:

$$l_{optimal} = \arg \max_{l \in c(i)} PDR_{multi-hop} \tag{17}$$

Formula 17 is used to calculate the data transmission rate of the routing section. However, for sparse networks, there are likely many link partitions in the network. If the scheme of carry-forward is adopted, the packet will have a higher transmission delay. In this case, connectivity will have a significant impact on network performance. Therefore, if all routing paths in the set $d(i)$ are disconnected, the path with the greatest connectivity is chosen as the best routing path, that is:

$$l_{optimal} = \text{argmax}_{l \in d(i)} p_l \tag{18}$$

Next Hop Option. After the next section is determined, the packet will be forwarded along this section and select the next hop [21, 22]. Most routing protocols choose greedy forwarding mode to select the next hop which can reduce the number of hops, loss and channel occupancy. However, because TRGR protocol is based on urban traffic trunk lines and the speed of vehicles is faster, the next hop node selected by greedy forwarding is usually located at the boundary of the transmission range of the sending node. It is likely that the selected vehicle node will drive out of the communication range of the sending node and cause packet loss. Therefore, compared with reducing hops and resource occupancy, priority should be given to ensuring the reliability of data packet transmission. For the problem of time delay, the introduction of workshop distance parameters to control can also be eliminated by calculating the formula in the later coordinated control of trunk lines, which has little impact on the whole system.

When choosing an intersection, a selection strategy requiring the next hop transmits data packets along the selected section. The location of adjacent nodes can be obtained by periodic exchange of beacon packets. In addition, to ensure the successful transmission of data packets, combined with the communication quality of nodes and the progress of single hop transmission, this protocol uses

AHP to ensure the quality of data packet transmission under trunk demand. Based on the physical characteristics of urban traffic trunk lines and the application scenarios of this routing protocol, the single hop transmission rate, channel fading and workshop distance are selected as the parameters for selecting the next hop.

Workshop Distance: Workshop distance [23] as a factor is based on the idea of greedy forwarding. The farther the next hop is from the sending node, the nearer the destination node is, the smaller the hops required, and the corresponding routing overhead will be reduced. However, because the next hop node farthest from the sending node is always located at the boundary of the transmission range of the node, it is easy to cause packet loss. So we introduce single hop transmission rate and channel fading as the measurement index. This section mainly describes the determination of workshop distance.

Assuming that there are m potential neighbor nodes in the transmission range of the sending node s, it is defined as a set of all neighbor nodes $N(s) = \{s_1, s_2, s_3, ..., s_m\}$. Neighbor nodes are filtered by conditions $t_{si} > T_{th}$ to indicate the average link connection time between the node s and its neighbor nodes calculated by Formula 21, and then the candidate nodes are identified separately. $d(s, i)$ represents the workshop distance between the node s and its neighbors, and a specified threshold value T_{th} for forwarding data packets.

Link connection time refers to the time when direct communication links between two nodes remain continuously available. For high-speed mobile vehicle network, it is an important basis for judging the quality of communication. The mean and variance of nodes velocity of node s and node i are defined respectively as v_s a_s v_i a_i. $D(t)$ represents the distance between vehicles, the initial value is $D(0) = d_0$. Because of the random mobility of the sum of nodes, the distance between vehicles can be considered as a $G/G/1$ queue, in which the movement of nodes in unit time can be considered as arrival queue, and the distance between nodes can be considered as departure queue.

$$
\begin{aligned}
p(x|d_0, t) &= Pr(x \le D(t) \le x + d_x|d_0) \\
&= \frac{1}{\sigma\sqrt{2\pi t}} \sum_{n=-\infty}^{\infty} [exp(\frac{\mu x_n^1}{\sigma^2} - \frac{[x-d_0-x_n^1-\mu t]^2}{2\sigma^2 t}) \\
&\quad - exp(\frac{\mu x_n^{''}}{\sigma^2} - \frac{[x-d_0-x_n^{''}-\mu t]^2}{2\sigma^2 t})]
\end{aligned}
\tag{19}
$$

Formula 19 calculates the connectivity of the road section. In the formula 20, the radius of the node is expressed as R. The cumulative distribution function (CDF) of link connection time between nodes can be expressed as:

$$
F_{s,i}(t) = 1 - \int_0^R p(x|d_0, t)dx
\tag{20}
$$

Therefore, the average link connection time can be deduced as:

$$
E(T_{s,i}) = \int_0^{+\infty} t \, dF_{s,i}(t)
\tag{21}
$$

Channel fading: For a transceiver system from the source node to the destination node, free space propagation is an ideal wireless signal propagation. It can be understood that the electromagnetic signal emitted by the source node reaches the destination node along a straight line after a certain attenuation. However, for the urban trunk scenario used in this paper, the electromagnetic wave emitted from the source node may reach the destination through a series of reflections. Nodes, due to the different phases of arriving at the destination node, cause the overall signal strength to decrease. This process is called channel fading.

Dedicated short-range Communications (DSRC) describes that the communication Range of a vehicle node in VANETs is usually several hundred meters, so small scale fading is considered. A large number of researchers have confirmed that the nakagami-m channel model has a good fit for the measured experimental data. We define the probability that a packet can be sent from the source node s to the node i after channel fading is C_{si}

$$C_{si} = 1 - \frac{m^m}{\Gamma(m)w^m} \int_0^{t_r} z^{m-1}e^{-(\frac{m}{w})^z}dz \tag{22}$$

$$t_r = \frac{T_p}{R^2}G \tag{23}$$

$$G = \frac{G_t G_r \lambda}{(4\pi)^2 l} \tag{24}$$

w denotes the average receiving power, t_r is the threshold of the received signal, T_p is the transmitting power, and the G_t G_r are the antenna gain of the transmitting node and the receiving node, respectively. m denotes the attenuation factor, which represents the severity of channel fading, and is related to the workshop distance required in the previous section. According to the application scenario of the protocol, m is 1.5 at that time (in meters). Therefore, the above formula can be simplified as follows:

$$C_{si}(d) = 1 - \frac{1.837d^3}{\Gamma(1.5)} \int_0^{\frac{1}{R^2}} z^{0.5}e^{-1.5x^2 z}dz \tag{25}$$

Analytic Hierarchy Process: AHP is a mathematical tool that can deal with non-quantitative multi-criteria decision-making problems. It is proposed by Professor Satie, an American operations researcher, to transform complex problems into hierarchical sub-problems to estimate the importance of each metric. In the next hop selection, this method can be used to calculate the weights of the next hop node and select the next hop node with the largest comprehensive evaluation value [24].

As shown in Fig. 2, The first step is to establish a hierarchical structure model. One of the candidate nodes of scheme level is selected as the next hop node to complete the decision.

The second step is to establish the judgment matrix. The weight of each factor is determined by the judgment matrix A. n is the total number of influencing factors in the criterion layer. In this agreement, n takes 3. Elements in A are

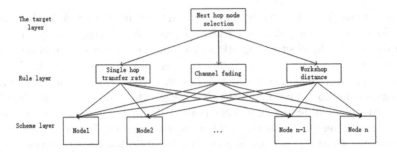

Fig. 2. Hierarchical structure model

expressed as the importance of elements compared with elements. The judgment matrix has the following properties:

$$a_{ij} = \frac{1}{a_{ji}} \tag{26}$$

$i = 1, 2, 3$ and $j = 1, 2, 3$ represent three elements workshop distance, single hop transmission rate and channel fading, respectively. Establish the judgment matrix as follows:

$$A = \begin{pmatrix} 1 & \frac{1}{7} & \frac{1}{5} \\ 7 & 1 & 3 \\ 5 & \frac{1}{3} & 1 \end{pmatrix} \tag{27}$$

Using the formula $b_{ij} = \frac{a_{ij}}{\sum\limits_{i=1}^{n} a_{ij}}$, the normalized matrix is obtained.

$$B = \begin{pmatrix} 0.0769 & 0.0968 & 0.0476 \\ 0.5385 & 0.6774 & 0.7143 \\ 0.3846 & 0.2258 & 0.2381 \end{pmatrix} \tag{28}$$

For each row of normalized matrix B, the eigenvectors obtained by summing the eigenvectors are as follows: normalizing the eigenvectors, the weights obtained can be obtained by calculating the formulas as follows:

$$w_i = \frac{W_i}{\sum\limits_{i=1}^{n} W_i} \tag{29}$$

In conclusion, the weights of workshop distance, single hop transmission rate and channel fading are 0.2213, 1.9302 and 0.8485, respectively. Therefore, the composite index is

$$M = 0.0738d(s, i) + 0.6434PDR_{one-hop} + 0.2828C_{si} \tag{30}$$

The third step is matrix consistency test. When the N-order judgment matrix satisfies consistency, its eigenvalues have and only have one value of N. The maximum characteristics of the matrix are computed in Formula 31 and the consistency index is calculated in Formula 32.

$$\lambda = \frac{\sum (Aw)_i}{nw_i} \tag{31}$$

$$C.I. = \frac{\lambda - N}{N - 1} \tag{32}$$

Table 2. Standard Values Mean Random Consistency Index RI

Matrix order	1	2	3	4	5	6	7
RI		0.00	0.00	0.58	0.90	1.12	1.24	1.32

The average consistency index R.I. can be obtained by Table 2, and the value of the third-order matrix R.I. is 0.58. Consistency ratios (C.R) can be obtained as shown in Formula 33. When C.R. is less than 0.1, matrix consistency is satisfied. After calculating the above process, the judgment matrix C.R. of this protocol is 0.0356, so the judgment matrix satisfies the consistency test, so the weight set by Formula 29 is reasonable.

$$C.R. = \frac{C.I.}{R.I.} \tag{33}$$

3 Simulation Results

3.1 Introduction of Simulation Tool

OPNET [25,26] is a leading global communication network simulation software, which can simulate wired and wireless networks. OPNET uses C++ programming language and its own core functions to simulate specific communication protocols. The graphical simulation interface also reduces the difficulty of network communication simulation. At last, the drawing function of MATLAB software is used to import OPNET software to simulate the data, remove the erroneous data which is obviously deviated, and compare the results.

3.2 Simulation Environment and Parameter Settings

The simulation parameters of OPNET are shown in Table 3. The simulation area is 10 km * 10 km, the number of vehicles is 100, the main road is two-way 6 lanes, and the other lanes are two-way 2 lanes. Vehicle node trajectory is set by OPNET's Define Trajectory function, and traffic lights are set at each intersection.

Table 3. OPNET simulation communication parameters settings

Parameter	Parameter values
Simulation area size	10000 m * 10000 m
Number of lanes	The main road has six lanes in two directions and the other two lanes in two directions
Number of signal lights	20 group
Minimum residence time of vehicles at intersections	3 s
Maximum residence time of vehicles at intersections	30 s
Minimum vehicle speed	30 km/h
Maximum vehicle speed	60 km/h
Simulation time	3600 s
Maximum transmission range of nodes	100 m
Packet size	128 bits
Packet type	TIP
Packet generation rate	1–10 packets/s
Beacon size	20 bytes
Beacon cycle	2 s
Channel capacity	2 Mbps
Mac protocol	802.11p

3.3 Simulation Index

In order to verify the performance of the TRGR, the meaning of three indicators is introduced.

- **End-to-End Delay (EED):** End-to-end delay refers to the time difference between sending data packets from source node and receiving data packets from destination node.
- **Packet Delivery Ratio (PDR):** Packet delivery rate is defined as the ratio of the total number of packets received at the destination to the total number of packets generated by the source vehicle.
- **Routing Overheads (RO):** Routing overhead is defined as the ratio between the total byte size of control packets and the cumulative size of packets forwarded to the target and control packets.

3.4   Simulation Results and Performance Analysis

In this section, according to the above simulation settings, the simulation results of different routing protocols obtained by OPNET simulation software are imported into MATLAB software, and are compared and analyzed from three aspects: end-to-end delay, data packet delivery rate and routing overhead.

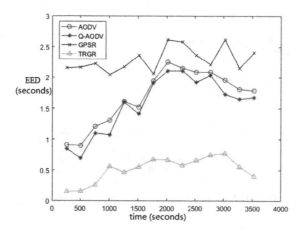

Fig. 3. Transmission delay of four routing protocols

As shown in Fig. 3, the transmission delay of the four routing protocols increases with the increase of traffic flowing into the road network in about 2500 s. For GPSR protocol, without considering vehicle traffic, the data packet may encounter local maximum or data congestion, which results in poor performance in terms of delay. Compared with the other three protocols, TRGR has the best performance. This is because it adopts an adaptive intersection selection scheme, which determines the intersection one by one according to the collected link connectivity and delay information. For TRGR, when choosing the next section, the connection section with the smallest delay is preferred. In addition, if there are no connected sections, the section with the greatest connectivity is selected as the next section to forward data packets. The protocol minimizes the use of carry-forward strategy and reduces the transmission delay.

Figure 4 shows the packet delivery rates of the four routing protocols as the simulation proceeds. In VANETs, due to the mobility and density of nodes, there may be some link partitions in the network. Then, the packets sent should be stored and carried until the next hop is found. Due to the size limitation of the buffer, upcoming new packages may be deleted when the buffer is full. Therefore, when the number of vehicles entering the road network increases near 2500 s, the data packet delivery rate of the four routing protocols decreases, especially the GPSR protocol. TRGR protocol is based on guaranteeing data packet delivery rate in section selection and next hop selection. According to the collected connectivity and delivery rate, appropriate weights are allocated for each road, and then relay nodes are determined one by one. This protocol can greatly reduce data packet loss.

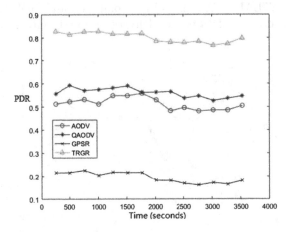

Fig. 4. Delivery rate of four routing protocols

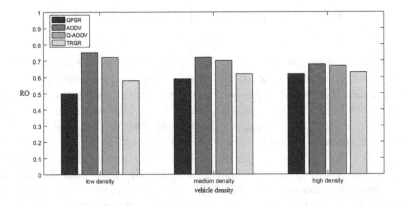

Fig. 5. Routing overhead comparison under different vehicle densities

According to Fig. 5, the effects of different vehicle densities on the routing overhead of the four routing protocols are shown. The routing overhead of AODV protocol and Q-AODV protocol is higher than that of GPSR protocol under different vehicle density conditions. For the GPSR protocol and TRGR protocol, vehicles can obtain their own location information through vehicle-mounted GPS and exchange information with other vehicles. The source node places the location information of the destination node into the header of the packet, without establishing and maintaining the update routing table, so the routing overhead is relatively low.

4   Summary

According to the actual demand of coordinated control of urban trunk lines, this chapter creatively proposes TRGR protocol. Routing forwarding is carried out according to the sequence of the next hop after choosing the section first. Under the condition of high density network, we use the index of data packet delivery rate as the criterion of road section selection, and propose a calculation method of data packet delivery rate under two-way multi-lane road network. In sparse networks, the next segment with higher link connectivity will be selected first. There are many factors to be considered when choosing the next hop. According to the single hop transmission rate, channel fading and workshop distance as evaluation indicators, the next hop selection of neighbor nodes is judged using analytic hierarchy process. Finally, the simulation software OPNET is used to evaluate the end-to-end delay, data packet delivery rate and routing overhead of TRGR protocol. The results show that TRGR protocol performs better than other protocolsGPSR, AODV and Q-ARDV in these three aspects in urban trunk network environment. In the future, we will take into account the mobile devices held by pedestrians to communicate information. In addition, we will optimize the weight determination method in AHP.

Acknowledgment. This work was supported by the National Key Research and Development Program of China (2017YFE0121400), the National Natural Science Foundation of China (61571338, U1709218,61672131), the Key Program of NSFC-Tongyong Union Foundation (U1636209), the key research and development plan of Shaanxi province (2017ZDCXL-GY-05-01, 2019ZDLGY13-04, 2019ZDLGY13-07), the Xi'an Key Laboratory of Mobile Edge Computing and Security (201805052-ZD3CG36) and the 111 Project of China (B08038) and the Innovation Fund of Xidian University (5001-20109195456).

References

1. Chen, C., Xiang, H., Qiu, T., Wang, C., Zhou, Y., Chang, V.: A rear-end collision prediction scheme based on deep learning in the Internet of vehicles. J. Parallel Distrib. Comput. **117**, 192–204 (2018)
2. Chen, C., Pei, Q., Li, X.: A GTS allocation scheme to improve multiple-access performance in vehicular sensor networks. IEEE Trans. Veh. Technol. **65**(3), 1549–1563 (2015)
3. Paier, A.: The end-to-end intelligent transport system (ITS) concept in the context of the European cooperative ITS corridor. In: 2015 IEEE MTT-S International Conference on Microwaves for Intelligent Mobility (ICMIM), pp. 1–4. IEEE, April 2015
4. Yang, Y., Li, D., Duan, Z.: Chinese vehicle license plate recognition using kernel-based extreme learning machine with deep convolutional features. IET Intel. Transp. Syst. **12**(3), 213–219 (2017)
5. Schiller, M., Dupius, M., Krajzewicz, D., Kern, A., Knoll, A.: Multi-resolution traffic simulation for large-scale high-fidelity evaluation of VANET applications. In: Behrisch, M., Weber, M. (eds.) Simulating Urban Traffic Scenarios. LNM, pp. 17–36. Springer, Cham (2019). https://doi.org/10.1007/978-3-319-33616-9_2

6. Walker, A., Radenkovic, M. GPSR-TARS: congestion aware geographically targeted remote surveillance for VANETs. In 2017 International Conference on Selected Topics in Mobile and Wireless Networking (MoWNeT), pp. 1–6. IEEE, May 2017

7. Saleet, H., Langar, R., Naik, K., Boutaba, R., Nayak, A., Goel, N.: Intersection-based geographical routing protocol for VANETs: a proposal and analysis. IEEE Trans. Veh. Technol. **60**(9), 4560–4574 (2011)

8. Mouhib, I., Smail, M., El Ouadghiri, M.D., Naanani, H.: Network as a service for smart vehicles: a new comparative study of the optimized protocol Q-AODV and GPSR protocol. In: 2016 International Conference on Engineering and MIS (ICEMIS), pp. 1–5. IEEE, September 2016

9. Anamalamudi, S., Sangi, A.R., Alkatheiri, M., Ahmed, A.M.: AODV routing protocol for Cognitive radio access based Internet of Things (IoT). Future Gener. Comput. Syst. **83**, 228–238 (2018)

10. Liu, C.C., Wang, T.Y., Yu, G.Z.: Using AHP, DEA and MPI for governmental research institution performance evaluation. Appl. Econ. **51**(10), 983–994 (2019)

11. Wang, R., Xu, Z., Zhao, X., Hu, J.: V2V-based method for the detection of road traffic congestion. IET Intel. Transp. Syst. **13**(5), 880–885 (2019)

12. Chen, C., Jin, Y., Pei, Q., Zhang, N.: A connectivity-aware intersection-based routing in VANETs. EURASIP J. Wirel. Commun. Netw. **2014**(1), 42 (2014)

13. Hafeez, K.A., Zhao, L., Ma, B., Mark, J.W.: Performance analysis and enhancement of the DSRC for VANET's safety applications. IEEE Trans. Veh. Technol. **62**(7), 3069–3083 (2013)

14. Liu, L., Chen, C., Qiu, T., Zhang, M., Li, S., Zhou, B.: A data dissemination scheme based on clustering and probabilistic broadcasting in VANETs. Veh. Commun. **13**, 78–88 (2018)

15. Al Najada, H., Mahgoub, I.: Anticipation and alert system of congestion and accidents in VANET using big data analysis for intelligent transportation systems. In: 2016 IEEE Symposium Series on Computational Intelligence (SSCI), pp. 1–8. IEEE, December 2016

16. Ou, L., Qin, Z., Liao, S., Hong, Y., Jia, X.: Releasing correlated trajectories: towards high utility and optimal differential privacy. IEEE Trans. Dependable Secure Comput. (2018)

17. Chen, C., Qiu, T., Hu, J., Ren, Z., Zhou, Y., Sangaiah, A.K.: A congestion avoidance game for information exchange on intersections in heterogeneous vehicular networks. J. Netw. Comput. Appl. **85**, 116–126 (2017)

18. Duan, Z., Yang, Y., Zhang, K., Ni, Y., Bajgain, S.: Improved deep hybrid networks for urban traffic flow prediction using trajectory data. IEEE Access **6**, 31820–31827 (2018)

19. Reis, A.B., Sargento, S., Neves, F., Tonguz, O.K.: Deploying roadside units in sparse vehicular networks: what really works and what does not. IEEE Trans. Veh. Technol. **63**(6), 2794–2806 (2013)

20. Wang, B., Sun, Y., Li, S., Cao, Q.: Hierarchical matching with peer effect for low-latency and high-reliable caching in social IoT. IEEE Internet Things J. **6**(1), 1193–1209 (2018)

21. Wang, B., Sun, Y., Sheng, Z., Nguyen, H.M., Duong, T.Q.: Inconspicuous manipulation for social-aware relay selection in flying Internet of Things. IEEE Wirel. Commun. Lett. (2019)

22. Wang, B., Sun, Y., Nguyen, H.M., Duong, T.Q.: A novel socially stable matching model for secure relay selection in D2D communications. IEEE Wirel. Commun. Lett. (2019). https://doi.org/10.1109/LWC.2019.2946828

23. Chen, C., Liu, X., Qiu, T., Liu, L., Sangaiah, A.K.: Latency estimation based on traffic density for video streaming in the internet of vehicles. Comput. Commun. **111**, 176–186 (2017)

24. Asghari, M., et al.: Weighting Criteria and Prioritizing of Heat stress indices in surface mining using a Delphi Technique and Fuzzy AHP-TOPSIS Method. J. Environ. Health Sci. Eng. **15**(1), 1 (2017)

25. Zhang, Y., Zhang, Y., Li, X., Wang, C., Li, H.: Simulation platform of LEO satellite communication system based on OPNET. In: Fourth Seminar on Novel Optoelectronic Detection Technology and Application, vol. 10697, p. 106975C. International Society for Optics and Photonics, February 2018

26. Pahlevan, M., Obermaisser, R.: Evaluation of time-triggered traffic in time-sensitive networks using the OPNET simulation framework. In: 2018 26th Euromicro International Conference on Parallel, Distributed and Network-based Processing (PDP), pp. 283–287. IEEE, March 2018

Sub-array Based Antenna Selection Scheme for Massive MIMO in 5G

Hassan Azeem[1], Liping Du[1(✉)], Ata Ullah[2], Muhammad Arif Mughal[1], Muhammad Muzamil Aslam[1], and Muhammad Ikram[1]

[1] School of Computer and Communication Engineering, University of Science and Technology Beijing (USTB), Beijing 10083, China
hassanazeem.gcu@gmail.com, dlp2001@ies.ustb.edu.cn,
cr3tiv3mac@hotmail.com, muzamil34410@qq.com,
ikramislam94@yahoo.com
[2] Department of Computer Science, National University of Modern Languages (NUML), Islamabad 44000, Pakistan
aullah@numl.edu.pk

Abstract. With rapidly increased throughput demand, operators are rapidly improving coverage and capacity with cost effective techniques in wireless communication network. Developments in technology enables advanced antenna system to be scalable across 5G and future wireless networks. Massive MIMO based advance antenna selection techniques provide powerful and affordable methods that are effective approaches for coverage and capacity of consumers. Prerequisites of an optimal communication system grow quickly, and therefore operators require more facilities to meet their needs. It is necessity to serve many operators and various devices at the same time in the integrated zone, while providing fast speed and consistent performance, makes it the boosting technology yard to meet the requirements of the 5G era. In this paper, we propose a Sub-array based Antenna Selection Scheme (SASS) for massive MIMO based on sub-array switching architecture which is beneficially helpful to achieve optimal throughput, energy efficiency and capacity. Moreover, SASS is cost effective technique which reduces the overall cost of system including computational, communication, and hardware impairments. We have validated our work using MATLAB and results are compared for spectral efficiency when number of antennas and Signal to noise ratio are varied. Results prove the dominance of SASS over counterparts.

Keywords: Massive MIMO · Antenna selection · Sub-array switching architecture · Optimal throughput

1 Introduction

Massive MIMO was firstly introduced by Marzetta and it is also known as communication system of large scale antennas [1]. Massive MIMO is promising technology and has showed significant results with spectral efficiency and collective capacity [2–5]. Fundamental conception of massive MIMO systems is connecting the base station with

© Springer Nature Singapore Pte Ltd. 2019
H. Ning (Ed.): CyberDI 2019/CyberLife 2019, CCIS 1138, pp. 38–50, 2019.
https://doi.org/10.1007/978-981-15-1925-3_3

massive network of large antennas utilizes the sources with same time and frequency [6]. It discusses the main fundamentals of massive MIMO and some other terminologies of MIMO systems. Furthermore, because higher frequencies, in some cases digital signal processing (DSP) delays as compare to analog signal processing demands [7]. The key challenges of mm-waves technology has no appropriately huge coverage because of mm-waves propagation nature [8, 9] and it contains another issue regarding environment of non-line-of-slight (NLoS) which is not supported for mobility [8, 10]. In traditional 4G MIMO technology normally utilize maximum 12 antennas for data transmission, on the other hand more than hundred antennas are utilizing at the BS in present MIMO systems. Massive MIMO system considers as measurable change and with many new opportunities and possibilities in data transmission system. The emphasis of massive MIMO technologies is to target emitted energy concerning the proposed objectives to minimize the interferences of intra and inter cells [2]. In digital linear processing under favorable propagation, massive MIMO technology attains extreme data communication rates [1, 3, 11].

Massive MIMO innovation is capable; it likewise faces a few difficulties. Massive MIMO frameworks need to manage pilot defilements, which are instigated by the predetermined number of symmetrical pilots produced by the base stations [1, 12]. Massive MIMO frameworks depend on TDD plots because of the certification of channel correspondence. Be that as it may, the vulnerability in the simple parts of the radio frequency chains (RFCs) may unbalance the channel correspondence, requiring an alignment [13, 14]. Massive MIMO is fairly unique in relation to everything showing up in past versatile correspondence measures, requesting for real changes in the structure of base stations [2, 4].

The main problem in modern communication system is requirement for high speed data transmission and bandwidth is always at the forefront. Advanced wireless communication systems are rapidly deploying from multiuser MIMO to massive MIMO, increasing the need to integrate multiple antennas on user devices. For improvement of massive MIMO antenna technology and its efficiency assessment for wireless communication have attracted many researchers that enhance gain, channel capacity, bandwidth, polarization diversity, and reduced linkage between the elements. These systems further require that multi-elemental antenna (MEA) be reduced to the minimum so that they could replace in intensive and reliable user equipment and support multiple recycling operations in different parts of the world. Antenna design technology has huge gap in developing an optimal antenna systems for next-generation networks by improving impedance bandwidth and multi-channel capacity. Another significant issue identified with the arrangement of massive MIMO frameworks is the expense of the base station. The expansion in the quantity of receiving antennas at the base station gives an expanding in the quantity of RFCs also, bringing about restrictively high power utilization and base station's expense [15].

This paper presents an antenna selection technique for massive MIMO that involves sub-array switching architecture. Sub-array switching architecture based antenna selection technique focuses on the work of massive MIMO antenna for next generation. The rapid deployment of the Internet has led to broadband studies such as wireless local area networks (WLANs) and global wireless communication systems. From a system standpoint, an increasing amount of data and improved quality of information should be achieved to meet excessive demand for existing capabilities. Massive multiple-input-multiple-output (MIMO) antenna technology is one of the most useful ways to increase reliability and improve wireless channel capacity. Because the bandwidth is strongly dependent on the side connection between the antenna elements and its design, for such systems, it is important to point out the distinctive characteristics of the antenna in the radiation wave propagation. The sub-array switching architecture based antenna technique for massive MIMO system is one of the most effective ways to increase system reliability and enhance wireless bandwidth.

Rest of the paper is organized as follows; Sect. 2 explores the literature review related to massive MIMO and antenna selection in M-MIMO. Section 3 is about system model and proposed algorithm. Section 4 discusses about results and analysis. In Sect. 5, we present the conclusion and future directions.

2 Literature Review

In this section we explore the implemented antenna selection (AS) techniques. Several antenna selection techniques have been deployed in previous decades. At first, BAB and Greedy based antenna selection techniques are discussed. Furthermore, antenna selection optimization using limited channel state information (CSI), convex optimization based antenna selection spatially correlated with channel estimation antenna selection techniques are discussed. Moreover, computational, cashing and communication based schemes and capacity based reduced complexity selection (CBRCS) scheme are also explored.

2.1 BAB Search Based Antenna Selection

One of the optimal antenna selection techniques is BAB [16]. A. H. Land and A. G. Doig firstly recommended the Branch and bound approach [17]. The term "branch and bound" initially happened in crafted by J. D. Little on the voyaging sales rep issue [18, 19]. It's a scheming plan universally for distinct and overall improvement problems, just as technical advancement. The BAB scheme involves of an orderly description of competitor schedules by approaches for public interval look: the planning of applicant schedules is assumed as framing a conventional tree with the filled group at the origin. The scheming probes parts present tree, which state to sub-sets of the planning set.

The BAB technique is grouped up and down according to the search objective [16]. Top BAB starts the search procedure from an empty set and gradually increases the size of the order until you receive one. In contrast, the downstream BAB has the reverse method, which gradually reduces the proportion to an empty one. Obviously, when only a small number of elements are selected, the BAB is more attractive. It means, only the higher BAB to find the maximum value, not the minimum value of the objective function when selecting a beam. Instead, to find the minimum value, we need the upward monopoly objective function for the BAB to the top. Below, we give the equivalent function of the equation, which increases uniformly with respect to the size of the boundary.

Downward BAB is the extent of set that will be diminished continuously from the root hub to the terminal hubs as appeared in Fig. 1 for instance of choosing two out of five components. Names close to every hub in Fig. 1 demonstrates which component is expelled [20]. Let $Jn(Xs)$, $s > n$, indicate a downwards upper bound of all n-component subsets of Xs, that is $Jn(Xs) \geq \max Xn \subseteq Xs\ J(Xn)$. Let the lower bound of $J(Xn)$ as B: $B \leq J(Xn)$ (1) At that point it pursues that $Jn(Xs) < B \Rightarrow J(Xn) < J(Xn)$, $\forall Xn \subseteq Xs$ (2). Condition (2) shows that if the upper bound of Xs is not exactly the lower bound of the ideal arrangement. At that point the subset Xs can be pruned without assessment [20] as shown in Fig. 1.

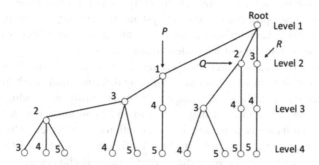

Fig. 1. Downward BAB Search

Upward BAB is the measure of set that will be expanded progressively from the root hub to the terminal hubs as appeared in Fig. 2 for instance of choosing two out of five components. Marks close to every hub, in Fig. 2 show in which component is included. Let $Jn(Xs)$, $s < n$, signify an upward upper bound of all n-component subsets of Xs, that is $Jn(Xs) \geq \max Xn \supseteq Xs\ J(Xn)$ (3). Let the lower bound of $J\,(Xn)$ still as B: $B \leq J(Xn)$ (4) At that point it pursues that $Jn(Xs) < B \Rightarrow J(Xn) < J(Xn)$, $\forall Xn \supseteq Xs$ (5). Condition (5) demonstrates that if the upper bound of Xs is not exactly the lower bound of the ideal arrangement, at that point the subset Xs can be pruned without assessment [20] as shown in Fig. 2.

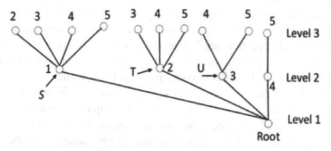

Fig. 2. Upward BAB Search

2.2 Greedy Search Based Antenna Selection

A greedy algorithm is a logical model that surveys the problem resolving experiential of building the in the neighborhood finest selection at every phase with the determined of conclusion a universal finest. In several difficulties a greedy approach isn't frequently create a finest resolution but nevertheless a greedy experimental can produce close by best clarifications that estimated a universally finest key in a sensible extent of instance [21]. For instance, a greedy approach for the TSP is the pursuing experimental: "At each step of the journey visit the nearest unvisited city." This experimental no need to propose to discover a finest explanation but it concludes in a sensible amount of stages: observing a best clarification to such a complicated difficulty usually expects excessively several phases. In numerical optimization, greedy systems in best way to explain combined issues consuming the characteristics of Metroid's and provide consistency estimates to optimization issues with the sub-integrated arrangements [22].

Methodology of greedy search based antenna selection is basically a competitor group, since through an answer has made with any determination, work that picks the finest contender for enhancing the arrangement. Attainability, work which utilized to decide whether an approximation may utilize to add to an answer with target work, which doles out an incentive to an answer, or a fractional arrangement which will show when we have found total arrangement [23]. A calculation is intended to accomplish ideal (productive) answer for a given issue. In greedy method, choices are produced using the given arrangement space. As remains greedy, the nearest arrangement that appears to give an ideal arrangement is picked. Greedy algorithms attempt to locate a limited ideal arrangement, which may in the end lead to all around improved arrangements. Be that as it may, for the most part voracious calculations don't give comprehensively enhanced arrangements [24]. Antenna selection strategies have enormous writing, a large portion of the work is accomplished for determination of antenna selection in MIMO framework or OFDM based MIMO framework. Antenna selection has been considered for MIMO with a small number of antenna selections in [25]. In [26], calculation for antenna selection in MIMO is demonstrated in which the best subset of receive antennas is chosen to boost the channel limit.

In [27], vitality proficient receives antenna selection calculation dependent on arched enhancement for massive MIMO framework has been given. In those choice criteria is to boost the vitality productivity. For that one condition is given that is, if the channel limit of the cell is bigger than a specific edge then the quantity of transmit and receive antennas, the subset of transmit receiving antennas and servable portable terminals are together advanced to boost vitality proficiency. In that, reenactment result indicates receive antennas determination utilizing given calculation shows better execution contrasting and no receive antenna selection. In [28], a model is given for antenna selection in massive MIMO systems. This framework model uses channel limit condition to make a scan for just the primary ideal receiving antennas and does not require a comprehensive inquiry to locate the staying ideal receiving antennas. It is important to send the channel state information (CSI) about the chose segment vectors of the channel from the beneficiary to transmitter as a piece of model prerequisite. In [29], transmit receiving antenna selection is given in the downlink of massive MIMO framework. In [30], a novel antenna selection consolidating plan is given for spatially correlated massive MIMO uplinks with imperfect channel estimation systems.

2.3 RF Switching Architectures in Antenna Selection

Accepting wire assurance is a different data in massive MIMO development, which uses RF changes to pick a fair sub-set of gathering mechanical assemblies. Antenna selection can moderate the need on the amount of RF handsets, as such being charming for massive MIMO structures. In massive MIMO antenna selection structures, RF changing models ought to be intentionally considered. RF switches are the fundamental RF-essential bits in the massive MIMO antenna selection structures. In large scale MIMO antenna selection structures, a totally trading RF mastermind are commonly acknowledged, which supports to pick any subset of accepting wires as showed up in Fig. 3, imply as Full-array switching (FAS). Regardless, as the MIMO estimation grows, full array RF organize ends up being less power capable and even infeasible to execute, as appeared in Fig. 4, to which we allude as Sub-array switching (SAS) [31].

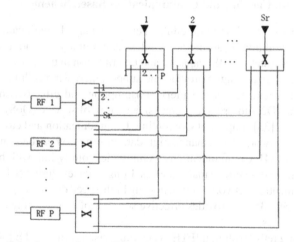

Fig. 3. Antenna Selection to choose (P < Sr) receiving antennas with the full-array RF exchanging.

In [32] antenna selection schemes are proposed, the capacity based reduced complexity selection (CBRCS) scheme is evaluated. We compared our results with CBRCS where we focused on massive MIMO antenna selection for next generation.

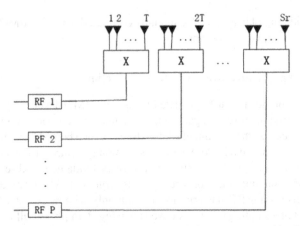

Fig. 4. Antenna Selection to choose (P < Sr) receiving antennas with the sub-array RF exchanging. For SAS, each sub-exhibit has M receive antennas.

2.4 Computation, Caching and Communication Based Schemes

In last few decades, Wireless and cellular networks played significant development and remarkable revolution regarding 5G wireless communication and made significant enhancement in base station (BS) capacity and up gradation in users perspective respectively. 5G networks are aiming to provide and support increasing cellular users in terms of high bandwidth, less delays & battery consumption and effective communication including IoTs and D2D to provide progressive quality of service (QoS) with software defined networking [33]. Impact of Computing, Communication and cashing (CCC) in terms of wireless networks to exchange high data information enables smart technologies including 5G [34]. Communications technologies jointly link with heterogeneous (HetNets) in terms of delivering smart services. Long-Term evolution (LTE) is advanced wireless communication network for transferring high-speed data among several cellular GSM based devices [35]. It provides effective services of broadcasting and covers fast data travelling devices.

Advanced long term evolution (LTE-A) is improvised form of LTE [36] with extension of bandwidth that supports > 100 MHz involving spatial multiplexing (both uplink & downlink), enhanced coverage and optimal throughput. Computing basically represents the ability of processing parameters or units e.g. microprocessors & microcontrollers and other software applications in terms of internet of things (IoTs). There are several good candidates of real-time operating systems RTOS-IoT based wireless communications networks [37]. Caching in wireless communications involves the devices transmitting data in delay applications scenarios. It improves efficiency of data transformation and boosts up information processing in pre-loaded data processed for constantly use. Caching is beneficially provides cellular and wireless communications applications regarding internet of things and its position fluctuates over offloading of communication and computation [38].

3   Proposed Subarray Based Antenna Selection Scheme

We present a Sub-array based Antenna Selection Scheme (SASS). It considered that sub-array switching architecture is unique antenna selection technique to improve performance of M-MIMO based systems, it is necessity of present widely expected data traffic efficiently and smartly because next level of advanced progressive standard for cellular communication which enhances the requirement of wireless communication at superior level of data communication. Sub-array switching architecture is one of the promising techniques to solve the data traffic issues in cellular networks which are now very important to meet the future innovation with extremely high data rate wireless connectivity. Here we consider a system with transmitting antennas Tr and receiving antennas Rr, where the data is distributed into Tr sub-antennas, which are transmitted by Tr transmit antennas respectively. Let $X \in \mathbb{C}Tt \times 1$ denote the transmit signal, then the received signal is $Y = HX + N$ where $H \in \mathbb{C}Tr \times Rt$ is the channel matrix, and $N \in \mathbb{C}Tr \times 1$ is the additive Gaussian noise. H and N have independent and identically distributed (i.i.d) entries according to $\mathbb{C} N (0,1)$ and $\mathbb{C} N (0, N0)$, respectively.

Multiple antennas are randomly scattered in each cell that potentially enhances the coverage and capacity of selected zone. Each antenna performs as base station which massively boosts the quality of service and significantly reduces computational complexity. Innovative array architecture is considered, in which the array is subdivided into one and uses an RF switch for each cluster multiple elements are used to optimize switching at the sub-level based massive MIMO system has less complexity and easy in adaptability.

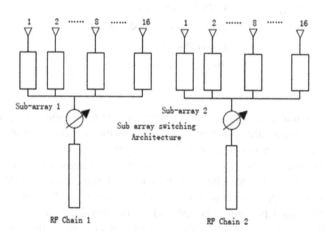

Fig. 5. Sub-array Switching Architecture based antenna selection

3.1   Sub-array Switching Architecture Based Antenna Selection

Sub-array switching architecture based antenna selection technique is a promising addition in massive MIMO systems. It mainly consists of base station which further subdivided in macro base stations and these macro base stations are furthermore divided in different branches, these branches contain massive antenna arrays. As shown in Fig. 5. Multiple antennas are sub-divided and scattered into several sub-arrays. In selecting method, Sub-array switching operates on grouped antenna arrays which begin by comparing a section in the focused array with the target value. If the target value matches the element, its position in the array is returned and the algorithm eliminates the half in which the target value cannot exists in each iteration. The suspension of unwanted signals entering the antenna mass from unknown directions is solved by adapting and interconnecting the subsets of elements from the carrier network and sub-arrays periodically positioned at array aperture, are equipped with radio frequency shifters, which increase the ratio between the desired signal strength and the direction of their arrival, and the full force of the received antenna. List of notations is illustrated in Table 1.

Table 1. List of Notations

Notation	Description
m	Level of sub-array
C_{TA}	Count of total transmitting antennas.
A_{ST}	Antenna Subset
CH_{MAT}	Channel Matrix
$[CH_{MAT}]_{A_{ST}}$	Channel submatrix (row) indexed by A_{ST}
$DMSV$	Decomposed Minimum Scalar Value
$D_{A_{ST}}$	DMSV of a matrix M
Q	Number of child nodes at $m + 1$ level
B_L	Lower Bound
TS_{TS}	Timestamp taken from TS

Sub-array switching architecture information is used to develop progressively theoretical information structures as follows; (i) when selecting an antenna in array, it rapidly adopts the easiest approach instead of visiting every node; (ii) in peak hours to meet the high data traffic load Sub-array switching architecture technique operates smartly. The normal stature of the tree approaches square foundation of the quantity of N antennas in a quicker way as compared to counterparts. In algorithm 1, a sub-array based antenna selection mechanism is presented. In steps (2) to (4), the value of m is identified which is level of sub-array where parent nodes are located. It increments the value of m until it equals the C_{TA} which is count of total transmitting antennas. After matching the desired value, it is iterated for Q values which equals number of child nodes at $m + 1$ level where only one antenna is selected out of each subset. Next, update the A_{ST} which is antenna subset for showing sub-arrays of antennas. After that, Decomposed Minimum Scalar Value (DMSV) for a matrix M is represented as $D_{A_{ST}}$ calculated for $[CH_{MAT}]_{A_{ST}}$ which is channel submatrix (row) indexed by A_{ST}. In this case, CH_{MAT} represents channel matrix. If the value of $D_{A_{ST}}$ is less than the lower bound B_L then the related values will be ignored to process the values at leaf nodes. In case of larger values of $D_{A_{ST}}$, new value of B_L is updated with $D_{A_{ST}}$ value.

In massive MIMO, the state of antennas can be computed first and then a few important calculations at prominent points can be cached for future calculations and then final antenna selection can be allocated to requesting users as per demand of spectrum and its technology. Previously cached values can be reused in future calculations. In case of modifications in configurationally values at child nodes, these cached values recalculated. It will reduce the computational cost and time.

Algorithm 1: Sub-Array based Antenna Selection Algorithm

1: Initialize $B_L = 0$, and $m = 0$.

2: **While** *m Not equals C_{TA} do*

3:   Incr m by 1

4: **End While**

5: **For** *i = 1 to Q do*

6:   Update Antenna Subset A_{ST}

7:   Calculate DMSV $D_{A_{ST}}$ of $[CH_{MAT}]A_{ST}$

6:   If $D_{A_{ST}} > B_L$ **then**

7:   Set $B_L = D_{A_{ST}}$ and $S = A_{ST}$

8:   End If

9: **End For**

10: return the final set

---

4 Results and Analysis

This section explores the results on MATLAB simulations performance of massive MIMO systems based on sub-array switching architectures. As we have antenna subset AST and channel matrix CHMAT containing m level of sub-array where parent nodes are located and $m + 1$ level where only one selected antenna is selected in each subset. Figure 6 elucidates the spectral efficiency (SE) for M-MIMO systems where total $N = 64$ antennas are considered. There are 10 users represented as U and number of RF chains represented as K which is set to 10. Spectral efficiency for i^{th} user is the sum of K sum-rate S_R values calculated as $S_R = \log_2(1 + SINR_K)$. In this case SINR is signal to interference plus noise ratio. Results show that for SNR = 10 dB, SE values are 25.16 and 28.11 bits/s/Hz for capacity based reduced complexity selection (CBRCS) and SASS respectively. Figure 7 illustrates the SE for a set of BS transmit antenna. Results show that for 40 antennas, SE values are 23.26 bits/s/Hz and 25.41 bits/s/Hz for CBRCS and SASS respectively. Proposed SASS improves SE due to sub-array based antenna selection.

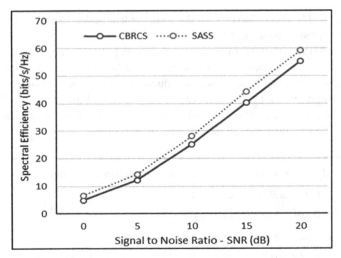

Fig. 6. Spectral Efficiency for variations in SNR

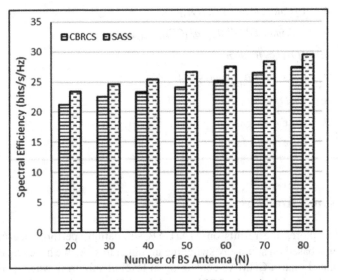

Fig. 7. Spectral Efficiency for a set of BS transmit antennas

5 Conclusion

Antenna selection is a fundamental factor in massive MIMO systems and smart decision in selecting antenna reduced the complexities in wireless communication systems.
We designed an efficient communication system based on novel antenna selection technique for massive MIMO system based on Sub-array switching architecture which effectively helpful to obtained optimal throughput. The proposed antenna selection algorithms
improved the system performance and enhanced system capacity significantly. Based on

achieved results, we attained better coverage and enhanced Quality of service (QoS). It has analyzed that SASS antenna selection technique is much faster than prior adopted techniques and progressively boosted the overall system performance. Moreover, it is proved to be a decent option to improve the energy efficiency and also helps to reduce the computational complexity and cost effective large scale multiple antenna systems. It enhances spectral efficiency when number of antennas and SNR values are varied. Results prove the supremacy of SASS as compared to counterparts.

References

1. Marzetta, T.L.: Noncooperative cellular wireless with unlimited numbers of base station antennas. IEEE Trans. Wireless Commun. **9**, 3590 (2010)
2. Health, R.W., Boccardi, F., Lozano, A., Marzetta, T.L., Popovski, P.: Five disruptive technology directions for 5G. IEEE Commun. Mag. **52**(2), 74–80 (2014)
3. Rusek, F., et al.: Scaling up MIMO: opportunities and challenges with very large arrays. arXiv preprint arXiv:1201.3210 (2012)
4. Andrews, J.G., et al.: What will 5G be? IEEE J. Sel. Areas Commun. **32**, 1065–1082 (2014)
5. Larsson, E.G., Edfors, O., Tufvesson, F., Marzetta, T.L.: Massive MIMO for next generation wireless systems. arXiv preprint arXiv:1304.6690 (2013)
6. Marzetta, T.L., Yang, H.: Fundamentals of Massive MIMO. Cambridge University Press, Cambridge (2016)
7. Alkhateeb, A., El Ayach, O., Leus, G., Heath, R.W.: Channel estimation and hybrid precoding for millimeter wave cellular systems. IEEE J. Sel. Top. Sign. Process. **8**, 831–846 (2014)
8. Roh, W., et al.: Millimeter-wave beamforming as an enabling technology for 5G cellular communications: theoretical feasibility and prototype results. IEEE Commun. Mag. **52**, 106–113 (2014)
9. Sayeed, A.M., Brady, J.H., Luo, F., Zhang, J.: Millimeter-wave MIMO transceivers: theory, design and implementation. In: Signal Processing for 5G: Algorithms and Implementations, IEEE-Wiley (2016)
10. Qian, L., Hyejung, J., Pingping, Z., Geng, W.: * 5G millimeter-wave communication channel and technology overview. In: Signal Processing for 5G: Algorithms and Implementations, pp. 354–371 (2016)
11. Bangerter, B., Talwar, S., Arefi, R., Stewart, K.: Networks and devices for the 5G era. IEEE Commun. Mag. **52**, 90–96 (2014)
12. Lu, L., Li, G.Y., Swindlehurst, A.L., Ashikhmin, A., Zhang, R.: An overview of massive MIMO: benefits and challenges. IEEE J. Sel. Top. Sig. Process. **8**, 742–758 (2014)
13. Wei, H., Wang, D., Zhu, H., Wang, J., Sun, S., You, X.: Mutual coupling calibration for multiuser massive MIMO systems. IEEE Trans. Wirel. Commun. **15**, 606–619 (2015)
14. Edfors, O., Liu, L., Tufvesson, F., Kundargi, N., Nieman, K.: Massive MIMO for 5G: theory, implementation and prototyping. In: Signal Processing for 5G: Algorithms and Implementations, pp. 189–230. Wiley (2016)
15. Björnson, E., Hoydis, J., Kountouris, M., Debbah, M.: Massive MIMO systems with non-ideal hardware: energy efficiency, estimation, and capacity limits. IEEE Trans. Inf. Theory **60**, 7112–7139 (2014)
16. Cao, Y., Kariwala, V.: Bidirectional branch and bound for controlled variable selection: part I. Principles and minimum singular value criterion. Comput. Chem. Eng. **32**, 2306–2319 (2008)
17. Land, A.H., Doig, A.G.: An automatic method for solving discrete programming problems. In: Jünger, M., et al. (eds.) 50 Years of Integer Programming 1958-2008, pp. 105–132. Springer, Berlin (2010). https://doi.org/10.1007/978-3-540-68279-0_5

18. Little, J.D., Murty, K.G., Sweeney, D.W., Karel, C.: An algorithm for the traveling salesman problem. Oper. Res. **11**, 972–989 (1963)
19. Balas, E., Toth, P.: Branch and bound methods for the traveling salesman problem. Carnegie-Mellon Univ Pittsburgh Pa Management Sciences Research Group (1983)
20. Gao, Y., Khaliel, M., Zheng, F., Kaiser, T.: Rotman lens based hybrid analog–digital beamforming in massive MIMO systems: Array architectures, beam selection algorithms and experiments. IEEE Trans. Veh. Technol. **66**, 9134–9148 (2017)
21. Cormen, T.H., Leiserson, C.E., Rivest, R.L., Stein, C.: Introduction to Algorithms, Chap. 11. The MIT Press, McGraw-Hill Book Company, Cambridge (2001)
22. Gutin, G., Yeo, A., Zverovich, A.: Traveling salesman should not be greedy: domination analysis of greedy-type heuristics for the TSP. Discrete Appl. Math. **117**, 81–86 (2002)
23. Bang-Jensen, J., Gutin, G., Yeo, A.: When the greedy algorithm fails. Discrete Optim. **1**, 121–127 (2004)
24. Bendall, G., Margot, F.: Greedy-type resistance of combinatorial problems. Discrete Optim. **3**, 288–298 (2006)
25. Molisch, A.F., Win, M.Z.: MIMO systems with antenna selection-an overview. IEEE Microwave Mag. **5**, 46–56 (2004)
26. Berenguer, I., Wang, X., Krishnamurthy, V.: Adaptive MIMO antenna selection via discrete stochastic optimization. IEEE Trans. Signal Process. **53**, 4315–4329 (2005)
27. Bibo, H., Yuanan, L., Gang, X., Fang, L., Feng, N., Jingchao, W.: Antenna selection for downlink transmission in large scale green MIMO system. In: 2014 4th IEEE International Conference on Network Infrastructure and Digital Content, pp. 312–316 (2014)
28. Al-Shuraifi, M., Al-Raweshidy, H.: Optimizing antenna selection using limited CSI for massive MIMO systems. In: Fourth edition of the International Conference on the Innovative Computing Technology (INTECH 2014), pp. 180–184 (2014)
29. Gao, X., Edfors, O., Liu, J., Tufvesson, F.: Antenna selection in measured massive MIMO channels using convex optimization. In: 2013 IEEE Globecom Workshops (GC Wkshps), pp. 129–134 (2013)
30. Mi, D., Dianati, M., Muhaidat, S., Chen, Y.: A novel antenna selection scheme for spatially correlated massive MIMO uplinks with imperfect channel estimation. In: 2015 IEEE 81st Vehicular Technology Conference (VTC Spring), pp. 1–6 (2015)
31. Gao, Y., Vinck, H., Kaiser, T.: Massive MIMO antenna selection: Switching architectures, capacity bounds, and optimal antenna selection algorithms. IEEE Trans. Signal Process. **66**, 1346–1360 (2017)
32. Lee, B.-J., Ju, S.-L., Kim, N.-i., Kim, K.-S.: Enhanced transmit-antenna selection schemes for multiuser massive MIMO systems. In: Wireless Communications and Mobile Computing, vol. 2017 (2017)
33. Farhady, H., Lee, H., Nakao, A.: Software-defined networking: a survey. Comput. Networks **81**, 79–95 (2015)
34. Bouras, M.A., Ullah, A., Ning, H.: Synergy between Communication, Computing, and Caching for Smart Sensing in Internet of Things. Procedia Comput. Sci. **147**, 504–511 (2019)
35. Crosby, G.V., Vafa, F.: Wireless sensor networks and LTE-A network convergence. In: 38th Annual IEEE Conference on Local Computer Networks, pp. 731–734 (2013)
36. Ghosh, A., Ratasuk, R., Mondal, B., Mangalvedhe, N., Thomas, T.: LTE-advanced: next-generation wireless broadband technology. IEEE Wirel. Commun. **17**, 10–22 (2010)
37. Dunkels, A., Gronvall, B., Voigt, T.: Contiki-a lightweight and flexible operating system for tiny networked sensors. In: 29th Annual IEEE International Conference on Local Computer Networks, pp. 455–462 (2004)
38. Wang, S., Zhang, X., Zhang, Y., Wang, L., Yang, J., Wang, W.: A survey on mobile edge networks: Convergence of computing, caching and communications. IEEE Access **5**, 6757–6779 (2017)

A Green SWIPT Enhanced Cell-Free Massive MIMO System for IoT Networks

Meng Wang, Haixia Zhang[⊠], Leiyu Wang, and Guannan Dong

Shandong Provincial Key Laboratory of Wireless Communication Technologies, Shandong University, Jinan 250100, China
wangmengsdu@mail.sdu.edu.cn, haixia.zhang@sdu.edu.cn

Abstract. This paper investigates the downlink performance of a green simultaneous wireless information and power transfer (SWIPT) enhanced cell-free massive multiple-input multiple-output (MIMO) system where the IoT devices are served by the virtual massive MIMO constituted by a large amount of distributed single antenna access points (APs). On the premise of that all the IoT devices can decode the information and harvest energy from the received signals, the closed-form expressions of downlink spectral efficiency (SE) and harvested energy of this system are firstly derived under the time switching (TS) and power splitting (PS) scheme. After that, to maximize the lowest SE of the IoT devices, a joint optimization problem which takes into account the power control coefficients and energy harvesting coefficients simultaneously is formulated and an optimal bisection based algorithm is proposed to solve it. Besides, the network load management problem is also studied to reduce the fronthaul burden and an AP selection method which considers the imbalance of AP transmission ratio is proposed. Simulation results show that compared with the benchmark methods, our proposed schemes can guarantee higher transmission rates for IoT devices as well as a reduction of load for fronthual links.

Keywords: IoT · SWIPT · Cell-free massive MIMO

1 Introduction

In recent years, the Internet of Things (IoT) has been considered as one of the most promising network paradigms for the next generation communication systems due to its huge potential application values. On one hand, the prevalence of IoT devices can gather accurate cities' environment parameters for data analyzing and processing, accelerating the development of smart cities. On the other hand, the ubiquitous deployment of IoT devices in plants can also further improve real-time sensing and decision-making abilities, enabling the intelligent manufacturing [1–3]. However, the IoT network performance severely suffers from the limitation of IoT devices' low battery capacity. Moreover, the widespread distribution of IoT devices and the terrible conditions where they had been

© Springer Nature Singapore Pte Ltd. 2019
H. Ning (Ed.): CyberDI 2019/CyberLife 2019, CCIS 1138, pp. 51–64, 2019.
https://doi.org/10.1007/978-981-15-1925-3_4

deployed make the work of replacing them unrealistic. Thus, how to prolong the IoT devices' life becomes a crucial problem.

As an attractive solution to improving the IoT devices' sustainability, simultaneous wireless information and power transfer (SWIPT) technology which absorbs the radio frequency signals power for information decoding (ID) and energy harvesting (EH) has drawn considerable attention [4–6]. According to the different receiver protocols, two schemes, i.e. time switching (TS) scheme and power splitting (PS) scheme have been investigated [4,5]. Particularly, the TS receiver periodically switches between ID and EH mode in the whole transmission phase with a TS ratio, while the PS receiver splits the received signals into two streams with a PS ratio for ID and EH. Nevertheless, the efficiency of information decoding and energy harvesting strongly depends on the distance between the transmitters and receivers. The long distance among them will inevitably reduce the transmission rates and power storage [7,8]. In order to narrow the distance between the transmitters and receivers, the cell-free system which puts its antennas in distributed manner, has been deemed as an effective method. The distributed antennas in cell-free system not only can be deployed close to the IoT devices to reduce the signal attenuation, but also constitute virtual massive multiple-input multiple-output (MIMO) system to further improve the devices' energy efficiency (EE) and spectral efficiency (SE). Fully understanding the SWIPT enhanced cell-free system performance in IoT scenarios is very important. Actually, the SWIPT has been introduced firstly into the cell-free massive MIMO system recently [9]. The focus of their study is to analyze the users' rate-energy tradeoff performance under the TS scheme. However, the system SE performance analysis and investigation under the PS scheme is absent. The problem of maintaining the reliable transmission rates while satisfying the energy requirement for all IoT devices is also ignored. Besides, due to the widespread of distributed APs, huge expenditure will be consumed on the fronthaul links in IoT cell-free system and how to reduce the fronthaul burden is still an open problem.

Motivated by the aforementioned problems, we consider a downlink SWIPT enhanced cell-free massive MIMO system in which the IoT devices will be served by the distributed APs. The closed-form expressions in terms of SE under TS and PS scheme are firstly investigated. To maximize the lowest SE of IoT devices, a joint optimization problem which takes account of power control coefficients and TS/PS ratio coefficients simultaneously is formulated and a bisection based iterative algorithm is proposed to solve it. After that, the fronthaul load management is also studied to alleviate the system burden and an AP selection method is proposed at the same time.

The remainder of this paper is organized as follows. In Sect. 2, the system model and transmission model are described, and the main assumptions required for analysis are introduced. Section 3 looks into the SE and energy analysis, and the exact expressions of the achievable SE and the harvested energy of each IoT device are derived. Section 4 presents the problem formulation and proposes corresponding solution to solve the joint optimization problem. Numerical results

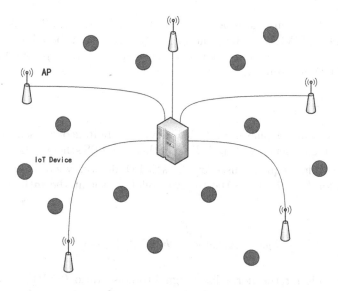

Fig. 1. System model of SWIPT enhanced cell-free massive MIMO system.

are shown in Sect. 5. Finally, Sect. 6 provides an overview of the results and concludes the paper.

2 System and Transmission Model

2.1 System Model

We consider a downlink (DL) SWIPT enhanced cell-free massive MIMO system where K single-antenna IoT devices are simultaneously served by M single-antenna APs in the same frequency band. All APs and IoT devices are geographically distributed over a wide area, as shown in Fig. 1. A central processing unit (CPU) is connected to all the APs via fronthaul links. The channel coefficient vector between the kth IoT device and the mth AP is modeled as

$$\mathbf{g}_{mk} = \sqrt{\beta_{mk}}\mathbf{h}_{mk}, \tag{1}$$

where β_{mk} denotes the large-scale fading and $\mathbf{h}_{mk} \sim \mathcal{CN}(0,1)$ is a complex Gaussian random variable which represents the small-scale fading.

For TS scheme, to support wireless information transmission (WIT) and wireless energy transfer (WET), the entire coherent time slot T_s is divided into three intervals, i.e. T_p, αT_d and $(1-\alpha)T_d$. In the first interval T_p, the IoT devices send orthogonal pilot sequences to the APs and each AP estimates the channel to all IoT devices. We assume the length of T_p for each IoT device is equally fixed because in such scenarios the distance between APs and IoT devices should be small. In the DL transmission phase, the system performs WIT and WET in the αT_d and $(1-\alpha)T_d$ intervals respectively, where $\alpha \in (0,1)$. For PS scheme, the channel estimation is also performed in the first intervals. The difference is that the WIT and WET will be simultaneously executed in the DL transmission phase, and the signal stream is split by a power splitting ratio ρ for ID and EH, where $\rho \in (0,1)$ accordingly.

2.2 Uplink Training

In the uplink training phase, all K IoT devices simultaneously send mutually orthogonal pilot sequences of length T_p in both TS and PS schemes. Let $\sqrt{T_p}\varphi_k \in \mathbb{C}^{T_p \times 1}$ be the pilot sequence used by the kth IoT device, where $\|\varphi_k\|^2 = 1$ and $\varphi_k^H \varphi_{k'} = 0$ for $k' \neq k$. Then the received pilot vector at the mth AP can be given by

$$\mathbf{y}_{p,m} = \sqrt{T_p P_p} \sum_{k=1}^{K} g_{mk}\varphi_k + \mathbf{w}_{p,m} , \tag{2}$$

where P_p stands for the normalized signal-to-noise ratio (SNR) of each pilot symbol and $\mathbf{w}_{p,m}$ is a vector of additive noise at the mth AP whose elements are i.i.d. $\mathcal{CN}(0, 1)$ random variables. The minimum mean square error (MMSE) estimation of g_{mk} is of the form

$$\hat{g}_{mk} = c_{mk}\breve{y}_{p,mk}, \tag{3}$$

where $\breve{y}_{p,mk} = \varphi_k^H \mathbf{y}_{p,m}$ is the projection of $\mathbf{y}_{p,m}$ onto φ_k, and

$$c_{mk} = \frac{\sqrt{T_p P_p}\beta_{mk}}{T_p P_p \beta_{mk} + 1}. \tag{4}$$

2.3 Downlink Transmission

In the DL transmission phase, the M APs use conjugate beamforming to transmit the precoded signals to the K IoT devices. The transmitted signal from the mth AP can be expressed as

$$x_m = \sqrt{P_d} \sum_{k=1}^{K} \sqrt{\eta_{mk}} \hat{g}_{mk}^* q_k, \tag{5}$$

where P_d is the normalized DL SNR and q_k which satisfies $\mathbb{E}\left\{|q_k|^2\right\} = 1$, is the data symbol intended for the kth user. Moreover, η_{mk} are power control coefficients that satisfy the power constraint at each AP

$$\mathbb{E}\left\{|x_m|^2\right\} \leq P_d, \tag{6}$$

which can be rewritten as

$$\sum_{k=1}^{K} \eta_{mk}\gamma_{mk} \leq 1, \tag{7}$$

where γ_{mk} is defined as $\gamma_{mk} = \mathbb{E}\left\{|\hat{g}_{mk}|^2\right\}$, and expressed as

$$\gamma_{mk} = \frac{T_p P_p \beta_{mk}^2}{T_p P_p \beta_{mk} + 1}. \tag{8}$$

Signal Model Under Time Switching Scheme. The signal received by the kth IoT device in the DL transmission phase under TS scheme for ID and EH is

$$y_k = \sqrt{P_d} \sum_{m=1}^{M} \sum_{k'=1}^{K} \sqrt{\eta_{mk'}} g_{mk} \hat{g}_{mk'}^* q_k + \mathbf{w}_{d,k} + \mathbf{n}_{TS}, \tag{9}$$

where $\mathbf{w}_{d,k}$ is additive $\mathcal{CN}(0,1)$ noise at the kth IoT device and $\mathbf{n}_{TS}$ is additive $\mathcal{CN}(0, \sigma_{TS}^2)$ noise brought by the TS receiver.

Signal Model Under Power Splitting Scheme. The signal received by the kth IoT device in the DL transmission phase under PS scheme for ID is

$$y_k = \sqrt{\rho_k} \left(\sqrt{P_d} \sum_{m=1}^{M} \sum_{k'=1}^{K} \sqrt{\eta_{mk'}} g_{mk} \hat{g}_{mk'}^* q_k + \mathbf{w}_{d,k} \right) + \mathbf{n}_{PS}, \tag{10}$$

and the signal received by the kth IoT device in the DL transmission phase under PS scheme for EH is

$$y_k = \sqrt{\varepsilon_k} \left(\sqrt{P_d} \sum_{m=1}^{M} \sum_{k'=1}^{K} \sqrt{\eta_{mk'}} g_{mk} \hat{g}_{mk'}^* q_k + \mathbf{w}_{d,k} \right) + \mathbf{n}_{PS}, \tag{11}$$

where $\varepsilon_k = 1 - \rho_k$ and $\mathbf{n}_{PS}$ is additive $\mathcal{CN}(0, \sigma_{PS}^2)$ noise brought by the PS receiver.

Thus, the energy harvested by each IoT device under TS/PS scheme can be expressed as (12) and (13) respectively

$$E_k = \xi \left(1 - \alpha_k\right) T_d P_d \left\{ \mathbb{E}\left[\left| \sum_{m=1}^{M} \sqrt{\eta_{mk}} g_{mk} \hat{g}_{mk}^* \right|^2 \right] + \sum_{k' \neq k}^{K} \mathbb{E}\left[\left| \sum_{m=1}^{M} \sqrt{\eta_{mk'}} g_{mk} \hat{g}_{mk'}^* \right|^2 \right] \right\}, \tag{12}$$

$$E_k = \xi \left(1 - \rho_k\right) T_d P_d \left\{ \mathbb{E}\left[\left| \sum_{m=1}^{M} \sqrt{\eta_{mk}} g_{mk} \hat{g}_{mk}^* \right|^2 \right] + \sum_{k' \neq k}^{K} \mathbb{E}\left[\left| \sum_{m=1}^{M} \sqrt{\eta_{mk'}} g_{mk} \hat{g}_{mk'}^* \right|^2 \right] \right\}. \tag{13}$$

3   Spectral Efficiency and Energy Analysis

In this section, we study the energy-rate tradeoff of the DL SWIPT enhanced cell-free massive MIMO system at a finite M. To do so the exact expressions of the achievable SE and the harvested energy of each IoT device are consequently derived.

3.1 Achievable Spectral Efficiency Analysis

In this subsection, we derive the closed-form expressions for the downlink SE under TS/PS scheme by using the worst-case Gaussian technique [10].

For the kth IoT device under TS scheme, the achievable SE can be expressed as

$$R_k = \alpha_k \log_2 \left(1 + \text{SNIR}_k \right), \tag{14}$$

where SINR_k stands for the signal to interference plus noise ratio of the kth IoT device, which can be expressed in the form of

$$\text{SINR}_k = \frac{|\text{DS}_k|^2}{\text{BU}_k + \sum_{k' \neq k}^{K} \text{UI}_{kk'} + \sigma_{TS}^2 + 1}, \tag{15}$$

where DS_k, BU_k, $\text{UI}_{kk'}$ represent the effect of the desired signal, the beamforming gain uncertainty, the interference caused by the k'th IoT device and can be given below

$$\text{DS}_k \triangleq \sqrt{P_d} \mathbb{E} \left\{ \sum_{m=1}^{M} \sqrt{\eta_{mk}} g_{mk} \hat{g}_{mk}^* \right\}, \tag{16}$$

$$\text{BU}_k \triangleq \sqrt{P_d} \sum_{m=1}^{M} \sqrt{\eta_{mk}} g_{mk} \hat{g}_{mk}^* - \text{DS}_k, \tag{17}$$

$$\text{UI}_{kk'} \triangleq \sqrt{P_d} \sum_{m=1}^{M} \sqrt{\eta_{mk'}} g_{mk} \hat{g}_{mk'}^*. \tag{18}$$

Thus, by the similar method in [11] the achievable rates for the DL SE under TS scheme is

$$R_k =$$
$$\alpha_k \log_2 \left(1 + \frac{P_d \left(\sum_{m=1}^{M} \sqrt{\eta_{mk}} \gamma_{mk} \right)^2}{P_d \sum_{k'=1}^{K} \sum_{m=1}^{M} \eta_{mk'} \gamma_{mk'} \beta_{mk} + 1 + \sigma_{TS}^2} \right). \tag{19}$$

Similar steps are applied to obtain the achievable rates for the DL SE under PS scheme, which can be written as

$$R_k =$$
$$\log_2 \left(1 + \frac{\rho_k P_d \left(\sum_{m=1}^{M} \sqrt{\eta_{mk}} \gamma_{mk} \right)^2}{\rho_k \left(P_d \sum_{k'=1}^{K} \sum_{m=1}^{M} \eta_{mk'} \gamma_{mk'} \beta_{mk} + 1 \right) + \sigma_{PS}^2} \right). \tag{20}$$

3.2   Achievable Harvested Energy Analysis

In this subsection, the achievable harvested energy at the kth IoT device under TS scheme is derived using (12)

$$
E_k =
$$
$$
(1 - \alpha_k) T_d P_d \left[\left(\sum_{m=1}^{M} \sqrt{\eta_{mk}} \gamma_{mk} \right)^2 + \sum_{k'=1}^{K} \sum_{m=1}^{M} \eta_{mk'} \gamma_{mk'} \beta_{mk} \right].
$$
(21)

Similarly, the achievable harvested energy at the kth IoT device under PS scheme using (13) is

$$
E_k =
$$
$$
(1 - \rho_k) T_d P_d \left[\left(\sum_{m=1}^{M} \sqrt{\eta_{mk}} \gamma_{mk} \right)^2 + \sum_{k'=1}^{K} \sum_{m=1}^{M} \eta_{mk'} \gamma_{mk'} \beta_{mk} \right].
$$
(22)

4   Problem Statement and Joint Optimization

In this section, we propose a joint optimization problem which selects the power coefficients $\{\eta_{mk}\}$ as well as the energy harvesting coefficient $v_k \in \{\alpha_k, \rho_k\}$ to maximize the smallest SE of all IoT devices under energy harvesting constraints. We also propose an AP selection method to reduce the fronthaul payload data transmission burden.

4.1   Problem Formulation

With the preliminaries above, the optimization problem can be formulated as

$$
(\text{P0}): \max_{\{v_k, \eta_{mk}\}} \min_k \ R_k, \tag{23a}
$$

$$
\text{s. t.} \quad \sum_{k=1}^{K} \eta_{mk} \gamma_{mk} \leq 1, \quad m = 1, \cdots, M, \tag{23b}
$$

$$
E_k \geq Q_k \quad k = 1, \cdots, K, \tag{23c}
$$

$$
\eta_{mk} \geq 0 \quad k = 1, \cdots, K, m = 1, \cdots, M, \tag{23d}
$$

$$
0 \leq v_k \leq 1, \tag{23e}
$$

where the constraint (23c) means the harvested energy at the kth IoT device must reach a given threshold to fulfill the energy request of the battery. We propose a bisection based iterative algorithm presented below to solve the problem.

Algorithm 1. Bisection Based Iterative Algorithm

1: For each IoT device, set initial v_{min}, v_{max} as $0, 1$. Choose a tolerance $\epsilon > 0$.
2: **repeat**
3:     Let $v_k = \frac{(v_{min}+v_{max})}{2}$.
4:     Set the interval (t_l, t_u) that contains the optimal value t^*. Choose a tolerance $\lambda > 0$.
5:     **repeat**
6:         Check the feasibility of the midpoint $t = \frac{(t_l+t_u)}{2}$.
7:         **if** t is feasible. **then**
8:             Let $t_l = t$.
9:         **else**
10:            Let $t_u = t$.
11:        **end if**
12:    **until** $t_u - t_l < \epsilon$.
13:    Output the optimal value $\left\{\eta_{mk}^{(*)}\right\}$ and E_k.
14:    **if** $E_k \geq Q_k$. **then**
15:        Let $v_{min} = v_k$.
16:    **else**
17:        Let $v_{max} = v_k$.
18:    **end if**
19: **until** $v_{max} - v_{min} < \epsilon$.
20: Output the optimal solution $\{\eta_{mk}^*, v_k^*\}$.

Time Switching Scheme. Specifically, for the case that adopts TS scheme, the optimization problem is written as

$$(P1): \max_{\{\alpha_k, \eta_{mk}\}} \quad \min_k \quad R_k(\alpha_k, \eta_{mk}), \tag{24a}$$

$$\text{s. t.} \quad \sum_{k=1}^{K} \eta_{mk}\gamma_{mk} \leq 1, \quad m = 1, \cdots, M, \tag{24b}$$

$$E_k \geq Q_k \quad k = 1, \cdots, K, \tag{24c}$$

$$\eta_{mk} \geq 0 \quad k = 1, \cdots, K, m = 1, \cdots, M, \tag{24d}$$

$$0 \leq \alpha_k \leq 1, \tag{24e}$$

The problem is not jointly convex with respect to $\{\eta_{mk}\}$ and $\{\alpha_k\}$. Nevertheless, for any optimization problems, it is possible to tackle the problem over some of the variables and then over the remaining ones [12]. Therefore, we can optimize η_{mk} with fixed α_k and optimize α_k with fixed η_{mk}, respectively.

The constraint (24c) implies we can always diminish α_k to obtain a larger rates as long as the harvested energy is greater than the threshold Q_k.

Thus problem (P2) is obtained with fixed α_k, which can be written as

$$(\text{P2}) : \max_{\{\eta_{mk}\}} \quad \min_{k} \quad \text{SINR}_k, \tag{25a}$$

$$\text{s. t.} \quad \sum_{k=1}^{K} \eta_{mk}\gamma_{mk} \le 1, \quad m = 1, \cdots, M, \tag{25b}$$

$$\eta_{mk} \ge 0 \quad k = 1, \cdots, K, m = 1, \cdots, M, \tag{25c}$$

where $\text{SINR}_k = \dfrac{P_d\left(\sum_{m=1}^{M} \sqrt{\eta_{mk}}\gamma_{mk}\right)^2}{P_d \sum_{k'=1}^{K} \sum_{m=1}^{M} \eta_{mk'}\gamma_{mk'}\beta_{mk}+1+\sigma_{TS}^2}$ according to (20).

The above problem is quasi-concave. Thus, we can formulate the problem into the equivalent form

$$(\text{P3}) : \max_{\{\eta_{mk}\},t} \quad t, \tag{26a}$$

$$\text{s. t.} \quad \text{SINR}_k \ge t, \quad k = 1, \cdots, K, \tag{26b}$$

$$\sum_{k=1}^{K} \eta_{mk}\gamma_{mk} \le 1, \quad m = 1, \cdots, M, \tag{26b}$$

$$\eta_{mk} \ge 0 \quad k = 1, \cdots, K, m = 1, \cdots, M, \tag{26c}$$

We can see that for a fixed t, the domain of constraints in (26) is convex and thus we can determine whether is feasible or not for a given t. The explicit solution to this problem can be found in [11].

Power Splitting Scheme. Similarly, the optimization problem for the case that adopts PS scheme can be expressed as

$$(\text{P4}) : \max_{\{\rho_k,\eta_{mk}\}} \quad \min_{k} \quad R_k\left(\rho_k, \eta_{mk}\right), \tag{27a}$$

$$\text{s. t.} \quad \sum_{k=1}^{K} \eta_{mk}\gamma_{mk} \le 1, \quad m = 1, \cdots, M, \tag{27b}$$

$$E_k \ge Q_k \quad k = 1, \cdots, K, \tag{27c}$$

$$\eta_{mk} \ge 0 \quad k = 1, \cdots, K, m = 1, \cdots, M, \tag{27d}$$

$$0 \le \rho_k \le 1, \tag{27e}$$

where $\text{SINR}_k = \dfrac{\rho_k P_d\left(\sum_{m=1}^{M} \sqrt{\eta_{mk}}\gamma_{mk}\right)^2}{\rho_k\left(P_d \sum_{k'=1}^{K} \sum_{m=1}^{M} \eta_{mk'}\gamma_{mk'}\beta_{mk}+1\right)+\sigma_{PS}^2}$ according to (21).

We can see that if we set $\{\eta_{mk}\}$ fixed, the function $\text{SINR}_k\left(\rho_k\right) = \frac{\rho_k a}{\rho_k b+c}$ is monotonically increasing within $(0, 1)$, and the constraint $E_k\left(\rho_k\right) \ge Q_k, \quad k = 1, \cdots, K$ is a convex set, which indicates the feasibility of a bisection based algorithm. Thus, the Algorithm 1 is reasonably applicable to the PS scheme as well.

AP Selection Method. The payload data transmission from CPU to the APs through the fronthaul links generates huge power consumption [13]. In order to relieve the fronthaul burden while maintaining a relatively good performance, we propose an AP selection method that turns on and off certain data stream dedicated to the kth IoT device from the mth AP. To measure the fronthaul expense, we define fronthaul usage as $FU = \frac{\sum_{k=1}^{K} M_k}{MK} \times 100\%$, where M_k is the number of APs that transmit signal to the kth IoT device.

We can find the mathematic form of the signal from the APs can be interpreted as the accumulation of the average transmit power for each IoT device (7), that is $p_{mk} = P_d \eta_{mk} \gamma_{mk}$. Thus, we define AP transmission ratio $p(m,k) = \frac{\eta_{mk} \gamma_{mk}}{\sum_{m=1}^{M} \eta_{mk} \gamma_{mk}}$ to represent the ratio of the power given by the mth AP to the total power collected by the kth IoT device. Corresponding algorithm is described as

---

Algorithm 2. AP Selection Algorithm

1: Perform Algorithm 1 to obtain $\{\eta_{mk}, v_k\}$.
2: Compare the $p(m,k)$ with a given threshold p_{th}.
3: If $p(m,k) \leq p_{th}$, set $p_{mk} = 0$.
4: Perform Algorithm 1 with the newly updated $\{p_{mk}\}$.
5: Output the optimal value $\{\eta_{mk}^*, v_k^*\}$.

5　Numerical Results

We assume that all M APs and K IoT devices are randomly distributed in a square area of 100×100 m^2. We use COST Hata model to formulate the large scale fading coefficients β_{mk} which can be written as

$$10 \log_{10}(\beta_{mk}) = -136 - 35 \log_{10}(d_{mk}) + X_{mk}, \tag{28}$$

where d_{mk} is the distance between AP m and IoT device k in kilometers and $X_{mk} \sim \mathcal{N}(0, \sigma_{shad}^2)$ with $\sigma_{shad}^2 = 8$ dB. The transmit powers of downlink data and pilot symbols are 400 mW and 200 mW, respectively. The corresponding normalized transmit SNRs P_p and P_d can be obtained by dividing these powers by the noise power, which is given by

$$\text{noise power} = B \times k_B \times T_0 \times \text{NF}, \tag{29}$$

where $k_B = 1.381 \times 10^{-23}$ (Joule per Kelvin), $\text{NF} = 9$ dB are Boltzmann constant and $T_0 = 290$ (Kelvin) is the noise temperature. The bandwidth B is 20 MHz and the carrier frequency is 1.9 GHz. The length of all sequences is $T_p = K$.

Firstly, the rate-energy tradeoff for TS scheme and PS scheme with power control are shown in Fig. 2 where the vertical coordinate is the ratio of the sum energy collected by all IoT devices to the maximum quantities of energy collected by all IoT devices as $\mathbf{v} = \mathbf{1}$, and the horizontal coordinate is the sum

SE represented by average achievable rates (bits per second per Hertz). We set the low and high receiver noise respectively, and the tradeoff of TS scheme shows a linear relation between energy and rate as α varies from 0 to 1. Besides, the tradeoff of PS scheme with low receiver noise shows a box-shaped curve as in [14], proving the optimal power control can reach the upper bound defined in [14].

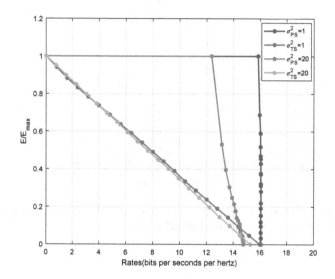

Fig. 2. The tradeoff between the average achievable sum rates and the normalized harvested energy. Here, $M = 60, K = 20, \sigma_v^2 = 1, 20$ respectively.

Besides the rate-energy tradeoff of the system, we next look into the performance in terms of achievable rates under TS scheme and PS scheme shown in Figs. 3 and 4 respectively. Both schemes compare the proposed optimal algorithm with the heuristic uniform power control method [15], the full power control method [15], and the AP selection method. For the convenience of simulation, we set $Q_k = 0.4E_{kmax}$ for all K IoT devices, where E_{kmax} is the energy harvested by the kth IoT device as $\rho_k = 1$. We can see the optimal power transmission shows about a 1.92 fold of improvement over the full power transmission. We set $p_{th} = 1/M$ as the threshold for AP selection method. The simulation shows a slight reduction of 1.28 fold over the optimal power transmission because it is equivalent to solving the problem with a sparse channel estimation matrix and this leads to the performance reduction. We can see that the performance under TS scheme is scaled down by roughly 40% compared to PS scheme because of the harvested energy threshold $Q_k = 0.4E_{kmax}$, which corresponds to the rate-energy tradeoff under low receiver noise in Fig. 3.

Finally, the fronthaul usage versus the average achievable rates of per IoT device is shown in Fig. 5. The dashed lines are the average achievable rates of IoT device without AP selection method, while the curves are that with AP

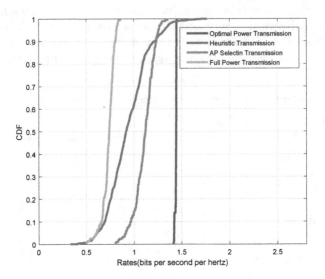

Fig. 3. CDFs of the average DL achievable rates of different transmission methods under TS scheme. Here, $M = 60, K = 20, \sigma^2_{TS} = 1$.

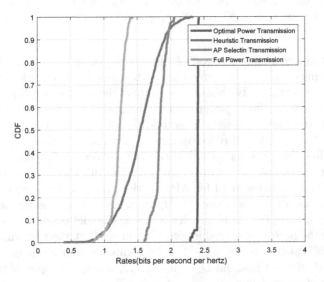

Fig. 4. CDFs of the average DL achievable rates of different transmission methods under PS scheme. Here, $M = 60, K = 20, \sigma^2_{PS} = 1$.

selection method under different thresholds p_{th}. We can see that it only requires about 40% fronthaul usage to achieve 90% of the per IoT device rates of the optimal transmission. The slope of the curves becoming diminished indicates as the distance grows, far-off devices are more difficult to get effective services from certain AP due to path loss and it is energy efficient to turn off the far-off devices' data streams in the corresponding fronthaul links.

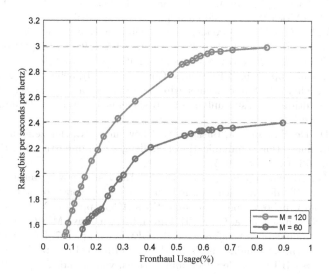

Fig. 5. Per IoT device rates versus fronthaul usage, where $K = 20$, $Q_k = 0$ for all K IoT devices.

6   Conclusion

This paper considered a green SWIPT enhanced cell-free massive MIMO system for IoT networks. The closed-form expressions of the system's achievable per IoT device SE and harvested energy were firstly derived. An optimal bisection based iterative algorithm that jointly optimized the power control coefficients and energy harvesting coefficients were proposed. Moreover, an AP selection method were proposed that saved huge fronthaul usage while maintaining good performance. The results showed the proposed algorithms guaranteed an uniformly good transmission rates, energy supplement for IoT devices as well as a reduction of fronthaul burden in the green SWIPT enhanced cell-free massive MIMO system.

Acknowledgment. This work is supported by the National Natural Science Foundation of China under Grant No. 61671278, 61801266.

References

1. Albreem, M.A.M., et al.: Green Internet of Things (IoT): an overview. In: 2017 IEEE 4th International Conference on Smart Instrumentation, Measurement and Application (ICSIMA), pp. 1–6, November 2017
2. Huang, J., Meng, Y., Gong, X., Liu, Y., Duan, Q.: A novel deployment scheme for green Internet of Things. IEEE Internet Things J. **1**(2), 196–205 (2014)
3. Dlodlo, N., Gcaba, O., Smith, A.: Internet of Things technologies in smart cities. In: 2016 IST-Africa Week Conference, pp. 1–7, May 2016
4. Zhang, R., Ho, C.K.: Mimo broadcasting for simultaneous wireless information and power transfer. IEEE Trans. Wirel. Commun. **12**(5), 1989–2001 (2013)
5. Liu, L., Zhang, R., Chua, K.: Wireless information and power transfer: a dynamic power splitting approach. IEEE Trans. Commun. **61**(9), 3990–4001 (2013)
6. Wang, X., Zhai, C.: Simultaneous wireless information and power transfer for downlink multi-user massive antenna-array systems. IEEE Trans. Commun. **65**(9), 4039–4048 (2017)
7. Mishra, D., De, S., Jana, S., Basagni, S., Chowdhury, K., Heinzelman, W.: Smart RF energy harvesting communications: challenges and opportunities. IEEE Commun. Mag. **53**(4), 70–78 (2015)
8. Mishra, D., Alexandropoulos, G.C., De, S.: Harvested power fairness optimization in MISO SWIPT multicasting IoT with individual constraints. In: 2018 IEEE International Conference on Communications (ICC), pp. 1–6, May 2018
9. Shrestha, R., Amarasuriya, G.: SWIPT in cell-free massive MIMO. In: 2018 IEEE Global Communications Conference (GLOBECOM), pp. 1–7, December 2018
10. Daza, L., Misra, S.: Fundamentals of Massive Mimo (Marzetta, T., et al.: 2016) [book reviews]. IEEE Wireless Communications, vol. 25, no. 1, pp. 9–9, February 2018
11. Ngo, H.Q., Ashikhmin, A., Yang, H., Larsson, E.G., Marzetta, T.L.: Cell-free massive mimo versus small cells. IEEE Trans. Wirel. Commun. **16**(3), 1834–1850 (2017)
12. Boyd, S., Vandenberghe, L., Faybusovich, L.: Convex optimization. IEEE Trans. Autom. Control **51**(11), 1859–1859 (2006)
13. Tombaz, S., Monti, P., Farias, F., Fiorani, M., Wosinska, L., Zander, J.: Is backhaul becoming a bottleneck for green wireless access networks? In: 2014 IEEE International Conference on Communications (ICC), pp. 4029–4035, June 2014
14. Zhou, X., Zhang, R., Ho, C.K.: Wireless information and power transfer: architecture design and rate-energy tradeoff. IEEE Trans. Commun. **61**(11), 4754–4767 (2013)
15. Nayebi, E., Ashikhmin, A., Marzetta, T.L., Yang, H., Rao, B.D.: Precoding and power optimization in cell-free massive mimo systems. IEEE Trans. Wirel. Commun. **16**(7), 4445–4459 (2017)

Non-orthogonal Multiple Access
in Coordinated LEO Satellite Networks

Tian Li[1,2(✉)], Xuekun Hao[3], Guoyan Li[1], Hui Li[1], and Xinwei Yue[4]

[1] The 54th Research Institute of China Electronics Technology Group Corporation,
Shijiazhuang 050081, Hebei, China
t.li@ieee.org
[2] Beijing University of Posts and Telecommunications, Beijing 100876, China
[3] CETC Advanced Mobile Communication Innovation Center, Shanghai, China
[4] School of Information and Communication Engineering, Beijing Information
Science and Technology University, Beijing 100101, China

Abstract. Non-orthogonal multiple access (NOMA) has been widely considered to improve the spectral efficiency in terrestrial wireless networks. In this paper, we extend the idea to satellite networks and propose a NOMA-based scheme for coordinated low Earth orbit satellite systems, where the beam-edge user is supported by two satellites. By exploiting the difference of the equivalent downlink channel gains, users located both at the beam-center and beam-edge can be served simultaneously using NOMA. It is shown that the NOMA-based cooperative method is capable of providing a higher system capacity while guarantee the rate quality of the beam-edge user.

Keywords: Cooperative transmission · Non-orthogonal multiple access · Satellite networks

1 Introduction

Recent years have witnessed extensive studies focusing on key technologies for 5G&Beyond [1–5]. Among them, non-orthogonal multiple access (NOMA) has been developed as a promising technology to improve the achievable rate without occupying additional resources [6, 7]. In NOMA, users' signals are enabled to be transmitted in the same time and frequency domain with the help of different power allocation levels [8].

Recognizing such an advantage, NOMA began to be applied in satellite communication systems [9–12]. In [10], the original NOMA was investigated in satellite networks, where the outage probability and average symbol error rate were also analyzed. Simulation results confirm the performance advantage of the NOMA-based scheme. By employing multiple antennas at satellites, a two-user transmission method using NOMA is studied in [11]. In this work, recognizing the difference of user link gains, a beamforming-aided superposition coding (SC) approach is proposed to enhance the throughput performance. Inspired

H. Ning (Ed.): CyberDI 2019/CyberLife 2019, CCIS 1138, pp. 65–78, 2019.
https://doi.org/10.1007/978-981-15-1925-3_5

by NOMA strategies studied in terrestrial coordinated systems [13], cooperative transmission methods have also been developed in satellite networks [14–16]. To fully exploit the potential of the strong user, Yan *et al.* in [14] developed a two time-slot cooperative NOMA scheme. In the first time-slot, the satellite transmits a superposed signal to a pair of users. After splitting the signal at the stronger user side, the corresponding information will be pushed forward to the weak user in the second time-slot. The superiority of the proposed method is validated in terms of the ergodic sum rate. For the application of NOMA in coordinated satellite-terrestrial networks, Zhu *et al.* in [15] proposed a NOMA-aided beamforming scheme, where a user pairing algorithm is further studied to enhance the system capacity. However, the NOMA-aided method investigated for coordinated system in [15] is only considered for the terrestrial network, which might not help to understand the performance in coordinated satellite networks.

In this paper, a NOMA-based cooperative satellite transmission approach is proposed for the purpose of enhancing system spectral efficiency, where a two-point coordinated low Earth orbit (LEO) satellite networks is set up carefully. To improve the quality-of-service for the beam-edge user, two satellites transmit the same signals in a cooperative way. Furthermore, by applying SC and successive interference cancellation (SIC) receivers, beam-center and beam-edge users are enabled to be served simultaneously. Simulation results demonstrate that the proposed method is capable of providing a higher achievable rate than that of the conventional frequency division multiple access (FDMA)-based cooperative approach.

The rest of the paper is organized as follows. Section 2 first presents the system model for coordinated LEO satellite networks. Then, the conventional cooperative transmission using FDMA is analyzed where the achievable sum rate is also derived. In Sect. 3, a NOMA-based transmission method is proposed and analyzed for the coordinated system. For performance evaluation purposes, the ergodic sum rate is derived for the two schemes in Sect. 4. Simulation results are presented and discussed in Sect. 5 and conclusions are drawn in Sect. 6.

Notation: The statistical expectation is presented by $\mathbb{E}[\cdot]$. While $\mathcal{CN}(a, c)$ denotes the distribution of circularly symmetric complex Gaussian (CSCG) random variables with mean a and variance c.

2   System Model

2.1   Channel Model in LEO Satellite Systems

Throughout the paper, we consider a coordinated LEO satellite system constitutes two satellites and three users. As shown in Fig. 1, Users 1 and 2, named the beam-center user, locate at the center of satellite beams. For LEOs 1 and 2, User e can be seen as the beam-edge user. We assume that satellites and users are equipped with reflector and single antennas, respectively.

In general, the signals from satellites to ground users may experience large-scale (mainly free-space path loss) and small-scale fadings [17]. Hence, the down-link channel from LEO i to User j can be modeled as

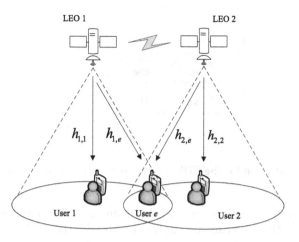

Fig. 1. System model for the two-point coordinated LEO satellite network.

$$h_{i,j} = \rho_{i,j} \sqrt{G_r} \sqrt{\left(\frac{\lambda}{4\pi d_{i,j}}\right)^2} \sqrt{\eta_{i,j} G_t}, \tag{1}$$

where $\rho_{i,j} \sim \mathcal{CN}(0,1)$ represents the channel coefficient. For the factors in the free-space path loss, λ denotes the wavelength, $d_{i,j}$ is the distance between LEO i and User j. The main lobe gains of transmit and receive antennas are presented by G_t and G_r, respectively. Since an attenuation of the antenna gain may exist at the beam edge, an attenuation factor $0 < \eta_{i,j} \leq 1$ is needed to make the channel precisely modeled.

2.2   Cooperative Transmission in FDMA Scheme

For conventional FDMA-based transmission scheme in coordinated LEO satellite systems, as shown in Fig. 2, Users 1 and 2 are allocated in the same frequency band which is orthogonal to User e. In this paper, we assume that the bandwidth assigned to Users 1 and 2 is β Hz, respectively, where the bandwidth for User e is $1 - \beta$ Hz at one satellite. Then, signals arrived at Users 1 and 2 are given by

$$y_1 = h_{1,1}x_1 + h_{2,1}x_2 + n;$$
$$y_2 = h_{2,2}x_2 + h_{1,2}x_1 + n. \tag{2}$$

Similarly, the signal received by User e can be written as

$$y_e = h_{1,e}x_e + h_{2,e}x_e + n$$
$$= (h_{1,e} + h_{2,e})x_e + n, \tag{3}$$

where $n \sim \mathcal{CN}(0, \sigma^2)$ denotes the background noise at users. The x_1, x_2, and x_e represent the signals for Users 1, 2, and e, respectively.

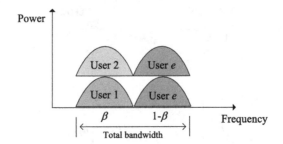

Fig. 2. An illustration for the FDMA-based scheme in the coordinated system.

Generally, the transmit power for Users 1 and 2 is the same, i.e., $P = \mathbb{E}[x_i]$, $i \in \{1, 2\}$. Then, the following assumption can be derived in detail.

Assumption 1. *Considering the pattern of spot beams and the much longer travel distance for signals from adjacent satellites, we have $|h_{i,i}x_i|^2 \gg |h_{j,i}x_j|^2$, $i, j \in \{1, 2\}$.*

According to Assumption 1, (2) can be approximated as

$$
\begin{aligned}
y_1 &= h_{1,1}x_1 + h_{2,1}x_2 + n \\
&\approx h_{1,1}x_1 + n; \\
y_2 &= h_{2,2}x_2 + h_{1,2}x_1 + n \\
&\approx h_{2,2}x_2 + n.
\end{aligned} \tag{4}
$$

To obtain a better performance, the maximal ratio combining (MRC) algorithm is applied at receivers. Thus, the signal-to-interference-plus-noise ratios (SINRs) at Users 1, 2, and e can be calculated as

$$
\mathrm{SINR}_1 = \frac{|h_{1,1}|^2 P}{\beta \sigma^2};
$$

$$
\mathrm{SINR}_2 = \frac{|h_{2,2}|^2 P}{\beta \sigma^2};
$$

$$
\mathrm{SINR}_e = \frac{(|h_{1,e}|^2 + |h_{2,e}|^2)P_e}{(1 - \beta)\sigma^2}, \tag{5}
$$

where P_e represents the signal power for User e. According to the Shannon formula [17], the achievable rates of users are derived as

$$
R_1 = \beta \log_2(1 + \frac{|h_{1,1}|^2 P}{\beta \sigma^2});
$$

$$
R_2 = \beta \log_2(1 + \frac{|h_{2,2}|^2 P}{\beta \sigma^2});
$$

$$
R_e = (1 - \beta) \log_2 \left(1 + \frac{(|h_{1,e}|^2 + |h_{2,e}|^2)P_e}{(1 - \beta)\sigma^2}\right). \tag{6}
$$

Finally, the achievable sum rate can be computed as $R = R_1 + R_2 + R_e$.

From (6), it can be found that the frequency resource has not been fully exploited. Thanks to the different link gains among users, we propose a NOMA-based transmission approach for the coordinated LEO satellite network.

3 The Proposed NOMA-Based Transmission Method in the Coordinated LEO Satellite Network

In the proposed scheme, like the strategy described above in cooperative FDMA, Users 1 and 2 will receive signals from LEOs 1 and 2, respectively. However, both satellites also transmit signals for User e in the same frequency band by using SC, as shown in Fig. 3. Hence, the signals at the three users are given by

$$y_1' = h_{1,1}(x_1 + x_e) + h_{2,1}(x_2 + x_e) + n;$$
$$y_2' = h_{2,2}(x_2 + x_e) + h_{1,2}(x_1 + x_e) + n;$$
$$y_e' = h_{1,e}(x_1 + x_e) + h_{2,e}(x_2 + x_e) + n. \tag{7}$$

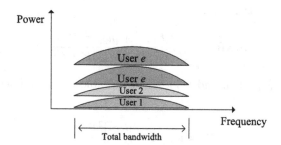

Fig. 3. An illustration for the NOMA-based scheme in the coordinated system.

According to Assumption 1, (7) can be further expressed as

$$y_1' = h_{1,1}x_1 + (h_{1,1} + h_{2,1})x_e + h_{2,1}x_2 + n$$
$$\approx h_{1,1}x_1 + (h_{1,1} + h_{2,1})x_e + n;$$
$$y_2' = h_{2,2}x_2 + (h_{2,2} + h_{1,2})x_e + h_{1,2}x_1 + n$$
$$\approx h_{2,2}x_2 + (h_{2,2} + h_{1,2})x_e + n;$$
$$y_e' = (h_{1,e} + h_{2,e})x_e + h_{1,e}x_1 + h_{2,e}x_2 + n. \tag{8}$$

Note that in satellite NOMA, the transmission power for User e is always higher than that for the other two users [14]. The component $h_{i,j}x_e$ in y_j', $i, j \in \{1, 2\}$ could not be ignored. In this case, SIC receivers are deployed at Users 1

and 2 to provide useful information after decoding and cancelling the signal for
User e. Thus, the SINRs at User 1 can be calculated as

$$\text{SINR}'_{e1} = \frac{(|h_{1,1}|^2 + |h_{2,1}|^2)P_e}{|h_{1,1}|^2 P + \sigma^2};$$

$$\text{SINR}'_1 = \frac{|h_{1,1}|^2 P}{\sigma^2}. \tag{9}$$

Also, we can derive the SINRs at User 2 as

$$\text{SINR}'_{e2} = \frac{(|h_{2,2}|^2 + |h_{1,2}|^2)P_e}{|h_{2,2}|^2 P + \sigma^2};$$

$$\text{SINR}'_2 = \frac{|h_{2,2}|^2 P}{\sigma^2}. \tag{10}$$

In (9) and (10), SINR'_1 and SINR'_2 represent the SINR for Users 1 and 2,
respectively. Moreover, SINR'_{e1} and SINR'_{e2} denote the SINR for the decoded
signals of User e at Users 1 and 2, respectively. Since the signal power of User
e is much higher than that of Users 1 and 2, we can obtain the useful information directly at User e by regarding x_1 and x_2 as interferences. Thus, the
corresponding SINR is given by

$$\text{SINR}'_{ee} = \frac{(|h_{1,e}|^2 + |h_{2,e}|^2)P_e}{|h_{1,e}|^2 P + |h_{2,e}|^2 P + \sigma^2}. \tag{11}$$

According to (9), (10), and (11), the achievable rates for Users 1 and 2 are
given by

$$R'_1 = \log_2(1 + \frac{|h_{1,1}|^2 P}{\sigma^2});$$

$$R'_2 = \log_2(1 + \frac{|h_{2,2}|^2 P}{\sigma^2}). \tag{12}$$

To make sure the information for Users 1 and 2 can be correctly decoded,
the maximum transmission rate for User e has to be upper-bound as

$$R'_e = \log_2(1 + \min\{\text{SINR}'_{e1}, \text{SINR}'_{e2}, \text{SINR}'_{ee}\}). \tag{13}$$

Consequently, we have the achievable sum rate calculated as $R' = R'_1 + R'_2 + R'_e$.

4   Performance Analysis

Different from the work presented in [12], we evaluate the performance in terms
of the ergodic rate in this section as small-scale fading is involved in the channel
model.

4.1 Ergodic Sum Rate for FDMA Scheme

We assume the transmit power at one satellite is denoted by $\bar{P}$. By defining α the power factor for User e, we have $P_e = \alpha\bar{P}$ and $P = (1 - \alpha)\bar{P}$, $0 < \alpha < 1$. Let $\gamma = \bar{P}/\sigma^2$ be the transmit signal-to-noise ratio (SNR). The achievable rates in (6) can be further expressed as

$$R_1 = \beta \log_2 \left(1 + \frac{|h_{1,1}|^2(1-\alpha)\bar{P}}{\beta\sigma^2}\right)$$

$$= \beta \log_2(1 + |h_{1,1}|^2 \frac{(1-\alpha)}{\beta}\gamma);$$

$$R_2 = \beta \log_2 \left(1 + \frac{|h_{2,2}|^2(1-\alpha)\bar{P}}{\beta\sigma^2}\right)$$

$$= \beta \log_2(1 + |h_{2,2}|^2 \frac{(1-\alpha)}{\beta}\gamma);$$

$$R_e = (1-\beta) \log_2 \left(1 + \frac{(|h_{1,e}|^2 + |h_{2,e}|^2)\alpha\bar{P}}{(1-\beta)\sigma^2}\right)$$

$$= (1-\beta) \log_2(1 + (|h_{1,e}|^2 + |h_{2,e}|^2)\frac{\alpha}{1-\beta}\gamma). \tag{14}$$

Since $\rho_{i,j} \sim \mathcal{CN}(0,1)$ is assumed in (1), we have $h_{i,j} \sim \mathcal{CN}(0, \eta_{i,j} \sin^2\theta_{i,j}L)$, where $L = G_r(\lambda/(4\pi d))^2 G_t$, d is the orbit height, and $\theta_{i,j}$ is the elevation angle from User j to LEO i. Note that Users 1 and 2 locate at the beam center while User e locates at the beam edge. Then, it can be derived that $h_{i,i} \sim \mathcal{CN}(0, L)$ and $h_{i,e} \sim \mathcal{CN}(0, \eta \cos^2(\phi/2)L)$, where ϕ represents the beam-width. With $\mathbb{E}[\log_2(1 + X)] = \frac{1}{\ln 2}\int_0^\infty \frac{1-F_X(x)}{1+x}dx$ [18], where X is a random variable and $F_X(x)$ represents the cumulative distribution function, we have the following theorem.

Theorem 1. Let $\bar{L} = \cos^2(\phi/2)L$. The ergodic sum rate for FDMA-based coordinated LEO satellite network is $\mathbb{E}[R] = \mathbb{E}[R_1] + \mathbb{E}[R_2] + \mathbb{E}[R_e]$, where

$$\mathbb{E}[R] = -2\beta\frac{1}{\ln 2}e^{\frac{\beta}{L\gamma(1-\alpha)}} \mathrm{Ei}\left(-\frac{\beta}{L\gamma(1-\alpha)}\right)$$

$$+ (1-\beta)\frac{1}{\ln 2}\left[-e^{\frac{1-\beta}{\eta\bar{L}\gamma\alpha}} \mathrm{Ei}\left(-\frac{1-\beta}{\eta\bar{L}\gamma\alpha}\right)\right.$$

$$\left. + \frac{1-\beta}{\eta\bar{L}\gamma\alpha}e^{\frac{1-\beta}{\eta\bar{L}\gamma\alpha}} \mathrm{Ei}\left(-\frac{1-\beta}{\eta\bar{L}\gamma\alpha}\right) + 1\right]. \tag{15}$$

Proof. See Appendix A.

4.2 Ergodic Sum Rate for the Proposed Approach

To obtain analytical expressions for NOMA-based cooperative transmission, we reuse some critical conclusions discussed in Appendix A. Since the achievable rates for Users 1 and 2 derived in NOMA scheme are similar to those presented in (6), we can easily obtain the ergodic rates for Users 1 and 2 as

$$\mathbb{E}[R_1'] = \mathbb{E}[R_2'] = -\frac{1}{\ln 2} e^{\frac{1}{L\gamma(1-\alpha)}} \text{Ei}\left(-\frac{1}{L\gamma(1-\alpha)}\right). \tag{16}$$

From the achievable rate in (13), it can be seen that the ergodic rate for User e is not easy to derive. For tractable analysis, we give the following lemma under the background of satellite networks.

Lemma 1. *Consider the noise power is always low in satellite communications [15]. The achievable rate of User e can be calculated as*

$$R_e' = \log_2(1 + \min\{\text{SINR}_{e1}', \text{SINR}_{e2}', \text{SINR}_{ee}'\})$$
$$= \log_2(1 + \text{SINR}_{ee}'). \tag{17}$$

Proof. For notational simplicity, we use $|h|^2$ and $|\bar{h}|^2$ to represent $|h_{i,j}|^2$ for $i = j$ and $i \neq j$, $i, j \in \{1, 2\}$, respectively in this proof. Then, SINR_{e1}' and SINR_{e2}' are lower-bounded by

$$\text{SINR}_{e1}' = \text{SINR}_{e2}' = \frac{(|h|^2 + |\bar{h}|^2)P_e}{|h|^2 P + \sigma^2}$$

$$> \frac{|h|^2 P_e}{|h|^2 P + \sigma^2} = \frac{|h|^2 \alpha}{|h|^2(1-\alpha) + 1/\gamma}. \tag{18}$$

Moreover, SINR_{ee}' in (11) can be further expressed as

$$\text{SINR}_{ee}' = \frac{(|h_{1,e}|^2 + |h_{2,e}|^2)P_e}{|h_{1,e}|^2 P + |h_{2,e}|^2 P + \sigma^2}$$

$$= \frac{(|h_{1,e}|^2 + |h_{2,e}|^2)\alpha}{(|h_{1,e}|^2 + |h_{2,e}|^2)(1-\alpha) + 1/\gamma}. \tag{19}$$

Since γ is relative high, we have

$$\text{SINR}_{e1}' = \text{SINR}_{e2}' > \frac{|h|^2 \alpha}{|h|^2(1-\alpha) + 1/\gamma} \approx \frac{\alpha}{1-\alpha} \tag{20}$$

and

$$\text{SINR}_{ee}' \approx \frac{\alpha}{1-\alpha}. \tag{21}$$

Consequently, it can be derived from the above that $\min\{\text{SINR}_{e1}', \text{SINR}_{e2}', \text{SINR}_{ee}'\} = \text{SINR}_{ee}'$.

Based on Lemma 1, the ergodic sum rate for NOMA-based scheme can be found in the following results.

Theorem 2. *The ergodic sum rate for NOMA-based coordinated LEO satellite network is* $\mathbb{E}[R'] = \mathbb{E}[R_1'] + \mathbb{E}[R_2'] + \mathbb{E}[R_e']$, *where*

$$\mathbb{E}[R'] = -\frac{2}{\ln 2} e^{\frac{1}{L\gamma(1-\alpha)}} \operatorname{Ei}\left(-\frac{1}{L\gamma(1-\alpha)}\right)$$

$$+ \frac{1}{\ln 2} \int_0^{\frac{\alpha}{1-\alpha}} \frac{\left(\frac{x}{\eta \bar{L}(\alpha\gamma - x(1-\alpha)\gamma)} + 1\right) e^{-\frac{x}{\eta \bar{L}(\alpha\gamma - x(1-\alpha)\gamma)}}}{1 + x} dx. \tag{22}$$

Proof. See Appendix B.

5   Results

In this part, simulation results are presented to validate performance advantages of the NOMA-based cooperative transmission scheme. We assume the system works at the frequency of $f = 1\,\mathrm{GHz}$ and the orbit height is $d = 700\,\mathrm{km}$ [9]. The satellite beam-width is $\phi = 30°$. Moreover, the main lobe gains for the satellite and user antennas are set to $G_t = 47\,\mathrm{dB}$ and $G_r = 0\,\mathrm{dB}$. Specifically, the satellite beam attenuation for the edge user is assumed to be $\eta = 0.5$.

In this paper, the following simulations are considered to validate the advantage of the proposed method and the correctness of the derived expressions:

- "NOMA-based": the ergodic sum rate of the proposed scheme using (22);
- "FDMA-based": the ergodic sum rate of FDMA-based scheme using (15);
- "NOMA-based sim": the ergodic sum rate of the proposed scheme using numerical simulation;
- "FDMA-based sim": the ergodic sum rate of FDMA-based scheme using numerical simulation.

Ergodic sum rates for NOMA-based and FDMA-based schemes are shown in Fig. 4. We set the transmit SNR $\gamma = 115\,\mathrm{dB}$. Without the loss of generality, $\beta = 0.5$ is assumed throughout the simulations. Note that the power allocation factor α for User e varies from 0.6 to 0.9 for fairness reasons. From Fig. 4, one can observe that the ergodic sum rate achieved by applying NOMA is higher than that by FDMA, which confirms the advantage of the proposed scheme. As expected, curves derived by using theoretical expressions closely follow those by numerical simulations, which validates the correctness of Theorems 1 and 2. It is noteworthy that the performance degrades with the increasing α, since the transmission rate for the edge user could not be apparently improved. More details can be found in Fig. 5.

Figure 5 shows the ergodic rates for User e and the sum of Users 1 and 2 under different transmission strategies, respectively. As expected, the channel capacity of User e grows with α because more power is allocated. On the contrary, the sum rate of Users 1 and 2 degrades accordingly. It is also shown in Figs. 4 and 5 that an optimal ergodic sum rate could be found by designing a proper power allocation algorithm. Overall, the proposed NOMA-based approach outperforms the conventional FDMA-based method on the aspect of ergodic sum rate. In

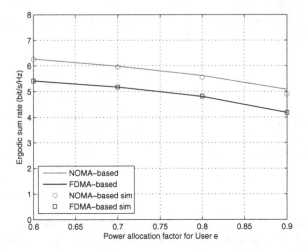

Fig. 4. The ergodic sum rate versus power allocation factor for User e.

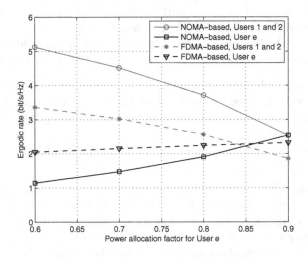

Fig. 5. The ergodic rate of different users versus power allocation factor for User e.

practical systems, the proposed method can be used to improve the capacity when transmitted signals experience obviously different attenuations.

In Fig. 6, we illustrate the ergodic sum rate versus γ with $\alpha = 0.7$ and 0.6. It is observed that ergodic rates increase with a better transmit condition. Note that an enhanced performance can be expected by setting $\alpha = 0.6$ as more power is allocated to strong users. Again, marginal performance differences between theoretical and numerical results confirm the correctness of expressions for the ergodic sum rate.

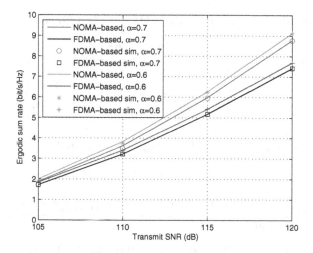

Fig. 6. The ergodic sum rate versus transmit SNR.

6   Conclusion

In this paper, we proposed a NOMA-based transmission approach for coordinated LEO satellite networks where the beam-edge user can be served by two satellites simultaneously. Thanks to the gaps among equivalent channel gains, users at different locations are enabled to be served in the same time and frequency resource. Particularly, the ergodic sum rates for NOMA and FDMA schemes are derived for performance analysis. Simulation results validate the advantage of the novel NOMA-based method.

Acknowledgment. This work was supported by the Project funded by China Postdoctoral Science Foundation (2019M661051), the Key Research and Cultivation Project at Beijing Information Science and Technology University (5211910924), the National Natural Science Foundation of China (61571338), the key research and development plan of Shaanxi province (2017ZDCXL-GY-05-01), and the Xi'an Key Laboratory of Mobile Edge Computing and Security (201805052-ZD3CG36).

Appendix

A   Proof of Theorem 1

Since $|h_{i,i}|^2 \sim \exp(1/L)$ and $|h_{i,e}|^2 \sim \exp(1/(\eta\bar{L}))$, we have

$$\mathbb{E}[R_1] = \mathbb{E}[R_2] = \beta\frac{1}{\ln 2}\int_0^\infty \frac{1 - (1 - e^{-\frac{\beta}{L\gamma(1-\alpha)}x})}{1+x}dx. \tag{23}$$

Based on the result derived in [19] (Page 341, line 2), (23) can be derived as

$$\mathbb{E}[R_1] = \mathbb{E}[R_2] = \beta \frac{1}{\ln 2} \int_0^\infty \frac{e^{-\frac{\beta}{L\gamma(1-\alpha)}x}}{1+x} dx$$

$$= -\beta \frac{1}{\ln 2} e^{\frac{\beta}{L\gamma(1-\alpha)}} \operatorname{Ei}\left(-\frac{\beta}{L\gamma(1-\alpha)}\right), \tag{24}$$

where $\operatorname{Ei}(x) = -\int_{-x}^\infty e^{-r}/r \, dr$.

From (14), the ergodic rate of User e can be calculated as

$$\mathbb{E}[R_e] = \mathbb{E}[(1-\beta)\log_2(1+(|h_{1,e}|^2+|h_{2,e}|^2)\frac{\alpha}{1-\beta}\gamma)]$$

$$= (1-\beta)\frac{1}{\ln 2}\int_0^\infty \frac{1-\Pr((|h_{1,e}|^2+|h_{2,e}|^2)\frac{\alpha}{1-\beta}\gamma \le x)}{1+x} dx. \tag{25}$$

Let $Y \triangleq |h_{1,e}|^2 + |h_{2,e}|^2$. Then, the probability density function of Y can be expressed as $f_Y(y) = (\frac{1}{\eta \bar{L}})^2 y e^{-y/(\eta \bar{L})}$. From this, we have

$$\Pr((|h_{1,e}|^2+|h_{2,e}|^2)\frac{\alpha}{1-\beta}\gamma \le x)$$

$$= \int_0^{(1-\beta)x/(\alpha\gamma)} \left(\frac{1}{\eta \bar{L}}\right)^2 y e^{-y/(\eta \bar{L})} dy$$

$$= 1 - \left(1 + \frac{(1-\beta)x}{\eta \bar{L} \alpha \gamma}\right) e^{-\frac{(1-\beta)x}{\eta \bar{L} \alpha \gamma}}. \tag{26}$$

Substituting (26) into (25), it can be derived that

$$\mathbb{E}[R_e] = (1-\beta)\frac{1}{\ln 2}\int_0^\infty \frac{(1+\frac{(1-\beta)x}{\eta \bar{L} \alpha \gamma})e^{-\frac{(1-\beta)x}{\eta \bar{L} \alpha \gamma}}}{1+x} dx$$

$$= (1-\beta)\frac{1}{\ln 2}\left(\int_0^\infty \frac{e^{-\frac{(1-\beta)x}{\eta \bar{L} \alpha \gamma}}}{1+x} dx + \int_0^\infty \frac{\frac{(1-\beta)x}{\eta \bar{L} \alpha \gamma}e^{-\frac{(1-\beta)x}{\eta \bar{L} \alpha \gamma}}}{1+x} dx\right)$$

$$= (1-\beta)\frac{1}{\ln 2}\left[-e^{\frac{1-\beta}{\eta \bar{L} \gamma \alpha}}\operatorname{Ei}\left(-\frac{1-\beta}{\eta \bar{L} \gamma \alpha}\right)\right.$$

$$\left. + \frac{1-\beta}{\eta \bar{L} \gamma \alpha}e^{\frac{1-\beta}{\eta \bar{L} \gamma \alpha}}\operatorname{Ei}\left(-\frac{1-\beta}{\eta \bar{L} \gamma \alpha}\right) + 1\right]. \tag{27}$$

Then, we can derive the result in Theorem 1.

The proof is completed.

B   Proof of Theorem 2

According to Lemma 1, the ergodic rate for User e is given by

$$\mathbb{E}[R_e'] = \mathbb{E}[\log_2(1+X)] = \frac{1}{\ln 2}\int_0^\infty \frac{1-F_X(x)}{1+x} dx, \tag{28}$$

where $X = \frac{(|h_{1,e}|^2+|h_{2,e}|^2)\alpha\gamma}{(|h_{1,e}|^2+|h_{2,e}|^2)(1-\alpha)\gamma+1}$.

Since $Y \triangleq |h_{1,e}|^2 + |h_{2,e}|^2$, which has been defined in above, $F_X(x)$ can be further expressed as

$$F_X(x) = \Pr(X \le x) = \Pr\left(\frac{Y\alpha\gamma}{Y(1-\alpha)\gamma+1} \le x\right)$$

$$= \Pr(Y(\alpha\gamma - x(1-\alpha)\gamma) \le x). \tag{29}$$

In (29), when $\alpha\gamma - x(1-\alpha)\gamma > 0$, i.e., $x < \frac{\alpha\gamma}{(1-\alpha)\gamma}$, we have

$$\Pr(Y(\alpha\gamma - x(1-\alpha)\gamma) \le x) = \Pr\left(Y \le \frac{x}{\alpha\gamma - x(1-\alpha)\gamma}\right)$$

$$= \int_0^{\frac{x}{(\alpha\gamma-x(1-\alpha)\gamma)}} \left(\frac{1}{\eta\bar{L}}\right)^2 y e^{-y/(\eta\bar{L})} dy$$

$$= \left(-\frac{x}{\eta\bar{L}(\alpha\gamma - x(1-\alpha)\gamma)} - 1\right)e^{-\frac{x}{\eta\bar{L}(\alpha\gamma-x(1-\alpha)\gamma)}} + 1. \tag{30}$$

For the case of $\alpha\gamma - x(1-\alpha)\gamma < 0$, it can be found that $\Pr(Y(\alpha\gamma - x(1-\alpha)\gamma) \le x) = \Pr(Y \ge \frac{x}{\alpha\gamma-x(1-\alpha)\gamma}) = 1$ as Y is always a positive random variable. Thus, (29) becomes

$$F_X(x) = \begin{cases} \left(-\frac{x}{\eta\bar{L}(\alpha\gamma-x(1-\alpha)\gamma)} - 1\right)e^{-\frac{x}{\eta\bar{L}(\alpha\gamma-x(1-\alpha)\gamma)}} + 1, \\ \qquad\qquad\qquad\qquad\qquad x < \frac{\alpha\gamma}{(1-\alpha)\gamma} \\ 1, \qquad\qquad\qquad\qquad x > \frac{\alpha\gamma}{(1-\alpha)\gamma}. \end{cases} \tag{31}$$

Substituting (31) into (28), we have

$$\mathbb{E}[R'_e] = \frac{1}{\ln 2} \int_0^{\frac{\alpha}{1-\alpha}} \frac{\left(\frac{x}{\eta\bar{L}(\alpha\gamma-x(1-\alpha)\gamma)} + 1\right)e^{-\frac{x}{\eta\bar{L}(\alpha\gamma-x(1-\alpha)\gamma)}}}{1+x} dx. \tag{32}$$

From (16) and (32), we can derive Theorem 2.
The proof is completed.

References

1. Bai, L., Li, T., Liu, J., et al.: Large-scale MIMO detection using MCMC approach with blockwise sampling. IEEE Trans. Commun. **64**(9), 3697–3707 (2016)
2. Ning, H., Liu, H., Ma, J., et al.: Cybermatics: cyber-physical-social-thinking hyperspace based science and technology. Future Gener. Comput. Syst. **56**, 504–522 (2016)
3. Ning, H., Liu, H.: Cyber-physical-social-thinking space based science and technology framework for the Internet of Things. Sci. China Inf. Sci. **58**(3), 1–19 (2015)
4. Liu, L., Chen, C., Qiu, T., et al.: A data dissemination scheme based on clustering and probabilistic broadcasting in VANETs. Veh. Commun. **13**, 78–88 (2018)
5. Chen, C., Jin, Y., Pei, Q., et al.: A connectivity-aware intersection-based routing in VANETs. EURASIP J. Wirel. Commun. Netw. **2014**(1), 42 (2014)

6. Chen, Y., Bayesteh, A., Wu, Y., et al.: Toward the standardization of non-orthogonal multiple access for next generation wireless networks. IEEE Commun. Mag. **56**(3), 19–27 (2018)
7. Liu, Y., Xing, H., Pan, C., et al.: Multiple-antenna-assisted non-orthogonal multiple access. IEEE Wirel. Commun. **25**(2), 17–23 (2018)
8. Saito, Y., Kishiyama, Y., Benjebbour, A., et al.: Non-orthogonal multiple access (NOMA) for cellular future radio access. In: Proceedings of IEEE Vehicular Technology Conference, Dresden, Germany (2013)
9. Zhang, X., Zhu, L., Li, T., et al.: Multiple-User transmission in space information networks: architecture and key techniques. IEEE Wirel. Commun. **26**(2), 17–23 (2019)
10. Yan, X., Xiao, H., Wang, C., et al.: Performance analysis of NOMA-based land mobile satellite networks. IEEE Access **6**, 31327–31339 (2018)
11. Shi, X., Thompson, J.S., Safari, M., et al.: Beamforming with superposition coding in multiple antenna satellite communications. In: Proceedings of IEEE ICC Workshops, pp. 705–710 (2017)
12. Caus, M., Vazquez, M.A., Perez-Neira, A.: NOMA and interference limited satellite scenarios. In: Proceedings of IEEE ACSSC, pp. 497–501. Pacific Grove (2017)
13. Choi, J.: Non-orthogonal multiple access in downlink coordinated two-point systems. IEEE Commun. Lett. **18**(2), 313–316 (2014)
14. Yan, X., Xiao, H., An, K., et al.: Hybrid satellite terrestrial relay networks with cooperative non-orthogonal multiple access. IEEE Commun. Lett. **22**(5), 978–981 (2018)
15. Zhu, X., Jiang, C., Kuang, L., et al.: Non-orthogonal multiple access based integrated terrestrial-satellite networks. IEEE J. Sel. Areas Commun. **35**(10), 2253–2267 (2017)
16. Yan, X., An, K., Liang, T., et al.: The application of power-domain non-orthogonal multiple access in satellite communication networks. IEEE Access **7**, 63531–63539 (2019)
17. Bai, L., Zhu, L., Choi, J., et al.: Cooperative transmission over Rician fading channels for geostationary orbiting satellite collocation system. IET Commun. **11**(4), 538–547 (2017)
18. Yue, X., Liu, Y., Kang, S., et al.: Exploiting full/half-duplex user relaying in NOMA systems. IEEE Trans. Commun. **66**(2), 560–575 (2018)
19. Gradshteyn, I.S., Ryzhik, I.M.: Table of Integrals, Series and Products, 7th edn. Academic, New York (2007)

Multi-sensor Data Fusion Based on Weighted Credibility Interval

Jihua Ye$^{(\boxtimes)}$, Shengjun Xue, and Aiwen Jiang$^{(\boxtimes)}$

College of Computer and Information Engineering, Jiangxi Normal University,
Nanchang 330022, China
yjhwcl@163.com, flix_2001@163.com

Abstract. The Dempster-Shafter combination rule often get wrong results when dealing with severely conflicting information. The existing typical improvement methods are mostly based on the similarity of attributes such as evidence distance, similarity and information entropy attribute as evidence weight correction evidence itself. Ultimately, the final weights of the evidences are applied to adjust the bodies of the evidences before using the Dempster's combination rule. The fusion results of these typical methods are not ideal for some complex conflict evidence. In this paper, we propose a new improved method of conflict evidence based on weighted credibility interval. The proposed method considers the credibility degree and the uncertainty measure of the evidences which respectively based on the Sum of Absolute Difference among the propositions and the credibility interval lengths. Then the original evidence is modified with the final weight before using the Dempster's combination rule. The numerical fusion example has verified that the proposed method is feasible and improved, in which the basic probability assignment (BPAs) to identify the correct target is 99.21%.

Keywords: Weighted credibility interval · Sum of absolute difference ·
Credibility interval length · Data fusion

1 Introduction

Dempster-Shafer evidence theory is an uncertainty reasoning method, which was firstly proposed by Dempster, and had been developed by Shafer. There are many advantages in Dempster's evidence theory. Firstly, a complete prior probability model is not required. After that, it can effectively deal with uncertain information, which is why it is widely applied in various fields [1, 2]. When combining highly conflicting evidence data, Dempster-Shafter evidence theory [3] often get counter-intuitive results. The improved method mainly divided into two categories, the first type is to revise the Dempster's combination rule, and the second type is to preprocess the body of evidence. The first kind of research work mainly includes Smets's method, Dubois and Prade's rule and Yager's combination rule [4]. However, the new combination rules tend to destroy the good mathematical properties of the original composition rules. More importantly, they cannot handle counter-intuitive error caused by sensor failure. Many researches tend

© Springer Nature Singapore Pte Ltd. 2019
H. Ning (Ed.): CyberDI 2019/CyberLife 2019, CCIS 1138, pp. 79–91, 2019.
https://doi.org/10.1007/978-981-15-1925-3_6

to preprocess the evidence that to solve the problem of highly contradictory evidence fusion.

The second kind of research work mainly considers the uncertainty measure and the credibility degree, which include Murphy's average method [5], Deng [6] proposed a method based on weighted average of the masses based on the evidence distance, Zhang's method [7], Yuan's method [8] based on vector space and the belief entropy. Xiao [9] proposed a multi-sensor fusion method combining belief divergence and Deng's entropy, which has a good effect. Most of these improvement methods only consider the similarity between evidences, and rarely consider the influence of evidences themselves. For complex evidence cases, there is still some room for improvement to the final fusion result.

In practical applications, when a sensor obtains information from a signal source, there is no uncertainty in the process of information generation and transmission. While in the aspect of information reception, with the reliability of the sensor, there is uncertainty in the information source obtained by different sensors.

In order to solve this problem, a new uncertainty parameter that is credibility interval lengths, is first proposed to measure the influence of proposition on final weight. Based on that, a new multi-sensor data fusion method is proposed. This method not only considers the credibility of the evidence, but also considers the influence of the uncertainty measure of the evidence on the weight. The credibility interval length of evidence interval $[Bel(A), Pl(A)]$ is used to represent the uncertainty degree of evidence itself. For each evidence, calculate the difference between the basic probability assignment and the proposition assignment under other evidence. Then the Sum of Absolute Difference indicates the support of other evidences, thus make a dent in the conflict degree between the evidences. We calculate two correction parameters based on the credibility interval lengths and the Sum of Absolute Difference among the propositions. Then the original evidence is modified with the final weight before using the Dempster's combination rule. The flow chart of the proposed method is shown in Fig. 1. To summarize, the primary contributions of this paper are listed as follows:

(1) We propose a new uncertainty parameter, the length of the interval, to measure the impact of the proposition on the final weight. Based on this, a new multi-sensor data fusion method is proposed.
(2) We validated the effectiveness of the proposed method in two real evidence cases and outperformed the existing excellent fusion methods.

The rest of this paper is organized as follows. In Sect. 2, the theoretical basis and improvement work of this paper are introduced, including a brief introduction to DS evidence theory and a new uncertainty parameter. Based on credibility interval lengths, an improved method for dealing with conflict evidence is proposed. The effectiveness of the proposed method is illustrated in Sect. 3 on two practical cases of target recognition and fault diagnosis. Finally, the conclusions and the next research work are given in Sect. 4.

2 Multi-sensor Data Fusion Based on Weighted Credibility Interval

2.1 Dempster-Shafer Evidence Theory

Dempster-Shafer evidence theory is an uncertainty reasoning method for dealing with uncertain problems. It is able to reason flexibly and has good effects even if there is no complete prior probability, or no knowledge of prior knowledge. And it is widely applied in dealing with uncertain problems. The basic model framework of Dempster-Shafer evidence theory is introduced in Fig. 1.

Definition 1 (Frame of discernment). In Dempster-Shafer evidence theory, a complete set of mutually exclusive events is defined as a frame of discernment which is described as $\Theta = \{\theta_1, \theta_2, \ldots, \theta_n\}$. Where the set of all subsets in the identification framework is defined as 2^θ.

Definition 2 (Basic Probability Assignment Function). For a frame of discernment Θ, the basic probability assignment function is a mapping from 2^θ to [0, 1], which satisfies the following condition, $m(\phi) = 0, \sum_{A \in \Theta} m(A) = 1$. The basic probability of any proposition A in the set 2^θ is assigned m(A), indicating the support of the inference model to proposition A.

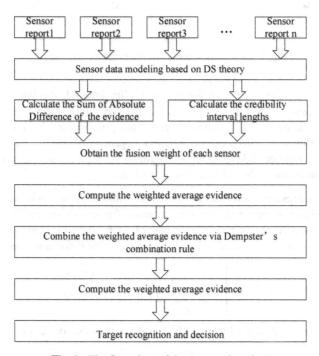

Fig. 1. The flow chart of the proposed method.

Definition 3 (Belief function). For the proposition A which $A \subseteq \theta$ in the recognition framework, the trust function Bel: $2^\theta \to [0, 1]$, and is defined as,

$$\text{Bel}(A) = \sum_{B \subseteq A} m(B). \tag{1}$$

Where Bel(A) is called the trust degree of proposition A, which indicates the overall credibility of proposition A in all evidences. Meanwhile, the plausibility function is defined as,

$$Pl(A) = 1 - Bel(\bar{A}) = \sum_{B \cap A \neq \phi} m(B). \tag{2}$$

Where $\bar{A} = \theta - A$. Obviously, $Pl(A)$ is greater than or equal to $Bel(A)$. The credibility interval consists of the belief function and the plausible function. And the relationship among them is shown in Fig. 2.

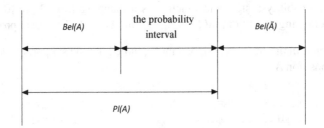

Fig. 2. Schematic diagram of the evidence interval.

Define interval [Bel(A), Pl(A)] for the credibility interval, the interval is neither support the proposition A, nor refuse to it. But the credibility interval lengths reflects the relevance difference of proposition A and other propositions. When the length at 0, namely the Bel(A) = Pl(A), evidence theory degenerates into the Bayes reasoning method. The average confidence interval length of each piece of evidence is used to represent the influence of the evidence on the weight.

Definition 4 (Dempster's combination rule). Assuming m_1 and m_2 is two sets of basic probability assignment under the same recognition framework Θ, and the two sets of basic probability assignment are independent of each other, the combination rule is represented by symbols $\oplus$, and $m = m_1 \oplus m_2$, the specific definition is as follows.

$$m(A) = \begin{cases} 0 & A = \Phi \\ \frac{1}{1-k} \sum_{\cap A_i = A} \prod_{j=1}^{n} m_j(A_i) & A \neq \Phi \end{cases} \tag{3}$$

Where k is the conflict coefficient, which is defined as the formula (4).

$$k = \sum_{\cap A_i = \Phi} \prod_{j=1}^{n} m_j(A_i), 0 \leq k \leq 1. \tag{4}$$

2.2 Improved Method Based on Credibility Interval

Similar to the improvement ideas [5–7], the proposed method does not change the Dempster-Shafter combination rule, but focuses on solving the problem that the evidence theory cannot integrate the evidence of serious conflicts effectively. So the evidences of multiple sensors can be better applied to Dempster-Shafter combination rule. The improvement of the proposed method is to propose a weight based on credibility interval to represent the influence of the evidence itself on the final weight. And the propositional Sum of Absolute Difference of evidence is used to represent the similarity between evidence. The final weight is composed of these two modified parameters is used to preprocess the original evidence and weaken the degree of conflict between the evidences. For each body of evidence, we calculate the absolute difference between each proposition and other propositions, and measure the trust degree between evidences with the difference sum. On the other hand, we calculate the propositional credibility interval lengths of evidence. The longer the interval is, the more relevance different it is with other propositions, and the greater the weight is. For the convenience of assignment calculation, the complement set between the regions is used as the direct parameter. The reliability of the evidence itself is measured by the length of the average confidence interval. The final adjustment weight is obtained by modifying the two parameters, and then it is combined with the original evidence to obtain the weighted average evidence. Finally, the final fusion result is obtained by using the Dempster-Shafter evidence combination rule.

Under the frame of discernment Θ, $m_i(.)(1 \leq i \leq n)$ is the basic probability assignment of the sensor i. The method based on weighted credibility interval is calculated as follows.

Step 1: The evidence interval matrix can be calculated by formula (5), and the range of evidence can be calculated by formula (6). The complement of credibility interval is used as the direct calculation parameter.

$$EIM_i = [Bel(.), Pl(.)]. \tag{5}$$

$$ROE_i(.) = 1 - |Pl(.) - Bel(.)|. \tag{6}$$

Step 2: The Sum of Absolute Difference between each proposition and the others can be calculated by

$$Suad_i(A) = \sum_{A \subseteq 2^\theta, i \neq j} |m_i(A) - m_j(A)|. \tag{7}$$

Step 3: The average value of each raw on two matrices ROE and $Suad$, which is denoted as two adjustment parameters.

$$\tilde{R}OE_i = \frac{ROE_i(.)}{2^\theta}, 1 \leq i \leq n. \tag{8}$$

$$\tilde{S}uad_i = \frac{Suad_i(.)}{2^\theta}, 1 \leq i \leq n. \tag{9}$$

Step 4: Support for evidence is expressed as Sup which is calculated by multiplying the two parameters obtained in step 3, then squaring, and taking the reciprocal.

$$\text{Sup}_i = \frac{1}{\left(\tilde{R}OE_i * \tilde{S}uad_i\right)^2}, 1 \leq i \leq n. \tag{10}$$

Step 5: The support of evidence is normalized as the final weight adjustment.

$$w_i = \tilde{S}\text{up}_i = \frac{Sup_i}{\sum\limits_{s=1}^{n} Sup_s}, 1 \leq i \leq n. \tag{11}$$

Step 6: The weighted average evidence can be calculated by

$$m(\{.\}) = \sum m_i(\{.\}) * W_i \tag{12}$$

Step 7: The modified evidence is combined n-1 times by Dempster's combination rule.

$$\tilde{m}(A) = \oplus_{i=1}^{n-1}(m_w). \tag{13}$$

Where, m_w is the weighted average evidence calculated in step 6.

3 Examples

In order to verify the improvement of the proposed method on the conflict evidence problem, experiments are carried out through MALTAB simulation, and the experimental results are compared and analyzed with the recent studies.

3.1 Example of Target Recognition

An evidence data from reference [6] is shown in Table 1. This is a multi-sensor based target recognition problem, an object set is defined as $\Theta = \{A, B, C\}$. The evidence information is independently collected by five different types of sensors, whose evidences are processed as basic probability assignments (BBAs).

Step 1: The evidence interval matrix can be calculated by formula (5), and the range of evidence can be calculated by formula (6).

$$EIM = \begin{pmatrix} [0.41, 0.41] \ [0.29, 0.29] \ [0.30, 0.30] \ [0.71, 0.71] \\ [0.00, 0.00] \ [0.90, 0.90] \ [0.10, 0.10] \ [0.10, 0.10] \\ [0.58, 0.93] \ [0.07, 0.07] \ [0.00, 0.35] \ [0.93, 0.93] \\ [0.55, 0.90] \ [0.10, 0.10] \ [0.00, 0.35] \ [0.90, 0.90] \\ [0.60, 0.90] \ [0.10, 0.10] \ [0.00, 0.35] \ [0.90, 0.90] \end{pmatrix}$$

$$ROE = \begin{bmatrix} 1 & 1 & 1 & 1 \\ 1 & 1 & 1 & 1 \\ 0.65 & 1 & 0.65 & 0.07 \\ 0.65 & 1 & 0.65 & 0.10 \\ 0.7 & 1 & 0.70 & 0.10 \end{bmatrix}$$

Table 1. The BBAs of a multi-sensor based target recognition.

BBAs	$\{A\}$	$\{B\}$	$\{C\}$	$\{A, C\}$
$S_1 : m_1(.)$	0.41	0.29	0.30	0.00
$S_2 : m_2(.)$	0.00	0.90	0.10	0.00
$S_3 : m_3(.)$	0.58	0.07	0.00	0.35
$S_4 : m_4(.)$	0.55	0.10	0.00	0.35
$S_5 : m_5(.)$	0.60	0.10	0.00	0.30

Step 2: The Sum of Absolute Difference between each proposition and others can be calculated by formula (7).

$$Suad = \begin{bmatrix} 0.91 & 1.21 & 1.1 & 1 \\ 2.14 & 3.04 & 0.50 & 1 \\ 0.8 & 1.11 & 0.4 & 0.75 \\ 0.77 & 1.02 & 0.4 & 0.75 \\ 0.86 & 1.02 & 0.4 & 0.70 \end{bmatrix}$$

Step 3: The adjustment parameters are denoted as $\tilde{R}OE$ and $\tilde{S}$uad. It is equal to the average of the two rows of the matrix.

$$\tilde{R}OE_1 = 1; \tilde{R}OE_2 = 1; \tilde{R}OE_3 = 0.5925; \tilde{R}OE_4 = 0.6; \tilde{R}OE_5 = 0.625.$$

$$\tilde{S}uad_1 = 1.055; \tilde{S}uad_2 = 1.67; \tilde{S}uad_3 = 0.765; \tilde{S}uad_4 = 0.735; \tilde{S}uad_5 = 0.745.$$

Step 4: Support for evidence is expressed as Sup which is calculated by multiplying the two parameters obtained in step 3, then squaring, and taking the reciprocal.

$$Sup_1 = 0.8984; Sup_2 = 0.3586; Sup_3 = 4.8674; Sup_4 = 5.1419; Sup_5 = 4.6124;$$

Step 5: The support degree of evidence is normalized as the final weight adjustment.

$$\tilde{S}up_1 = 0.8984; \tilde{S}up_2 = 0.3586; \tilde{S}up_3 = 4.8674; \tilde{S}up_4 = 5.1419; \tilde{S}up_5 = 4.6124;$$

Step 6: The weighted average evidence can be calculated as follows.

$$m(\{A\}) = 0.5534$$
$$m(\{B\}) = 0.1196$$
$$m(\{C\}) = 0.0192$$
$$m(\{A, C\}) = 0.3078$$

Step 7: The modified evidence is combined 4 times by Dempster's combination rule, and the results of target A is $m_f(\{A\}) = ((((m \oplus m) \oplus m) \oplus m) \oplus m)(\{A\}) = 0.9023$. The same is available, $m_f(\{B\}) = 0.1201$, $m_f(\{A, C\}) = 0.0986$, The Combination results of different fusion algorithms on the evidence of target recognition shown in Table 2 and Fig. 3.

Table 2. Combination results of different fusion algorithms on the evidence of target recognition

Method	$\{A\}$	$\{B\}$	$\{C\}$	$\{A, C\}$	Target
Dempster [3]	0	0.1422	0.8578	0	C
Murphy [5]	0.9620	0.0210	0.0138	0.0032	A
Deng et al. [6]	0.9820	0.0039	0.0107	0.0034	A
Zhang et al. [7]	0.9886	0.0002	0.0072	0.0032	A
Yuan et al. [8]	0.9886	0.0002	0.0072	0.0039	A
Xiao [9]	0.9895	0.0002	0.0061	0.0043	A
Proposed method	0.9921	0.0001	0.0021	0.0058	A

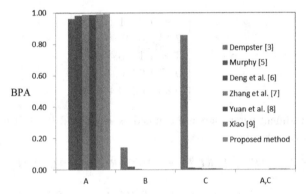

Fig. 3. The comparison of BBAs generated by different methods in Example 1.

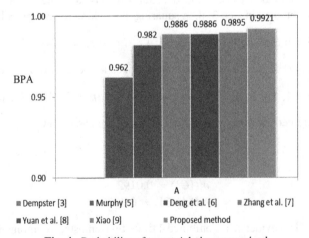

Fig. 4. Probability of target A being recognized

It can be seen that target A can be correctly identified in proposed method from Table 2 and Fig. 3. Dempster's rule cannot deal with the conflict of evidence 2 and other evidence in this case, and reaches the wrong target C. While Murphy's method [3], Deng et al.'s method [6], Zhang et al.'s method [7], Yuan et al. [8], Xiao [9] and the proposed method have certain effect on the processing of conflict evidence, identifying the correct target A.

As shown in Fig. 4, the proposed method present superior performance with a higher probability (99.21%) on target A compared with the best results in the comparison method. The numerical fusion example is verified that the proposed method is feasible and improved, of which weighted credibility interval can reduce the conflict of evidence.

3.2 Example of Fault Diagnosis

For the problem of fault diagnosis [10, 11], an evidence data from reference [10] as shown in Table 3. This is a multi-sensor based fault diagnosis problem, a object set is defined as $\Theta = \{F_1, F_2, F_3\}$. The evidence information is independently collected by three independently distributed sensors, whose evidence data are processed as basic probability assignments (BBAs).

Table 3. The BBAs of a multi-sensor based fault diagnosis.

BBAs	$\{F_1\}$	$\{F_2\}$	$\{F_2, F_3\}$	$\{F_1, F_2, F_3\}$
$S_1 : m_1(.)$	0.41	0.29	0.30	0.00
$S_2 : m_2(.)$	0.00	0.90	0.10	0.00
$S_3 : m_3(.)$	0.58	0.07	0.00	0.35

Step 1: The evidence interval matrix can be calculated by formula (5), and the range of evidence can be calculated by formula (6).

$$EIM = \begin{pmatrix} [0.60, 0.80] \ [0.10, 0.40] \ [0.20, 0.40] \ [1, 1] \\ [0.05, 0.15] \ [0.80, 0.95] \ [0.85, 0.95] \ [1, 1] \\ [0.70, 0.80] \ [0.10, 0.30] \ [0.20, 0.30] \ [1, 1] \end{pmatrix}$$

$$ROE = \begin{bmatrix} 0.80 \ 0.70 \ 0.80 \ 1 \\ 0.90 \ 0.85 \ 0.90 \ 1 \\ 0.90 \ 0.80 \ 0.90 \ 1 \end{bmatrix}$$

Step 2: The Sum of Absolute Difference between each proposition and others can be calculated by formula (7).

$$Suad = \begin{bmatrix} 0.65 \ 0.70 \ 0.05 \ 0.20 \\ 1.20 \ 1.40 \ 0.10 \ 0.10 \\ 0.75 \ 0.70 \ 0.05 \ 0.10 \end{bmatrix}$$

Step 3: Two adjustment parameters which is denoted as $\tilde{R}OE$ and $\tilde{S}$uad can be calculated. They are equivalent to the average values of two rows.

$$\tilde{R}OE_1 = 0.825; \; \tilde{R}OE_2 = 0.9125; \; \tilde{R}OE_3 = 0.9.$$

$$\tilde{S}uad_1 = 0.4; \; \tilde{S}uad_2 = 0.7; \; \tilde{S}uad_3 = 0.4.$$

Step 4: Support for evidence is expressed as Sup which is calculated by multiplying the two parameters obtained in step 3, then squaring, and taking the reciprocal.

$$Sup_1 = 0.4746; \; Sup_2 = 0.1267; \; Sup_3 = 0.3988;$$

Step 5: The support calculated in step 4 is used as the dynamic support of evidence. The total support of evidences can be weighted by it and the static support which represent the importance of sensors and the reliability of sensor distribution and importance. $\tilde{S}up_i = Sup_i * sw_i$. Where the static support parameter is known by the sensor distribution.

$$sw_1 = 1.0000; \; sw_2 = 0.2040; \; sw_3 = 1.0000.$$

$$\tilde{S}up_1 = 0.4746; \; \tilde{S}up_2 = 0.0258; \; \tilde{S}up_3 = 0.3988;$$

Step 6: The support of evidence is normalized as the final weights W_i

$$W_1 = 0.5278; \; W_2 = 0.0287; \; W_3 = 0.4435.$$

Step 7: The weighted average evidence can be calculated by

$$m(\{\cdot\}) = \sum m_i(\{\cdot\})^* Rpe_i.$$

$$m(\{F_1\}) = 0.6285$$
$$m(\{F_2\}) = 0.1201$$
$$m(\{F_1, F_2\}) = 0.0986$$
$$m(\{\Theta\}) = 0.1528$$

Step 8: The modified evidence is combined with 2 times by Dempster's combination rule, and the results of target A is $m_f(\{F_1\}) = (m \oplus m) \oplus m(\{F_1\}) = 0.9023$. The same is available, $m_f(\{F_2\}) = 0.1201$, $m_f(\{F_1, F_2\}) = 0.0986$, $m_f(\{\Theta\}) = 0.1528$. Combination results of different fusion algorithms on the evidence of fault diagnosis are shown in Table 4 and Fig. 5.

As shown in Table 4, the fault type F_1 can be correctly diagnosed with the proposed method. Fan and Zuo's method [10], Yuan et al. [11] and Xiao [9] can correctly diagnose fault F_1, while the wrong diagnosis results is obtained by Dempster's method [3]. This suggests that improved methods of weakening the degree of conflict in the original evidence have a good effect in dealing with the problem of evidence of serious conflict. Compared with the comparison method, the method proposed in this paper has a highest probability of diagnostic target support (90.23%), which is because weighted credibility

Table 4. Combination results of different fusion algorithms on the evidence of fault diagnosis

Method	$\{F_1\}$	$\{F_2\}$	$\{F_2, F_3\}$	$\{F_1, F_2, F_3\}$	Target
Dempster [3]	0.4519	0.5048	0.0336	0.0096	F_2
Fan et al. [10]	0.8119	0.1096	0.0526	0.0259	F_1
Yuan et al. [11]	0.8948	0.0739	0.0241	0.0072	F_1
Xiao [9]	0.8973	0.0688	0.0254	0.0080	F_1
Proposed method	0.9023	0.0674	0.0235	0.0068	F_1

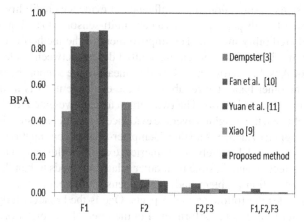

Fig. 5. The comparison of BBAs generated by different methods in Example 2.

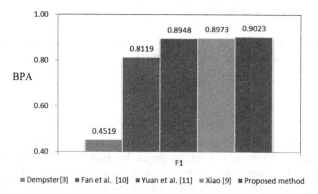

Fig. 6. Probability of fault F_1 being diagnosed

interval and the Sum of Absolute Difference are good measures of the reliability, and the conflict degree between the evidence can be reduced.

As shown in Fig. 6, the proposed method present superior performance with a higher probability (90.23%) on target A compared with the best results in the comparison method. The numerical fusion example is verified that the proposed method is feasible and improved, of which weighted credibility interval can reduce the conflict of evidence. This is because Sum of Absolute Difference can succinctly measure the mutual support between evidences, and credibility interval lengths can effectively measure the reliability of the evidence itself through the distribution of propositions. The proposed method considers both the effect of the two on the final weight to be effective.

4   Conclusion

The comprehensive consideration of the influence of evidence credibility and evidence uncertainty on weight, this paper proposes a new multi-sensor data fusion method based on the weighted credibility interval. The improvement of the method is to propose the weighted credibility interval to weaken the conflict degree between evidences. Specifically, the Sum of Absolute Difference is used to measure the support between any two evidences. On the other hand, the reliability of the evidence itself is measured by the average credibility interval lengths. The two parameters are weighted to obtain the final adjustment weight, and the weighted average evidence is calculated using the adjustment weight and the original evidence. And then Dempster-Shafter combination rule is used to get the final fusion result. Ultimately, the numerical examples illustrate that the proposed method is more effective and feasible than other related methods to handle the conflicting evidence combination problem under multi-sensor environment. Further research work mainly includes the following two aspects. One is the broader study to compare the generalization of uncertainty information in the interval parameter processing. For example, the distance of evidence [12], the joint performance of information entropy [8, 13] and similarity [14]. Second, the use of more practical and difficult verification cases is also of practical significance.

Acknowledgment. This work was supported by the Nation Natural Science Foundation of China (NSFC) under Grant No. 61462042 and No. 61966018.

References

1. Gravina, R., Alinia, P., Ghasemzadeh, H., Fortino, G.: Multi-sensor fusion in body sensor networks: state-of-the-art and research challenges. Inf. Fusion **35**, 68–80 (2017)
2. Fu, C., Xu, D.-L.: Determining attribute weights to improve solution reliability and its application to selecting leading industries. Ann. Oper. Res. **245**, 401–426 (2014)
3. Dempster, A.P.: Upper and lower probabilities induced by a multivalued mapping. Ann. Math. Stat. **38**(2), 325–339 (1967)
4. Yager, R.R.: On the Dempster-Shafer framework and new combination rules. Inf. Sci. **41**(2), 93–137 (1987)
5. Murphy, C.K.: Combining belief functions when evidence conflicts. Decis. Support Syst. **29**(1), 1–9 (2000)

6. Deng, Y., Shi, W., Zhu, Z., Liu, Q.: Combining belief functions based on distance of evidence. Decis. Support Syst. **38**(3), 489–493 (2004)
7. Zhang, Z., Liu, T., Chen, D., Zhang, W.: Novel algorithm for identifying and fusing conflicting data in wireless sensor networks. Sensors **14**(6), 9562–9581 (2014)
8. Yuan, K., Xiao, F., Fei, L., Kang, B., Deng, Y.: Conflict management based on belief function entropy in sensor fusion. Springerplus **5**(1), 638 (2016)
9. Xiao, F.: Multi-sensor data fusion based on the belief divergence measure of evidences and the belief entropy. Inf. Fusion **46**, 23–32 (2019)
10. Fan, X., Zuo, M.J.: Fault diagnosis of machines based on D-S evidence theory. Part 1: D-S evidence theory and its improvement. Pattern Recognit. Lett. **27**(5), 366–376 (2006)
11. Yuan, K., Xiao, F., Fei, L., Kang, B., Deng, Y.: Modeling sensor reliability in fault diagnosis based on evidence theory. Sensors **16**(1), 113 (2016)
12. Jiang, W., Zhuang, M., Qin, X., Tang, Y.: Conflicting evidence combination based on uncertainty measure and distance of evidence. SpringerPlus **5**(1), 12–17 (2016)
13. Wang, J., Xiao, F., Deng, X., Fei, L., Deng, Y.: Weighted evidence combination based on distance of evidence and entropy function. Int. J. Distrib. Sens. Netw. **12**(7), 3218784 (2016)
14. Fei, L., Wang, H., Chen, L., Deng, Y.: A new vector valued similarity measure for intuitionistic fuzzy sets based on OWA operators. Iran. J. Fuzzy Syst. **15**(5), 31–49 (2017)

A Locality Sensitive Hashing Based Collaborative Service Offloading Method in Cloud-Edge Computing

Wenmin Lin[1], Xiaolong Xu[2(✉)], Qihe Huang[2], Fei Dai[3], Lianyong Qi[4], and Weimin Li[5]

[1] School of Computer Science and Technology, Hangzhou Dianzi University, Hangzhou, China
linwenmin@hdu.edu.cn
[2] School of Computer and Software,
Nanjing University of Information Science and Technology, Nanjing, China
njuxlxu@gmail.com
[3] School of Big Data and Intelligence Engineering, Southwest Forestry University, Yunnan, China
daifei@swfu.edu.cn
[4] School of Information Science and Engineering, Qufu Normal University, Jining, China
qilianyong@gmail.com
[5] School of Computer Engineering and Technology, Shanghai University, Shanghai, China
wmli@shu.edu.cn

Abstract. Benefiting by the big data produced by ever increasing IoT devices, big data services are gaining popular attention in many areas. However, general IoT terminals are unable to execute these services due to the exponentially growing data and the limited computing resources. And a possible solution is to execute the services on remote cloud data centers. However, transferring all data to remote cloud for process brings huge energy consumption and congestion on the backends under high load conditions. The development of edge servers makes it possible to handle some simple tasks on edge servers. Towards this end, it is imperative to design a collaborative service offloading scheme to process data of complex big data services on both edge servers and clouds. In this paper, to protect user's privacy and quickly decide offloading destination for big data services, we propose a locality sensitive hashing based allocating strategy called Loyal. Loyal relies on E2LSH technique to hash and encrypt the sensitive data information. In addition, Loyal is able to retrieve suitable service that can be offloaded to the ES in a short time. Finally, the performance of Loyal is presented by simulation experiment.

Keywords: Edge computing · Service offloading · Locality sensitive

H. Ning (Ed.): CyberDI 2019/CyberLife 2019, CCIS 1138, pp. 92–104, 2019.
https://doi.org/10.1007/978-981-15-1925-3_7

1   Introduction

Over the last decades, data is accumulated and increasing rapidly (35 ZB data will be produced over the world by 2020), which spawning various types of big data services [1–3]. Moreover, the Internet of Things (IoT) has greatly expanded the supply of information for these services. In IoT, billions of devices that satisfy a variety of purposes are interconnected, gathering and handling big data [4,5]. However, due to the exponentially growing data and the limited computing resources of these smart devices, relying solely on IoT is hard to meet the efficient operation of complex systems Cloud Computing, on the other hand, enables big data services to be carried out in the cloud data centers. By offloading services to the cloud, the execution time of each service can be sharply reduced [6,7]. However, communication with the cloud centers will bring about huge energy consumption on the backhauls under high load conditions. What's more, the data transmission delay between the data sources and the cloud centers is too long for time-critical or real-time complex systems.

Fortunately, Edge Computing (EC), which can make use of computing resources in network edges, is promoted as an efficient solution to alleviate above problems [8,9]. EC supports handling data on edge servers without transferring to remote cloud center. Nevertheless, edge server's computing ability is relatively limited, and therefore it is unable to process intensive offloading requests under high loads, setting off the advantages of low transmission delay. For example, if an edge server is overburdened, its will hold big data processing requests and queue up requests until the edge server becomes idle. Towards this end, it is meaningful to offload services queuing up in the edge server to the cloud server in favor of the less execution time. More specifically, the ES will compare the transmission time to the cloud server with the queuing time in the ES and then decide where the big data service should be executed.

Thus, collaborative methodology to combine edge computing and cloud computing can be an effective computing platform for complex systems [10–12]. Cloud-edge computing platform brings more opportunities to execute the big data services with offloading strategies. It is still a challenge to conduct offloading due to the inherent privacy disclosure risk. Generally, encrypting the data is a common method to solve the data privacy issue. But in the cloud-edge computing platform, it is suitable to apply hash strategy to protect privacy of big data services, since hashed passwords are more secure than encrypted passwords because they can't be decrypted. In this paper, a locality sensitive hashing (E2LSH) based allocating strategy called Loyal is proposed to realize the effect big data services offloading.

The main contributions of this paper are listed as follows:

- Model the transmission delay and execution consumption for big data services in complex systems and calculate the average waiting time of each service in the edge server.
- Devise a collaborative offloading methodology where combining edge computing with cloud computing to avoid excessive queuing time, decrease the time latency and reduce energy expenditure.

The rest of this paper is structured as follows. In Sect. 2, we introduce the related work of this paper. In Sect. 3, we present the system model. In Sect. 3.5, the optimization problem is formulated. Section 4 investigates the offloading strategy Loyal based on E2LSH. Experiment results are presented in Sect. 5 and the whole paper is concluded in Sect. 6.

2   Related Work

Researches on addressing the design of an edge-cloud computing platform that inspired our work has recently started. E2LSH is a k-neighbor algorithm, which was first proposed to address image indexing problem by dimension reduction and hash processing. [13] efficiently finished the face recognition by making use of false positive images comparison technology based on E2LSH. Kim et al. [14] applied the sublinear time, scalable locality-sensitive hashing and majority discrimination to the problem of predicting critical events based on physiological waveform time series. Tanaka et al. [15] focused on a fundamental problem of feature-based localization in multi robot systems, using exact Euclidean locality sensitive hashing and extending the algorithm of Monte Carlo localization. Chafik et al. [16] proposed multidimensional indexing methods based on the approximation approach and used two criteria evaluating the E2LSH performances, namely average precision and CPU time using a database of one million image descriptors.

Prior to our work, there has been some researchers like Guo et al. [17] bent their effort for both time-aware and energy-aware offloading methods in the edge-cloud computing platform, but nobody tried to apply E2LSH strategy to offloading strategies. The private experiment has proved E2LSH's accuracy and low latency of processing high-dimensional retrieval. By establishing hash index table of big data services for each ES and generating its appropriate feature vector to conduct index lookup, there is a great hope to quickly determine which services can be offloaded to the corresponding ESs or cloud server.

In [18], a hierarchical edge-cloud architecture is proposed and its latency advantage is demonstrated over flat edge-cloud using formal analysis and simulation. Reference [19] presents the edge-cloud system as an enabler for the Industry 4.0 and smart factory paradigms, where the computation and storage capacity highlighted are larger for higher tiers, and also describes existing challenges such as the infrastructure energy and cost. Ceselli et al. [20] studied the design of an edge-cloud network where the placement decision for cloudlets among the available sites is made (edge, aggregation, or core nodes).

3   System Model

3.1   Complex System Model

Assume a complex system consists of N terminals providing big data services (BDS), J edge computational access points and a cloud computing center. An edge computational access point is composed of a base station (BS) and an edge server (ES). We assume that each BS is connected with densely deployed BDSs via wireless channel, and every BS is integrated with an edge server ES, which can provide relatively rich computational resources to execute offloaded services. Besides, data could be delegated to cloud computing center through backhaul links. As is well documented, the backhaul link suffers from congestion and its transmission speed is slow.

We use $\mathcal{N} = \{1, ..., N\}$ to represent a terminals set, each terminal having only one BDS requesting for being offloaded onto an ES at every interval due to terminal device's limited capabilities. We consider that each BDS arrives simultaneously and have the same form. In addition, the computing logic is atomic and can't be split. As to a certain terminal $i(i \in \mathcal{N})$, its BDS is represented by $T_i = (w_i, m_i)$ where w_i denotes the size of input data and m_i represents the CPU cycles executing this BDS needs.

The set of the edge computational access points is donated by $\mathcal{J} = \{1, ..., J\}$. For the sake of convenience, we use j ($\mathcal{J} = \{1, ..., J\}$) to represent the j-th BS with ES and define the j-th ES's total computation ability as F_j^{edge}, whose measurement is the number of one second's CPU cycles. What's more, we consider every edge server has s virtual machines and each virtual machine (VM) possesses the same computation capacity. Hence, the computation ability of each VM in j-th ES is expressed as

$$f_j^{edge} = \frac{F_j^{edge}}{s}.$$ (1)

In this system, one BDS is processed by only one VM in ES. Certainly, BDS can be exchanged between base stations within a couple of hops via optical fiber. Moreover, we construct a queuing model at the edge server. Hence, if a BDS's waiting time is too long, instead of being executed at the edge server, the BDS can be delegated to the cloud computing center and then the result of its computation will be returned as soon as it is accomplished.

3.2   Queuing Model of Big Data Service in the ES

We consider each edge server has virtual machines $\{Z_1, ..., Z_s\}$. For the j-th ES, its number of the arrival offloaded BDSs is assumed to follow Poisson distribution, and the parameter is denoted by λ_j. Similarly, the service time at j-th edge server is supposed to obey negative exponential distribution with a parameter of μ_j. Then queuing model can be constructed at edge server. According to the queuing theory, when the whole system tends to be steady-state, its service intensity of a single virtual machine can be described by

$$\rho_j = \frac{\lambda_j}{\mu_j},$$ (2)

and the whole system's service intensity is

$$\rho_{s_j} = \frac{\lambda_j}{s \cdot \mu_j} \quad (\rho_{s_j} < 1).$$ (3)

The probability that the number of the BDSs in the j-th ES equals k is represented by P_{k_j}. Then we can accord queuing model to calculate the probability size of the j-th edge server being idle,

$$P_{0_j} = \frac{1}{\sum\limits_{n=0}^{s-1} \frac{1}{n!} \cdot \rho_j^n + \frac{\rho_{s_j}}{s!(1-\rho_{s_j})}}, \tag{4}$$

Correspondingly, we can calculate the probability that the number of the BDSs at the j-th ES equals k ($k \geq 1$) in the light of the size of P_{0_j},

$$P_{k_j} = \begin{cases} \frac{(\lambda_j/\mu_j)^k}{k!} \cdot P_{0_j}, & (1 \leq k \leq s) \\ \frac{(\lambda_j/\mu_j)^k}{s! s^{(k-s)}} \cdot P_{0_j}, & (k > s) \end{cases}. \tag{5}$$

In some cases, the number of BDSs at the ES is smaller than the total number of virtual machines, so the queue length will be zero and the offloaded BDS can be executed as soon as it is transferred to ES. In other cases, the offloaded BDSs are assumed to wait for execution and queue up at the ES. We derive the average queue length of BDSs from formula (1) and formula (2) in the j-th ES by evaluating the integration and taking the limit,

$$L_j = \sum_{k=s+1}^{\infty} (k-s) \cdot P_{k_j} = \frac{P_{0_j} \cdot \rho_j^{\ s}}{s!} \sum_{k=s+1}^{\infty} (k-s) \cdot \rho_{s_j}^{k-s}$$

$$= \frac{P_{0_j} \cdot \rho_j^{\ s}}{s!} \cdot \frac{d}{d\rho_{s_j}} \left(\sum_{k=1}^{\infty} k \cdot \rho_{s_j}^k \right) = \frac{P_{0_j} \cdot \rho_j^{\ s} \cdot \rho_{s_j}}{s!(1-\rho_{s_j})^2}, \tag{6}$$

Based on formula (4), we eventually deduce the average waiting time in the j-th edge server,

$$t_j^{que} = \frac{L_j}{\lambda_j} = \frac{P_{0j} \cdot \rho_j^{\ s}}{s!(1-\rho_{s_j})} \cdot \frac{1}{s \cdot \mu_j - \lambda_j}. \tag{7}$$

3.3   Wireless Communication Model of BDS Between BSs and Terminals

Each terminal is associated with only one BS, and the uplink communication between a terminal and its connected BS is allocated on the same spectrum resource of B hertz. In this paper, to properly analyze the offloading strategy and obtain more insightful results, we consider different BSs share the wireless channel orthogonally. We propose a decision variable $x_{i,j}$ to express the relationship between the i-th BDS and the j-th BS,

$$x_{i,j} = \begin{cases} 1, & ith \ BDS \ is \ associated \ jth \ BS, \\ 0, & otherwise. \end{cases} \tag{8}$$

Note that each BDS is connected with only one BS, so the $x_{i,j}$ is subject to

$$\sum_{i}^{N} \sum_{j}^{J} x_{i,j} = 1, \tag{9}$$

Then, let y_i represent the ordinal number that meets $x_{i,j} = 1$, meaning the y_i-th BS is the associated BS of i-th BDS. We further define the channel gain between the i-th terminal and the y_i-th BS is $M_{(i,y_i)}$, which is a random variable. Define $p_{(i,y_i)}$ as the transmission power from the i-th terminal to the y_i-th BS. On the basis of above definition, we express the uplink transmission rate from the i-th terminal to its connected y_i-th BS according to the Shannon formula,

$$r_{i,y_i} = B \cdot \log_2(1 + \frac{p_{i,y_i} \cdot M_{i,y_i}}{p_0}). \tag{10}$$

where p_0 is the power of Gaussian white noise, which is considered to be a constant during the transmission. It can be seen from formula (10) that the high transmission power will decrease latency but increase the cost of more energy consumption. Without loss of generality, we neglect the downlink communication from the BSs to the terminals on account of the much smaller output size compared to the BDS's input size. In our future work, we can expand our research to analyze the situation in downlink communication and more complex communication resource sharing scenarios, which can make our model more detailed and accurate.

3.4   Latency and Energy Model of BDSs in the ES and Cloud

As noted in the Sect. 3.1, we use a two-tuple $T_i = (w_i, m_i)$ to represent an offloaded BDS of the i-th terminal, and the terminals are incapable of processing BDSs due to its limited computing resources. The terminal sends the BDS offloading request and then it will be selected by a ES within a couple of hops. Afterwards, the BDS will be offloaded to its connected ES, and the ES will decide where the BDS should be offloaded, the suitable ES or the cloud. Denote a decision variable $\theta_i \mid 0, 1$ to express the offloading decision of the i-th BDS, where $\theta_i = 0$ means the i-th BDS will be offloaded to the cloud center through backhaul link and $\theta_i = 1$ means the BDS is offloaded to the suitable ES via optical fiber.

In the following discussion, we present the offloading model in terms of the time latency and energy consumption respectively by the wireless transmission, the execution at edge server and the execution at cloud computing center. Following the research in [25], we consider a quasi-static scenario, in which the terminal set N keeps unchanged through the whole BDSs offloading.

The Transmission via Wireless Channel. Each BDS is supposed to directly transmit its BDS to the associated BS via wireless channel without any local computation. The time latency for the i-th BDS to transmit its BDS to the associated y_i-th BS is given by

$$t_i^{tran,mobile} = \frac{w_i}{r_{i,y_i}} \tag{11}$$

The energy expenditure in data transmission from the i-th BDS to the y_i-th BS is expressed as

$$e_i^{tran,mobile} = t_i^{tran} \cdot p_{i,y_i}, \tag{12}$$

Execution in Edge Server. For the offloading decision $\theta_i = 1$, the associated BS chooses to offload the i-th BDS to a suitable ES within serval hops in order to maintain satisfactory Quality-of-Service (QoS). The hop distance between two BSs is calculated by Manhattan distance. Define $h_{(y_i,j)}$ as the number of hops between the y_i-th BS and the j-th BS. θ is the transmission delay per hop, which is a constant for each BDS and BS in our paper. Then the transmission time between j-th BS and y_i-th BS can be expressed as

$$t_{y_i,j}^{tran,edge} = (\theta + \frac{w_i}{r^{fiber}}) \cdot h_{y_i,j}. \tag{13}$$

Please be noted that if the i-th BDS is chosen by y_i-th ES to offload its BDS, the transmission delay between BSs will not exist due to the number of interval hops is zero.

The execution time and energy consumption that i-th BDS is computed at j-th ES need can be respectively expressed as

$$t_{i,j}^{exe,edge} = \frac{m_i}{f_j^{edge}} + t_j^{que}, \tag{14}$$

$$e_{i,j}^{exe,edge} = w_i \cdot \eta_j^{edge}. \tag{15}$$

where t_j^{que} is the queuing time at the j-th ES and η_j^{edge} is the execution power of the j-th ES.

Execution in Cloud Computing Center. For the offloading decision $\theta_i = 0$, the associated BS chooses to offload the i-th BDS to the cloud center in favor of the richer computing resources. The total time offloading and executing i-th BDS at cloud need is expressed as

$$t_i^{cloud} = wan + \frac{m_i}{f^{cloud}} + \frac{w_i}{b}, \tag{16}$$

where f^{cloud} is the computation ability of cloud center, b is the transmission rate from BS to cloud and wan is the transmission delay which results from the congestion on the backhaul lines.

Define η^{cloud} as the execution power of the cloud center, and then the energy expenditure executing i-th BDS at cloud center is calculated as

$$e_i^{cloud} = w_i \cdot \eta^{cloud}. \tag{17}$$

3.5   Problem Formulation

As mentioned in Sect. 2, there are numberous big data services in complex systems needing to be offloaded in favor of the richer computing resources at edges or cloud, and this amount will continuously grow along with the upgrading big data technology. In order to better optimize the QoS in complex system, it's necessary to minimize both the energy consumption and time delay. In the above discussion, we have analyzed two kinds offloading destinations for big data services in complex system, ES and Cloud. The total time and energy expenditure of the i-th service in this system can be expressed as

$$T_i^{total} = t_i^{tran,mobile} + \sigma_i \cdot (t_{y_i,j}^{tran,edge} + t_{i,j}^{exe,edge}) + (1 - \sigma_i) \cdot t_i^{cloud}, \qquad (18)$$

$$E_i^{total} = e_i^{tran,mobile} + \sigma_i \cdot e_{i,j}^{exe,edge} + (1 - \sigma_i) \cdot e_i^{cloud}. \qquad (19)$$

In this paper, we aim at minimizing the system time delay and the executing expenditure of all big data services. To this end, the optimization problem and following constraints can be formulated as

$$P_1 \min \sum_{i=1}^{N} (t_i^{tran,mobile} + \sigma_i \cdot (t_{y_i,j}^{tran,edge} + t_{i,j}^{exe,edge}) + (1 - \sigma_i) \cdot t_i^{cloud}), \qquad (20)$$

$$P_2 \min \sum_{i=1}^{N} (e_i^{tran,mobile} + \sigma_i \cdot e_{i,j}^{exe,edge} + (1 - \sigma_i) \cdot e_i^{cloud}), \qquad (21)$$

$$P_2 \min \sum_{i=1}^{N} (e_i^{tran,mobile} + \sigma_i \cdot e_{i,j}^{exe,edge} + (1 - \sigma_i) \cdot e_i^{cloud}), \qquad (22)$$

$$s.t. \sum_{j=1}^{s} f_j^{edge} \leq F_j^{edge}, f_j^{edge} \geq 0, \qquad (23)$$

$$\sum_{i=1}^{N} (1 - \sigma_i) \cdot f^{cloud} \leq F^{cloud}, f^{cloud} \geq 0, \qquad (24)$$

$$f_j^{edge} \leq f^{cloud}, \forall j \in \mathcal{J}. \qquad (25)$$

4   An E2LSH Based Cloud-Edge Collaborative Offloading Strategy Loyal

The key challenge of the problem is to properly and quickly decide where to offload these big data services, cloud center or edge server. Although traditional heuristic algorithm (e.g. genetic algorithm) can reasonably accomplish the overall allocation of offloading big data services, its time complexity is too high to meet the time limitation. Moreover, chances are that most existing offloading algorithms will leakage the privacy information since they need personal information as parameters during execution. To finish deciding where to offload these services in a short time and better protect the privacy of users, we propose an E2LSH based offloading strategy named Loyal in following discussion.

4.1 E2LSH Strategy

The Euclidean Locality Sensitive Hashing (E2LSH) is a technique which could further reduce time complexity of the traditional LSH method. E2LSH relies on p-stable distribution to map vector from higher dimension to lower dimension. Generally, E2LSH is confirmed as an effective method to solve the nearest problem, for there is a great probability that the similar vectors will be put into the same hash bucket by E2LSH. That is to say, these near vectors will keep their similarity after hash processing and still be close.

Definition 1 (Locality Sensitive Hashing Family (LSH Family)). A locality sensitive family H $(H = h : S^d -> R)$ is a group of functions which can map the vectors from d dimensions in domain S to real number space. The family H is called (r_1, r_2, p_1, p_2) sensitive if it satisfies

$$f \, \|v - q\| < r_1, then \, P\,[h_m(v) = h_m(q)] > p_1, \tag{26}$$

$$f \, \|v - q\| > r_2, then \, P\,[h_m(v) = h_m(q)] < p_2. \tag{27}$$

for any $v, q \mid S^d, h_m \mid H$, where P implies the probability that the hash code of vector v is the same with $q's$ hash code, with $r_2 > r_1$ and $0 < p_1 < p_2 < 1$. Then we can say that the family H is locality sensitive and the possibility of the hashed vectors having the same hash value is positively correlated with the previous distance between them.

4.2 Loyal Design

Loyal relies on E2LSH strategy to determine which big data services can be offloaded to the ES. After constructing the feature vectors of the services connected to its agreeable ES, these vectors will be hashed and mapped into hash buckets in L hash tables. Then each ES generates its feature vector according to its corresponding ability, and quickly retrieves the near vector in hash tables to gain suitable big data services. In addition, we will compare the queuing time in the ES with the offloading time to cloud, choosing less than s (the number of VMS in each ES) big data services executed in the ES and offloading the other to the cloud server.

Algorithm 1. Offloading strategy Loyal

Require: $s, w_i, m_i, f_j^{edge}, \eta_j^{edge}, f^{cloud}, \eta^{cloud}, B, [Lim]_{hop}$
Ensure: T^{total}, E^{total}
1: Set $T^{total} = 0$ and $E^{total} = 0$
2: **for** $j = 1. J$ **do**
3:     set $cnt = 0$
4:     Generate feature vector V' based on the j-th $ES's$ ability
5:     **for** $i = 1. I$ **do**
6:         Calculate V' 's index $T_1(V')$ and $T_2(V')$ by the i-th LSH family
7:         Find index meeting $T_1 == T_1(V')\&\&T_2 == T_2(V')$ in the i-th LSH Index
    Table
8:         **for** $k = 1.$ length (index) **do**
9:             **if** $t_{index(k)}^{tran,edge} + \frac{m_{index(k)}}{f_j^{edge}} < t_{index(k)}^{cloud}$ **then**
10:                 Delete k in index
11:                 Update T^{total} and E^{total} by the cost in the cloud server
12:             **else**
13:                 Update $cnt = cnt +$ length(index)
14:                 Update T^{total} and E^{total} by the cost in the ES
15:             **end if**
16:         **end for**
17:         **if** $cnt \leq s$ **then**
18:             Break
19:         **end if**
20:     **end for**
21: **end for**

---

5 Experiment

In this section, we compare the performance of our scheme with three similar schemes: all2cloud, time-greedy and energy-greedy, respectively, to prove the feasibility of our proposal.

Figure 1 compares the executing energy consumption in the ESs through time-greedy, energy-greedy and LSH strategies. In this simulation, we gradually increase the number of services from 1000 to 4500 with the roughly same size. We can observe that the energy-greedy keeps lowest executing energy consumption due to its good stability of decreasing energy expenditure. However, in the majority of cases, Loyal consumes the highest executing energy in the ESs among three offloading strategies because it takes other energy expenditure (e.g. transmission energy consumption) into account and the executing energy occupies a relatively tiny percentage of the total consumption in the ESs.

As shown in Fig. 2, time-greedy expends the most energy in the ESs while energy-greed keeps consuming the least energy. And Loyal holds moderate energy consumption, verifying the above conclusion to some degree.

Figure 3 illustrates the comparison of energy expenditure in the cloud by Time-greedy, Energy-greedy and Loyal with different number of big data services. When these services are offloaded to the cloud, Loyal consumes less energy

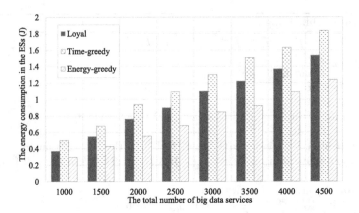

Fig. 1. Executing energy consumption in the ESs with the number of big data services.

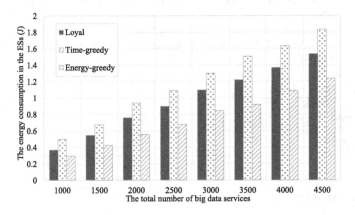

Fig. 2. Energy consumption in the ESs with the number of big data services.

(including transmission and execution) than Energy-greedy strategy from Fig. 3, because Loyal reasonably chooses the suitable services to be offloaded to the cloud data center on the basis of overall situation, not limited to only decreasing the production of energy in the edge layer.

6   Conclusion

To achieve the goal of decreasing time latency and reducing energy consumption in the cloud-edge computing while protecting user's data privacy, we designed Loyal. Firstly, we construct the feature vector of each big data service and the conduct hash process based on E2LSH technique, which will greatly ensure data integrity. Then we can quickly retrieve the suitable service for each ES to receive offloading requests. Afterwards, rest services that may not suitable be executed at ES will be offloaded to the cloud center. In the future, we will focus on

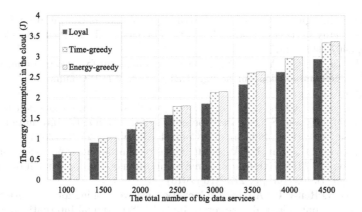

Fig. 3. Energy consumption in the ESs with the number of big data services.

transferring Loyal to the real scene of IoT, taking account of more unique features of complex systems.

References

1. Puyun, B., Miao, L.: Research on analysis system of city price based on big data. In: 2016 IEEE International Conference on Big Data Analysis (ICBDA), Hangzhou, pp. 1–4 (2016)
2. Norman, M.D.: Complex systems engineering in a federal IT environment: lessons learned from traditional enterprise-scale system design and change. In: 2015 Annual IEEE Systems Conference (SysCon) Proceedings, Vancouver, pp. 33–36 (2015)
3. Wang, H., Zhong, D., Zhao, T., Ren, F.: Integrating model checking with SysML in complex system safety analysis. IEEE Access **7**, 16561–16571 (2019)
4. Chze, P.L.R., Leong, K.S.: A secure multi-hop routing for IoT communication. In: 2014 IEEE World Forum on Internet of Things (WF-IoT), Seoul, pp. 428–432 (2014)
5. Fox, J., Donnellan, A., Doumen, L.: The deployment of an IoT network infrastructure, as a localised regional service. In: 2019 IEEE 5th World Forum on Internet of Things (WF-IoT), Limerick, pp. 319–324 (2019)
6. Bahrami, M.: Cloud computing for emerging mobile cloud apps. In: 2015 3rd IEEE International Conference on Mobile Cloud Computing, Services, and Engineering, San Francisco, pp. 4–5 (2015)
7. Zhang, C., Green, R., Alam, M.: Reliability and utilization evaluation of a cloud computing system allowing partial failures. In: 2014 IEEE 7th International Conference on Cloud Computing, Anchorage, pp. 936–937 (2014)
8. Muñoz, R., et al.: Integration of IoT, transport SDN, and edge/cloud computing for dynamic distribution of IoT analytics and efficient use of network resources. J. Lightwave Technol. **36**(7), 1420–1428 (2018)
9. Mao, Y., You, C., Zhang, J., Huang, K., Letaief, K.B.: A survey on mobile edge computing: the communication perspective. IEEE Commun. Surv. Tutor. **19**(4), 2322–2358 (2017)

10. El Haber, E., Nguyen, T.M., Assi, C.: Joint optimization of computational cost and devices energy for task offloading in multi-tier edge-clouds. IEEE Trans. Commun. **67**(5), 3407–3421 (2019)
11. Ren, J., Yu, G., Cai, Y., He, Y.: Latency optimization for resource allocation in mobile-edge computation offloading. IEEE Trans. Wirel. Commun. **17**(8), 5506–5519 (2018)
12. Li, H., Hao, W., Chen, G., Liao, X.: Large-scale documents reduction based on domain ontology and E2LSH. In: Proceedings of the 11th IEEE International Conference on Networking, Sensing and Control, Miami, pp. 24–29 (2014)
13. Su, L., Zhang, F., Ren, L.: An adult image recognition method facing practical application. In: 2013 Sixth International Symposium on Computational Intelligence and Design, Hangzhou, pp. 273–276 (2013)
14. Kim, Y.B., O'Reilly, U.: Analysis of locality-sensitive hashing for fast critical event prediction on physiological time series. In: 2016 38th Annual International Conference of the IEEE Engineering in Medicine and Biology Society (EMBC), Orlando, pp. 783–787 (2016)
15. Tanaka, K., Kondo, E.: A scalable localization algorithm for high dimensional features and multi robot systems. In: 2008 IEEE International Conference on Networking, Sensing and Control, Sanya, pp. 920–925 (2008)
16. Chafik, S., Daoudi, I., Ouardi, H.E., Yacoubi, M.A.E., Dorizzi, B.: Locality sensitive hashing for content based image retrieval: a comparative experimental study. In: 2014 International Conference on Next Generation Networks and Services (NGNS), Casablanca, pp. 38–43 (2014)
17. Guo, H., Liu, J.: Collaborative computation offloading for multiaccess edge computing over fiber-wireless networks. IEEE Trans. Veh. Technol. **67**(5), 4514–4526 (2018)
18. Tong, L., Li, Y., Gao, W.: A hierarchical edge cloud architecture for mobile computing. In: Proceedings 35th Annual IEEE International Conference on Computer Communications (INFOCOM), pp. 1–9 (2016)
19. Dao, N.N., Lee, Y., Cho, S., Kim, E., Chung, K.S., Keum, C.: Multi-tier multi-access edge computing: the role for the fourth industrial revolution. In: Proceedings International Conference on Information and Communication Technology Convergence (ICTC), pp. 1280–1282 (2017)
20. Ceselli, A., Premoli, M., Secci, S.: Mobile edge cloud network design optimization. IEEE/ACM Trans. Netw. **25**(3), 1818–1831 (2017)

A Testbed for Service Testing: A Cloud Computing Based Approach

Qinglong Dai[1]([⊠]), Jin Qian[2], Jianwu Li[3], Jun Zhao[4], Weiping Wang[5],
and Xiaoxiao Liu[1]

[1] Beijing Union University, No. 97, Beisihuan East Road, Beijing, People's Republic of China
xxtqinglong@buu.edu.cn
[2] Dareway Software Co., Ltd., No. 300, Gangxing 1st Road, Jinan,
Shandong, People's Republic of China
[3] H3C Technologies Co., Limited, No. 8 Guangshun South Road, Beijing,
People's Republic of China
[4] China Telecom Beijing Research Institute, Penglaiyuan South Road, Beijing,
People's Republic of China
[5] University of Science and Technology Beijing, No. 30, Xueyuan Road, Beijing,
People's Republic of China

Abstract. Although simulation is an important tool in studying new emerged networks, tests on a real system are still the most ultimate way to validate the capability of a newly emerged network. Traditional tests have their drawback, because of the coupling of test service and test devices. The arise of new network architecture, industry 4.0 and the internet of things, also brings new challenges. Therefore, a testbed for service testing and data process based on cloud computing is proposed. In this testbed, a test is realized as a service hosted by one or several cloud hosts, i.e., test as a service (TaaS). The relationship of different kinds of services in a testbed is modeled, based on the Lotka-Volterra model. The testbed implementation is demonstrated, based on cloud computing, using a typical video test as an example. It is proved that our testbed for service testing and data process is practicable and effective.

Keywords: Testbed · Cloud computing · Service test · Visualization

1 Introduction

Rapid progress have been made on seeking network architecture with better performance, such as scalability, flexibility, controllability, mobility and security support. A lot of promising future network architectures, like software-defined networking (SDN), information-centric networking (ICN), network virtualization, cloud computing, have emerged in the past few years.

Although analysis and simulation are important tools in studying the behavior of data traffic and analyzing new protocols and algorithms, experiments or tests have played a most important role in advancing networking research. New research ideas must be finally validated on a real testbed. To realize the experiment, a variety of testbeds have

© Springer Nature Singapore Pte Ltd. 2019
H. Ning (Ed.): CyberDI 2019/CyberLife 2019, CCIS 1138, pp. 105–120, 2019.
https://doi.org/10.1007/978-981-15-1925-3_8

been built in academia, national laboratories and industry, including GENI, PlanetLab, FIRE, CENI, etc. [1, 2]. By conducting experiments on these testbeds, researchers can well validate their ideas for developing better future network architectures.

Before a network is deployed, it has to pass a network test. The network test is conducted by the means of service accomplishment. In a traditional test, the test is bound with the tested network. This means that the network test must be executed in a place where the network is. If the test place is changed, the entire network and test device have to be dismantled and redeployed in the new place. Plenty of time, human resources and costs are taken. On the other hand, if an error appears in a network test, the network test has to stop, until the error is settled.

Cloud computing is a system for enabling ubiquitous, convenient, on-demand network access to a shared pool of configurable computing resources (e.g., networks, servers, storage, applications, and services) that can be rapidly provisioned and released with minimal management effort or service provider (SP) interaction [3]. Cloud computing makes it possible that SPs access resources on-demand to host services via a network, rather than having to provide a resource on their own [4]. The resources sliced from the resource pool are assembled as a virtual network (VN) to host a specific service [5]. SPs and users need not think about the place where physical device is. And if one device has a problem, the sliced resource can be replaced by another rival device via a resource pool. Cloud computing is thought of as a feasible way to overcome the drawback of traditional tests.

With the rise of new network architecture, industry 4.0 and the internet of things, the data generated by service in-network service test has great growth [6]. How to obtain useful knowledge from the huge data, is the new problem. If the data scale is small, the data process can be finished by human resources. When the data scale is large enough, the data process must be finished with the help of big data.

The contributions of this paper are listed as follows:

1 A testbed for service testing and data process based on cloud computing is proposed. In this testbed, a test is realized as a service hosted by one or several cloud hosts, i.e., test as a service (TaaS). The testbed is built on the OpenStack platform. The relationships of infrastructure provider, resource pool, cloud manager, service and so on, are analyzed.
2 The relationships of different kinds of services in testbed are modeled based on the Lotka-Volterra model. In this model, the competition and promotion of services are presented.
3 The testbed implementation is demonstrated, including testbed environment, cloud host allocation, video test, and other data visualization. Using a typical video test as an example, cloud computing-based service testing and data process are displayed.

The rest of this paper is organized as follows. Section 2 reviews related work on cloud computing and data process. Section 2.2 proposes a testbed for service testing and data process. The relationship between the services in the testbed is modeled in Sect. 3. Section 4 represents the testbed implementation. Finally, Sect. 5 concludes the paper.

2 Related Work

Around cloud computing and data processes, a lot of works have done.

2.1 Cloud Computing

Cloud computing is thought of as a promising way to be applied in the testbed, because of the cost reduction brought by decoupling service from infrastructure. Note that, virtualization is the key technology in cloud computing.

Due to the scale and complexity of shared resources, it is often hard to analyze the performance of new scheduling and provisioning algorithms on actual cloud testbeds. Therefore, simulation tools are becoming more and more important in the evaluation of cloud computing research. The most popular simulation tool on cloud computing is CloudSim. CloudSim is used to study the cloud computing infrastructure schedule mechanics, in an emulation way [7]. Garg et al. extended CloudSim with a scalable network and generalized application model in a data center. Their work allowed a more accurate evaluation of scheduling and resource provisioning policies to optimize the performance of a cloud infrastructure [8].

Meanwhile, a lot of works on cloud computing are verified in the real system. Dai et al. proposed a network virtualization based seamless networking scheme for heterogeneous networks, including hierarchical model, service model, service implementation and dynamic bandwidth assignment. Then, the performance changes after network virtualization, like the ability of key switch and the time consumed by ping, were evaluated in a real experiment environment. The proposed networking scheme was helpful for cloud computing research [9, 10]. Khazae et al. developed an analytical model, in which task service times were modeled with a general probability distribution, but the model also accounted for the deterioration of performance due to the workload at each node. The model allowed for the calculation of important performance indicators such as mean response time, waiting time in the queue, queue length, blocking probability, probability of immediate service, and the probability distribution of the number of tasks in the system [11]. Note that their work did not consider the existence of different level SPs in cloud computing.

Ameixieira et al. presented a real-world testbed for research and development in vehicular networking that has been deployed successfully in the seaport of Leixoes in Portugal. The testbed allowed cloud-based deployment, remote network control and distributed data collection from moving container trucks, cranes, towboats, patrol vessels, and roadside units, thereby enabling a wide range of experiments and performance analyses. After describing the testbed architecture and its various modes of operation, they gave concrete examples of its use and offered insights on how to build effective testbeds for wireless networking with moving vehicles [12].

2.2 Data Process

With the development of industry and scientific research, the volume and generated velocity of the data in the test are explosively increasing. Thus, the data process is closely related to big data.

Based on specific data compression requirements, Yang et al. proposed a novel scalable data compression approach based on calculating similarity among the partitioned data chunks. With real-world meteorological big sensing data experiments on a cloud platform, they demonstrated that the proposed scalable compression approach based on data chunk similarity could significantly improve data compression efficiency with affordable data accuracy loss [13].

The primary focus of data process technologies is currently in storage, simple processing, and analytical tasks. Big data initiatives rarely focus on the improvement of end-to-end processes. To address this mismatch, Der et al. advocated a better integration of data science, data technology, and process science. They discussed the interplay between data science and process science and related process mining to big data technologies, service orientation, and cloud computing [14].

Based on the data process result, visualization is a feasible and effective way to demonstrate knowledge. The most common visualization tools are Highcharts and Echarts [15, 16]. Highcharts can visualize based on scalable vector graphics and good at 3D visualization. Meanwhile, Echarts is able to visualize based on HTML 5's canvas technology. Normally, Echarts visualization is more user-friendly and interactive. Besides, in order to realize 3D visualization, EhartsX is developed [17].

2.3 Basic Architecture of Testbed

A testbed is able to decouple service from physical infrastructures, by exploiting cloud computing. This provides a feasible method to solve the problems faced by the traditional test. The basic architecture of testbed for testing and data process is shown in Fig. 1.

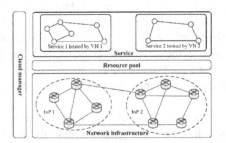

Fig. 1. Basic architecture of testbeds

There are three layers in the basic architecture of testbeds. The bottom layer is network infrastructures, which are the contents provided by different infrastructure providers (InPs). In order to disengage the services from the physical network resources which have complex characteristics, virtualization is used to abstract the physical infrastructures to virtual resources. Virtual resources are an independently manageable partition of all the physical resources and inherit the same characteristics as the physical resources. The abstracted virtual resources are stored in a resource pool and managed by a cloud manager. The capacity of the virtual resources is not infinite but limited by the capacity of network physical resources.

The middle layer in basic architecture is a resource pool which is the heart of testbed. From resource pool, a portion of virtual resources is allocated to an SP as a form of VN according to its virtual resources requirement. The SP loads the specific service on the allocated virtual resources, which means that different services may be hosted by the same physical node or physical link.

The top layer in basic architecture shows two independent VNs for different kinds of services. A VN can be seen as the collection of allocated cloud hosts and links. Actually, a VN can be deployed upon another VN. In testbed, a VN can be used to provide a test environment for a network test and also can be used as the computing environment for the data process.

2.4 OpenStack

Cloud computing in the testbed is realized via OpenStack, which is a free and open-source software platform for cloud computing. OpenStack consists of seven main modules. The modules and their interaction are shown in Fig. 2.

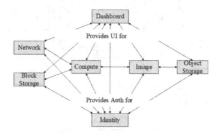

Fig. 2. OpenStack modules interaction

Dashboard module provides cloud manager and users with a direct graphical interface to access, provision and automatical deployment of cloud-based resources.

Network module can manage network and IP address. Through the network module, the network would not be a bottleneck or limiting factor in a cloud deployment and gives users self-service ability, even over network configurations.

Compute module is a cloud computing fabric controller, which is designed to manage and automate pools of computer resources. Compute module can widely exploit available virtualization technologies and high-performance computing.

Image module provides discovery, registration and delivery services for disk and server images. Image module can store disk and server images in different kinds of back-ends. Through a standard REST interface for querying information about images, clients are able to migrate images to new servers.

Object storage module is a scalable redundant storage system. Objects and files are written to the multiple disks that spread throughout cluster servers, with object storage module responsible for ensuring data replication and integrity across the cluster servers. Cluster servers used for storage can scale simply by adding new servers. Once a server or disk fails, the object storage module replicates its content from other active nodes to new locations.

Block storage module provides persistent block-level storage devices for users with computing instance. Block storage module manages the creation, attaching and detaching of the block devices to servers. Block storage volumes are fully integrated into compute and dashboard modules, allowing for cloud users to manage their own storage needs.

Identity module provides a central mapped directory of users and the services that they can access. It acts as a common authentication system across the entire OpenStack and can integrate with existing backend directory services. It supports multiple forms of authentication, including standard user name and password credentials, token-based system and Amazon web service style.

2.5 Test Achievement

The test achievement in the testbed is shown in Fig. 3. The test achievement in the testbed is called test as a service.

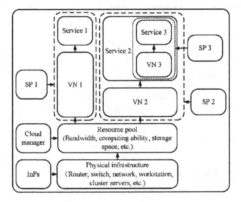

Fig. 3. Test achievement

Here, physical infrastructures provided by multiple InPs, such as a router, switch, network, workstation, cluster servers and etc., are collected and abstracted into virtual resources which are centrally managed by cloud managers. The virtual resources stored in the resource pool include bandwidth, computing ability, storage space and so on. According to the resource request from SPs, the cloud manager would allocate appropriate resources. These allocated resources would be used by SPs to form a VN and then host a test service. The VNs are composed of cloud hosts, virtual links and storage space. The resource request is generated according to the user's request.

Thus, a test service conducted by a VN is isolated from another testing service and the service test is decoupled from underlying infrastructures. This means that the test services can be executed, no matter where test devices are and the tested network is. Even when an error appears in the test, the test resource can be re-allocated by the cloud manager, without affecting the test execution.

The process of service implementation is as follows. In order to get a test service, users firstly make a service request to SPs. After receiving the service request, SPs will verify whether SPs can provide this service or not. If yes, according to test service's features (i.e., delay-sensitive, throughput-sensitive, or others), SPs make a request which contains needed virtual resources to cloud manager; Otherwise, such as too many subscribers, SPs return a service rejection to users.

When a cloud manager receives the virtual resources request from SPs, it checks available virtual resources. If available virtual resources are enough to fulfill the request, then the cloud manager gives these virtual resources' disposition to requested SP and updates available virtual resources; Otherwise, the cloud manager returns a virtual resource insufficiency message to users via SPs. Using allocated virtual resources, SPs form a VN to host users' requested service, until this VN's duration expires. Then, users make a service remove request to SPs. After receiving service remove request, SPs will release occupied virtual resources. At last, the cloud manager will update available virtual resources at a fixed time interval.

In Fig. 3, virtual resources are allocated to two SPs, i.e., SP 1 and SP2. SP 1 assembles obtained virtual resources into VN 1 to host Service 1. SP 2 assembles obtained virtual resources into VN 2 to host Service 2. Note that a user of Service 2 further exploits one portion of resources to form VN 3 and host Service 3. This user is actually SP 3. Here, Service 1 and Service 2 have the same level. Service 1 has one level higher compared to Service 3, so does Service 2.

3 Lotka-Volterra Model Based Service Relationship

3.1 Service Model

In the testbed that we constructed, one kind of service may be hosted by several VNs. One service is seemed as a product of TaaS. A kind of service can be seen as a species. One kind of service can compete for resource with another kind of service, like Service 1 and Service 2 in Fig. 3. One kind of service can also promote another kind of service, like Service 2 and Service 3 in Fig. 3.

We assume that there are M kinds of service. They are all hosted by testbed. The number of these M kinds of service are successively $N_1(t), N_2(t), ..., N_m(t), ..., N_M(t)$. The number of services varies as the time, i.e., t, change. If there is only one kind of service hosted by testbed, the maximum service number of the mth service is K_m. According to the Lotka-Volterra model, for the mth service

$$\frac{dN_m(t)}{dt} = r_m N_m(t) \left(1 - \frac{N_m(t)}{K_m} - \sum_{i \neq m} \alpha_{im} \frac{N_i(t)}{K_m} \right) \tag{1}$$

where r_m is the ideal increase rate for the mth service, α_{im} is the ith service's competition coefficient. The physical meaning of α_{im} is that the existence of $N_i(t)$ would occupy α_{im}

$N_m(t)$'s resource space. Therefore, there is

$$
\begin{cases}
\dfrac{dN_1(t)}{dt} = r_1 N_1(t)\left(1 - \dfrac{N_1(t)}{K_1} - \displaystyle\sum_{i\neq 1} \alpha_{i1}\dfrac{N_i(t)}{K_1}\right) \\[3mm]
\dfrac{dN_2(t)}{dt} = r_2 N_2(t)\left(1 - \dfrac{N_2(t)}{K_2} - \displaystyle\sum_{i\neq 2} \alpha_{i2}\dfrac{N_i(t)}{K_2}\right) \\[2mm]
\vdots \\[1mm]
\dfrac{dN_m(t)}{dt} = r_m N_m(t)\left(1 - \dfrac{N_m(t)}{K_m} - \displaystyle\sum_{i\neq m} \alpha_{im}\dfrac{N_i(t)}{K_m}\right) \\[2mm]
\vdots \\[1mm]
\dfrac{dN_M(t)}{dt} = r_M N_M(t)\left(1 - \dfrac{N_M(t)}{K_M} - \displaystyle\sum_{i\neq M} \alpha_{iM}\dfrac{N_i(t)}{K_M}\right)
\end{cases}
\tag{2}
$$

3.2 Model Analysis

For the situation of three kinds of services, Eq. (2) becomes

$$
\begin{cases}
\dfrac{dN_1(t)}{dt} = r_1 N_1(t)\left(1 - \dfrac{N_1(t)}{K_1} - \alpha_{21}\dfrac{N_2(t)}{K_1} - \alpha_{31}\dfrac{N_3(t)}{K_1}\right) \\[3mm]
\dfrac{dN_2(t)}{dt} = r_2 N_2(t)\left(1 - \dfrac{N_2(t)}{K_2} - \alpha_{12}\dfrac{N_1(t)}{K_2} - \alpha_{32}\dfrac{N_3(t)}{K_2}\right) \\[3mm]
\dfrac{dN_3(t)}{dt} = r_3 N_3(t)\left(1 - \dfrac{N_3(t)}{K_3} - \alpha_{13}\dfrac{N_1(t)}{K_3} - \alpha_{23}\dfrac{N_2(t)}{K_3}\right)
\end{cases}
\tag{3}
$$

We define that, if service A's SP and service B's SP have the same level, the relationship between service A and service B is denoted as A-B. If service B's SP is based on service A's SP, the relationship between service A and service B is denoted as A(B). Therefore,

Table 1. Typical service relationship

Situation	Relationship	α_{BA}, α_{CA}	α_{AB}, α_{CB}	α_{AC}, α_{BC}
1	A-B-C	$\alpha_{BA} > 0, \alpha_{CA} > 0$	$\alpha_{AB} > 0, \alpha_{CB} > 0$	$\alpha_{AC} > 0, \alpha_{BC} > 0$
2	A(C)-B	$\alpha_{BA} > 0, \alpha_{CA} < 0$	$\alpha_{AB} > 0, \alpha_{CB} > 0$	$\alpha_{AC} < 0, \alpha_{BC} > 0$
3	A-B(C)	$\alpha_{BA} > 0, \alpha_{CA} > 0$	$\alpha_{AB} > 0, \alpha_{CB} < 0$	$\alpha_{AC} > 0, \alpha_{BC} < 0$
4	A(B(C))	$\alpha_{BA} < 0, \alpha_{CA} < 0$	$\alpha_{AB} < 0, \alpha_{CB} < 0$	/

Four situations are listed in Table 1. Situation 1 has one level of SPs. Situation 2 has two-level SPs, so does situation 3. Situation 4 has three levels of SPs. Actually, other relationships are the variant of these 4 situations in Table 1. The SP relationship in Fig. 3 is situation 3.

3.3 Numerical Simulation

In this subsection, a numerical simulation is conducted using Matlab. A specific situation in Sect. 3.2, i.e. situation 2 A(C)-B, is selected. Because it has the same level of SPs and the different levels of those at the same time. Storage space is selected as a measurement index of service occupied resources.

The simulation result is shown in Fig. 4. The baseline is available storage space, which is equal to storage space. There seems to be no regularity for a single service, no matter Service A, B, and C. However, the sum of occupied storage spaces of competing Service A and B is nearly equal to baseline. And the tiny difference between baseline and sum of Service A and Service B is the storage space occupied by the SP of the entire testbed. Because of a similar reason, there is also a tiny difference between the occupied storage space of Service A and that of Service C.

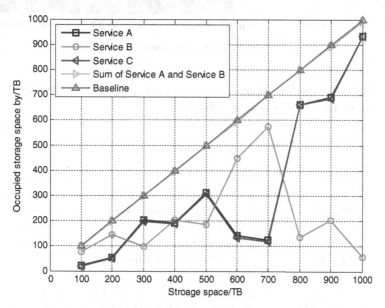

Fig. 4. Occupied storage space by all the services in situation 2

4 Testbed Implementation

The testbed for service testing and data process is shown in this section, through real implementations. Using a video service test in this testbed as an example, cloud computing-based service testing and data process are displayed. Furthermore, other data visualization based on ready-made data is also shown. Because this testbed is built as a commercial product towards the china market, the words in this testbed are Chinese characters.

4.1 Testbed Environment

The testbed environment is displayed in Fig. 5. The testbed is composed of two parts: the back end and front end. The back end is cluster servers which are responsible for data storage, computing and so on. The front end is workstations that are responsible for providing users with interactive interfaces. All the devices in testbed are connected via a network.

(a)Testbed architecture   (b)Back end cluster servers   (c)Front end workstation

Fig. 5. Testbed environment

The devices used in testbed are the server, switch, cable and etc. The detail of the devices used in the testbed is listed in Table 2.

Table 2. Equipment parameters in testbed

Device	Version	Parameter	Application
Server	Inspnr Xeon NF8560M2	Memory: Intel DDR3 8 GB*32; Hard Disk: 300 GB*10; Basic frequency: 1.86 GHz	Provide data storage and computing ability
	Inspur Xeon NF5280M3	Memory: Intel DDR3 8 GB*24; Hard Disk: 300 GB*24; Basic frequency: 2 GHz	
Switch	TP-LINK TL-SG7016T	Backplane bandwidth: 32 Gbps: Packet forwarding rate: 1488000 pps; Port number: 16	Devices connection
Cable	RJ45 twisted pair	Data transmission rate: 1000 Mbps	
Rack	ToTen	Size: 2 m*0.6 m*1 m	Devices placement
Workstation	HP XW4600	Memory: Intel DDR3 4 GB: Hard Disk: 500 GB; Basic frequency: 2.67 GHz	Client

OpenStack is the most important software in the testbed. It is responsible for realizing cloud computing. At the same time, software like Wireshark, Apache Tomcat and etc., are also employed in the testbed. The detail of the software used in the testbed is listed as Table 3.

Table 3. Software parameters in testbed

Name	Version	Application
Wireshark	1.10.8.2	For test data access
Apache Tomcat	7.0.34	Storage for handled data; Storage for data visualization page
Baidu Echarts	3.0	Data visualization tool
Google Chrome	35.0.1916.114	Brower for client work station
Eclipse	3.6.1	Web page edit environment
OpenStack	Juno	Cloud computing platform

4.2 Cloud Host Allocation

Figure 6 is the testbed main page which provides the function interfaces for test service users. The first one in the first row is the entrance for OpenStack used in this testbed. After successful OpenStack login, cloud hosts can be allocated to host different kinds of test services. The others in the first row are the tested video service, captured packets and visualized captured data. The second and third row are other data visualizations based on ready-made data.

Fig. 6. Testbed main page   **Fig. 7.** OpenStack login page

Figure 7 is the OpenStack login page used in this testbed. If login successfully, SPs are able to allocate and abort cloud hosts for test services. SPs can also look up the basic information of allocated cloud hosts, which are shown in Fig. 8.

Fig. 8. Cloud host list

Cloud host allocation is shown in Fig. 9. SPs have to configure basic resource, security, and network in order, before a cloud host is allocated.

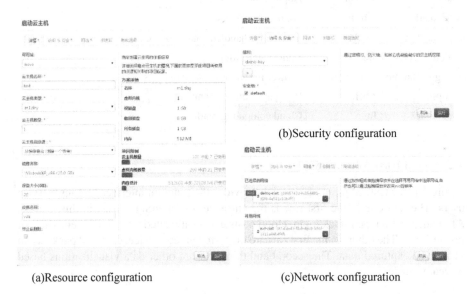

(a)Resource configuration   (c)Network configuration

(b)Security configuration

Fig. 9. Cloud host allocation

4.3   Video Test

Using a typical video test as an example, cloud computing-based service testing and data process are displayed. Through the method in Section V.B, two cloud hosts are allocated. One cloud host is used as a video server, and another is used as a video client. The video service test is a simple server-client video transmission which is shown in Fig. 10. Here, the video service can be seemed as the product of TaaS.

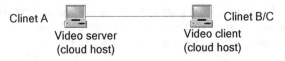

Fig. 10. Video service test based on cloud test

The video server is mapped upon client A, the video client is mapped upon client B. These are shown in Fig. 4(c). In the process of video service test, the place of host video client can be changed from client B to client C, without any influence on the video service test. The tested video is shown in Fig. 11.

The captured packets are shown in Fig. 12. The result of the video service test based on captured packets analysis, i.e., video test result, is visualized in Fig. 12. This video test result visualization in Fig. 13 is dynamically changed as time.

视频业务应用测试

Fig. 11. The video used in the test

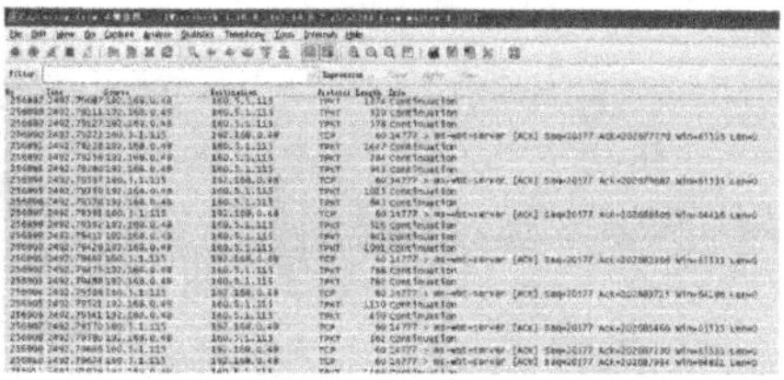

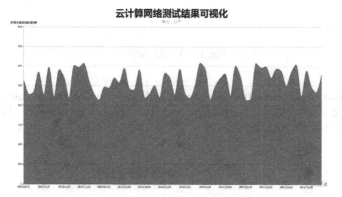

Fig. 12. Captured packets

云计算网络测试结果可视化

Fig. 13. Video test result visualization

4.4 Other Data Visualizations

In order to display the diverse functions of this testbed, based on ready-made data, other data visualizations are shown.

From a lot of text files, the most common keywords are shown as a word cloud in Fig. 14. The higher a key word's occurrence frequency is, the bigger it is.

词云

Fig. 14. Word cloud

Based on the data of global airline data, a world airline visualization is shown in Fig. 15. Each line is an actual airline with its own source and destination. The denseness of red lines in Fig. 15 stands for how busy the airport is. We can see that the busiest areas are the east coast of the USA, Europe, and eastern China.

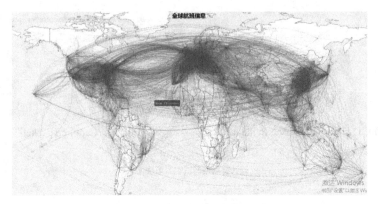

Fig. 15. World airline

The China backbone network is shown in Fig. 16, based on the data from ChinaNet's official site. We can see that the backbone network is deployed all over china and connected to a foreign core node.

Fig. 16. China backbone network

The user distribution of a web site is shown in Fig. 17. It is clear that most users can be located in the main cities such as Beijing, Shanghai, Guangzhou and etc.

Fig. 17. Website visitor distribution

5 Conclusions

In this paper, in order to overcome the drawback of the traditional test in a real system, a testbed for service testing and data process based on cloud computing was proposed. In this testbed, a test was realized as a service hosted by one or several cloud hosts, i.e., test as a service. The relationship of different kinds of service in testbed was modeled based on the Lotka-Volterra model. The testbed implementation was demonstrated based on cloud computing, using a typical video test as an example. It was proved that our testbed for service testing and data process was practicable and effective.

Acknowledgement. This work was supported by the Support Project of High-Level Teachers in Beijing Municipal Universities in the Period of the 13th Five-Year Plan (CIT&TCD 201704069) and China Computer Federation (CCF) Opening Project of Information System (No. CCFIS2019-01-01).

References

1. Wang, A., Iyer, M., Dutta, R., et al.: Network virtualization: technologies, perspectives, and frontiers. J. Lightwave Technol. **31**(4), 523–537 (2013)
2. Berman, M., Chase, J.S., Landweber, L., et al.: GENI: a federated testbed for innovative network experiments. Comput. Netw. **61**, 5–23 (2014)
3. Mell, P., Grance, T.: SP 800-145. The NIST Definition of Cloud Computing. Cloud Computing (2011)
4. Gangadharan, G.R.: Open source solutions for cloud computing. Computer **50**(1), 66–70 (2017)

5. Rimal, B.P., Maier, M.: Workflow scheduling in multi-tenant cloud computing environments. IEEE Trans. Parallel Distrib. Syst. **28**, 290–304 (2016)

6. Georgakopoulos, D., Jayaraman, P.P., Fazia, M., et al.: Internet of Things and edge cloud computing roadmap for manufacturing. IEEE Cloud Comput. **3**(4), 66–73 (2016)

7. Buyya, R., Ranjan, R., Calheiros, R.N.: Modeling and simulation of scalable Cloud computing environments and the CloudSim tool kit: challenges and opportunities. In: International Conference on High Performance Computing & Simulation (HPCS 2009), pp. 1–11 (2009)

8. Garg, S.K., Buyya, R.: NetworkCloudSim: modelling parallel applications in cloud simulations. In: Cloud Computing, pp. 105–113 (2011)

9. Dai, Q., Shou, G., Hu, Y., et al.: A general model for hybrid fiber-wireless (FiWi) access network virtualization. In: IEEE International Conference on Communications (ICC 2013), pp. 858–862 (2013)

10. Dai, Q., Zou, J., Shou, G., et al.: Network virtualization based seamless networking scheme for fiber-wireless (FiWi) networks. China Commun. **11**(5), 1–16 (2014)

11. Khazaei, H., Misic, J., Misic, V.B., et al.: Performance of cloud centers with high degree of virtualization under batch task arrivals. IEEE Trans. Parallel Distrib. Syst. **24**(12), 2429–2438 (2013)

12. Ameixieira, C., Cardote, A., Neves, F., et al.: Harbornet: a real-world testbed for vehicular networks. IEEE Commun. Mag. **52**(9), 108–114 (2014)

13. Yang, C., Chen, J.: A scalable data chunk similarity based compression approach for efficient big sensing data processing on cloud. IEEE Trans. Knowl. Data Eng. **29**, 1144–1157 (2016)

14. Der Aalst, W.M., Damiani, E.: Processes meet big data: connecting data science with process science. IEEE Trans. Serv. Comput. **8**(6), 810–819 (2015)

15. Baidu. Echarts Official Site[EB/OL]. http://echarts.baidu.com/index.html

16. Jianshukeji. Highcharts Official Site[EB/OL]. https://www.highcharts.com/

17. Baidu. Echarts-X Official Site[EB/OL]. http://echarts.baidu.com/echarts2/x/doc/index.html

Improve the Efficiency of Maintenance for Optical Communication Network: The Multi-factor Risk Analysis via Edge Computing

Yucheng Ma[1(✉)], Yanbin Jiao[2], Yongqing Liu[2], Hao Qin[3], Lanlan Rui[1], and Siqi Yang[1]

[1] State Key Laboratory of Networking and Switching Technology,
Beijing University of Posts and Telecommunications, Beijing, China
{mayucheng,llrui,yangsiqi}@bupt.edu.cn
[2] State Grid Information & Telecommunication Group Co., Ltd., Beijing, China
{jiaoyanbin,liuyongqing}@sgitg.sgcc.com.cn
[3] Anhui Jiyuan Software Co., Ltd., Hefei, China
qinhao@sgitg.sgcc.com.cn

Abstract. Optical communication networks carry people's communication services. As the structure of communication networks becomes more complex, and the volume of services is growing. It is greater to ensure that the network is functioning properly. Therefore, people are deeply concerned about the maintenance of communication networks. The nodes in the network are an important component of the network. Therefore, it is necessary to conduct the risk analysis on nodes and perform intelligent maintenance based on the results. The normal operation of nodes in a communication network is influenced by many aspects. In order to access the risk of nodes, the risk analysis algorithm proposed in this paper considers network topology, service type, maintenance, environmental factors and the load of the node. Then intelligent maintenance is performed based on the results of the node's risk analysis, and the edge computing server is introduced to perform the risk analysis on the node in time. So as to improve the timeliness of updating information. According to the results of simulation, maintenance personnel can get more accurate information about the nodes or aspects that need to be maintained, and because of the use of edge computing structures, the method has become more efficient.

Keywords: Optical communication network · Edge computing · Intelligent maintenance · Risk analysis

1 Introduction

The amount of data has increased dramatically, which has led to the rapid development of communication networks. Not only is the scale of communication networks growing, but also optical communication networks carry more and more communication services. Effective maintenance of the communication network is important for its normal operation. Communication networks have complex operating environments. The timeliness of

© Springer Nature Singapore Pte Ltd. 2019
H. Ning (Ed.): CyberDI 2019/CyberLife 2019, CCIS 1138, pp. 121–129, 2019.
https://doi.org/10.1007/978-981-15-1925-3_9

maintenance and the knowledge of maintenance personnel will affect the normal operation of the communication network [1]. Nodes in a communication network are very important to the operation of the entire network. Therefore, it is necessary to ensure that the node can be maintained in time. And the cause of the failure of the network node may be different, and it is necessary to analyze the risk of nodes in the network in time and perform intelligent maintenance based on the results. In order to solve the above problems, this paper has conducted research.

We propose a network node risk analysis based on edge computing. The architecture of edge computing is utilized to perform the risk analysis for network nodes. When the service supported in the communication network changes rapidly, the risk analysis result of the node is quickly obtained.

If a node that support a large number of services fails, it will have a serious impact on the stable operation of the network. In order to effectively evaluate the risks that each node may bring during the operation of the network. We propose a risk analysis of node algorithm based on multi-factor. Since the running state of the node is affected by the complicated situation. The assessment of the importance of maintenance takes into account several factors, which mean whether a node should pay more attention to maintenance. It also combined with the load of the node to analyze the risk of the node, which can reflect the real-time operation of the node. Depending on the result of the risk analysis, intelligent maintenance can be implemented. For different factors that may cause node failure, we can intelligently choose more accurate maintenance.

2   Related Work and Problem

For the maintenance architecture of communication networks, reference [2] proposed a cloud-based architecture. The system collects relevant resources and sends the information to the cloud server for processing and analysis. Reference [3] proposed method that includes the concept of edge computing. It migrates some of the services to servers close to the edge nodes, which can satisfy the demand of the system's real-time response. For the risk analysis of nodes in a communication network, the importance of nodes in the network may determine the degree to which nodes need to pay attention. Reference [4] considered the importance of nodes by considering the degree of nodes and the average degree of direct neighbors. Reference [5] used a node deletion method based on the number of spanning trees to evaluate the importance of nodes in the network. The number of spanning trees is a parameter used to evaluate network reliability. But the node deletion method also cannot fully consider the connection relationship of all nodes. As in [6], a node contraction method was proposed which assumes the importance of the node by shrinking the degree of network contraction. Reference [7] considered the importance of nodes based on network cohesion. In addition to the network structure, this method also considers the properties of the node. But there is a lack of operational factors of network.

Usually, the maintenance system of the communication network adopts a cloud service architecture [2]. In this architecture, the system collects network information and sends it to the cloud server for analysis. Maintenance personnel can perform maintenance according to the results of the analysis. But the size of the network can be large. The factors that cause the node to failure may also be different, and the load on the nodes is constantly changing. These conditions can put severe stress on the cloud server and make the maintenance information update untimely. Therefore, we introduce an edge computing maintenance architecture to migrate the node risk analysis tasks undertaken by the cloud server to the servers of the edge nodes. This method can effectively respond quickly to changes in the operating state of the node and analyze risk. The data are sent to the central server after processing at the edge node, which can effectively reduce the flow of data and the pressure of the central server.

Risk analysis of nodes includes the importance of maintenance and risk estimates of the nodes. According to the above description, we can know that most methods evaluate the importance of nodes through the topology of the network. Some methods consider other factors. However, the operating environment and operating state of the node are not considered. In order to satisfy the operation of a communication network, more factors about the actual situation should be considered. Therefore, this paper identifies four main influencing factors, and comprehensively evaluates the importance of maintenance.

The edge server will update the overall structure of the network synchronously from the data center, and it collects operational data for the nodes in the network. The Edge Server evaluates the importance of four aspects based on relevant information. Finally, the data of node importance will be centralized in the data center. Maintenance personnel will choose to pay more attention to maintenance of the priority nodes and aspects.

3 Risk Analysis of Nodes Based on Multi-factors

The multi-factor based the risk analysis of node was proposed to evaluate the importance of maintaining the nodes in the network from the four levels of network structure, service, the maintenance situation and environment. Then the load of node and the importance of maintenance is applied to estimate the node risk.

3.1 The Importance Based on Network Structure

In the communication network, the structure of network is very important and fundamental for the importance of the evaluation node. At this level, the concept of network cohesion and node contraction is chosen to assess the importance of the nodes. According to the definition of network cohesion, it can be expressed as:

$$\partial(WG) = \frac{1}{N * \frac{\sum_{i,j \in V, i \neq j} d_{ij}}{N(N-1)}} = \frac{N-1}{\sum_{i,j \in V, i \neq j} d_{ij}} \tag{1}$$

In (1), d_{ij} is the shortest length of path that is from node i to node j. Firstly, the degree of cohesion $\partial(WG_0)$ of the initial network can be computed. Then the nodes that connected to the node i directly transform one node. Therefore we can calculate the degree of cohesion $\partial(WG_i)$ of new network. Finally, the node importance W_i about node i can be obtained:

$$W_i = 1 - \frac{\partial(WG_0)}{\partial(WG_i)} \tag{2}$$

Equation (2) evaluates node importance by calculating the rate of change of network cohesion. This approach can fully consider the importance of nodes in the network. But some nodes in the network may have important significance.

3.2 The Importance Based on Service

Quality of service is a very important goal for communication networks [10]. Set services with high quality of service requirements to a higher priority, and the more important service is, the more attention and maintenance the node supporting it needs.

Considering the optical communication network service process comprehensively, the importance of the services can be considered from the three aspects, including Delay, Delay variation and information loss, According to the ITU-T Rec G1010, several typical parameters of importance of communication service are evaluated by the 1–5 scale method. Each indicator is scored according to various types of services, and then weighted sum calculation and logarithmic normalization are performed with the corresponding delay, delay variation and information loss indicators to obtain the importance of each service. In the calculation of the importance of service that supported by node i, the maximum value of all the importance levels of services supported by this node is taken as the node importance. Therefore, P_i of the node depends on the type and level of the supported service. The equation is as follows:

$$P_i = \max_{j=1,2,3...n} (I_j) \tag{3}$$

I_j is the importance of the service supported by node i, and n is the service type.

3.3 The Importance Based on Maintenance Situation

In the network, nodes will have a certain probability of failure during operation, and whether the node can recover in time, it has an important impact on the operation of the network. The operation of a network is affected by nodes with high probability of failure, which require more maintenance. The classical equipment failure probability model can used. The running equipment enters the Stop state due to the failure, and the failed equipment is repaired and re-entered into the running state. The model does not consider changes in the external environment. Where λ, μ are the failure rate and the repair rate, respectively, defined as:

$$\lambda = \frac{\text{Number of failures during the operation period}}{\text{Total time of node commissioning}} \tag{4}$$

$$\mu = \frac{\text{Number of fixes during the outage period}}{\text{Total node outage time}} \tag{5}$$

By counting the failure rate and the repair rate, the failure probability of the node can be evaluated. The probability P(t) of the node failure within the time t is:

$$P(t) = e^{-\lambda t} \tag{6}$$

According to the failure probability of the node, the importance of the node based on the probability of failure can be obtained. The calculation formula is as follows:

$$M_i = 1 - e^{-\lambda_i} \tag{7}$$

λ_i is the average failure probability of the node.

3.4 The Importance Based on Environmental Factors

Generally, the coverage area of communication networks is very wide. There may be some special environmental impacts in different areas. Therefore, this situation requires protection against specific environmental factors. Firstly, calculate the fault level of the cable. The environment factors can be classified into three types. **Suppose:** α indicates the temperature, humidity and other factors, β indicates disasters such as ice, snow, and γ indicates Man-made harm. When evaluating the fault level of a specific optical cable, α, β, γ are respectively scored (range 1 to 5), and then the addition calculation and logarithmic normalization are performed to obtain the fault level of the optical cable line: L_k. Then the damage caused by the cable failure to the service is B_j:

$$B_j = I_j \sum_{k \in r} L_k \tag{8}$$

I_j is the importance of services j and L_k is the failure level of the cable through which services j passes. r is a collection of optical cables through which service i passes.

The node importance based on environmental factors is:

$$E_i = \max_{i=1,2,3\ldots n} \left(1 - (B_j)_i\right) \tag{9}$$

E_i is the degree of environmental impact on node i. $(B_j)_i$ is the degree of harm assumed by the service of node i.

3.5 Risk Analysis of Nodes in the Network

According to the above four aspects, we can get the comprehensive importance of node i. k_1, k_2, k_3, k_4 are the weights of each aspect:

$$N_i = k_1 * W_i + k_2 * P_i + k_3 * M_i + k_4 * E_i \tag{10}$$

N_i represents the comprehensive importance in the network.

The node and the link of network support related services. Once the critical nodes are damaged and failed, it will not only directly affect the service quality of the communication system, but also have a huge impact on the performance of the communication network [8, 9]. Therefore, it is necessary to analyze the risk of nodes in the network to evaluate the impact of failed node. I_k is the k-type service importance, n_k is the number of k-th services, and k is the service type carried by the link connected to the node. ω_q is the amount of information of the link q.

$$I(i, j) = \begin{cases} \sum_{k=1}^{n} n_k I_k, & Node\ i\ is\ directly\ connected\ to\ node\ j \\ 0, & Node\ i\ is\ not\ directly\ connected\ to\ node\ j \end{cases} \tag{11}$$

$$R_i = N_i \sum_{j}^{n} I(i, j) \tag{12}$$

N_i is the above-mentioned network node importance combined with various factors (service level, maintenance and environment), and R_i is the risk degree of node i.

4 Performance Evaluation

Firstly, we create the abstract structure of the simulated communication network. The nodes in abstract network have the function of initiating, forwarding, and receiving services. In addition, the simulation produced a number of services and scored the parameters of the quality of service. The number of failures and the time to failure are simulated for each node. For each link, three environmental impact types are scored. Assuming that each node has an edge server, it is responsible for the risk analysis of the node. The central server is responsible for storing and statistic data.

In order to compare the impact of the structure of cloud and edge computing on the above methods, we randomly generate 200 services, and the calculation time of the two structures is calculated separately. When the number of services is small, the calculation time of the two structures is similar. When the number of services increases, the computing time based on the cloud service structure increases rapidly. However, the calculation time based on the edge computing increases slowly. It shows that when the number of services is large, the structure of edge computing can make the above method more efficient.

Table 1 shows the importance evaluation and ranking of nodes in the network. Firstly, the importance of the node in terms of network structure, service, the maintenance situation and environment factors is calculated, and finally the comprehensive importance of the node is obtained.

According to the Table 1, node 1 is the highest ranked. We can find that node 1 has a high degree of node importance in the network structure, the service layer and the environment layer. The location of node 1 that supported many services with high importance is at the center of this network, therefore it connected with the many surrounding nodes, and the links supported large traffic so that this node also has a high value in the risk analysis, and maintenance personnel need to reduce it via reducing traffic. Although this

Table 1. Ranking of the importance evaluation for each level

Node ID	Network structure level	Service level	Maintenance level	Environment level	Comprehensive importance
1	0.950	0.979	0.865	0.906	0.950
2	0.860	0.979	0.993	0.906	0.942
5	0.805	0.911	0.996	0.868	0.889
8	0.435	0.979	0.998	0.906	0.814
18	0.398	0.979	0.994	0.906	0.802
7	0.426	0.979	1.000	0.833	0.800
4	0.426	0.979	0.997	0.833	0.799
3	0.692	0.748	0.998	0.878	0.783
9	0.398	0.911	1.000	0.868	0.766
10	0.100	0.979	0.986	0.906	0.711
88	0.679	0.647	0.918	0.794	0.708
15	0.100	0.979	0.995	0.783	0.693
6	0.398	0.748	0.976	0.878	0.690
14	0.100	0.911	0.990	0.886	0.677
16	0.100	0.911	0.995	0.852	0.673
13	0.100	0.935	0.918	0.833	0.669
11	0.100	0.748	0.998	0.868	0.602
12	0.100	0.748	1.000	0.827	0.596
17	0.100	0.647	0.998	0.878	0.557

node's importance about the maintenance situation is not high, the overall importance of the node is still the highest. So node 1 needs more maintenance.

In order to verify the reasonableness of the proposed algorithm. The algorithm proposed in this paper was compared with the node contraction method and the node deletion method. As in Table 2. The node deletion algorithm and the node contraction method cannot evaluate the importance of the node very well. Because, many nodes have the same importance. But, the algorithm proposed in this paper considers many factors, and the results show that it can distinguish each node well.

According to the results of the comparison, the three methods have the same ranking in the evaluation of the importance for some nodes, such as 1, 2, and 5. These nodes are at the core of the network with many important services, and the complex operating environment. But most of the nodes have different ranking results. For example node 3 has a higher ranking in the results of the node contraction algorithm. However, this node is not ranked high in the multi-factor based algorithm. By analyzing Table 1, node 3 is

Table 2. Comparison of different node evaluation algorithms

Node ID	Multi-factor based algorithm		Node contraction algorithm		Node deletion algorithm	
	Node Importance	Ranking	Node Importance	Ranking	Node Importance	Ranking
1	0.950	1	0.950	1	1.000	1
2	0.942	2	0.860	2	1.000	2
3	0.783	8	0.692	4	1.000	4
4	0.800	7	0.426	7	0.727	11
5	0.889	3	0.805	3	1.000	3
6	0.690	13	0.100	13	1.000	8
7	0.800	6	0.398	9	1.000	6
8	0.814	4	0.435	6	1.000	5
9	0.766	9	0.398	10	1.000	7
10	0.711	10	0.100	17	0.000	13
11	0.602	17	0.100	12	0.000	15
12	0.596	18	0.100	15	0.000	12
13	0.669	16	0.100	18	0.000	19
14	0.677	14	0.100	19	0.000	16
15	0.693	12	0.398	11	0.000	18
16	0.673	15	0.100	14	0.000	17
17	0.557	19	0.100	16	0.000	14
18	0.802	5	0.426	8	0.750	10
88	0.708	11	0.679	5	0.909	9

not very important at the service level and the environment level. Therefore, the overall importance of the node 3 is less than the node 8 and node 18, although these two nodes are less important in the network structure. The node deletion algorithm and the node contraction algorithm only focus on the importance of nodes in the network structure. The multi-factor based key node identification algorithm can comprehensively consider various factors. Therefore, this method can give maintenance suggestions based on the estimation results of different aspects.

5   Conclusion

Efficient maintenance is critical to the communications network. In this study, we propose a multi-factor based risk analysis approach. This method includes an assessment of the maintenance importance of the node and a risk estimate. The evaluation of the importance of node maintenance takes into account the four factors. The real-time load situation of the nodes was used to estimate the risk. The introduction of edge computing makes the method improve the processing efficiency and transmission capacity of information. According to the simulation, maintenance personnel can perform intelligent maintenance on specific aspects depending on the evaluation of node importance, which makes the maintenance result more efficient. The Risk analysis can reflect the extent of damage that a node can cause in real time so that maintenance personnel can take action to reduce risk and prevent losses.

Acknowledgment. The work is supported by State Grid ICT Industry Group ICT Research Institute "2019 Standardization Research Innovation Seed Fund Project", the National Natural Science Foundation of China (61302078), the 863 Program (2011AA01A102).

References

1. Zhao, Y., Chen, L., Li, Y.: Efficient task scheduling for many task computing with resource attribute selection. China Commun. **11**(12), 125–140 (2014)
2. Lojka, T., Bundzel, M., Zolotova, I.: Service-oriented architecture and cloud manufacturing. Acta Polytechnica Hungarica **13**(6), 25–44 (2016)
3. Qiao, X., Ren, P., Dustdar, S.: A new era for web AR with mobile edge computing. IEEE Internet Comput. **22**(4), 46–55 (2018)
4. Ai, J., Li, L., Su, Z., Jiang, L., Xiong, N.: Node-importance identification in complex networks via neighbors average degree. In: 2016 Chinese Control and Decision Conference (CCDC), Yinchuan, pp. 1298–1303 (2016)
5. Chen, Y., Hu, A., Hu, J., et al.: Determination of the most important nodes in communication networks. High Technol. Lett. **14**(1), 21–24 (2004)
6. Tan, Y., Wu, J., Deng, H.: Node shrinkage method for node importance evaluation in complex networks. Syst. Eng. - Theory Pract. **26**(11), 79–83 (2006)
7. Huang, H., Jiang, W., Tang, Y.: Evaluation method of node importance in power communication network. Electr. Autom. **40**, 238(04), 47–49 (2018)
8. Liu, Z., Wu, J.: Mobile satellite network virtual mapping algorithm based on node risk. In: 2016 International Conference on Network and Information Systems for Computers (ICNISC), Wuhan, pp. 76–79 (2016)
9. Waqar, A., Osman, H., Usman, P., Junaid, Q.: Reliability modeling and analysis of communication networks. J. Netw. Comput. Appl. **78**, 191–215 (2017)
10. Meng, Y.: QoS management system of smart grid communication network based on SDN technology. In: Proceedings of 2017 2nd International Conference on Materials Science, Machinery and Energy Engineering (MSMEE 2017), p. 9. Research Institute of Management Science and Industrial Engineering (2017)

Edge Computing for Intelligent Transportation System: A Review

Qian Li[1], Pan Chen[1], and Rui Wang[1,2,3(✉)]

[1] School of Computer and Communication Engineering, University of Science
and Technology Beijing (USTB), Beijing 100083, China
wangrui@ustb.edu.cn
[2] Institute of Artificial Intelligence, University of Science
and Technology Beijing, Beijing 100083, China
[3] Shunde Graduate School, University of Science and Technology Beijing,
Foshan 528300, China

Abstract. To meet the demands of vehicular applications, edge computing as a promising paradigm where cloud computing services are extended to the edge of networks can enable ITS applications. In this paper, we first briefly introduced the edge computing. Then we reviewed recent advancements in edge computing based intelligent transportation systems. Finally, we presented the challenges and the future research direction. Our study provides insights for this novel promising paradigm, as well as research topics about edge computing in intelligent transportation system.

Keywords: Edge computing · Intelligent transportation system · Vehicular networks

1 Introduction

Cloud computing is a resource-rich and large-scale service system. It is an ideal strategy to use cloud computing systems as a platform for processing big data [1]. But, cloud computing systems also bring a lot of problems, such as increasing the delay of requests, serious waste of resources, and unstable connection [2]. Edge computing is a distributed computing structure that divides complex computing tasks into multiple small element units and assigns them to local nearby devices. As an extension of cloud computing, edge computing itself has the characteristics of dynamics, low latency, mobility, and location awareness [3]. Each node in the edge computing is not isolated, and they can perform allocation and cooperation to complete the same task request. At the same time, edge computing is not a substitute for cloud computing and both of them are not isolated. On the contrary, they complement each other and constitute a three-tier architecture, that is, equipment, edge computing, and cloud computing to jointly solve many challenges in the development of the Internet of Things [4]. Its architecture is shown in Fig. 1.

In this paper, we first discuss the research progress of intelligent transportation based on edge computing in the second section. In the third section, we summarize

© Springer Nature Singapore Pte Ltd. 2019
H. Ning (Ed.): CyberDI 2019/CyberLife 2019, CCIS 1138, pp. 130–137, 2019.
https://doi.org/10.1007/978-981-15-1925-3_10

the challenges faced by intelligent transportation based on edge computing. Then, we summarize the future development direction of intelligent transportation based on edge computing in the fourth section. Finally, we conclude this paper.

Fig. 1. Edge computing architecture diagram

2 Research on Edge Computing in Intelligent Transportation System

In this section, we will discuss the research progress from the three perspectives of resource management and relay framework, deep reinforcement learning and 5G automotive system, security message authentication.

2.1 Resource Management and Relay Framework

In terms of resource management, [5] conducts research on spectrum resource management. First, the authors used the logarithmic and linear utility functions to develop three aggregated network utility maximization issues. Then, linear programming relaxation and first-order Taylor series approximation are used, and the alternating concave search (ACS) algorithm is designed to solve the three non-convex network maximization problems. [6] proposes an alternating direction method of multipliers (ADMM) scheme. After the scheme adds a set of local variables to each in-vehicle user equipment (UE), the initial variable optimization problem is transformed into a consistency problem with separable targets and constraints. The consistency problem can be further broken down into a set of sub-problems, and the sub-problems are distributed to the UE to be solved in parallel.

In order to efficiently add mobile vehicles to information dissemination and select appropriate relays to meet the diverse transmission tasks, the researchers have done a lot of research work and will introduce the following. In [7], the researchers developes a vehicle content communication framework based on edge computing. In terms of incentives, the framework develops a new theoretical model of auction games that evaluates the resource contributions of idle vehicles by using a Bayesian Nash equalization algorithm based on game theory. [8] studies the problem of optimal deployment and dimensioning (ODD) of in-vehicle networks supported by edge computing, minimizing the cost of developing ODD problems in the form of integer linear programming (ILP), while considering infrastructure deployment, overall network organization, coverage requirements and network latency. The above research work is summarized in Table 1.

Table 1. Summary of research work on resource management and relay framework

Application scenario	Research work		Application Technology	Experimental program
Resource management	[5]	Dynamic spectrum resource management framework	ACS algorithm, etc.	Linear programming relaxation; first-order Taylor series approximation; ACS algorithm
	[6]	ADMM solution	ADMM algorithm, etc.	Decompose the consistency problem into a set of sub-problems and processing them in parallel
Relay framework	[7]	Vehicle content communication framework	Bayesian Nash Equilibrium Algorithm Based on Game Theory etc.	Game theory as an incentive mechanism; content cache to ECD; ECD selection relay
	[8]	ODD problem optimization	ILP, coupling, decoupling	ILP minimizes the cost of ODD problems; develops coupling and decoupling models to compare

2.2 Deep Reinforcement Learning and 5G Automotive System

The Internet of vehicle has a certain deployment structure and specific connections. It is easier to capture multi-dimensional data from the environment, including vehicle behavior, mission information and network status. [9] proposes a DRL-based offloading decision-making model based on the effective use of deep reinforcement learning to assist the vehicle interconnection network to make better computing and offloading decisions, combined with knowledge-driven (KD) method. In [10], the author uses D2D communication and heterogeneous network HetNets to cooperate with the base station (BS) and the roadside unit (RSU) to increase the on-board caching capability. On the other hand, a deep Q-learning multi-time scale framework was developed, and the parameters required for compute, communication, and caching were optimally configured to determine the best choice for idle vehicle and edge nodes.

Researchers have also proposed introducing 5G technology into in-vehicle systems to improve the quality of in-vehicle application services. In [11], the author introduces the concept of Follow Me Edge Cloud (FMeC) to maintain the requirements of 5G automotive systems by using the Mobile Edge Computing (MEC) architecture. Based on the LTE-based V2I communication type, the author introduces its SDN/OpenFlow-based architecture and a set of algorithm-based mobile sensing frameworks, which enable autopilot QoS requirements in 5G networks, meeting autopilot delay requirements.

[12] proposes a new in-vehicle network architecture that integrates 5G mobile communication technology and software-defined networks. On the basis of the traditional 5G vehicle network, the author combines software-defined network (SDN) with cloud computing and edge computing technology, and divides the network into application layer, control layer and data layer that can divide control and data functions to improve the vehicle network. The above research work is summarized in Table 2.

Table 2. Summary of research work on vehicle network based on deep reinforcement learning and 5G automotive system

Application scenario	Research work		Application Technology	Experimental program
Deep reinforcement learning	[9]	DRL-based offloading decision model	Knowledge driven algorithm, A3C algorithm	Mobility to judge task delay; online learning KD service; exploring A3C algorithm optimization offloading decision
	[10]	Deep Q-learning multi-time scale framework	D2D communication, heterogeneous network HetNets, deep Q-learning	D2D communication, heterogeneous network HetNets improve cache computing power; optimal configuration of Q-learning parameters
5G automotive system	[11]	Follow Me Edge Cloud Architecture	FMeC Architecture, 5G, Mobile Perception	Follow Me Edge Cloud Architecture for Automated Driving
	[12]	SDN-based 5G automotive system	5G, SDN etc.	Hierarchical structure processing control and data request, SDN, cloud computing, edge computing, 5G combination

2.3 Secure Message Authentication

The communication protocol in VANET should satisfy the anonymity, that is, the vehicle should communicate with all entities through pseudo-identities rather than real identity. In [13], the authors incorporated a novel edge computing concept into VANET's message authentication process. In this scenario, the RSU acts as a cloud for the vehicle, and the system selects a number of edge computing vehicles (ECVs) to assist the RSU in verifying the message signature sent by nearby vehicles and then transmitting the results to the RSU based on the vehicle's limited computing power. Finally, the RSU verifies the results sent from the ECV, obtains the legitimacy of these messages, and broadcasts information about legality to the vehicle through the filter at the end.

[14] advocates the introduction of edge computing in the IoV, referred to as F-IoV. The author proposes a privacy-preserved pseudonym (P3) scheme in F-IoV and introduces an edge computing node that is deployed in different regions to directly manage the pseudonyms of passing vehicles. [15] proposes a novel edge computing based anomaly detection, coined edge computing based vehicle anomaly detection (EVAD). Edge-based sensors obtain data through Fourier transform formula, and the time domain characteristics, that is, the correlation between different onboard sensors and the frequency domain characteristics of the sensor data, are used to determine whether an abnormality has occurred in the vehicle.

3   Challenges

3.1   The Vehicular Environment Is Dynamic and Uncertain

The dynamic and uncertainty of vehicle environment is mainly reflected in the varying network topologies, wireless channel states, and computing workload. These uncertainties bring additional challenges to task offloading. Sun et al. [16] consider the task offloading among vehicles and then they designed an adaptive learning-based task offloading (ALTO) algorithm in order to minimize the average delay. The algorithm is based on the multi-armed bandit theory. Vehicles can learn the delay performance of their neighboring vehicles while offloading tasks. The proposed algorithm is proved under both synthetic scenario and realistic highway scenario.

3.2   Lack of Effective Incentives

The deployment of Vehicular fog computing (VFC) still confronts several critical challenges, such as the lack of efficient incentive. Zhou et al. [17] investigated the task assignment and computation resource allocation problem in VFC from a contract-matching integration perspective. An incentive mechanism based on contract is proposed to promote resource sharing among vehicles. At the same time, a stable matching algorithm based on pricing is proposed to solve the task assignment problem.

3.3   Security and Privacy

In the research of vehicle edge computing, due to the distributed and heterogeneous characteristics of edge nodes, some original security mechanisms related to cloud computing are not fully applicable to edge computing, and they bring new challenges to authentication and access control etc. Cui et al. [13] put forward a valid message authentication scheme for the redundancy in the communications security mechanism in the VANET authentication problem, this scheme introduce the edge of computing concept into the VANET message authentication, the roadside unit can efficiently authenticate messages from nearby vehicles and broadcast the authentication results to the vehicles within its communication range, in order to reduce the redundant certification, improve the efficiency of the whole system.

4 Future Research Direction

4.1 Task Offloading and Resource Allocation

In the existing research on vehicle edge computing, most of the research on task offloading is to design an offloading architecture, jointly consider the network, communication and computing resources, and design the resource allocation algorithm, so as to achieve the minimum energy consumption, delay or the system cost. For the problem of computing offloading decision, the decision algorithm is designed, aiming at the minimum delay or energy consumption. But these algorithms lack of performance optimization and analysis, how to design a low complexity approximate optimal algorithm is a worthy research topic. In addition, there are dependencies between subtasks in some complex applications, which makes task scheduling more complicated and brings additional challenges to the framework. Designing a task scheduling algorithm with reasonable computational complexity and good performance is a field worthy of research. Many current studies are conducted in some simple hypothetical scenarios. In the future, we need to consider more realistic situations, such as unpredictable vehicle mobility and dynamic resource utilization, to find more feasible solutions.

4.2 Incentive Mechanism

In vehicle network, content dissemination usually relies on roadside infrastructure and mobile vehicles to deliver and disseminate content. Due to vehicle mobility, selfishness and limited communication capabilities of infrastructure, how to effectively motivate vehicles to participate in content dissemination is a challenging task. To solve this problem, Hui et al. [7] put forward a novel edge computing-based content dissemination framework in which the contents are uploaded to an edge computing device(ECD). A two-stage relay selection algorithm is designed based on the selfishness and transmission capability of the vehicle, to help edge computing devices selectively transport content by vehicles to infrastructure (V2I) communication. Experiments show that the framework can transmit content to vehicles more effectively than traditional methods and bring more benefits to content providers. Therefore, the vehicle edge computing system needs the assistance of nearby vehicles or roadside units. How to motivate vehicles to participate in the vehicle edge computing is another direction of future research.

4.3 Collaborative Computing

In the internet of vehicles environment, each edge node has certain computing resources, such as vehicles and roadside units. They can cooperate with each other to complete computing tasks, which can not only improve the utilization rate of resources, but also improve the completion efficiency of tasks. Vehicle-to-everything (V2X) technology enables collaboration between roadside units and vehicles. They share their information and task offloading through V2X communication to achieve the purpose of collaborative computing. However, there are still some challenges with the technology [18], for example, how different vehicles work together to make perception and planning decisions; how to dynamically balance the cost of infrastructure sensors and on-board sensors; and how to share traffic information in real time. Therefore, how to design a collaborative strategy so that edge nodes can coordinate task scheduling is a future research direction.

4.4 Security and Privacy

Privacy and security are very important in vehicle network. Data generated by vehicles are very sensitive to privacy, such as vehicle location information, user entertainment preference information, etc., all of them need to have a good protection mechanism [19]. Offloading vehicle tasks to nearby vehicles or roadside units can also lead to information leakage, raising security and privacy issues. In addition, the deployment, operation and maintenance of cloud computing are operated by specific service providers, while the deployment and maintenance of nodes in edge computing are operated by third-party developers or even directly by users. Therefore, ensuring the security and privacy protection of vehicle edge computing is a future research direction.

5 Conclusion

Edge computing is viewed as a promising technology to provide massive connectivity and support delay-sensitive applications for the vehicle network. In this paper, we first briefly introduced the edge computing and then we reviewed recent advancements in edge computing based intelligent transportation systems (ITS). Last, we presented the challenges and the future research direction for ITS based on edge computing. These open issues are expected to be a prime focus for researchers in the next decade.

Acknowledgment. This work was supported in part by the National Key Research and Development Program of China under Grant No. 2016YFC0901303.

References

1. Sahni, Y., Cao, J., Zhang, S., Yang, L.: Edge mesh: a new paradigm to enable distributed intelligence in internet of things. IEEE Access **5**, 16441–16458 (2017)
2. Ning, Z., Wang, X., Huang, J.: Mobile edge computing-enabled 5G vehicular networks: toward the integration of communication and computing. IEEE Veh. Technol. Mag. **14**(1), 54–61 (2019)
3. Swarnamugi, M., Chinnaiyan, R.: IoT hybrid computing model for intelligent transportation system (ITS). In: 2nd International Conference on Computing Methodologies and Communication (ICCMC), Erode, pp. 802–806 (2018)
4. Liu, K., Xu, X., Chen, M., Liu, B., Wu, L., Lee, V.C.S.: A hierarchical architecture for the future internet of vehicles. IEEE Commun. Mag. **57**(7), 41–47 (2019)
5. Peng, H., Ye, Q., Shen, X.: Spectrum management for multi-access edge computing in autonomous vehicular networks. IEEE Trans. Intell. Transp. Syst. Early Access, 1–12 (2019)
6. Zhou, Z., Feng, J., Chang, Z., Shen, X.: Energy-efficient edge computing service provisioning for vehicular networks: a consensus ADMM approach. IEEE Trans. Veh. Technol. **68**(5), 5087–5099 (2019)

7. Hui, Y., Su, Z., Luan, T.H., Cai, J.: Content in motion: an edge computing based relay scheme for content dissemination in urban vehicular networks. IEEE Trans. Intell. Transp. Syst. **20**(8), 3115–3128 (2019)

8. Yu, C., Lin, B., Guo, P., Zhang, W., Li, S., He, R.: Deployment and dimensioning of fog computing-based internet of vehicle infrastructure for autonomous driving. IEEE Internet of Things J. **6**(1), 149–160 (2019)

9. Qi, Q., et al.: Knowledge-driven service offloading decision for vehicular edge computing: a deep reinforcement learning approach. IEEE Trans. Veh. Technol. **68**(5), 4192–4203 (2019)

10. Tan, L.T., Hu, R.Q.: Mobility-aware edge caching and computing in vehicle networks: a deep reinforcement learning. IEEE Trans. Veh. Technol. **67**(11), 10190–10203 (2018)

11. Aissioui, A., Ksentini, A., Gueroui, A.M., Taleb, T.: On enabling 5G automotive systems using follow me edge-cloud concept. IEEE Trans. Veh. Technol. **67**(6), 5302–5316 (2018)

12. Ge, X., Li, Z., Li, S.: 5G software defined vehicular networks. IEEE Commun. Mag. **55**(7), 87–93 (2017)

13. Cui, J., Wei, L., Zhang, J., Xu, Y., Zhong, H.: An efficient message-authentication scheme based on edge computing for vehicular ad hoc networks. IEEE Trans. Intell. Transp. Syst. **20**(5), 1621–1632 (2019)

14. Kang, J., Yu, R., Huang, X., Zhang, Y.: Privacy-preserved pseudonym scheme for fog computing supported internet of vehicles. IEEE Trans. Intell. Transp. Syst. **19**(8), 2627–2637 (2018)

15. Guo, F., et al.: Detecting vehicle anomaly in the edge via sensor consistency and frequency characteristic. IEEE Trans. Veh. Technol. **68**(6), 5618–5628 (2019)

16. Sun, Y., et al.: Adaptive learning-based task offloading for vehicular edge computing systems. IEEE Trans. Veh. Technol. **68**(4), 3061–3074 (2019)

17. Zhou, Z., Liu, P., Feng, J., Zhang, Y., Mumtaz, S., Rodriguez, J.: Computation resource allocation and task assignment optimization in vehicular fog computing: a contract-matching approach. IEEE Trans. Veh. Technol. **68**(4), 3113–3125 (2019)

18. Liu, S., Liu, L., Tang, J., Yu, B., Wang, Y., Shi, W.: Edge computing for autonomous driving: opportunities and challenges. Proc. IEEE **107**, 1697–1716 (2019)

19. Khattak, H.A., Islam, S.U., Din, I.U., Guizani, M.: Integrating fog computing with VANETs: a consumer perspective. IEEE Commun. Stand. Mag. **3**(1), 19–25 (2019)

Consensus Performance of Traffic Management System for Cognitive Radio Network: An Agent Control Approach

Muhammad Muzamil Aslam[1], Liping Du[1(✉)], Zahoor Ahmed[2], Hassan Azeem[1], and Muhammad Ikram[1]

[1] School of Computer and Communication Engineering,
University of Science and Technology Beijing, Beijing, China
lpdu200@163.com
[2] Department of Automation, Shanghai Jiaotong University, Shanghai, China

Abstract. The Spectrum sharing is an important topic in Cognitive Radio Sensor Networks (CRSNs). Bio-inspired consensus-based schemes can provide lightweight and efficient solutions to ensure spectrum sharing fairness in CRSNs. This paper studies the consensus performance of traffic management system in Cognitive Radio Networks (CRNs). Research focused on significant topics of spectrum management in scalable CR ad hoc networks such as mobility, sharing, and allocation driven by local control. First of all, CRN is analyzed and is considered as a network of multiagent systems with a directed graph. Secondly, this multiagent problem is transformed into the multi-input and multi-output model of cognitive radio users in the frequency domain. Thirdly, the consensus condition of CR users is proposed based on their information sharing in the network. Moreover, the communication delay is also considered and a delay margin criterion is derived to guarantee the performance of information sharing for CR users. From simulation results, the use of proposed schemes shows low complexity and effectiveness for spectrum sharing processes in randomly deployed CRSNs also shows the effectiveness of the proposed method.

Keywords: Cognitive Radio Network · Agent control · Consensus · Primary user · Secondary user

1 Introduction

The key concept and methodology of distributed Artificial Intelligence (AI), led to the emergence of agent, and multi-agent AI which have grown and spread quickly since their inception around the mid of 1980s. The establishment of agent, and multiagent AI have become a promising research area, and their applications are giving greater results and ideas from several other fields such as AI, sociology, computer science, organization behavior, economics, philosophy and management science. The agent AI system has progressed to a new definition of distributed AI (DAI) which is the study, and applications of multiagent systems and construction where the system is interacts with other intelligent

© Springer Nature Singapore Pte Ltd. 2019
H. Ning (Ed.): CyberDI 2019/CyberLife 2019, CCIS 1138, pp. 138–145, 2019.
https://doi.org/10.1007/978-981-15-1925-3_11

agents to follow some set of targets or do some errands. The multi-agent AI system is an accomplish area of AI that is widely used for controlling Cognitive Radio Network (CRN). With the development of modern control tactics especially hierarchical control methods for crowded management problems, conventional control methods approach based on functional decay are pre-valid in both practical applications and theoretical studies [1]. Many recent researches mostly focus on improving hierarchical structures, optimized algorithm, and analytical modeling which are effective for CRN, as it can be seen in various wireless management systems (WMS). However, the working decay systems are helpful and fruitful for several CRN management problems, but hurdles occur with their development, maintenance, operation, upgrading and expansion, and cost which are habitually excessive and occasionally needless, especially in quick arrival age of connectivity. Therefore, we have to rethink control systems again and start the investigation on usage of simple target-oriented agents for CRN systems.

The Wireless Network and management are well known as agent-based control because of its real, and distributed nature and busy alternating mode. In the future CRN technology, we will see independent intelligent agents among the number of users and control centers via wireless and ad-hoc networks. These agents will help collect the right information at the right time in real-time and will make a smart decision so that our Cognitive radio system will be "intelligent".

Currently, there is rise in more trainings on the application of agent-based approaches in the transportation sector (e.g. the agents for executing future carpooling, distributed control, scheduling, and traffic simulation) [2, 3]. Although these issues are important, they are not still systematically involved core intelligent transportation system (ITS) issues. In this paper, we applied the idea of agent-based control for CRN systems based on control theory [4], to users. The Consensus Performance problem of Traffic Management system in CRN was discussed. We applied basic graph Theory in problem formation, and performance analysis of CRNs. The consensus criteria of CRNs under Communication delay together with the simulations are provided for at the end of the proposed work.

2 Basic Graph Theory

Here, the communication topology of the networks of agents is represented by a directed graph $A = (M, N)$ with a set of nodes $M = \{1, 2, 3, \ldots\ldots, n\}$ and edges $N \in M * M$. The surrounding agents are represented by $\{y \in M : (x, y) \in N\}$. According to [5], a simple consensus algorithm to reach a contract about the state of n integrator agent with dynamics $m_x = n_x$ can be represented as a linear system on a graph in nth order.

$$m_x(t) = \sum_{y \in K_x} \left(m_x(t) - m_y(t) \right) + C_x(t), \ m_x(0) = G_x \varepsilon \overset{\ast}{R}, C_x(t) = 0 \qquad (1)$$

Dynamic graph with state M, m, values of m changes regarding network dynamics $m_x(t)$. M is a distinct state of the system that changes with time, m is also known to be information flow. The collective dynamic of groups of agents following protocol Eq. 1. can be

$$m_x = -L_x \qquad (2)$$

where $L = L_{xy}$ is the Laplacian graph of the network and its elements are given below.

$$l_{xy} = \begin{cases} -1, & y \, \varepsilon \, K_x \\ |K_x|, & y = x \end{cases} \tag{3}$$

where $|K_x|$ represents the number of surroundings nodes x. Figure 1 represents the equivalent forms of consensus algorithm in Eqs. (1) and (2) for agents with a scaler state.

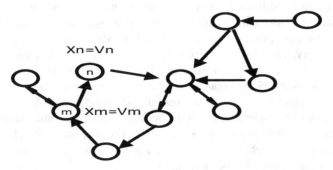

Fig. 1. Collective system

Regarding the Laplace graph definition in Eq. 3, all rows sum of W is zero because of $\sum_y w_{xy} = 0$. Therefore, W always has zero eigenvalue $Y_1 = 0$. Here zero eigenvalue corresponds to eigenvector $\mathbf{1} = (1, \ldots \ldots, 1)^T$, here $\mathbf{1}$ is because of null space of $W (W\mathbf{1} = 0)$. In the hand (2) is a state in the form of $x^* = (d, \ldots \ldots \ldots d)^T = d\mathbf{1}$ in which all nodes satisfy. To check on analytical tools from algebraic theory [6], later on, we express that x^* is a unique equilibrium of Eq. 2 for graphs that are connected. Here, for connected network, equilibrium $x^* = (d, \ldots \ldots \ldots d)^T$ is worldwide exponentially stable, and Furthermore, the consensus value is $\propto = \frac{1}{n} \sum_x G_x$.e.g. if one network has $n = 10^6$ nodes, where every node can talk to $\log_{10} n = 6$ surroundings, to find the average value of start conditions of nodes is more intricate. In Eq. 1, there are several functions which can be calculated in a similar fashion using asynchronous and synchronous algorithms [5].

3 Problem Formation

In a network architecture there is permission of agents such as controllers to be interconnecting together with actuators and sensors. A universal network pyramid model has been studied in [7] research. When sensing and action agents are unified on the network at the 1st or 2nd level, we further suppose a typical network architecture e.g. in newly manufacturing system as shown in Fig. 2(a). Here network architecture has three various levels namely information system network, Discrete Event Cell (DEC) network and the continuous variable Device (CVD). The first level information system network is used to transmit nontimed-critical information's. e.g. hourly or daily production data, and for communicating with factory wide database. The middle level is a Discrete event cell network, which loads commands or updates the working config-uration for various subsystems or cells. The discrete event cell network messages are probability periodic, time-critical or sporadic. Analysis of control of discrete event cell network system e.g. manufacturing system and multi targets robotic systems have been studied using DEC network technique like finite state machines and Petri-nets, with or without timing parameters [8] to focus on the right research and safety problems of system operation in a sensible and optimal logical order. The lower level is continuous variable device network, which communicates physical signals like velocity, position, and temperature by the means of network messaging and coding [9]. Typical agents in a continuous variable device network consist of smart network controllers. Actuating or sensing, communication capabilities and data processing are the three main things of the network agents. Schematic diagram for transmission of message among network agents is shown in Fig. 2(b). Precisely, there are three main features of smart sensors, those are for intelligence, data acquisition and communication ability [10]. In a physical environ-ment and network medium to see different information's, there must be the capability of sensors to properly encode/share the information before transmitting it to the network.

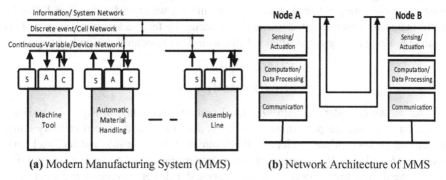

(a) Modern Manufacturing System (MMS)   (b) Network Architecture of MMS

Fig. 2. Network architecture of modern manufacturing system

Smart actuators that are also similar to smart sensors have the structures of actuation communication and intelligence. There should be ability in actuator to crack informa-tion's from network medium and send it to the physical devices. For further explanation of performance evaluation of network study [11].

4 Performance Analysis of Cognitive Radio Network

For performance analysis of (CRN) First, we study standard MIMO system, in which for control system design in Fig. 3 discrete-time is supposed to be first also with this linear time-invariant, multi-input and multi-output system with "A" states, "V" inputs and "z" outputs represented in Fig. 6 and explained below

$$m(l + 1) = X_A(l) + Qv(l) + U(l)$$
$$Z(l) = K_A(l) + Y(l) \tag{4}$$

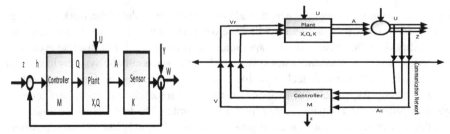

Fig. 3. General closed-loop system (MIMO).   **Fig. 4.** Control system of MIMO

In this "l" is representing time index which is connected with the time sampling T in discrete-time domain. $X \in \check{R}^{u*u}$, $Q \in \check{R}^{u*v}$ and $K \in \check{R}^{w*u}$ are system input and output matrices correspondingly, and $U(.)$ and $Y(.)$ are system disturbances and measurements. Noise correspondingly. $U(.) = [u1(.), \ldots \ldots, uf(.)]^T$ and $Y(.) = [y1(.), \ldots \ldots, yz(.)]^T$ are considered to be restricted e.g. $|ui(.)| \leq g_{iu}, i = 1, \ldots, f$ and $|yj(.)| \leq g_{jv}, j = 1, \ldots z$, in which g_{iu}'s and g_{jv}'s are well-known constants which are positive. For easy understanding, we consider that $\mathbf{K = I}$, that is, all the states are supposed measurable. For system, the feedback controller state (1) is possible to design using any normal MIMO, as following control design techniques

$$V(M) = M_k(l)$$
$$M[(z(l) - A(M) + Y(l))] \tag{5}$$

That is, we consider the MIMO control system represented in Fig. 3 is designed in a good way. So, the performance of the system Eq. 4 and system stability could be definite by appropriately selecting the sampling time T in (4), and feedback gain designing K in Eq. 5. We supposed dynamic system Eq. 4 and controller design Eq. 4 as baseline design framework. Next, we supposed a dispersed control architecture in which actuators, sensors and controllers and physically dispersed and swapping data through one communication network as represented in Fig. 3. If data have time delays in transmission and sampled asynchronously, the controller dynamics and system at the controller sampling immediate should be improved.

$$A(l + 1) = X_A(l) + Q_{v_r}(l) + U(l) \tag{6}$$

$$v(l) = M[z(l) - A_c(l) + U(l))] \tag{7}$$

In this $Ac(.)$ and $Vr(.)$ are delayed form of $A(.)$ and $V(.)$ correspondingly, that if for the mth element of $A(.)$, $d_a^m(l) = d^m(l - d^m)$ and for nth elements $V(.)$, $V_h^n(l) = V^n(l - h^j)$, in this s^i and a^j are representing resultant of the transmission delays, and mth and nth sensory sampling are mismatched at actuation data. For its explanation can study [12]. Here in paper we attentive on the network agent's performance analysis with general maximum input and maximum output controller in the network control design and learn the interaction between communication and control mechanism. Therefore, the controller feedback and estimator state are discussed as consider to be in system dynamics 1 and expanded on designing of Fig. 4 and framework of Eqs. 3 and 4.

5   Consensus Criteria of CRN Under Communication Delay

Consider a message send by neighbor **j** is received by agent **I** after T time delay, which is equal to network with time delay of uniform one-hop communication. Below in Eq. 8, given is algorithm of consensus

$$x_m(t) = \sum_{n \in Z_m} h_{mn} (x_m(t - \dot{T}) - x_m(t - \dot{T})) \tag{8}$$

To reach an average consensus for undirected graph G has been proposed in [5]. The algorithm

$$x_m(t) = \sum_{n \in Z_m} h_{mn} (x_n(t - \dot{T}) - x_m(t)) \tag{9}$$

It's not protective the average $\bar{a}(t) = (1/n) \sum_m x_n(t)$ in time for standard graph, with 0–1 weights for balance graph it turns out. Here weights $\bar{a}(t)$ is an invariant number along (8) solution. Here we can show collective dynamics of network in following form

$$x(t) = -L_x(t - \dot{T})$$

After taking Laplace transform of the above form, we can write it again as

$$A(s) = \frac{W(s)}{s} x(0) \tag{10}$$

Here function $W(s) = (In + I/s) \exp(-s\dot{T})L)^{-1}$. Here Nyquist criterion can be used for the verification of stability of H(s). Similar was introduced in [13] by Fax and Murray. It providing us time delay upper bound so that network stability maintained in the availability of time-delays. The algorithm in Eq. 8 asymptotically explains the average consensus problem $\forall$ initial state if and only if $0 \leq \dot{T} < \pi/2\gamma_n$. We can concern for this proof in [5]. Here $\gamma_n < 2\Delta$, in Eq. 8 an enough condition of average consensus algorithm is $\tau < \pi/4\Delta$. Networks having a large degree (with hub) generally consider as scale-free networks as in [14]. In [15] random graphs and small networks [16] are impartially vigorous to time delay since they do not have a large number of degrees. Finally, we can say that construction or buildup of engineering networks with hub is not so good experience for accomplishment consensus.

6   Simulation

Consider four Cognitive Radio Users in CRN after their traffic Management the communication topology in their information sharing is shown in Fig. 5.

If all users have integrator dynamics, then by using Eq. 2 the performance without delay is shown in Fig. 6(a). When delay has appeared in the network, it influences the performance of CRN network. Thus, consensus performance of the CRN could be seen in Fig. 6(b) after communication delay without the proposed method. Although, these responses are in consensus but their transient responses are very poor. Hence the settling time is very large. So, to improve the performance, the designer should follow

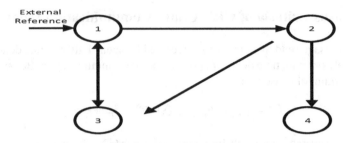

Fig. 5. CR users in CRN

this proposed method to minimized the effects of time delay and communication delay as well. Now if we consider that there is communication delay between all cognitive radio users, let a communication delay of 2 s between all users is the same then their Consensus Performance can be calculated by using an algorithm Eq. 9 shown in Fig. 6(d). This proposed method is also sufficient for large arbitrary communication delay. Thus, the consensus performance of the CR users with a large communication delay of 4 s as shown in Fig. 6(c).

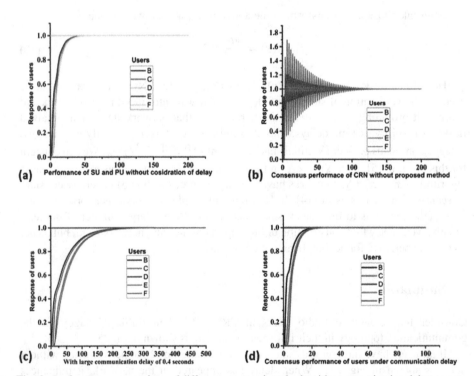

Fig. 6. Performance comparison of different proposed methods with communication delay or not

7   Conclusion

In this paper, we discussed the traffic management performance problems in Cognitive Radio Network. we studied information sharing performance of numerous supportive agents over one communication Network. Cognitive Radio Network has been analyzed and supposed as a network of multiagent systems with graphs. This multiagent problem was transformed into multi-input and multi-output model of CR users in the frequency domain. After that CR user with consensus, condition was considered on network information sharing base, furthermore, communication delay has been derived to guarantee the performance of information sharing for Users of Cognitive Radio Network. Simulation results were giving effective results of our proposed method.

References

1. Baskar, L.D., De Schutter, B., Hellendoorn, J., Papp, Z.: Traffic control and intelligent vehicle highway systems: a survey. IET Intel. Transport Syst. **5**(1), 38–52 (2011)
2. Bishop, R., et al.: Intelligent transportation, pp. 1–20 (2000)
3. Papageorgiou, M., Kiakaki, C., Dinopoulou, V., Kotsialos, A., Wang, Y.: Review of road traffic control strategies. Proc. IEEE **91**(12), 2043–2067 (2003)
4. Wang, F.Y., Wang, C.H.: Agent-based control systems for operation and management of intelligent network-enabled devices. In: Proceedings of the IEEE International Conference on Systems, Man and Cybernetics, vol. 5, pp. 5028–5033 (2003)
5. Olfati-Saber, R., Murray, R.: Consensus problems in networks of agents with switching topology and time-delays. Autom. Control. IEEE Trans. **49**(9), 1520–1533 (2004)
6. Hedman, S., Pong, W.Y.: Quantifier-eliminable locally finite graphs. Math. Log. Q. **57**(2), 180–185 (2011)
7. Definition, T.: Chapter 1 Introduction to Fieldbus Systems, pp. 1–28
8. Silva, M., Valette, R.: Petri nets and flexible manufacturing, no. June 2014 (1990)
9. Shi, X., Kim, K., Rahili, S., Chung, S.J.: Nonlinear control of autonomous flying cars with wings and distributed electric propulsion. In: Proceedings of the IEEE Conference on Decision and Control, vol. 2018-Decem, no. Cdc, pp. 5326–5333 (2019)
10. Edgar, T.F., et al.: Automatic control in microelectronics manufacturing: practices, challenges, and possibilities. Automatica **36**(11), 1567–1603 (2000)
11. Lian, F.L., Moyne, J., Tilbury, D.: Network design consideration for distributed control systems. IEEE Trans. Control Syst. Technol. **10**(2), 297 (2002)
12. Lian, F.L., Moyne, J., Tilbury, D.: Modelling and optimal controller design of networked control systems with multiple delays. Int. J. Control **76**(6), 591–606 (2003)
13. Fax, J.A., Murray, R.M.: Information flow and cooperative control of vehicle formations. IEEE Trans. Autom. Control **49**(9), 1465–1476 (2002)
14. Taani, D.S.M.Q., Al-Wahadni, A.M., Al-Omari, M.: The effect of frequency of toothbrushing on oral health of 14-16 year olds. J. Ir. Dent. Assoc. **49**(1), 5–20 (2003)
15. Erdos, P., Rényi, A.: On the evolution of random graphs. Struct. Dyn. Networks **9781400841**, 38–82 (2011)
16. Olfati-saber, R.: Olfati05-ACC.pdf, pp. 2371–2378 (2005)

Conclusion

References

CyberLife 2019: Cyber Philosophy, Cyberlogic and Cyber Science

An Efficient Concurrent System Networking Protocol Specification and Verification Using Linear Temporal Logic

Ra'ed Bani Abdelrahman[1], Hussain Al-Aqrabi[2]([✉]), and Richard Hill[2]

[1] Department of Computer Science, Loughborough University, Loughborough, UK
R.Bani-Abdelrahman@lboro.ac.uk
[2] Department of Computer Science, University of Huddersfield, Huddersfield, UK
{h.al-aqrabi,r.hill}@hud.ac.uk

Abstract. In critical computer-based systems, safety and reliability are of principal concern, especially when dealing with concurrent transactions on which mobile systems depend on, such as the emerging Internet of Things (IoT). We present a protocol to ensure safety and reliability of systems where concurrent modification of data on routers in a network is possible, by detecting cycles in the conflict graph and ensuring the system is free of any cycle in an effective manner. The existence of a cycle in a conflict graph means that the schedule of such concurrent transactions cannot be serialized. We use temporal logic in the representation of this protocol model to ensure the safety of systems. Administrative routing protocols benefit significantly from this protocol model.

Keywords: Concurrent transactions · Output schedules · Conflict graphs · Routing protocol

1 Introduction

Reliability and safety are two fundamental concerns in critical computer-based systems. The jobs running on these systems are not tolerant to errors in output. Concurrency makes these systems very difficult to build and verify, especially with the emergence of mobile Internet of Things devices that incorporate shared resources [2]. These resources and their concurrent transactions must leave the system in a consistent state, as having more than one transaction trying to access and update similar data items could leave the system in an inconsistent and unsafe state [3].

If we consider networking systems, we notice that the software used to make sure a certain packet is delivered from its source to its destination on routers is built according to certain routing algorithms. Routing protocols serve to specify the path by which data packets are delivered to the specified destination. Routing protocols are usually tested according to different situations and scenarios but in certain situations, which are not tested for, there is the possibility that a system

© Springer Nature Singapore Pte Ltd. 2019
H. Ning (Ed.): CyberDI 2019/CyberLife 2019, CCIS 1138, pp. 149–168, 2019.
https://doi.org/10.1007/978-981-15-1925-3_12

could malfunction or become insecure. Current Internet routing protocols differ according to the algorithms used.

This paper focuses on systems of routers where the routers are accessed by concurrent transactions. The transactions access and update the routers' data; it is vital that this data remains consistent. Because of the high concurrency of such systems, they are considered system critical, where their failure can lead to great losses. Since different transactions try to access routers at the same time, it becomes cumbersome to deal with them due to the importance of keeping the data being accessed correct.

A protocol is presented in this paper to ensure serializability. This protocol checks for cycles in the conflict graph, which represents the concurrent transactions accessing the data on routers. Due to the particular importance of temporal logic and model checker tools, it is used in this paper to specify and verify this protocol. Since routers are naturally ordered in some given manner, the transactions will tend to access them in an ordered manner. A transaction can access a set of routers, and indeed can skip routers in the set. This research considers the skipped routers to be a gap in the set. By knowing the size of the gap in the different transactions, the protocol presented in this paper will verify the system if the system is serializable or a non-serializable in a highly efficient manner, leading to a reliable and safe system.

Our major concern is an effective means to find if the path is cyclic which means that the scheduler has created a non-serializable schedule for accessing the routers. In this paper, we present a networking protocol, specified using temporal logic formulae, which is modelled to verify its properties, improving concurrency and assuring serializability.

In the next section, we will present a review of some of the previous work in this regard. In Sect. 3, we discuss concurrent transactions and serializability, in addition to gap theory, followed by an examination of the use of temporal logic in Sect. 4. Section 5 provides an overview of the proposed protocol, before an explanation of LTL in Sect. 6. In Sect. 7, we describe a more detailed specification of the protocol using in LTL. In Sect. 8 we demonstrate protocol verification using the NuSMV model checker, reporting the results achieved. Finally, we make some concluding remarks.

2   Related Work

Temporal logic is used to provide reasoning about a changing world [1], where the formula truth values may vary over time [6]. Facts about past, present, and future states can be expressed in the formula of temporal logic. The use of temporal logic for formal specification and verification of computer systems was introduced by Pnueli [19].

Verification of concurrent and reactive systems makes extensive use of temporal logic [23]. E-commerce, traffic control and IoT systems are examples where an error in this type of system is considered fatal. Model checkers for different types of temporal logic were developed with the ability to verify real-world systems in only a short period of time, given that these systems generally contain a massive number of states. Temporal logic is helpful in the specification of concurrent systems by describing the event ordering over time. With the exponential growth of technologies and concurrent systems, these systems have become more complicated and an understanding of their interactions more important. This means that the specification and verification of some of their properties are essential [15,21]. In a paper called "what good is temporal logic?" and which is a considered to be of particular significance in the field, Leslie Lamport emphasised that the main function of temporal logic is in modelling concurrent systems [16]. We use linear temporal logic (LTL) in the specification and verification of our novel routing protocol.

Concurrency is one of the topics that has been considered for decades and indeed has grown with new mobile technologies. The data accessed by any transaction should qualify and meet certain conditions upon completion of the transaction (when the transaction commits). Different algorithms are used in this field to accomplish this task. In [6], multi-step transactions can access data items infinitely many times. Partial-ordered temporal logic (POTL) was used to specify concurrent database transactions in [18]. The specification for this type of transaction, using quantified-propositional temporal logic (QPTL), was used in [12]. In [13], linear temporal logic (LTL) was used. A monadic fragment of first-order temporal logic was used to specify the concurrent transactions in [14]. All of these logics have an exponential space complexity and at worst are undecidable, with the exception of LTL. An advantage of LTL is the availability of model checkers. A disadvantage of these logics, (excluding LTL), is that it is impractical to demonstrate the basic serializability correctness condition [6].

Formal methods are considered to be powerful when used to build critical systems in order to make sure that the system is robust and secure. Due to the availability of powerful model checker tools, temporal logic can specify properties that are required, and to subsequently verify them. In [4], data items were accessed in such a way that a gap would exist. In this research, we show how we use temporal logic to specify and verify the proposed protocol.

3   Methodology

This paper focuses on specifying and verifying a protocol for unlimited number of multi-step transactions accessing a finite set of routers with different properties using temporal logic and a model checker. This work introduces a routing protocol that can efficiently detect any breach of serializability in the case where the routers are accessed by multi-step transactions with gaps. By calculating the size of a gap in different transactions, a cycle can be detected, as there will be no need to check all the different cycles of different sizes possible. This work builds upon prior research which (a) defined the gap and, (b) introduced a theorem to calculate the gap size [4]. As such, this article makes two contributions. First, this work is founded upon applying the theorem introduced in [4] to routing systems, thereby significantly improving administrative routing protocols in order to have consistent, reliable, and safe systems. We model the protocol based on temporal logic LTL specifications and use the NuSMV model checker to prove - or disprove - that the model satisfies the serializability property. Second, we specify and verify concurrent systems using temporal logic and this research illustrates how powerful temporal logic can be when dealing with concurrent systems in spite of their difficulties. The significance of temporal logic in computer science is clear, especially in the specification and verification of critical computer-based systems. Model checkers, such as NuSMV, used to model temporal logic properties, and their capabilities of dealing with a large number of states and verifying real-world systems, allow us to verify the correctness of the proposed protocol. To gain a fully automated verification, the NuSMV model checker is used to verify the protocol properties specified in the temporal logic.

The protocol presented will be modelled in terms of finite state transition systems whose specifications are expressed in LTL. The next step is to explore the state space of the state transition system, where it is possible to automatically check if the protocol satisfies its specifications or otherwise. The model checker will be used at the final stage, where it will produce a "true" result if the specifications of the required properties are satisfied. A counterexample is given by the model checker, which will indicate any potential source of error.

4   Concurrent Transactions and Serializability

A transaction is a sequence of operations performed on one or more databases which is representative of a single real-world transition. This section introduces some basics about concurrent data transactions and their histories. We will be concerned with an unlimited number of transactions creating an unlimited number of histories. If we capture the existing transactions at a certain point in time, they will be less than or equal to n transactions.

4.1   Concurrent Transactions and Histories

A serial schedule represents a schedule where, for all transactions participating in the schedule, every transaction operation is executed sequentially for all transactions. If this is not the case, the schedule is said to be a non-serial.

Definition 1. *The transaction is formed of a sequence of read and write steps that are represented as a totally ordered set on a set of data items $D_i = \{x_1, x_2, \ldots, x_m\}$, where every read step $r_i(x)$ comes before a write step $w_i(x), \forall x \in D_i$ such that*

$$T_i = r_i(x_1)w_i(x_1)...r_i(x_m)w_i(x_m). \tag{1}$$

is called a multi-step transaction, as in [22].

The set of multi-step transactions is denoted $T = \{T_i : i \in \mathbb{N}_1\}$, where $\mathbb{N}_1$ is the set of positive integers. If a step s_i of T_i in a history h occurs before step $s_{i'}$ of $T_{i'}$ in h, we write it as $s_i <_h s_{i'}$.

Two transactions are deemed to be conflicting if they require access to a shared data item and at least one of their operations is a write operation (upon the shared data item). This conflict may leave the database in an inconsistent state if the transactions are running concurrently. To avoid this, some form of ordering the execution of the steps of the concurrent transactions is needed. A schedule S is serializable if it is equivalent to some serial schedule of the same transactions [17]. A serializable schedule is 'correct' when it leaves the database in a correct state [17]. Conflict in serializable schedules provides higher efficiency where the transactions are executed concurrently, leading to higher throughput.

4.2 Multi-step Transactions with Gaps

To test conflict serializability in polynomial time, conflict graphs are commonly used [6,9,17,20]. A conflict in the graph occurs when there are two steps of different transactions on the same data item, one of which is a write step. We include gap theorem [4] in this research in addition to other definitions from the same work. We have used the order of the set of data items accessed by transactions, as defined by Definition 2.5 of [4], where if a set of data items $D' \subseteq D$ is denoted by $\{x_a, \ldots, x_b\}$, this will mean that $x_a <_D \ldots <_D x_b$. A gap is defined in [4] as follows:

Definition 2. *Assume that the transaction T_i accesses a set of data items D_i, where $D_i \subseteq D$ such that*

$$D_i = \{x_a, x_b, x_c\} \tag{2}$$

where $x_a <_D \ldots <_D x_b \ldots <_D x_c$. Then the gap G^i of the set D_i can be calculated as follows:

$$G^i = (c - a + 1) - (|D_i|) \tag{3}$$

where $(c - a + 1)$ is the number of elements in the sequence $x_a \ldots x_c$, and $|D_i|$ is the cardinality of the set D_i [4].

The maximum gap G is defined in [4] as follows:

Definition 3. *Let $T = \{T_i : 1 \le i \le n\}$ be a set of transactions that iterates an unlimited number of times to constitute $T = \{T_i : i \in \mathbb{N}\}$, and each $T_i \in T$ accesses a set of data items D_i. At any given point in time, there exist k transactions where $1 \le k \le n$, such that*

$$\bigcup_{i=1}^{k} D_i = \{x_a, x_{a+1}, \ldots, x_u\} \tag{4}$$

where $x_a <_D \ldots <_D x_a + 1 \ldots <_D x_u$. The maximum gap G can then be calculated as follows:

$$G = \begin{cases} 0, & \forall i, 1 \le i \le k, G^i = 0 \\ u - a - 1, & \exists i, 1 \le i \le k, G^i \ne 0. \end{cases} \tag{5}$$

In a conflict graph containing a cycle of length n, where data items are accessed by transactions without gaps (the maximum gap $G = 0$), there exist a cycle of length two as stated in Theorem 3.3 of [4]. We use the order of the set of data items accessed by transactions as defined in Definition 2.5 of [4], where if a set of data items $D' \subseteq D$ is denoted $\{x_a, \ldots, x_b\}$, this will mean that $x_a <_D \ldots <_D x_b$. Our work also benefits from the theoretical results of [4], especially from Theorem 3.4 which states the following:

Theorem 1. *Assume D to be an irreflexively totally ordered set of data items such that*

$$D = \{x_1, x_2, \ldots, x_{n-1}, x_n\}, \tag{6}$$

where it is accessed by a set of transactions T iterating an unlimited number of times such that $T = \{T_i | i \in \mathbb{N}_1\}$ and accesses the set D as per Definition 3. Assuming we have a maximum gap $G = n - 2$ in the set D denoted by G_{n-2} and that there is a cycle in the corresponding conflict graph $G(h)$, there then exists a cycle of length n, denoted by C_n, in the corresponding conflict graph $G(h)$.

The following example illustrates how we use these definitions and theorems. We assume that we have five transactions, $T_1, \ldots, T_5$, accessing the set of data items $D = \{x_1, x_2, \ldots, x_8\}$, as in Definition 2.5 of [4], as follows:

$$D_1 = \{x_2, x_3\} \qquad D_2 = \{x_3, x_4\},$$
$$D_3 = \{x_3, x_6\} \qquad D_4 = \{x_3, x_5\},$$
$$D_5 = \{x_4, x_5\}.$$

First we calculate the *gap* G^i using Eq. 3 in Definition 2 as follows:

$$G^1 = 3 - 2 - 2 + 1 = 0, \qquad G^2 = 4 - 3 - 2 + 1 = 0,$$
$$G^3 = 6 - 3 - 2 + 1 = 2, \qquad G^4 = 5 - 3 - 2 + 1 = 1,$$
$$G^5 = 5 - 4 - 2 + 1 = 0.$$

After finding the gaps, we find the maximum gap. Theorem 3.4 of [4] introduced a very efficient way to determine if there is a cycle caused by the transactions accessing data items. The theorem allows for greater efficiency in finding a cycle in a conflict graph. This means that if any of the transactions accessing the data items yield a gap, we can check for a cycle of length = *maximum gap* + 2. In the instance that we find this cycle in the conflict graph, this means that we do not have a serializable schedule; otherwise, the history is serializable. The maximum gap is $G = 3$ (G^3), hence the maximum cycle will be of length 5 (C_5). Building a precedence graph for the history h of the all transactions $T_1, \ldots, T_5$ will show that we have a cycle of length 5 (C_5). This significant finding is used when specifying and verifying our routing protocol, where we will be able to detect a cycle in an efficient manner. This will be used to maintain the serializability of the transactions. Serializability is checked by testing for the existence of a cycle of length $n + 2$ if the *maximum gap* equals n. Theorem 3.4 in [4] can be applied to many applications, as discussed in [4]. One of the more important applications discussed in [4] is that of booking a flight e-ticket through different agencies. Destinations are naturally ordered, and booking a ticket from any place to another, will present different options where a person can book a direct ticket, meaning that there will be no transit, and creating a gap. Another situation is when a ticket is booked that has multi-stop destinations. In the second situation, there could also be gaps. In the case where the ticket contains all the stops on the path from departure to destination arrival, then there is no gap. In our work, we use these findings, as per [4], in a routing protocol, as will be discussed in the next section.

5   Description of the Protocol

In our protocol, we assume that different routers connect to different systems, where packets are passed through some of them (or all of them) from sender to receiver. The router's job is to create a path where the packet will travel from source to destination. Routers are naturally ordered, where they are presented as the set D with its own information table. We denote each router by $r_i \in D$. Packets can travel through two or more routers according to the path it is set to travel through. A packet path can be set where the packet can reach the destination by going from the first and closest router (source), to the destination, without going through any other routers. Another scenario is when a path through which a packet must travel to reach its destination consists of more than two routers. The router table needs to be updated, through which data is kept consistent. The different concurrent transactions associated with sending data across a network require this concurrency where a transaction can check (read) the router or update (write on) the router. A scheduler needs to be used to avoid conflicting transactions from stopping the network or corrupting certain data while sending it across the network. An important class of routing protocols is that of *administrative* protocols, which we model herein.

The main two steps in the transactions we will be using in our protocol are (a) read step and, (b) write step, where accessing and updating the associated data occurs accordingly.

Since routers are naturally ordered in some manner, creating a path from router A to router F can include many choices from the other routers included in the path. In our model, we assume that there can be direct or multi-stop paths between two chosen routers. To represent this scenario, we assume that we have the set routers D, which here we will call *destinations*, where $|D| = k$ are ordered as per Fig. 1.

The set D contains all destinations starting from location (router) A to end at location F. The next location from location i is x_i such that $x_i \in D$, as per Fig. 1. The first option in creating a path is that the path runs directly from A to F without passing through other destinations. We call this a direct path without stop. Transaction T_1, which represents this case, accesses the set $D_1 = \{A, F\}$. The gap for the set D_1 is $G^1 = k - 2$. This path is illustrated as edge 1 in Fig. 2.

The second choice is to have a path from A to B then from B to F, which is represented by edges 2 and 3 in Fig. 2. A third choice might be to select a path from A to B, then from B to C, and finally from C to F. A fourth path which can be selected might be by going from A to B to C to D to E and then from E to F; this path contains all the stops from the initial point A to the final point F. Here we represent the read step of the transaction as accessing the destination (router), whilst the write step represents the modification on the router. The set D represents the ordered routers as defined in Definition 2.5 of [4]. The number of ignored destinations from the start to the end destination is represented by the gap, for which we can refer to Definition 2. The maximum gap illustrates the number of destinations not accessed in the path where there is an available path from A to F.

Fig. 1. Representation of the set of ordered routers.

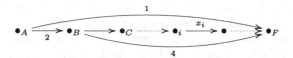

Fig. 2. Representation of the set of ordered routers with gaps.

6   Syntax and Semantics of LTL

This section introduces LTL syntax followed by LTL semantics.

6.1   Syntax of LTL

The syntax and semantics of LTL are standardised, and in this paper is used from [7,8,11]. The alphabet of LTL consists of a set of proposition symbols p_i, $i = 0, 1, 2, \ldots$ and read/write step propositional symbols $r_i(x_j), w_i(x_j)$, with $i \geq 1$ and $j \geq 1$, booleans $\neg, \vee, \wedge, \top, \bot$, and temporal operators X, F, G, U. Formulae in LTL are those generated by:

$$\phi ::= p_i | r_i(x_j) | w_i(x_j) | \neg \phi | \phi_1 \vee \phi_2 | \phi_1 \wedge \phi_2 | X\phi | F\phi | G\phi | \phi_1 U \phi_2$$

The symbols $\top$ and $\bot$ will also be used to denote true and false values, respectively. The symbols $\Rightarrow$ and $\leftrightarrow$ present their normal logical meanings.

6.2   Semantics of LTL

An interpretation for LTL, $I(s_a)$, at a given state $s_a \in S$ where S is a set of states, assigns truth values $p_i^{Is(a)}$, $r_i(x_j)^{I(s_a)}$ and $w_i(x_j)^{I(s_a)} (\in \{\bot, \top\})$ to the propositional symbols p_i, $r_i(x_j)$ and $w_i(x_j)$, respectively. A Kripke structure M, as defined in [7], is a triple $<S, R, I>$, where S is a set of states, $R \subseteq S \times S$ a transition relation such that, for all $s \in S$, there exists $s' \in S$ with $(s, s') \in R$. A path in M is an infinite sequence of states, $\pi = s_a, S_{a+1}, \ldots$, such that for every $b \geq a, (s_b, s_{b+1}) \in R$. We use π^a to denote the $suffix$ of π starting at s_a. As each state in a Kripke structure is required to have at least one successor, thus it follows that $\pi^a \neq \{\}$ for any state s_a. The semantics of an LTL formula ϕ is given by the truth relationship $M, s_a \models \phi$ which means that ϕ holds at state s_a in the Kripke structure M. Similarly, if ϕ is a path formula, $M, \pi \models \phi$ means that ϕ holds along path π in the Kripke structure M. The relation $\models$ is defined inductively as follows:

$M, s_a \models p_i$ iff $p_i^{I(s_a)} = \top$

$M, s_a \models r_i * (x_j)$ iff $r_i(x_j)^{I(s_a)} = \top$

$M, s_a \models w_i(x_j)$ iff $w_i(x_j)^{I(s_a)} = \top$

$M, s_a \models \neg\phi$ iff $M, s_a \not\models \phi$

$M, s_a \models \phi_1 \vee \phi_2$ iff $M, s_a \models \phi_1$ or $M, s_a \models \phi_2$

$M, s_a \models \phi_1 \wedge \phi_2$ iff $M, s_a \models \phi_1$ and $M, s_a \models \phi_2$

$M, s_a \models \mathbf{X}\phi$ iff $M, s_{a+1} \models \phi$

$M, s_a \models \mathbf{F}\phi$ iff there exists $k \geq a$ such that $M, s_k \models \phi$

$M, s_a \models \mathbf{G}\phi$ iff for all $k \geq a$ such that $M, s_k \models \phi$

$M, s_a \models \phi_1\mathbf{U}\phi_2$ iff there exists $c \geq a, M, s_c \models \phi_2$ and, for all $a \leq b < c, M, s_b \models \phi_1$.

7   Specification of Routing Protocol in LTL

Assuming we have four transactions T_1, T_2, T_3 and T_4 accessing four ordered sets of routers D_1, D_2, D_3 and D_4, respectively. These transactions are iterated to constitute an infinite history. Data items in the sets represent the different routers. We elected to represent our protocol using four different transactions accessing the ordered routers in a way where we can have a gap. The router item sets are as follows:

$$D1 = \{x_3, x_4\}, D_2 = \{x_3, x_4\}, D_3 = \{x_3, x_5\}, D_4 = \{x_4, x_5\}.$$

The transactions are as follows:

$$T_1 : \{begin_1, r_1(x_3), w_1(x_3), r_1(x_4), w_1(x_4), end_1\};$$

$$T_2 : \{begin_2, r_2(x_3), w_2(x_3), r_2(x_4), w_2(x_4), end_2\};$$

$$T_3 : \{begin_3, r_3(x_3), w_3(x_3), r_3(x_5), w_3(x_5), end_3\};$$

$$T_4 : \{begin_4, r_4(x_4), w_4(x_4), r_4(x_5), w_4(x_5), end_4\};$$

The reason for having four transactions T_1, T_2, T_3 and T_4 and datasets D_1, D_2, D_3 and D_4 is to ensure that we satisfy the conditions of the Theorem 1 where we can have a gap, as we have it here to be G^1 by skipping one data element (x_4) in transaction T_3, and can create cycles of sizes 2, 3 and 4. Accordingly, we can observe a cycle with a length of 3 and we will be checking only for this cycle. The transactions arrive at the scheduler S in the order T_1 then T_2 then T_3 Then T_4. The semantics of the formula φ are given by a truth relation $M, s_i \models \varphi$, where M is a structure for LTL which satisfies the properties we will present in this section. Given a state s_i and a path π, we will have a matching sequence of steps of reads and writes which becomes true in $s_i, s_{i+1}, \ldots$. Hence, π will create a history of the transactions $\{T_1, T_2 \ldots\}$ generated by the protocol and which begin at s_i. The resulting history h is:

$$h = r_1(x_3)w_1(x_3)r_3(x_3)w_3(x_3)r_2(x_3)w_2(x_3)r_2(x_4)$$
$$w_2(x_4)r_1(x_4)w_1(x_4)r_4(x_4)w_4(x_4)r_4(x_5)w_4(x_5)r_3(x_5)w_3(x_5)$$

We also add to the transaction propositions $begin_i$ and end_i, respectively, indicating when a transaction starts and finishes.

Using temporal logic operators, we encode the properties (P1)–(P6) of the protocol in LTL structure. In this section, we only present an example of each property, where the example is for describing the LTL formula in NuSMV. The remainder are found in the next section. The properties of the protocol are as follows:

(P1) No two reads without a write in-between
Any transaction which has completed a read step to one data item cannot read another data item without having it write to the first one before the second read. If $x <_D y$, which means that x precedes y in the data items domain, D, $r_i(y)$ cannot be executed before $w_i(x)$ [5]. This property is to maintain the structure of multi-step transactions. We can encode this property into the LTL formula as follows:

$$\sigma_1 = \bigwedge_{i \geq 1} \bigwedge_{x,y \in D_i, x <_D y} G[(r_i(x) \Rightarrow F(w_i(x) \wedge F(r_i(y))))]$$

Taking the case of T_1, this means that T_1 cannot read $x3$ and $x4$, where $x3 <_D x4$, without having first written to $x3$. This is encoded into LTL in NuSMV as:

```
LTLSPEC G (T1=r1x3 -> (F (T1=w1x3 & F (T1=r1x4))))
```

We can also write it in another way as:

```
LTLSPEC G (((T1=r1x4) & O(T1=r1x3)) -> O(T1=w1x3))
```

(P2) A write step happens if an item was read
A transaction T_i can only write to x if it has read x beforehand [5].

$$\sigma_2 = \bigwedge_{i \geq 1} \bigwedge_{x \in D_i} G[(w_i(x) \Rightarrow O(r_i(x)))]$$

Taking the case of T_1, this means that if T_1 accomplished a write step on x_3, it must have previously read x_3 to ensure that we have read and write steps to each data item x. This is encoded into LTL in NuSMV as follows:

```
LTLSPEC G ((T1=w1x3) -> O(T1=r1x3))
```

(P3) Step of read/write stays true until the next operation of the same transaction itself becomes true. If there is a read/write step, no changes to this will be made until the next operation in T_i becomes true, i.e., if $r_i(x)/w_i(x)$ is true, it is unchanged until the next step, where $x <_D y$, becomes true [5].

$$\sigma_3 = \bigwedge_{i \geq 1} \bigwedge_{x \in D_i} G[w_i(x) \Rightarrow \neg(r_i(x))] \bigwedge_{i \geq 1} \bigwedge_{x,y \in D_i, x <_D y} G[r_i(y) \Rightarrow \neg(w_i(x))]$$

If T_1 reads x_3, then $r_1(x_3)$ stays true until T_1 has written to x_3, at which point $r_1(x_3)$ becomes false and $w_1(x_3)$ becomes true. After that, if T_1 needs to read another data item, say x_4, then $w_1(x_3)$ becomes false and $r_1(x_4)$ becomes true. This property is encoded into LTL in NuSMV as follows:

```
LTLSPEC G(((T1=w1x3) -> !(T1=r1x3)) & G((T1=r1x4) -> !(T1=w1x3)))
```

(P4) Each successful state includes only one occurrence of a step.
This is adopted so as not to have two different false steps in a state, and after that the same steps are true subsequently [5].

$$\sigma_4 = \bigwedge_{\substack{i,i' \geq 1 \\ 1 \leq j,j' \leq m \\ i \neq i', j \neq j'}} G[\neg((\neg(r_i(x_j) \wedge \neg r_i'(x_j')) \wedge X r_i(x_j) \wedge r_i'(x_j')))$$

$$\wedge \neg((\neg r_i(x_j) \wedge \neg w_i'(x_j')) \wedge X(r_i(x_i) \wedge w_i'(x_j')))$$

$$\wedge \neg((\neg w_i(x_j) \wedge \neg w_i'(x_j')) \wedge X(w_i(x_j) \wedge w_i'(x_j')))]$$

This property emphasises the fact that, if, say, transaction T_1 reads item x_3, it cannot simultaneously write and read in the next step. Only one successful step can happen in each state. This is written in LTL for T_1 in NuSMV as follows:

```
LTLSPEC G ((T1=begin1) -> X!((T1=r1x3)&(T1=w1x3)))
LTLSPEC G ((T1=r1x3) -> X!((T1=w1x3)&(T1=r1x4)))
LTLSPEC G ((T1=r1x4) -> X!((T1=w1x4)&(T1=end1)))
```

(P5) Any transaction can read and write only once to a data item [5]. For all $x \in D_i$, a transaction T_i can only read data item x once and only write on data item x once.

$$\sigma_5 = \bigwedge_{i \geq 1} \bigwedge_{x \in D_i} G\neg[r_i(x) \wedge F(\neg r_i(x) \wedge F r_i(x)) \wedge U end_i]$$

$$\wedge (\bigwedge_{i \geq 1} \bigwedge_{x \in D_i} G\neg[w_i(x) \wedge F(\neg w_i(x) \wedge F w_i(x)) \wedge U end_i]))$$

This means that a transaction can only access the data item once for both the read and write steps in a given history. If transaction T_1 writes on data item x_3 having previously read x_3, it is not allowed to read x_3 again until transaction T_1 ends. This is encoded into LTL as follows:

```
LTLSPEC G ((T1=w1x3 & O(T1=r1x3)) -> (F!(T1=r1x3)))U(T1=end1)
```

(P6) The conflict graph of the routing protocol is serializable if there is no cycle in the conflict graph G of a history h that is generated by the protocol. Since we found a maximum gap $G = 1$ (G^1), we only need to check for a cycle of length 3 (C_3). If we find this cycle, the history is not serializable. We use the following LTL formula:

$$\sigma_6 = \bigwedge_{\substack{i,j,k \geq 1 \\ i \neq j \neq k}} \bigwedge_{\substack{x,y \in D_i \\ z \in D_j, D_k \\ y \in D_k, D_i}} ! \, G[(r_i(x) \lor w_i(x)) \Rightarrow F(w_j(x) \lor (w_j(x) \land (w_j(z) \lor r_j(z))))]$$

$$\Rightarrow F(w_k(x) \land w_k(z) \land w_k(y) \lor (w_k(x) \land (w_k(z) \land (w_k(y)))) \Rightarrow F(w_i(y)]$$

A cycle can be produced by three or more transactions where the first transaction conflicts with the second transaction, creating an edge from the first to the second transaction in the conflict graph, and where the second transaction conflicts with the third, similarly creating an edge from the second to the third transaction in the conflict graph. Finally, the third transaction conflicts with the first, creating an edge from the third to the first transaction in the graph. In this scenario, we will have a cycle of length 3 which matches our goal. This is encoded into LTL in NuSVM as follows:

```
LTLSPEC !G((T1=r1x3)->F(T3=w3x3)->F(T2=w2x3)->F(T1=w1x4))
```

This will check whether there is not a cycle of this form. This will detect the cycle $T_1T_3T_2T_1$ of length 3. As a result, the NuSMV model checker will give a counterexample stating that this is not satisfied, which means that it is not the case that we do not have this cycle, i.e., we have a cycle. The error is shown in Fig. 3 after executing the code and the specification in the NuSMV model checker. The error is given with a counter example by the NuSMV model checker showing where the error happens by specifying the states of the error and their trace. If we remove the ! from the beginning of the LTLSPEC, this will not cause an error, instead we will have a result that is *true*. The following LTL formula in NuSMV will check if there is a cycle of length 3 formed as $T_1T_2T_4T_1$:

```
LTLSPEC !G((T1=r1x3)->F(T2=w2x3)->F(T4=w4x4)->F(T1=w1x4))
```

The result in Fig. 4 shows the result given by the NuSMV model checker indicating an error and giving a counterexample as a result of running the code, which means that this cycle exists. The following will check if there is a cycle of length 3 formed as $T_2T_1T_4T_2$:

```
LTLSPEC !G((T2=r2x3)->F(T1=w1x3)->F(T4=w4x4)->F(T2=w2x4))
```

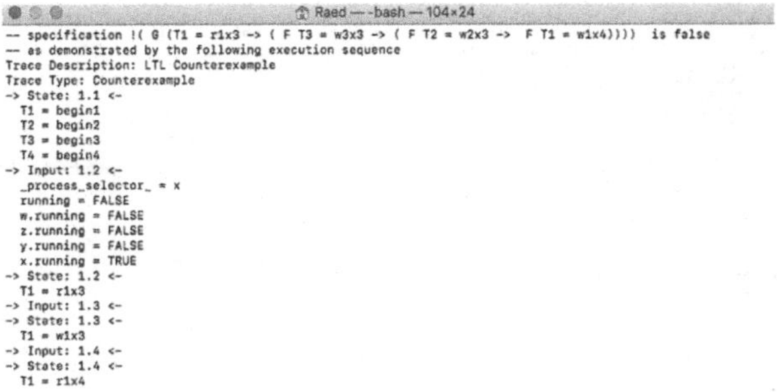

```
Raed — -bash — 104×24
-- specification !( G (T1 = r1x3 -> ( F T3 = w3x3 -> ( F T2 = w2x3 ->  F T1 = w1x4)))  is false
-- as demonstrated by the following execution sequence
Trace Description: LTL Counterexample
Trace Type: Counterexample
-> State: 1.1 <-
  T1 = begin1
  T2 = begin2
  T3 = begin3
  T4 = begin4
-> Input: 1.2 <-
  _process_selector_ = x
  running = FALSE
  w.running = FALSE
  z.running = FALSE
  y.running = FALSE
  x.running = TRUE
-> State: 1.2 <-
  T1 = r1x3
-> Input: 1.3 <-
-> State: 1.3 <-
  T1 = w1x3
-> Input: 1.4 <-
-> State: 1.4 <-
  T1 = r1x4
```

Fig. 3. Counterexample on cycle $T_1T_3T_2T_1$.

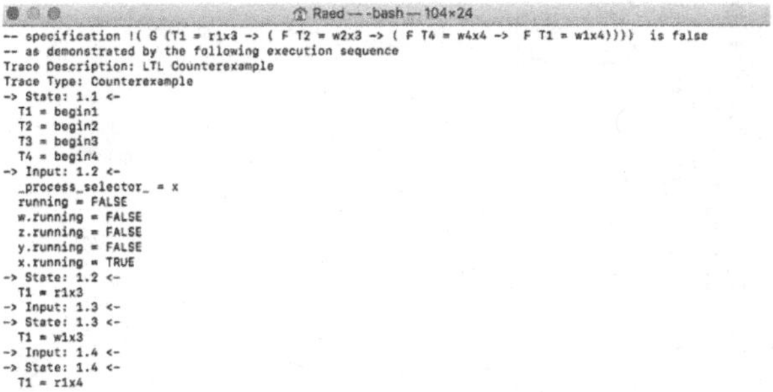

```
Raed — -bash — 104×24
-- specification !( G (T1 = r1x3 -> ( F T2 = w2x3 -> ( F T4 = w4x4 ->  F T1 = w1x4)))  is false
-- as demonstrated by the following execution sequence
Trace Description: LTL Counterexample
Trace Type: Counterexample
-> State: 1.1 <-
  T1 = begin1
  T2 = begin2
  T3 = begin3
  T4 = begin4
-> Input: 1.2 <-
  _process_selector_ = x
  running = FALSE
  w.running = FALSE
  z.running = FALSE
  y.running = FALSE
  x.running = TRUE
-> State: 1.2 <-
  T1 = r1x3
-> Input: 1.3 <-
-> State: 1.3 <-
  T1 = w1x3
-> Input: 1.4 <-
-> State: 1.4 <-
  T1 = r1x4
```

Fig. 4. Counterexample on cycle $T_1T_2T_4T_1$.

Figure 5 shows a result that indicates a counterexample is given, which means that this cycle exists. The following formula will check if there is a cycle of length 3 formed as $T_3T_2T_4T_3$:

```
LTLSPEC !G((T3=r3x3)->F(T2=w2x3)->F(T4=w4x4)->F(T3=w3x5))
```

Figure 6 shows a result that indicates a counterexample is given, which means that this cycle exists. The process is repeated for subsequent cases such as $T_3T_1T_4T_3$. The results shown in Figs. 3, 4, 5 and 6 shows the significance of using the protocol presented in this paper. The results in these figures checks for the existence of cycles using the LTL encoded into the NuSMV model checker. As the model checker checks the model and the specifications presented earlier, it gives an error of the form of a counterexample for each specification that caused an error. The specifications where used to check that "there is no cycle of such form of certain length" and the model checker finds that there is a cycle of these details, hence gives an error. The importance of the counterexample is to trace where the cycle starts and how it is formed.

Fig. 5. Counterexample on cycle $T_2T_1T_4T_2$.

Fig. 6. Counterexample on cycle $T_3T_2T_4T_3$.

8   Verification of the Routing Protocol with Gap Using the NuSMV Model Checker

We use NuSMV model checker [10] as a verification language introduced by model checking, to determine whether the protocol specifications expressed in LTL hold, or otherwise. The model checker produces a true result if the specification of the required property conforms with all system behaviours; otherwise, a counterexample will be given that represents the error source. We will first explain some keywords and variables used in the model. We used MODULE move (Tr, n, Ta, Tb, Tc). The variables (Tr, n, Ta, Tb, Tc) represent:

- Tr: transaction that is currently in process.
- n: an integer indicating the number of the transaction.
- Ta, Tb, Tc: other transactions that are waiting in the queue.
- T1, T2, T3, T4: transaction number one, two, three and four.
- r1x1: T1 reads item x1.
- w1x1: T1 writes on item x1.

8.1   Encoding into LTL

```
{MODULE move(Tr,n,Ta,Tb,Tc)
ASSIGN
next(Tr):=case
Tr= begin1 &n= 1 &(!(Tr=r1x3) & (!(Ta=r2x3)) & (!(Tb=r3x3))  ) : r1x3;
Tr= r1x3 &n= 1  : w1x3;
        Tr= w1x3 &n= 1 : r1x4;
        Tr= r1x4 &n= 1 : w1x4;
        Tr= w1x4 &n= 1 : end1;
        Tr= end1 : begin1;
        Tr= begin2 &n= 2 &(!(Tr=r2x3) & (!(Ta=r1x3)) &
        (!(Tb=r3x3))  ) : r2x3;
        Tr= r2x3 &n= 2 : w2x3;
        Tr= w2x3 &n= 2 : r2x4;
        Tr= r2x4 &n= 2 : w2x4;
        Tr= w2x4 &n= 2 : end2;
        Tr= end2 : begin2;
        Tr= begin3 &n= 3 &(!(Tr=r3x3) & (!(Ta=r1x3)) &
        (!(Tb=r2x3))  ) : r3x3;
        Tr= r3x3 &n= 3 : w3x3;
        Tr= w3x3 &n= 3 : r3x5;
        Tr= r3x5 &n= 3 : w3x5;
        Tr= w3x5 &n= 3 : end3;
        Tr= end3 : begin3;
    Tr= begin4 &n= 4 &(!(Tr=r4x4) & (!(Ta=r1x4)) &
    (!(Tb=r2x4))  ) : r4x4;
        Tr= r4x4 &n= 4 : w4x4;
        Tr= w4x4 &n= 4 : r4x5;
        Tr= r4x5 &n= 4 : w4x5;
        Tr= w4x5 &n= 4 : end4;
        Tr= end4 : begin4;
        TRUE : Tr;
        esac;
MODULE main
  VAR
        T1 : {begin1,r1x3,w1x3,r1x4,w1x4,end1};
        T2 : {begin2,r2x3,w2x3,r2x4,w2x4,end2};
        T3 : {begin3,r3x3,w3x3,r3x5,w3x5,end3};
        T4 : {begin4,r4x4,w4x4,r4x5,w4x5,end4};
        x: process move(T1,1,T2,T3,T4);
        y: process move(T2,2,T1,T3,T4);
```

```
        z: process move(T3,3,T1,T2,T4);
        w: process move(T4,4,T1,T2,T3);
  ASSIGN
        init(T1):= begin1;
        init(T2):= begin2;
        init(T3):= begin3;
        init(T4):= begin4;
        FAIRNESS (T1=end1)
        FAIRNESS (T2=end2)
        FAIRNESS (T3=end3)
        FAIRNESS (T4=end4)
----- This is h
  LTLSPEC F( (T1=r1x3)->X(T1=w1x3)->X(T3=r3x3)->X(T3=w3x3)->(T2=r2x3)->
  X(T2=w2x3)->X(T2=r2x4)->X(T2=w2x4)->X(T1=r1x4)->X(T1=w1x4)->X(T4=r4x4)->
  X(T4=w4x4)->X(T4=r4x5)->X(T4=w4x5)->(T3=r3x5)->X(T3=w3x5))
------No two reads without a write in-between for a transaction
    --T1
    LTLSPEC G (((T1=r1x4) & O(T1=r1x3)) -> O(T1=w1x3))
--T2
LTLSPEC G (((T2=r2x4) & O(T2=r2x3)) -> O(T2=w2x3))
--T3
LTLSPEC G (((T3=r3x5) & O(T3=r3x3)) -> O(T3=w3x3))
--T4
LTLSPEC G (((T4=r4x5) & O(T4=r4x4)) -> O(T4=w4x4))
-- another way of representing this ---
--T1
LTLSPEC G (T1=r1x3 -> (F (T1=w1x3 & F (T1=r1x4))))
--T2
LTLSPEC G (T2=r2x3 -> (F (T2=w2x3 & F (T2=r2x4))))
--T3
LTLSPEC G (T3=r3x3 -> (F (T3=w3x3 & F (T3=r3x5))))
 --T4
LTLSPEC G (T4=r4x4 -> (F (T4=w4x4 & F (T4=r4x5))))
---- A write step happens if item was read
---T1
LTLSPEC G ((T1=r1x3) -> O(T1=begin1))
LTLSPEC G ((T1=w1x3) -> O(T1=r1x3))
LTLSPEC G ((T1=r1x4) -> O(T1=w1x3))
LTLSPEC G ((T1=w1x4) -> O(T1=r1x4))
LTLSPEC G ((T1=end1) -> O(T1=w1x4))
--T2
LTLSPEC G ((T2=r2x3) -> O(T2=begin2))
LTLSPEC G ((T2=w2x3) -> O(T2=r2x3))
LTLSPEC G ((T2=r2x4) -> O(T2=w2x3))
    LTLSPEC G ((T2=w2x4) -> O(T2=r2x4))
```

```
LTLSPEC G ((T2=end2) -> O(T2=w2x4))
--T3
LTLSPEC G ((T3=r3x3) -> O(T3=begin3))
LTLSPEC G ((T3=w3x3) -> O(T3=r3x3))
LTLSPEC G ((T3=r3x5) -> O(T3=w3x3))
LTLSPEC G ((T3=w3x5) -> O(T3=r3x5))
LTLSPEC G ((T3=end3) -> O(T3=w3x5))
---T4
LTLSPEC G ((T4=r4x4) -> O(T4=begin4))
LTLSPEC G ((T4=w4x4) -> O(T4=r4x4))
LTLSPEC G ((T4=r4x5) -> O(T4=w4x4))
LTLSPEC G ((T4=w4x5) -> O(T4=r4x5))
LTLSPEC G ((T4=end4) -> O(T4=w4x5))
--- Step of read/write stays true until the next operation of
the same transaction becomes true
--T1
LTLSPEC G ((T1=begin1) -> X!((T1=r1x3)&(T1=w1x3)))
LTLSPEC G ((T1=r1x3) -> X!((T1=w1x3)&(T1=r1x4)))
LTLSPEC G ((T1=r1x4) -> X!((T1=w1x4)&(T1=end1)))
--T2
LTLSPEC G ((T2=begin2) -> X!((T2=r2x3)&(T2=w2x3)))
LTLSPEC G ((T2=r2x3) -> X!((T2=w2x3)&(T1=r2x4)))
LTLSPEC G ((T2=r2x4) -> X!((T2=w2x4)&(T2=end2)))
--T3
LTLSPEC G ((T3=begin3) -> X!((T3=r3x3)&(T3=w3x3)))
LTLSPEC G ((T3=r3x3) -> X!((T3=w3x3)&(T3=r3x5)))
LTLSPEC G ((T3=r3x5) -> X!((T3=w3x5)&(T3=end3)))
--T4
LTLSPEC G ((T4=begin4) -> X!((T4=r4x4)&(T4=w4x4)))
LTLSPEC G ((T4=r4x4) -> X!((T4=w4x4)&(T4=r4x5)))
LTLSPEC G ((T4=r4x5) -> X!((T4=w4x5)&(T4=end4)))
---A data item is only read once and written once by a transaction
LTLSPEC G ((T1=w1x3 & O(T1=r1x3)) -> (F!(T1=r1x3)))
LTLSPEC G ((T1=w1x4 & O(T1=r1x4)) -> (F!(T1=r1x4)))
LTLSPEC G ((T2=w2x3 & O(T2=r2x3)) -> (F!(T2=r2x3)))
LTLSPEC G ((T2=w2x4 & O(T2=r2x4)) -> (F!(T2=r2x4)))
LTLSPEC G ((T3=w3x3 & O(T3=r3x3)) -> (F!(T3=r3x3)))
LTLSPEC G ((T3=w3x5 & O(T3=r3x5)) -> (F!(T3=r3x5)))
LTLSPEC G ((T4=w4x4 & O(T4=r4x4)) -> (F!(T4=r4x4)))
LTLSPEC G ((T4=w4x5 & O(T4=r4x5)) -> (F!(T4=r4x5)))
--iteration of transactions
--T1
LTLSPEC G ((T1=r1x3) -> F(T1=r1x3))
LTLSPEC G ((T1=w1x3) -> F(T1=w1x3))
LTLSPEC G ((T1=r1x4) -> F(T1=r1x4))
```

```
LTLSPEC G ((T1=w1x4) -> F(T1=w1x4))
--T2
LTLSPEC G ((T2=r2x3) -> F(T2=r2x3))
LTLSPEC G ((T2=w2x3) -> F(T2=w2x3))
LTLSPEC G ((T2=r2x4) -> F(T2=r2x4))
LTLSPEC G ((T2=w2x4) -> F(T2=w2x4))
--T3
LTLSPEC G ((T3=r3x3) -> F(T3=r3x3))
LTLSPEC G ((T3=w3x3) -> F(T3=w3x3))
LTLSPEC G ((T3=r3x5) -> F(T3=r3x5))
LTLSPEC G ((T3=w3x5) -> F(T3=w3x5))
--T4
LTLSPEC G ((T4=r4x4) -> F(T4=r4x4))
LTLSPEC G ((T4=w4x4) -> F(T4=w4x4))
LTLSPEC G ((T4=r4x5) -> F(T4=r4x5))
LTLSPEC G ((T4=w4x5) -> F(T4=w4x5))
              --- no cycle
---x3  is accessed by T1, T3 and T2 respectively as in history h.
If a transaction reads, it should write on before other transactions read.
LTLSPEC G ((T1=w1x3) -> O!(T2=r2x3))
LTLSPEC G ((T1=w1x3) -> O!(T3=r3x3))
LTLSPEC G ((T3=w3x3) -> O!(T2=r2x3))
----x4 is accessed by T,2 T1, and T4 respectively, so we check here again.
LTLSPEC G ((T2=w2x4) -> O!(T1=r1x4))
LTLSPEC G ((T2=w2x4) -> O!(T3=r4x4))
LTLSPEC G ((T1=w1x4) -> O!(T4=r4x4))
----x5 is accessed by T4 and T3 respectively, so we check here again.
LTLSPEC G ((T4=w4x5) -> O!(T3=r3x5))
-- We can also check in another way as follows:
--X3
LTLSPEC G ((T2=r2x3 & O(T1=r1x3)) -> (F!(T1=w1x3)))
LTLSPEC G ((T3=r3x3 & O(T1=r1x3)) -> (F!(T1=w1x3)))
LTLSPEC G ((T3=r3x3 & O(T2=r2x3)) -> (F!(T2=w2x3)))
--X4
LTLSPEC G ((T2=r2x4 & O(T1=r1x4)) -> (F!(T1=w1x4)))
LTLSPEC G ((T4=r4x4 & O(T1=r1x4)) -> (F!(T1=w1x4)))
LTLSPEC G ((T4=r4x4 & O(T2=r2x4)) -> (F!(T2=w2x4)))
--X5
   LTLSPEC G ((T4=r4x5 & O(T3=r3x5)) -> (F!(T3=w3x5)))
-----The gap G=1 this means that we can have a cycle of length= 3
--- Here we check for this cycle only.
---these create the cycle if we put ! before G
--1321
LTLSPEC !G((T1=r1x3)->F(T3=w3x3)->F(T2=w2x3)->F(T1=w1x4))
--if we remove  "!"  before G it will execute and give TRUE,
```

```
i.e, does not check for cycle.
--- this is continuing to check of n length cycle
--1241
LTLSPEC G((T1=r1x3)->F(T2=w2x3)->F(T4=w4x4)->F(T1=w1x4))
--2141
LTLSPEC G((T2=r2x3)->F(T1=w1x3)->F(T4=w4x4)->F(T2=w2x4))
--3242
LTLSPEC G((T3=r3x3)->F(T2=w2x3)->F(T4=w4x4)->F(T3=w3x5))
--3143
LTLSPEC G((T3=r3x3)->F(T1=w1x3)->F(T4=w4x4)->F(T3=w3x5))
--4234
LTLSPEC G((T4=r4x4)->F(T2=w2x4)->F(T3=w3x5)->F(T4=w4x5))
FAIRNESS running
```

9 Conclusion

The importance of this contribution is represented in the different ways that routers can be accessed by an unlimited number of concurrent transactions, and in the detection of cycles according to the theoretical work of the gap theory presented in [4]. We have illustrated the specification and verification of a novel administrative protocol using LTL and NuSMV model checker. Determining if a conflict graph contains a cycle is very time consuming and difficult, especially when the number of transactions are unlimited. We have shown how to determine if a graph has a cycle, or indeed otherwise, by discovering the gap and the length of the cycle and checking only for a specific cycle of a specific length, rather than searching for all cycles possible. This work describes an excellent technique by which to maintain reliability and security in critical systems in a highly efficient manner. This is considered an advantage which can save a considerable amount of resources in different systems using administrative routing protocols, especially in the emerging area of pervasive communications and networking for IoT. Failure can be avoided in a much more straightforward manner by detecting cycles using the proposed protocol. We illustrate how powerful and important temporal logic is when specifying computer-based systems protocols. Despite the significant findings of this research, and indeed its different potential applications, it is limited to the order in which the transactions access the data items. This opens the door to further research in this area as regards different properties. We are now evaluating different applications in which we can apply gap theory, and further specify and verify protocols for those systems. We are also considering other situations where data is accessed in a different manner.

References

1. Abadi, M., Manna, Z.: Temporal logic programming. J. Symb. Comput. 8(3), 277–295 (1989)

2. Al-Aqrabi, H., Hill, R.: Dynamic multiparty authentication of data analytics services within cloud environments. In: 2018 IEEE 20th International Conference on HPCC, pp. 742–749. IEEE (2018)
3. Al Aqrabi, H., Liu, L., Hill, R., Antonopoulos, N.: A multi-layer hierarchical intercloud connectivity model for sequential packet inspection of tenant sessions accessing BI as a service. In: IEEE International Conference on HPCC, pp. 498–505. IEEE (2014)
4. Alshorman, R., Fawareh, H.: Reducing conflict graph of multi-step transactions accessing ordered data with gaps. Math. Comput. Sci. **40**(1), 1–8 (2013)
5. Alshorman, R., Hussak, W.: Multi-step transactions specification and verification in a mobile database community. In: 2008 3rd International Conference on Information and Communication Technologies: From Theory to Applications, ICTTA 2008, pp. 1–6. IEEE (2008)
6. Alshorman, R., Hussak, W.: A serializability condition for multi-step transactions accessing ordered data. Int. J. Comput. Sci. **4**, 13–20 (2009)
7. Alshorman, R., Hussak, W.: Specifying a timestamp-based protocol for multi-step transactions using LTL. World Acad. Sci. Eng. Technol. Int. J. Comput. Electr. Autom. Control Inf. Eng. **4**(11), 1716–1723 (2010)
8. Baier, C., Katoen, J.P., Larsen, K.G.: Principles of Model Checking. MIT Press, Cambridge (2008)
9. Bernstein, P.A., Hadzilacos, V., Goodman, N.: Concurrency Control and Recovery in Database Systems. Addison-Wesley Longman Publishing Co. Inc., Boston (1987)
10. Cimatti, A., Clarke, E., Giunchiglia, F., Roveri, M.: NuSMV: a new symbolic model checker. Int. J. Softw. Tools Technol. Transf. **2**(4), 410–425 (2000)
11. Clarke, E.M., Grumberg, O., Peled, D.: Model Checking. MIT Press, Cambridge (1999)
12. Hussak, W.: Serializable histories in quantified propositional temporal logic. Int. J. Comput. Math. **81**(10), 1203–1211 (2004)
13. Hussak, W.: Specifying strict serializability of iterated transactions in propositional temporal logic. Int. J. Comput. Sci. **2**(2), 150–156 (2007)
14. Hussak, W.: The serializability problem for a temporal logic of transaction queries. J. Appl. Non-Class. Log. **18**(1), 67–78 (2008)
15. Lamport, L.: Proving the correctness of multiprocess programs. IEEE Trans. Softw. Eng. SE **3**(2), 125–143 (1977)
16. Lamport, L.: What good is temporal logic? IFIP Congress. **83**, 657–668 (1983)
17. Papadimitriou, C.: The Theory of Database Concurrency Control. Computer Science Press, Rockville (1986)
18. Peled, D., Pnueli, A.: Proving partial order properties. Theor. Comput. Sci. **126**(2), 143–182 (1994)
19. Pnueli, A.: Applications of temporal logic to the specification and verification of reactive systems: a survey of current trends. In: de Bakker, J.W., de Roever, W.-P., Rozenberg, G. (eds.) Current Trends in Concurrency. LNCS, vol. 224, pp. 510–584. Springer, Heidelberg (1986). https://doi.org/10.1007/BFb0027047
20. Elmasri, R., Navathe, S.: Fundamentals of Database Systems, 4th edn. Addison-Wesley, Boston (2004)
21. Sistla, A.: Safety, liveness and fairness in temporal logic. Formal Aspects Comput. **6**(5), 495–511 (1994)
22. Sommerville, I.: Software Engineering. Addison-Wesley, Boston (2010)
23. Manna, Z., Pnueli, A.: The Temporal Logic of Reactive and Concurrent Systems: Specification, 1st edn. Springer, New York (1992). https://doi.org/10.1007/978-1-4612-0931-7

Performance Evaluation of Multiparty Authentication in 5G IIoT Environments

Hussain Al-Aqrabi[(⊠)] [iD], Phil Lane[iD], and Richard Hill[iD]

School of Computing and Engineering, University of Huddersfield, Huddersfield, UK
{h.al-aqrabi,p.lane,r.hill}@hud.ac.uk

Abstract. With the rapid development of various emerging technologies such as the Industrial Internet of Things (IIoT), there is a need to secure communications between such devices. Communication system delays are one of the factors that adversely affect the performance of an authentication system. 5G networks enable greater data throughput and lower latency, which presents new opportunities for the secure authentication of business transactions between IIoT devices. We evaluate an approach to developing a flexible and secure model for authenticating IIoT components in dynamic 5G environments.

Keywords: Internet of Things · 5G · Security · Authentication · Analytics

1 Introduction

Fifth generation (5G) networks are becoming more recognised as a significant driver of Industrial Internet of Things (IIoT) application growth [1,2]. Recent developments in the growth of wireless and networking technologies such as software-defined networking and hardware virtualisation have led to the next generation of wireless networks and smart devices. In comparison to 4G technologies, 5G is characterised by: higher bit rates (more than 10 gigabits per second), higher capacity, and very low latency [3], which is a significant asset for the billions of connected devices in the context of Internet of Things (and Industrial Internet of Things) domains. As such, emerging IoT applications and business models require new approaches to measuring performance, utilising criteria such as security, trustworthy, massive connectivity, wireless communication coverage, and very low latency for a large number of IoT devices.

The introduction of 5G infrastructure is particularly important for IoT and Industrial IoT (IIoT) as it supports increases in data throughput transmission rates, as well as reducing system latency.

As IoT devices proliferate and we delegate more tasks to them, there is a greater need to (a) exchange data, and (b) augment existing data transmission to facilitate improved trust mechanisms between objects. In any network, an increase in the size and volume of data packets that are transported can increase response times, which is undesirable in a highly interconnected environment

© Springer Nature Singapore Pte Ltd. 2019
H. Ning (Ed.): CyberDI 2019/CyberLife 2019, CCIS 1138, pp. 169–184, 2019.
https://doi.org/10.1007/978-981-15-1925-3_13

of smart objects. 5G performance facilitates reduced response times between communicating entities, which helps enhance the user experience.

Together, these characteristics directly support the closer cooperation of decoupled physical objects, and is the basis for a more connected future.

5G deployments in the millimeter wave region of the radio spectrum is one of the main enabling factor for improved network performance, albeit at a loss of transmission distance. Operating at ≈ 30 GHz offers some physical security due to the limited propagation ranges available. However, the industrial scenario of a malicious employee, stood beside some manufacturing equipment that is IIoT enabled, means that data can potentially be 'sniffed' and relayed to an external system.

Awareness and accessibility of technologies that can connect the operations of devices, with a view to permitting innovative ways of optimising operations and minimising waste through cooperation, is a key objective of the Industry 4.0 movement. Industrial organisations have mission-critical intellectual property (IP) that is central to their ability to compete effectively and maintain profits now and into the future. The detail and insight that describes such IP is contained within the myriad industrial operations that take place, and therefore the use of IIoT, which generally includes wireless networking technologies, is a concern to many who wish to protect their IP.

The traditional methods of ensuring secure network communications tend to depend on the checking of credentials against a central authority. This has proven satisfactory in many cases, though this is up to a point as systems grow to the point where the authentication mechanism itself can become a bottleneck. If we consider the potential number of connected parties that would require authentication in an IIoT environment, it is clear that a centralised authentication system cannot scale sufficiently without harming the operation of the whole system.

IIoT devices are dynamic, often mobile, and need to work and be trusted for variable amounts of time, usually to complete a particular transaction in a timely fashion. The potential requirements of IIoT are such that any architecture that is essential for the secure exchange of data, must also be scalable to meet what seems to be an inconceivable future demand.

1.1   Industrial Internet of Things in the 5G Era

The IIoT has the ability to provide intelligent services to users, while presenting privacy and security issues and perhaps new challenges to standards and governance bodies [3]. Research studies have focused on state-of-the-art research in several aspects of IIoT and 5G technologies, from academic and industrial perspectives [4,5]. The aim is to find a place for recent developments in theory, application, standardisation and the application of fifth-generation technologies in IIoT scenarios [6].

5G technology has the potential to expand IIoT capabilities, significantly beyond what is feasible with existing technology. A 5G wireless network will enable IIoT devices to communicate to a new level within smart environments, through connected 'smart sensors'. In addition, a 5G wireless network may also considerably expand the scope and scale of IIoT coverage by offering the fastest communication and capacity for business transactions [7].

Whilst IoT systems are often aimed at enhancing the quality of everyday life, including interconnecting users, smart home devices, and smart environments such as smart cities and smart homes, the IIoT is a domain of significant interest as it is through the enhanced coordination and optimisation possibilities of interconnected manufacturing operations that industrial organisations can become more competitive.

However, IIoT adoption is still developing in industry and faces several challenges, including new demands for product and solutions, and the transformation of business models. In some industries, such as healthcare, or traffic management, etc., the IIoT still must overcome several technical challenges such as flexibility, reliability and robustness [8].

5G-enabled IoT can make important contributions to the future of IIoT by connecting billions of IoT devices to generate a huge 'network of things' in which intelligent devices interact and share data without any human assistance. Presently, a heterogeneous application domain makes it very hard for IIoT to identify whether the individual system components are capable of meeting application requirements [9].

1.2   Key Challenges for 5G-IoT Architecture

While a lot of studies have been done on 5G IoT, there are still technical challenges to overcome. In this section, we will briefly review the major challenges for 5G IoT.

Due to the openness of network architectures and rapid communication network deployment of a wide variety of services, 5G IoT systems pose major challenges for information security, increased privacy concerns, trusted communications between devices, etc. Several researchers have contributed to strengthen authentication mechanisms [10–12] in 4G and 5G cellular networks and there are various cryptographic algorithms to address potential security and privacy problems for 4G and 5G networks [13–16]. The emerging interest in microservices architecture [17] emphasises the need to consider how trust can be engendered between ever-decreasing units of computation.

Although there are numerous heterogeneous mechanisms for secure communication, 5G IoT integrates a number of different technologies and this has an important impact upon IoT applications [18,19].

As the amount of devices in IoT networks increases every day, the management of these devices is becoming increasingly complex [1]. Due to the large amount of IoT devices, scalability of the network and network management is a significant problem in 5G-IoT. To manage the state information such a large number of devices IoT devices with satisfying performance is also a problem that needs to be addressed. Also, several current IoT applications comprise of overlaid deployments of IoT devices networks where both applications and devices are unable to communicate and share information.

These devices need to be able to flexibility connect the network at any time. Since IoT systems generate and/or process sensitive information, it must authenticate itself to obtain and deliver information to the gateway.

Furthermore, the capability and effectiveness to gather and distribute information in the physical globe is challenging. There are still several challenges remaining for 5G IoT that need to be addressed, such as the seamless interconnection between heterogeneous communication networks where a large amount of IoT heterogeneous devices are connected via a complex communication network with varying technology to communicate, and retrieve vital information with other intelligent networks or applications. High availability of IoT devices is essential for real-time monitoring systems, as they need to be accessible to monitor/collect data. IoT devices that may be compromised and vulnerable to hackers, physically harmed or stolen, resulting in service disruption, and it is not easy to locate an impacted node.

2   Dynamic Multiparty Authentication

Due to the rapid development of various emerging technologies and computing paradigms, such as Mobile Computing and the Internet of Things presents significant security and privacy challenges [8, 20].

With the explosive development of the Internet of Things applications, the shift from traditional communication facilities to the Internet is becoming increasingly crucial for group communication.

Many new Internet services and applications are emerging, such as cloud computing that allows users to elastically scale their applications, software platforms, and hardware infrastructure [21]. Cloud-based business systems are dynamic in a multi-tenancy setting and require likewise dynamic authentication relationships.

Therefore, the authentication frameworks can not be static. However, experience in the domain of Cloud computing helps the comprehension of how IoT applications can be subject to security threats, such as exploiting virtue, malware attacks, distributed attacks, and other known cloud challenges [9].

These cloud implementations increase the sharing of resources that can be made available by dividing solutions into distinct levels. Consequently, the increasing proliferation of services provided by IoT technologies also presents many security and privacy-related risks.

Within the shared domains of IoT cloud, the user becomes dynamic or the system may need to upgrade its product to remain up-to-date. However, the IoT application is subjected to increased security and privacy threats because an unauthorised user may be able to obtain access to highly delicate, consolidated business information [2].

In a complex and challenging application there is a need to delegate access control mechanisms securely to one or more parties, who can in turn, control the methods that enable multiple other parties to authenticate with respect to the services they wish to consume. The primary challenge of any multiparty application is the need to authenticate customers so that they can be granted controlled access to information and data resources hosted on the cloud [22,23].

The wider distribution of IoT nodes and the extent and nature of the data collected and transformed by such devices is a major challenge for security [24]. In the IoT domain, authentication permits the integration of various IoT devices deployed in various contexts. In view of the fact that services and organisations can adopt a collaborative process in an extremely vibrant and flexible manner, direct cross-realm authentication relationship is not simply a means of joining the two collaborating realms.

The lack of authentication path connecting two security realms will necessitate two security realms [25], when working together, to follow a more traditional and long route that will involve creating a mutual trust entailing entering into contractual agreements, multi-round cooperation and human intervention [22].

The primary reason for this lack of progress is due to serious concerns about the security, privacy, and reliability of these systems [9]. IoT is capable of monitoring all aspect of day-to-to life, including the above-mentioned concerns. Citizens, therefore, have legitimate concerns about privacy.

In addition, businesses are concerned with damage to their reputations due to data being handled by wrong hands, and the governments fear the consequences of security risks [26]. Multiparty authentication is a complex challenge in a multi-cloud environment.

These challenges increase in complexity when we consider the potential proliferation of devices in IoT systems. In general, such systems may be a one-to-one mapping between system access devices and the clouds themselves.

However, there are also several additional complications of numerous devices with varying degrees of functionality and capability. An example of such a device is a Wireless Sensor Network (WSN), which are often adaptive entities that may be applicable to the addition or removal of sensor nodes during operation.

Various reports predict a remarkable increase in the number of connected intelligent 'things' exceeding 20 billion by 2020 [27,28]. As we see the exponential growth of the connected devices, the predictions seem to be believable. If these predictions come true, then the demand for authentication of devices will be a major challenge to address, especially as there will be insufficient capacity to manually authenticate even a fraction of the devices and consequently, some automation will be mandatory.

A fundamental challenge in a complex distributed computing environment that emerges as a consequence of the IoT and other combinations of technologies such as cloud architectures and microservices, is the necessity to manage and ensure that the required authentication approvals are in place to enable effective, secure communications.

For instance, Service Oriented Architectures (SOA) enable software systems to re-use 'black box' functionality, by way of intra-service messaging that is facilitated by internet protocols. A more recent refinement of this is 'microservices' architecture where there is a consideration of the level of granularity of service that is offered by software.

These paradigms, although abstract, offer considerable opportunities for system architects to develop resilient functionality within software, especially since re-used code can be comprehensively tested and secured. This approach to software development emphasises the need to develop secure systems, particularly since the IoT is in many cases a manifestation of a Cyber Physical System, whereby physical actuation is controlled and governed by software. Such systems present risks as well as opportunities, giving rise to the importance of formal approaches to the design of such systems.

As such, secure communication between software components is essential [29], both in terms of ensuring that a particular message or instruction reaches the intended destination, but also that the complexity of out-sourced service functionality is provided with the appropriate authority to perform the task that is being requested.

The use of Single Sign On (SSO) [30] also allows the use of a key exchange technique to actually manage the provision of authentication credentials certified by a named authority. In addition, it eliminates the need for users to enter different security credentials multiple times.

However, despite the relative simplicity of the technique, it simply provides a secure method of key exchange is insufficient for the situation when we need multiple parties to be capable to establish certain trust each other in a dynamic, heterogeneous environment [31], and therefore SSO technique is lacking in this regard.

2.1   A Multiparty Authentication Model

Prior work [22], as described briefly below, describes a framework that addresses the challenges of obtaining the required authorisation agility in a dynamic multiparty environment.

This multiparty authentication model for dynamic authentication interactions is relevant when participants from various security realms want to access distributed operational data through a trusted manager. Al-Aqrabi et al. [22] addressed issues related to reliable, timely and secure data transfer processes needed for shared company data processing networks.

This scenario is directly transferrable to the situation where large numbers of sensor nodes are producing streamed data, that requires real-time processing for the purposes of signal conditioning [26], data cleansing, localised analytics processing, etc. For the sensor and computational nodes to work together in a service oriented way, there needs to be a mechanism where trusted access to data that is both in-transit and stored in a repository is feasible.

In addition, the multiparty authentication model can be used effectively to assist any distributed computing environment, for instance where cloud session users need to authenticate their session participants, and thus require a simplified authentication processes in multiparty sessions.

Therefore, we have developed and extended this work to support the development of, for example, specific use cases where the availability of 5G network infrastructure can allow new business opportunities through enhanced performance. Figure 1 shows the framework for a *Session Authority Cloud* that is applied as a certificate authority in this situation, although it could also be a remote cloud.

The *SAC*'s function is to control the individual sessions requested by any of the various parties (clouds). The *SAC* does not differentiate between clouds and does not depend on them and it retains general authority over any party wishing to join the system. The *SAC* retains authentication data for all tenants, including, for instance, root keys.

2.2   An Authentication Protocol

In this section, we introduce our proposed authentication protocol that addresses IoT application scenarios and data analytics applications that are accessed through IoT clouds, where participants from different security realms need to access distributed analytics services through a trusted principal.

This may apply especially when there are no direct authentication relationships in multiple IoT cloud systems between the people of different security realms and the distributed IoT cloud services.

Let A be the trusted principal by which the requesting user approaches the SAC. The session authentication approval protocol starts with a user, U who is a member of any security realm approved by the trusted principal. Providing access to IoT database objects in IoT clouds A and B is presumed, if the SAC authorises the request forwarded by the principal. This is also presumed that SAC will not accept any request that is not forwarded by the trusted principal.

The user who requests access is not a member of IoT $Cloud_A$ or a member of IoT $Cloud_B$. In principle, the user is a member of a security realm that is a different IoT cloud ($Cloud_C$) that the SAC trusts.

The principal should know who the user is, since the SAC mainly trusts the principal to accept the session request. ID_r is the requesting user's cloud membership key. ID_s is the requesting user's sub-domain membership and ID_{sess} is the session key allocated by the SAC to access IoT database files on IoT $Cloud_A$ and $Cloud_B$. IoT $Cloud_A$ and $Cloud_B$ will only open access to this key when authorised and forwarded by the SAC.

2.3   Algorithm: Protocol for Session Approval

Figure 1 shows the session approval protocol. First user U_A sends a request to create a new session in order to access IoT database objects in IoT $Cloud_A$ and $Cloud_B$. F sends a request for U_A's keys. U_A sends his/her certificate, which contains a root key and subdomain key, and the certificate is encrypted with U_A's private key. The multiparty session handler (F) generates a new session ID and sends it, along with U_A's request. SAC then verifies U_A identity and to approve a new session. SAC_DB uses U_A's public key. SAC also generates the key of the new session and then registers the session its session list.

Output A value in variable *Flag* to show that a session
is granted (*Flag* = 1) or denied (*Flag* = 0).

Steps

1. U_A to F: request Access to IoT cloud
2. F to U_A: request for the identity ID
3. U_A to F: U_A sends the certificate CA to F
4. F to SAC: session request sent to SAC
5. SAC to $SAC\text{-}DB\text{-}SH$: verifies U_A identity
6. SAC to IoT cloud: Flag indicating U_A is authenticated
   or not. Sends the SessionID and UserID to the
   IoT Cloud CA if authenticated.
7. IoT Cloud to SAC: stores the session ID and key in its
   registry and then sends a reply
8. SAC to F: sends a reply for session approval to F
   for authenticated user U_A.
9. F to U_A: approves the decision to grand session
   for authenticated user U_A. Flag = 1 and exit.

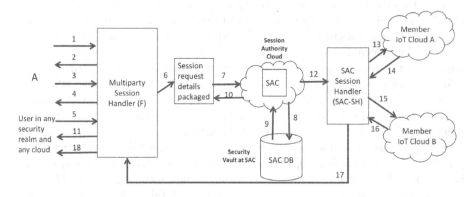

Fig. 1. Protocol for session approval.

Then it can verify U_A's identity. If U_A's identity is valid, then SAC generates
a session key and sends a request to access IoT clouds. After receiving a reply
from resources, SAC then sends a response of session approval with session key
and available resources list.

IoT $Cloud_A$ stores the session ID and key in its registry and then sends a
reply to SAC. SAC sends a reply for session approval to F. Then, F sends a
response for session approval to access to access IoT database objects in IoT
$Cloud_A$ and $Cloud_B$.

3   Simulation Approach

The key focus of the work reported here was an in-depth exploration of how the reduction in connection latency offered by 5G systems, in comparison to 4G systems, impacts the performance of our novel multiparty authentication protocol.

The simulation was built to enable the comparison of the delay experienced by a party seeking authorisation as the rate of authentication requests varied. The authentication delay is conditioned by two factors:

- The time taken for the authentication request to be transported by the mobile part of the system - there are two aspects to this parameter: the time taken to transmit the bits; and other sources of latency in the mobile system such as scheduling, resource allocation and routing
- The time taken for the authentication request to be processed by the authentication server. Earlier simulation work suggest that the time to service a request with reasonable hardware is of the order of 6 ms [22].

When reduced to its fundamental structure, the entire mobile radio system and authentication server systems can be modelled as two cascaded queues as shown in Fig. 2. Authentication requests are generated by a pool of devices with a Poisson distribution wholly defined by a generation rate.

These requests are queued and the service rate of this first M/M/1 queue is dependent on a combination of the mobile system latency mentioned above and the data-rate dependent time taken to transmit the authentication request package. A typical packet size is of the order of 1 kbit, and the transmission delay associated with transporting this is of the order of 0.2 ms at 5 Mbit/s, or 0.02 ms at 50 Mbit/s.

Compared to the overall latency of the end-to-end transmission, and typical service times of the authentication server, these times are negligible and are safely ignored for the rest of the simulation with the service rate of the queue modelling the radio part of the system being determined solely by a single parameter with the system data-rate being irrelevant in this context.

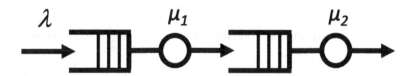

Fig. 2. Cascaded M/M/1 queue.

The authentication requests are then passed from the queue modelling the radio part of the system to a second queue which models the latency of the granting of the authentication request by setting an appropriate service rate for that queue. The implementation of such a system requires the utilization of an event-driven simulation approach, and in Python, the SimPy [32] package was selected as the framework for the simulation as, when combined with SimComponents.py [33], queue models can be assembled and executed with relative ease, and comprehensive performance data is readily accessible for further analysis.

4   Results

Figure 3 shows how the average (mean) delay experienced by an authentication request varies with the rate of requests for a number of different link latencies. The results in this figure are based on the 6 ms authentication delay discussed above. As expected, a higher request rate and/or greater link latency yields a greater average authentication delay. Figure 4 fixes the link latency at 5 ms – a typical, conservative, value for a 5G system – and explores how average authentication delay varies with request rate and the authentication request service time. Figure 5 presents the same information, but this time with an assumption of a 20 ms link latency – a value that is typical of a well performing 4G system. Again, both of these results follow the expected form.

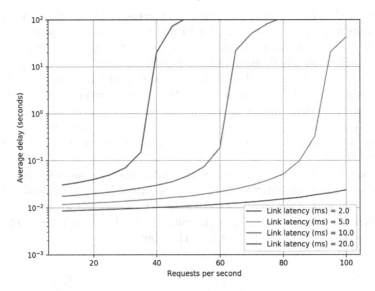

Fig. 3. Average (mean) authentication delay for a 6 ms authentication request service time

The previous three figures only look at the mean delay for an authentication request. In a real-world deployment, we would often have considerable interest in the range and distributions of the actual delays experienced by individual authentication requests. Figures 6 and 7 show the distribution of authentication requests as box-plots for a range of request rates. Figure 5 is for a 20 ms link latency, and Fig. 7 is for a link latency of 5 ms. The whiskers adopt the customary convention of defining points that fall outside of $\pm 1.5 \times IQR$ either side of the median as outliers.

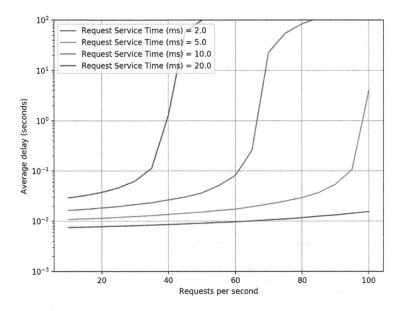

Fig. 4. Average (mean) authentication delay for a link latency of 50 ms

5   Discussion of Results

For most application scenarios, authentication requests will need to be provided in a relatively timely manner. If, for example, we impose an upper limit of 0.1 s for an authentication request to be serviced, then Fig. 3 shows that for a link latency typical of a 4G system – 20 ms – then only slightly over 30 requests per second can be accommodated.

This in contrast to the situation when the link latency is lower. With a latency of 2 ms, 100 requests per second can be accommodated with only ≈0.02 s of delay, and even with a link latency of 5 ms, 85 requests per second can be accommodated within a 0.1 s mean delay. These results show that the adoption of 5G is a key enabler of the enhanced multiparty authentication mechanism described in this research, as the link performance of 4G systems severely inhibits the usefulness of the authentication approach.

Figure 4 considers a 5G-like scenario with a 5 ms link latency and considers the impact of the time required to service an authentication request. This part of the investigation can help to inform the dimensioning of the authentication server hardware. Taking the same 0.1 s target for mean authentication delay, it can be seen that a service request time of 20 ms severely limits the overall system performance with only some 35 requests per second being accommodated.

Conversely, a service request delay of 10 ms yields an ability to support some 95 requests per second. In contrast, Fig. 5 shows that for a 4G-like scenario with a 20 ms link delay, the benefits achieved by improving the performance of the authentication server are limited. Even with a 5 ms service time, only

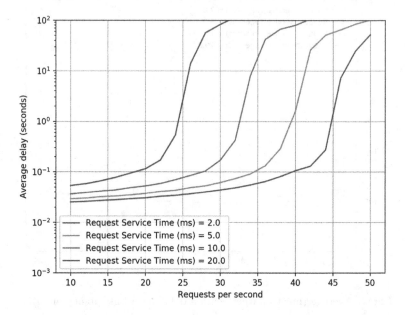

Fig. 5. Average (mean) authentication delay for a link latency of 20 ms

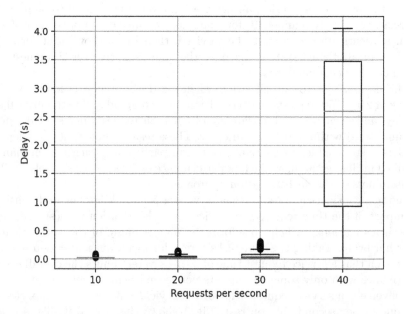

Fig. 6. Distribution of authentication delays for a 20 ms link latency

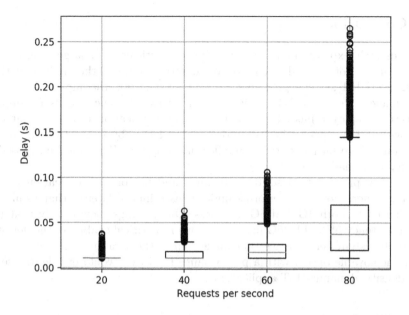

Fig. 7. Distribution of authentication delays for a 5 ms link latency

some 35 requests per second can be accommodated and importantly, the inherent latency of the radio link limits the overall throughput even if authentication server response times are pushed to levels that are not easily achievable in practice.

The lower latency offered by 5G amplifies the benefits offered by improving server performance, and excellent overall throughput can be achieved with server response times that are reasonably realisable.

Mean delays are only one part of the overall system performance assessment as they only allow us to quantify the service as experienced across the user base on an aggregated basis. The spread of delays experienced by devices attempting to authenticate are also of interest, as for a particular node attempting to authenticate, it is the delay experienced by that one node that is of importance. Figure 6 shows that for a 4G-like system with a link latency of 20 ms and an authentication response time of 6 ms, that even with 30 requests per second (where the mean response time was reasonable at under 0.1 s), a considerable number of requests had delays of significantly more than this, up to around 0.4 s as a worst case.

Figure 7 shows that for a 5G-like system then nearly all of the requests at a rate of 60 requests per second achieve an authentication time under 0.1 s, and at 80 requests per second, 75% of the requests are handled within 0.08 s, and all but a very few within 0.25 s. Again, this emphasises the key enabling role that 5G deployment will play in facilitating the deployment of real-world systems based on the enhanced authentication protocol described here.

6 Conclusions

This research explores issues around the authentication of large numbers of devices in an Iot or IIoT scenario. We describe some of the challenges that emerge as (I)IoT systems grow in scale, and consider how the improved network performance offered by 5G particularly in terms of latency opens up opportunities to deploy robust, flexible, dynamic authentication protocols that can improve the trust placed in (I)IoT devices, and thereby facilitate their use in situations where business critical intellectual property (IP) could be exposed if security were inadequate.

We then proceed to describe a dynamic and flexible authentication protocol and explore how this performs under constraints of latency that would be experienced in both 4G and 5G networks. Our findings show clearly that the reduced latency offered by 5G mobile systems is a critical enabler of the delivery of overall system performance that makes our authentication protocol a realistically deployable option with a performance level that would enable its use in high density, dynamic IIoT applications.

References

1. Kalra, S., Sood, S.K.: Secure authentication scheme for IoT and cloud servers. Pervasive Mob. Comput. **24**, 210–223 (2015)
2. Al-Aqrabi, H., Liu, L., Hill, R., Antonopoulos, N.: Cloud BI: future of business intelligence in the cloud. J. Comput. Syst. Sci. **81**, 85–96 (2015)
3. Ferrag, M.A., Maglaras, L., Argyriou, A., Kosmanos, D., Janicke, H.: Security for 4G and 5G cellular networks: a survey of existing authentication and privacy-preserving schemes. J. Netw. Comput. Appl. **101**, 55–82 (2018)
4. Xu, L.D., He, W., Li, S.: Internet of things in industries: a survey. IEEE Trans. Ind. Inf. **10**(4), 2233–2243 (2014)
5. Hosek, I.J.: Enabling technologies and user perception with integrated 5G-IoT ecosystem (2016)
6. Ishaq, I., et al.: IETF standardization in the field of the internet of things (IoT): a survey. J. Sens. Actuator Netw. **2**(2), 235–287 (2013)
7. Akpakwu, G.A., et al.: A survey on 5G networks for the internet of things: communication technologies and challenges. IEEE Access **6**, 3619–3647 (2017)
8. Elkhodr, M., Shahrestani, S., Cheung, H.: The internet of things: new interoperability, management and security challenges (2016). arXiv:1604.04824
9. Al-Aqrabi, H., Liu, L., Hill, R., Antonopoulos, N.: A multi-layer hierarchical inter-cloud connectivity model for sequential packet inspection of tenant sessions accessing BI as a service. In: Proceedings of 6th International Symposium on Cyberspace Safety and Security and IEEE 11th International Conference on Embedded Software and Systems, France, Paris, 20–22 March 2014, pp. 137–144. IEEE (2014)
10. He, D.: An efficient remote user authentication and key agreement protocol for mobile client and server environment from pairings. Ad Hoc Netw. **10**(6), 1009–1016 (2012)
11. Deng, Y., Fu, H., Xie, X., Zhou, J., Zhang, Y., Shi, J.: A novel 3GPP SAE authentication and key agreement protocol. In: Proceedings of International Conference Networks Infrastructure and Digital Content, pp. 557–561. IEEE (2009)

12. Karopoulos, G., Kambourakis, G., Gritzalis, S.: PrivaSIP: ad-hoc identity privacy in SIP. Comput. Stand. Interfaces **33**(3), 301–314 (2011)
13. Alrawais, A., Alhothaily, A., Hu, C., Cheng, X.: Fog computing for the internet of things: security and privacy issues. IEEE Internet Comput. **21**(2), 34–42 (2017)
14. Barni, M., et al.: Privacy-preserving fingercode authentication. In: Proceedings of the 12th ACM Workshop on Multimedia and Security - MMSec 2010, p. 231. ACM Press, New York
15. Ma, C.-G., Wang, D., Zhao, S.-D.: Security flaws in two improved remote user authentication schemes using smart cards. Int. J. Commun. Syst. **27**(10), 2215–2227 (2014)
16. Mahmoud, M., Saputro, N., Akula, P., Akkaya, K.: Privacy-preserving power injection over a hybrid AMI/LTE smart grid network. IEEE Internet Things J. **4**, 870–880 (2016)
17. Shadija, D., Rezai, M., Hill, R.: Towards an understanding of microservices. In: Proceedings of the 23rd International Conference on Automation and Computing, University of Huddersfield, 7–8 September 2017. IEEE (2017)
18. Bessis, N., Xhafa, F., Varvarigou, D., Hill, R., Li, M. (eds.): Internet of Things and Inter-cooperative Computational Technologies for Collective Intelligence. Studies in Computational Intelligence. Springer, Heidelberg (2013). https://doi.org/10.1007/978-3-642-34952-2
19. Hill, R., Devitt, J., Anjum, A., Ali, M.: Towards in-transit analytics for industry 4.0. In: IEEE International Conference on Internet of Things (iThings) and IEEE Green Computing and Communications (Green-Com) and IEEE Cyber, Physical and Social Computing (CPSCom) and IEEE Smart Data (SmartData). IEEE Computer Society (2013)
20. Sotiriadis, S., Bessis, N., Antonopoulos, N., Hill, R.: Meta-scheduling algorithms for managing inter-cloud interoperability. Int. J. High Perform. Comput. Netw. **7**(2), 156–172 (2013)
21. Baker, C., Anjum, A., Hill, R., Bessis, N., Kiani, S.L.: Improving cloud datacentre scalability, agility and performance using OpenFlow. In: Proceedings of the 4th International Conference on Intelligent Networking and Collaborative Systems (INCoS), pp 1–15. IEEE (2012)
22. Al-Aqrabi, H., Hill, R.: Dynamic multiparty authentication of data analytics services within cloud environments. In: 20th IEEE International Conference on High Performance Computing and Communications (HPCC-2018), Exeter, pp. 28–30. IEEE (2018)
23. Al-Aqrabi, H., Hill, R.: Dynamic multiparty authentication of data analytics services within cloud environments. In: Proceedings of the 20th International Conference on High Performance Computing and Communications, 16th International Conference on Smart City and 4th International Conference on Data Science and Systems, HPCC/SmartCity/DSS 2018, pp. 742–749. IEEE Computer Society (2018)
24. He, D., Zeadally, S., Wu, L., Wang, H.: Analysis of handover authentication protocols for mobile wireless networks using identity-based public key cryptography. Comput. Netw. **128**, 154–163 (2016)
25. Hada, S., Maruyama, H.: Session authentication protocol for web services. In: Proceedings of Symposium on Application and the Internet, pp. 158–165 (2002)
26. Cao, N., Nasir, S.B., Shreyas Sen, R.A.: Self-optimizing IoT wireless video sensor node with in-situ data analytics and context-driven energy-aware real-time adaptation. IEEE Trans. Circuits Syst. I Regul. Pap. **64**(9), 2470–2480 (2017)

27. Ndiaye, M., Hancke, G.P., Abu-Mahfouz, A.M.: Software defined networking for improved wireless sensor network management: a survey. Sensors **17**(5), 1–32 (2017)

28. Gartner: Gartner Says 6.4 Billion Connected Things Will Be in Use in 2016, Up 30 Percent From 2015? Gartner website. http://www.gartner.com/newsroom/id/3165317. Accessed 10 Nov 2015

29. Modieginyane, K.M., Letswamotse, B.B., Malekian, R., Abu-Mahfouz, A.M.: Software defined wireless sensor networks: application opportunities for efficient network management: a survey. Comput. Electr. Eng. **66**, 1–14 (2017)

30. Clercq, J.: Single sign-on architectures. In: Davida, G., Frankel, Y., Rees, O. (eds.) InfraSec 2002. LNCS, vol. 2437, pp. 40–58. Springer, Heidelberg (2002). https://doi.org/10.1007/3-540-45831-X_4

31. Xu, J., Zhang, D., Liu, L., Li, X.: Dynamic authentication for cross-realm SOA-based business processes. IEEE Trans. Serv. Comput. **5**(1), 20–32 (2012)

32. SimPy documentation. https://simpy.readthedocs.io/en/latest/index.html. Accessed Oct 2019

33. Grotto Networking, Basic Network Simulations and Beyond in Python. https://www.grotto-networking.com/DiscreteEventPython.html. Accessed Aug 2019

PAM: An Efficient Hybrid Dimension Reduction Algorithm for High-Dimensional Bayesian Network

Huiran Yan[1] and Rui Wang[1,2,3(✉)]

[1] School of Computer and Communication Engineering, University of Science and Technology Beijing (USTB), Beijing 100083, China
s20190706@xs.ustb.edu.cn
[2] Institute of Artificial Intelligence, University of Science and Technology Beijing, Beijing 100083, China
[3] Shunde Graduate School, University of Science and Technology Beijing, Foshan 528300, China

Abstract. In recent years, machine learning has been gradually widely applied to the big data in medical field, such as prediction and prevention of disorders. Bayesian Network has been playing an important role in machine learning and has been widely applied to the medical diagnosis field for its advantages in reasoning under uncertainty. But as the rise of the number of interest variables, the Bayesian Network structure search space is growing super-exponentially. Aiming at improving the efficiency of finding the optimize structure from the large search space of high-dimensional network, in this paper we propose a method, PAM, which is applied to Bayesian Network learning to constrain the search space of high-dimensional network. Several Experiments are performed in order to confirm our hypothesis.

Keywords: High dimensional Bayesian network · Dimension reduction · PCA · Apriori · Maximum information coefficient

1 Introduction

Bayesian network is usually used as a tool to model various disease prediction or classification problems [1–5] for its great performance in reasoning field. Application that Bayesian network was used as an auxiliary tool in Breast Cancer diagnosis has already existed [6]. This paper is based on the background of using Bayesian network to build the predicting model for Breast Cancer. Breast Cancer is one of the most dangerous common cancer among women, which threatens the health of women [7, 8]. If an accurate model for predicting Breast Cancer is built by Bayesian network, the Breast Cancer will be detected early, discovered early and treated early, which can greatly reduce the incidence of Breast Cancer. Former studies have proved that the incidence of Breast Cancer is the result of interaction of multiple causes [9]. In this case, the number of nodes (variables of interest) considered to build a Breast Cancer disease assessment model with

© Springer Nature Singapore Pte Ltd. 2019
H. Ning (Ed.): CyberDI 2019/CyberLife 2019, CCIS 1138, pp. 185–204, 2019.
https://doi.org/10.1007/978-981-15-1925-3_14

Bayesian network is very large. With the increase of the number of nodes in the network, the search space of Bayesian network structure and the complexity of the algorithm will increase exponentially. In view of the inefficiency of learning Bayesian network under high-dimensional variables, many scholars have proposed different methods [10, 11] to learn Bayesian network structure efficiently with high-dimensional variables. Reducing the dimension of variables is one of the most important ideas among these methods, which is very helpful to improve the Bayesian network structure learning efficiency and find the optimal structure.

Based on the background of building a Breast Cancer prediction model with Bayesian network, a hybrid dimension reduction method, PAM, is proposed in this paper to improve the learning efficiency of high-dimensional Bayesian network structure for the problem that Bayesian network structure learning efficiency is limited to the high-dimensional nodes set, that is, the structure learning efficiency of the Breast Cancer model is limited to the pathogenic factors of Breast Cancer. PAM consists of three algorithms, MIC, Apriori and PCA. Being different from other traditional methods [3, 12, 13] which use PCA to reduce dimension without other pre-processing steps, PAM is a kind of hybrid dimension reduction algorithm. Three novel contributions of PAM are as follows, (i) Instead of adopting only a single method, PAM is a hybrid dimension reduction method which absorbs respective advantages of Feature selection methods and Feature extraction methods. (ii) Features selection step is introduced before PCA, in which features irrelevant to classification result are eliminated to improve the quality of the Principal Components (PCs) found by PCA and to strengthen the relevance of these PCs to classification result. (iii) Features selection step consists of two algorithms which are from two different angle. Two features set relevant to classification result are found by the two different algorithms respectively through this step, we adopt the union of this two sets to make complementary advantages.

In Sect. 2, PCA, MIC and the concepts of support and frequent item set are briefly reviewed, and we introduce how these three algorithms mentioned above can be employed and combined to bring forward the PAM algorithm and reduce the dimension of the high-dimensional data in Sect. 3. Finally, in Sect. 4, several experiments are performed in order to confirm the feasibility of PAM algorithm in dimension reduction.

2 Preliminary

2.1 Maximum Information Coefficient

Maximum information coefficient (MIC) is a method to measure the strength of relationship based on mutual information [14]. By choosing the maximum normalized mutual information, MIC can greatly help the situation that the optimal network structure solution may be mistakenly deleted in searching for the optimal Bayesian network structure. At the same time, MIC can better find the relationship and non-linear relationship between variables [15].

Given m samples, a data set D, two variables X and Y, MIC divides the data set D into several empty boxes according to the values of X and Y, obtains an X*Y grid G which allows the occurrence of empty partitioned grids [14]. When G is divided into i*j grids (i rows and J columns), the maximum information coefficient between X and Y is defined as following (1): (B(m) = m 0.6)

$$MIC(X, Y|D) = \max_{i \times j < B(m)} \{ \arg \max I(X, Y, D|_G, i, j) / \log_2 \min(i, j) \} \quad (1)$$

Because the mutual information between two random variables is symmetrical, it is not difficult to deduce that the MIC of two random variables X and Y are still symmetrical, that is MIC(X, Y|D) = MIC(Y, X|D).

2.2 The Concept of Support and Frequent Item Set

Support: Support represents the ratio of the number of occurrences of a single item or a combination of items to the total number of transactions in the whole transaction set. For example, let X and Y be random variables. The degree of support for item X is:

$$support(X) = P(X) \quad (2)$$

In (2) P(X) represent the proportion of some transactions in which item X occurs in the total transactions. We make P(X $\cup$ Y) represent the proportion of the transactions in which item X and Y both occur in the total transactions, then the support of X $\cup$ Y is expressed as (3):

$$support(X \cup Y) = P(X \cup Y) \quad (3)$$

Frequent item set: If the support of an item set is greater than the minimum support threshold min_support set by user, then the item set can be called a frequent item set, and all non-empty subsets of this frequent item set must be also frequent. For example, we let the minimum support threshold min_support be 0.5, if support (X $\cup$ Y) >= 0.5, then X $\cup$ Y is a frequent item set, and X is also a frequent item set as well as Y. The Apriori algorithm we used in PAM is a very representative algorithm which finds the frequent item sets by generating a set of candidate items [16, 17]. The details of Apriori will be introduced in Sect. 3.

2.3 PCA

Principal Component Analysis (PCA) is a kind of linear dimension reduction method [18]. The key idea of PCA is derived from karhunen-loeve transform. A lower number of uncorrelated variables generated by the PCA algorithm is called principal component (PCs). Each principal component can be considered as a linear combination of the original correlated variables. The redundancy between new features generated by PCA is minimized. The PCA algorithm [19] can be expressed as (4).

$$X = TP^T + E \quad (4)$$

Where X is a $N \times m$ mean-variance scaled data matrix (N observations and m measured variables), T is the $N \times k$ scores matrix, $P \in R^{m \times k}$ is the m $\times$ k loadings matrix. k is the number of PCs we choose in the lower space. $E \in R^{N \times m}$ is the residual matrix.

3 PAM: A Hybrid Dimension Reduction Algorithm

3.1 Overview of Modules of PAM

Given a n dimensional data set D with m samples, the n dimensional variables set $X = \{x_1, x_2, .., x_n\}$ and g classes in class label $C = \{c_1, .., c_k\}$, aiming at reducing the original n-dimensional data set D to a dimension much smaller than n, we propose the PAM algorithm, a kind of hybrid method of feature selection and feature extraction for features dimension reduction, to improve the Bayesian network structure learning efficiency. The overall process of reducing dimension with PAM is shown in Fig. 1. The red dashed box in Fig. 1 indicates detailed design of the PAM, which is divided into three steps, or three modules.

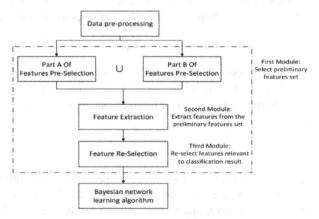

Fig. 1. Dimension reduction process of PAM. (Color figure online)

PAM is mainly divided into three modules: features pre-selection module, features extraction module and features re-selection module. In the first module, features pre-selection module, MIC combined with Apriori is used to select attributes set MAF in which attributes are very relevant to the classification result, the class label C. Some attribute columns of D whose attributes are not in MAF should be eliminated, which can give us a filtered data set D-MAF. Then D-MAF and MAF are inputted into the next module in PAM. In the second module, D-MAF is projected to a lower dimensional space by PCA. Then new low dimensional data set D-PCA and low dimensional features set PX are obtained in this step. Though the data redundancy is low after PCA, there may still exist some irrelevant features to C in PX, so the third module, features re-selection module, is introduced to eliminate these irrelevant features by MIC. From the third module, the target features set Final-X can be obtained. Features in Final-X are nodes, or variables of interest of the Bayesian network which we aim to build. We select the attribute columns not in Final-X from D-PCA and eliminate them, then we can get the target data set D-Final. Final-X and D-Final are inputted into vary kinds of learning algorithms to learn the Bayesian network. We introduce the details and function of three modules of PAM in Sects. 3.2, 3.3, and 3.4, respectively.

3.2 Pre-selection of Features

Select Features from First Angle. First of all, we introduce how to select features set FA by Apriori in PAM. The popular impression left by Apriori is that it is often used to find association rules and frequent item sets in Data Mining [20, 21]. However, in this paper we only use the function of finding frequent item sets of size 1 of Apriori to select features set FA.

As mentioned in Sect. 3.1, D has g classes, $C = \{C_1, ..., C_g\}$, then we divide D into g subsets $D = \{D_1, .., D_g\}$ according to the value of C, for example, all samples in D_i belong to class C_i, which can be formulate in (5).

$$D_i = D(C = C_i) \; (i = 1, \ldots, g) \tag{5}$$

Set a minimum support threshold min_support, we use Apriori to find frequent item set of size 1 Li in each subset Di. Apriori is an algorithm finding the frequent item sets by generating a set of candidate items [16]. The steps to find all frequent item sets are listed below 13:

1. Apriori algorithm first scan the transaction database and take all the items as candidate of size 1, C_1;
2. Apriori choose those whose frequencies are larger than or equal to the min_support as the frequent item sets L_1 of size 1;

After finishing finding frequent item sets of g subsets, we obtain $L = \{L_1, ..., L_g\}$. Each Li in L represents the frequent item set of $D_i (i = 1, 2, ..., g)$. We iterate over each Li in L in order, if there is some duplication between Li and other frequent item set in L, then we will remove the duplication part from all frequent item sets in L where the duplication part has ever occurred. For example, when Li is accessed, then we will traverse L from L_{i+1} to L_g, the duplication part between Li and others can be expressed by (6).

$$x = ((L_i \cap L_{i+1}) \cup, \ldots, (L_i \cap L_g)) \; (i = 1, \ldots g - 1) \tag{6}$$

So the duplication part should be removed as (7) from all item sets in L where the duplication part has ever occurred:

$$L = \left(F_1, \ldots, F_i = L_i - x, \ldots, L_g = L_g - x\right) \tag{7}$$

When finishing iterating over the first g − 1 frequent item sets in L in order, the g-th frequent item set L_g needs not to be accessed because L_g has become F_g at this point. Then new g unique frequent item sets $F = \{F_1, F_2, ..., F_g\}$ can be obtained. The contents of each frequent item set in F do not duplicate that of other frequent item sets in F, so each of the g unique frequent item sets in F, F_i, can be regard as a unique item set of Di, and these item sets in F can be regarded as an important basis for distinguishing g classes in D. We only keep the attribute names in each F_i and abandon the attribute values, and then we take a union of all the attributes of each F_i in F to form an attribute variables set. These attributes in this set are comparatively relevant to the class label C, and the values of them are strongly differentiated in different classes. The union of these frequent item sets $F_i(i = 1, 2, ..., g)$ in (8) in F is the features set FA selected by Apriori.

$$FA = F_1 \cup, ..., \cup F_i \cup, ..., \cup F_g \tag{8}$$

In summary, steps of finding attribute variables set by Apriori are listed as follows:

1. Divide D into g classes according to the value of C, $D = \{D_1, ... D_g\}$;
2. Set the minimum support threshold min_support to 0.5, and find frequent item set of size 1 L_i of each D_i by Apriori to form $L = \{L_1, ..., L_i\}$;
3. Iterate over each Li in L in order, $(i = 1, 2, ..., g)$. If the index of the current L_i is less than g, then continue to step (4), or skip to step (5) directly;
4. The duplication part between L_i and other frequent item sets is $x = ((L_i \cap L_{i+1}) \cup, ..., (L_i \cap L_g))$, so the duplication part should be removed as $L = (F_1, ..., F_i = L_i - x, ..., L_g = L_g - x)$ from all item sets in L where the duplication part has ever occurred. Then go back to step (3);
5. New g unique frequent item sets $F = \{F_1, F_2, ..., F_g\}$ are obtained, abandon attribute values of all the F_i in F and keep the attribute names of $F_i(i = 1, 2, ..., g)$;
6. The union of all item sets in F, $F_1 \cup ... \cup F_i \cup ... \cup F_g$, is the features set FA selected by Apriori.

Select Features from Second Angle. In Section A, Apriori is used to select features set FA. However the features selected by Apriori are from the frequent item sets angle, frequent item sets are those whose support is larger than or at least equal to minimum support threshold min_support. But image that, there may exist a number of special features whose support values are a little bit smaller than min_support in original data, which are also strong relevant to classification result and the class label C. Apriori may ignore these special features because it selects features only according to the support value. In this case, we select another features set FM from another angle that we select features set by measuring the strength of the relationship between these attributes and class label C. Then the union of FA and FM is the target features set MAF in the first dimension reduction module in PAM.

We use MIC to measure the correlation between each feature in the original features set X consisting of n attribute variables $X = \{x_1, ..., x_n\}$ and the class label C with g classes $C = \{c1, .., cg\}$. If the MIC value between a feature x_a and the class label C is large enough, then the relationship between x_a and C can be considered as strong, which also indicates that x_a has a great impact on the classification result. MIC divides the data set D into several empty boxes according to the values of x_a and C, and obtains a grid G which allows the occurrence of empty partitioned grids [14]. When G is divided into i*j grids (i rows and j columns), the MIC value between x_a and C is defined as following (9) according to (1):

$$MIC(x_a, C|D) = \max_{i \times j < B(m)} \left\{ \arg\max I(x_a, C, D|G, i, j) / \log_2 \min(i, j) \right\} \tag{9}$$

$I(x_a, C, D|_G, i, j)$ expresses the mutual information between x_a and C. As mentioned above, the data D has been discretized before it is inputted in to PAM, so the mutual information between x_a and C can be figured out as following (10):

$$I(x_a; C) = \sum_{x \in x_a} \sum_{c_b \in C} p(x, c_b) \log \frac{p(x, c_b)}{p(x)p(c_b)} = I(C; x_a) \qquad (10)$$

The highest mutual information between x_a and C is $I^*(x_a, C, D, i, j) = $ arg max $I(x_a, C, D|_G, i, j)$ according to the principle of MIC [14]. Then the characteristic matrix between x_a and C can be defined as (11).

$$Ma(x_a, C|D)_{i,j} = I^*(x_a, C, D, i, j)/\log_2 \min(i, j) \qquad (11)$$

According to (1), MIC between x_a and C can be simplified into (12).

$$MIC(x_a, C|D) = \max_{i \times j < B(m)} \{Ma(x_a, C|D)_{i,j}\} \qquad (12)$$

The strength of relationship between x_a and class label C can be measured by the value of $MIC(x_a, C|D)$.

In this module of PAM, besides Apriori, we also use MIC to select features. What is our idea of selecting features set by MIC is that if the MIC value between x_a and C is large, then x_a can be considered having a strong relationship with c, which also means that this feature x_a has an important impact on the classification result and this feature x_a can be selected. Given training set D, class label C, original attribute variables set $X = \{x_1, ..., x_n\}$, the detailed steps of selecting features set by MIC algorithm are as follows:

1. Calculate MIC value between class label C and each x_a in X, $MIC(x_a, allcase|D)$, $a = 1, ..., n$.
2. Sort the corresponding attribute variables according to MIC values in descending order.
3. Select an appropriate number of features in the step 2 to form FM.

Features in FM are all comparatively strong related to C selected by C. It is also feasible to select a little bit more attributes according to their MIC values, but if excessive attribute variables are introduced, such as taking all attribute variables instead of filtering, the performance of PCA in the next module will be affected, which can be confirmed in Part 4.

In the first module of PAM, the Apriori algorithm in Section A and the MIC algorithm in Section B select attribute variables set relevant to classification result from two different angle respectively, and then we take a union of the two features sets FA ∪ FM as the target features set MAF in first module. This step can eliminate a lot of irrelevant attribute variables and improve the efficiency of the later steps in PAM on one hand. On the other hand, taking a union of two features sets found by two algorithms from different angle can ensure the found features are selected comprehensively.

In summary, the target attribute variables set in first module of PAM is MA, the union of FA and FM. We let v represent the dimension of MAF (v << n), that is, MAF has v attribute variables. The original n dimension has been reduced to v dimension. Usually, v is less than a third of n. We eliminate some attribute columns whose attributes are not in MAF from D, which can give us a filtered data set D-MAF, then we use D-MAF and MAF as input to the second module of PAM.

3.3 Features Extraction

After the features pre-selection process in Sect. 3.1, we obtain data set D-MAF and attribute variables set MAF, which are used as the input of PCA in this Section. Data pre-processing including zero-mean and z-score which can normalize all the vectors to unity norm are performed on D-MAF before using PCA to reduce dimension.

When using PCA to reduce the dimension of original high-dimensional data, for example, the v dimensional data D-MAF in this paper, it is necessary to determine the number of PCs k (k << v) first and foremost. The number of PCs is often determined by the cumulative contribution rate. Generally, the cumulative contribution rate is supposed to be more than 80%. Here we take 90% as the criterion for choosing how many PCs we need. The cumulative contribution rate of PCs is determined by the value of eigenvalue. To get the eigenvalue, we need to figure out the covariance matrix at first. Each column in D-MAF represents an attribute variable so the v dimensional data set D-MAF has v columns. There are m samples in D-MAF, so D-MAF has m rows. The covariance matrix of $D\text{-}MAF_{m\times v}$ [22] can be expressed as (13).

$$\sum{}_{D-MAF} = \frac{1}{m-1}(\text{D-MAF})(\text{D-MAF})^T \tag{13}$$

Here we let the eigenvalues of $D-MAF_{m\times v}$ be $\lambda = \{\lambda_1 \geq \lambda_2 \dots \geq \lambda_j\}$ $(j \leq v)$, we can figure out the eigenvalues λ and eigenvectors υ according to (14).

$$\sum{}_{D-MAF} \upsilon = \upsilon\lambda \tag{14}$$

The cumulative contribution rate of the former j PCs of $D-MAF_{m\times v}$ can be figured out by $\frac{\sum_{i=1}^{j}\lambda_i}{\sum_{i=1}^{v}\lambda_i}$ [23]. When $\frac{\sum_{i=1}^{j}\lambda_i}{\sum_{i=1}^{v}\lambda_i}$ is larger than or equal to 90%, that is, when the cumulative contribution rate of the former j PCs of $D-MAF_{m\times v}$ is larger than or equal to 90%, j is the number of PCs we need, k. The bigger the corresponding eigenvalue of an eigenvector is, the more important the projection direction is. We sort these eigenvectors in a descending order according to their corresponding eigenvalues, and select the first k eigenvalues to form the projection matrix $P_{v\times k}$.

The attribute variables set MAF and data set $D-MAF_{m\times v}$ obtained by the first step of PAM are used as input of PCA. The steps of PCA feature extraction in this paper are as follows:

1. Data pre-processing including zero-mean and z-score which can normalize all the vectors to unity norm are performed on $D-MAF_{m\times v}$;
2. Find covariance matrix by $\sum_{D-MAF} = \frac{1}{m-1}(\text{D-MAF})(\text{D-MAF})^T$;
3. The eigenvalues matrix λ and eigenvectors matrix υ of covariance matrix are obtained according to (14);

4. Sort these eigenvectors in descending order according to their corresponding value of eigenvalues, and select the first k components satisfying the requirement that the cumulative contribution rate is larger than or equal to 90% to form the PCs matrix $P_{v \times k} = [v1, v2, \ldots, vk]$;

5. After projected by $P_{v \times k}$, the dimension of $D-MAF_{m \times v}$ is reduced to k from v, and the k-dimensional data set generated by PCA is $D-PCA_{m \times k} = D-MAF_{m \times v} \times P_{v \times k}$ $(k << v)$.

The original v dimensional attribute variables set is mapped to a lower dimensional space after the five steps above, and the redundancy between the new k features set which we use PX to represent is greatly reduced with the class capability well preserved at the same time. After PCA, we obtain a new lower dimensional data set $D-PCA_{m \times k}$ and new k dimensional features set PX (k << v). These k features in PX have lost the physical meaning after PCA, so we can't use the former attribute names to represent these new k features and we let PX = $(p_1, \ldots, p_k)$, pi represent the i-th new feature generated by PCA. However, the features extracted by PCA are not always strongly related to the class label C, so the logical third and last step of PAM algorithm uses MIC algorithm again to select features most relevant to C from the features set PX generated by PCA.

3.4 Features Re-selection

We select the MIC algorithm again in the third and final module in PAM to eliminate some features from PX, which are irrelevant to the classification result in further. After this step, a low dimensional features set Final-X strong relevant to class label C with low redundancy is obtained, which is used as the input of the following Bayesian network structure learning algorithm.

For building Bayesian network structure easier in the subsequent step, here we discretize data in $D-PCA_{m \times k}$. The mutual information between each feature $p_d(d = 1, 2, .., k)$ in PX and class label C can be computed as (15).

$$I(p_d; C) = \sum_{p \in p_d} \sum_{c_b \in C} p(p, c_b) \log \frac{p(p, c_b)}{p(p)p(c_b)} = I(C; p_d) \qquad (15)$$

The highest mutual information between p_d and C is $I^*(p_d, C, D-PCA, i, j) = \arg \max I(p_d, C, D - PCA|_G, i, j)$, then the characteristic matrix between p_d and C is $Ma(p_d, C|D-PCA)_{i,j} = I^*(p_d, C, D - PCA, i, j)/\log_2 \min(i, j)$, so the MIC between p_d and C can be figured out as (16) according to (12).

$$MIC(p_d, C|D-PCA) = \max_{i \times j < B(m)} \left\{ \frac{\arg \max (\sum_{p \in p_d} \sum_{c_b \in C} p(p, c_b) \log \frac{p(p, c_b)}{p(p)p(c_b)})}{\log_2 \min(i, j)} \right\}$$

$$(16)$$

Sort these k features $p_d(d = 1, 2, ..., k)$ in PX according to their corresponding MIC values with C, $MIC(p_d, C|D-PCA)$, in a descending order. The higher the corresponding MIC value of p_d is, the stronger the relation between p_d and C is and the more important impact p_d has on classification result.

First, we eliminate those whose MIC values are very close to zero, here other proper specific elimination rules are also acceptable. Because features may be the parent node or may be the child node of the class label C. Image that if we only select one feature besides C in this step to build Bayesian network, and this feature happens to be the child node of C, of course, this is an extreme situation. As mentioned above, Bayesian network uses DAG to represent the relations between nodes and to reason the classification result, if there is only one child node of the class label node in DAG and not any parent nodes of C, the reasoning accuracy will be low. In this case, a little bit more features can be chosen here to build DAG to preserve the classification accuracy. The exact number of chosen features can be decided by performing some specific experiments or by analyzing some graphs. In Sect. 4, we choose the second way, choosing features by analyzing graphs. But introducing excessive features is also not advisable, which not only weakens the dimension reduction efficiency but also reduces the learning speed of the subsequent structure learning algorithm. We choose p features from PX here.

After this step of PAM, we obtain the final p dimensional target features set Final-X. We eliminate columns whose features are not in Final-X from $D-PCA_{m \times k}$, then we obtain the p dimensional data set D-Final$_{m \times p}$. The low-dimensional data set D-Final$_{m \times p}$ and features set Final-X are used as the input of various of Bayesian network structure learning algorithms to learn the structure of Bayesian network efficiently. We choose SA algorithm combined with K2 algorithm as the Bayesian network structure learning algorithm in this paper. However, the structure learning algorithm is not the focus of this paper so it is not introduced in detail.

4   Experiments

Experiment data set: The background of this paper is based on building a Breast Cancer prediction model by the Bayesian network. So the experiment data set D we use is a breast cancer data set, consisting of real breast cancer samples and healthy sample, which are all from about 120,000 volunteer people in several cooperate hospitals in China. To protect privacy of these hospital, their names are not released without their permission in this paper. D has about 120,000 samples, 1 class label allcase with 2 classes {allcase = 0 healthy, allcase = 1 breast cancer}, and 122 attribute variables in D can make up a 122 dimensional attribute variables set X = {x1, x2, .., x122}. There are only 320 breast cancer samples in D, that is allcase = 1, and about 110,000 samples, allcase = 0. So the experiment data set D we use is a high-dimensional and extremely imbalanced data set with numerous samples. To avoid the over-fitting problem of the Bayesian network, some pre-processing operations such as data discretization and using SMOTE method to solve the imbalanced data problem should be performed on this data set D before it is used as input of our following experiments. After pre-processing the original breast cancer data D, there are about 240,000 samples in D, 122 attribute variables X = (x1, .., x122) and 1 class label allcase. Now the breast cancer samples (allcase = 1) to healthy samples (allcase $= 0$) is 1 to 1. We select randomly 15% data, about 35,000 samples, as the training set data TrainS from D. To ensure that the verify process is comprehensive and accurate, we select three different number of testing sets TestS1, TestS2, TestS3 from D, and 6% of data each time, about 14,000 samples each time. We verify each experiment with all three testing sets. The samples in training set to testing set is around 7 to 3.

Experiment design: Three experiments are designed to prove the feasibility of the proposed PAM algorithm in reducing dimension. A brief introduction of these experiment is given here, in part A, the first experiment, the features subset F1 processed by PAM algorithm is used to build Bayesian network B1. In the second experiment in part B, we skip the first step of PAM in which MIC combined with Apriori is used to select the preliminary subset of features, and we use MIC directly to sift out our target features subset F2 from the low dimensional features set generated by PCA, and learn Bayesian network B2 from F2. In the third experiment of part C, we directly build a Naive Bayes network B3 using the original 122 features of TrainS, and at the same time we use the features set which is generated by PAM on TrainS in the first experiment to build the Naive Bayes network B4 as a comparative experiment. Three experiments learn their Bayesian networks on the same training set TrainS, and verify their networks with same three testing sets TestS1, TestS2, TestS3. Here we evaluate these Bayesian networks B1, B2, B3, B4 by the accuracy rate. We verify each Bayesian network with all three testing sets, in other words, each network is tested three times with TestS1, TestS2, TestS3.

4.1 Experiments 1

In this part, the first experiment, the features subset F1 processed by PAM algorithm is used to build Bayesian network B1. We first use Apriori to select attribute variables set FA strong relevant to the class label allcase from the original 122 dimensional attribute variables set X in TrainS. With 2 classes in allcase, those whose values of allcase equal to 0 are the healthy samples and those whose values of allcase equal to 1 are the breast cancer samples. Here we divide the TrainS into 2 data subsets D $=$ (D0, D1) according to their allcase values. All samples in Di belong to class where the values of allcase are equal to i (i $=$ 0, 1). Apriori first finds frequent item sets in different data subsets, D0 and D1, respectively. We let the minimum support threshold min_support be 0.5 here. According to the steps of finding frequent item set of size 1 by Apriori [17], the frequent item set of size 1 L1 found in the training set D1 where values of allcase are equal to 1 can be regard as an item set containing high-frequency and common attribute variables in most of breast cancer samples. Likewise, the frequent item set of size 1 L0 found in the training set D0 where values of allcase are equal to 0 can also be regard as an item set containing high-frequency and common attribute variables in most of healthy samples. We delete the intersection of L0 and L1, the common features part in L0 and L1, then we can get L1-(L1 ∩ L0) and L0-(L1 ∩ L0), which represent the attribute variables set specific for breast cancer samples and the set specific for healthy samples respectively. Attribute variables in FA have a great impact on the risk of breast cancer. Given the limited space available, here we only show L0-L1 ∩ L0, L1-L1 ∩ L0 and FA in Table 1:

Table 1. Attribute variables set selected by Apriori.

Features set	Features and corresponding values
(L1-L1 ∩ L0)	['N74', 1], ['N81', 1], ['N201AEOE', 1], ['N208', 1], ['A', 1], ['C', 1], ['BAAAA', 1], ['fm', 1]
(L0-L1 ∩ L0)	['N74', 0], ['N81', 0], ['fm3', 2], ['N201AEOE', 0], ['N208', 0], ['A', 0], ['C', 0], ['BAAAA', 0], ['fm', 0], ['bmi', 1]
FA	['N74', 1], ['N81', 1], ['N201AEOE', 1], ['N208', 1], ['A', 1], ['C', 1], ['BAAAA', 1], ['fm', 1] ['N74', 0], ['N81', 0], ['fm3', 2], ['N201AEOE', 0], ['N208', 0], ['A', 0], ['C', 0], ['BAAAA', 0], ['fm', 0], ['bmi', 1]

Because only the attribute names of these items are need here instead of the attribute values, so FA is {N74, N81, fm3, N201AEOE, N208, A, C, BAAAA, fm, bmi}.

Ten attribute variables are selected by Apriori from the original 122 dimensional attribute variables set X to form FA. These 10 attribute variables found by Apriori are high-frequency, but the possibility that in original data there may exist a number of special features with support values a little bit smaller than min_support but are also strong relevant to classification result and the class label C can't be ruled out. These special attribute variables are easily eliminated in Apriori for their a little bit smaller support values. To ensure that the attribute variables set MAF found in the first module of PAM is as comprehensive and accurate as possible, beside Apriori, here we also use MIC to select attribute variables set FM from a different angle of measuring the strength of relations between attribute variables and class label. We compute MIC value between each attribute variable $x_a(a = 1, 2, .., 122)$ in X and allcase, $MIC(x_a, allcase|TrainS) = \max_{i \times j < B(m)} \{Ma(x_a, allcase|TrainS)_{i,j}\}$, according to the (1) and (12) and measure the strength of relation between each attribute variable x_a and the class label allcase according to the MIC value. To ensure the attribute variables set FM selected by MIC algorithm is as accurate as possible, three data subsets with different numbers are sampled randomly from TrainS, training set of 10% of TrainS (DataRatio = 0.1), training set of 50% of TrainS (DataRatio = 0.5) and training set of 100% of TrainS (DataRatio = 1). We compute MIC values on the three different subsets and arrange these 122 attribute variables according to their corresponding MIC values in a descending order, the following Table 2 shows part of these attribute variables (Space lacks for a detailed representation of all the 122 features).

From Table 2 we can see that the order and content of the first 18 attribute variables (underlined part in Table 2) are stable and unchanged, while the order of attribute variables in the latter part is affected by the values of DataRatio of three different data subsets, that is to say, the order of attribute variables in the latter part fluctuate greatly with different number of data subsets, so the first 18 attributes are selected by MIC algorithm to form the attribute variables set. Here it is also feasible to select more attributes than 18 according to the MIC values, but if excessive attributes are introduced, such as taking all 122 attributes instead of filtering, the performance of PCA in the next step will be affected, which can be confirmed in Experiment 2 of part B. So the FM in this

Table 2. Features sorted by the descending order of corresponding MIC values.

Different sizes of the training set	Variables sorted by the descending order of corresponding MIC values
DataRatio = 1	**N81, N72AEOE, age, C, BX, fm, N410, N74, N411, fm3, A, N655OE, N208, N204AEAA, height, bretum, mensage, BAAAA**, bmi, CAAAA, AAAAA, weight, N203, N5, N202, DAAAA, N122, EAAAA, N201AEOE, N207AEAE, family
DataRatio = 0.5	**N81, N72AEOE, age, C, BX, fm, N410, N74, N411, fm3, A, N655OE, N208, N204AEAA, height, mensage, bretum, BAAAA**, bmi, AAAAA, CAAAA, weight, N203, N5, N202, DAAAA, N122, EAAAA, N201AEOE, family, N207AEAE
DataRatio = 0.1	**N81, N72AEOE, age, C, BX, fm, N410, N74, N411, fm3, A, N208, N655OE, N204AEAA, height, bretum, mensage, BAAAA**, CAAAA, bmi, AAAAA, weight, N203, DAAAA, N5, N202, EAAAA, N122, N201AEOE, N207AEAE, bage, N8, family

part is {N81, N72AEOE, age, BX, C, fm, N410, N74, N411, fm3, A, N655OE, N208, N204AEAA, height, mensage, bretum, BAAAA}. The union of FA and FM, {N81, N72AEOE, age, BX, C, fm, N410, N74, N411, fm3, A, N655OE, N208, N204AEAA, height, mensage, bretum, BAAAA, N201AEOE, bmi}, is the attribute variables set MAF selected in the first module of PAM. Compared with 122 attributes in the original data set, the MAF has only 20 attribute variables. Some attribute columns of TrainS whose attributes are not in MAF should be eliminated, which can give us a filtered data set D-MAF. Then D-MAF and MAF are inputted into the next module in PAM.

In the second module of PAM, PCA is employed to extract features from D-MAF. Before the PCA features extraction step, data pre-processing operations including zero-mean and z-score which can normalize all the vectors to unity norm should be performed on D-MAF. Here we choose 90% as minimum threshold of the cumulative contribution rate. When the cumulative contribution rate of first j PCS arrives 90%, j is the target number of PCs. On our D-MAF data set, j is fourteen here. After PCA, we obtain a 14 dimensional data set D-PCA1 and features set PX with 14 features. Because features in the low-dimensional space have lost the original physical meaning after PCA, so the names of original attribute variables can no longer represent these new features. Here we use simple numbers (from 0 to 13) to express these 14 new features, ['0', '1', '2', '3', '4', '5', '6', '7', '8', '9', '10', '11', '12', '13']. However, features extracted by PCA are not always strongly related to the class label allcase, so the logical third and last step of PAM algorithm are introduced to use MIC algorithm again to select features most relevant to allcase from the features set PX generated by PCA. MIC in this module selects features in much the same way as what is done in the first module of PAM. To ensure the features selected by MIC here in this module is as accurate as possible, three data subsets with different numbers are also sampled randomly from D-PCA1 when computing the MIC values, training set of 10% of D-PCA1 (DataRatio = 0.1), training set of 50% of D-PCA1 (DataRatio = 0.5) and training set of 100% of D-PCA1 (DataRatio = 1). MIC value between each feature and class label allcase is computed on the three different subsets, and then we sort these 14 features according to their corresponding MIC values in a descending order, the following Table 3 and Fig. 2 show the order of these features.

Table 3. Fourteen features sorted by their corresponding MIC values in a descending order.

The number of training samples	Several features sorted by the descending order of corresponding MIC values
DataRatio = 1	0, 1, 11, 12, 8, 2, 4, 13, 10, 5, 9, 6, 3, 7
DataRatio = 0.5	0, 1, 11, 12, 8, 2, 4, 13, 10, 5, 9, 6, 3, 7
DataRatio = 0.1	0, 1, 11, 12, 8, 2, 4, 13, 10, 5, 9, 3, 6, 7

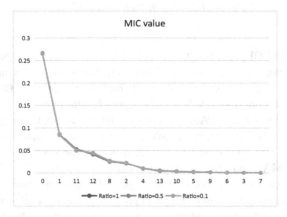

Fig. 2. MIC values between 14 features and allcase in 3 different data sets.

From Table 3 we can see that even if computed in three different sizes of data sets, order of these 14 features is still same with no fluctuations occurring, which means that the strength of relations between these 14 features and allcase measured by MIC are basically correct and stable here. Figure 2 shows that the difference between the MIC value of the first feature '0' and the subsequent features is large, which means that '0' is the most relevant feature to allcase on the one hand. On the other hand, aiming at this phenomenon, two assumptions can be made here that '0' may be the parent or child of allcase in Bayesian network. If '0' is the child of allcase, only taking '0' and allcase as nodes to build Bayesian network will lower the accuracy of classification. In this case, we can not only take '0' as Final-X, and more features should be taken to form Final-X. The difference between the MIC value of '12' and value of it is subsequent feature '8' is also large, besides, '8' and its subsequent features can be ignored because their MIC values are less than 0.03, which are too small to be chosen. Here other proper threshold value can also be accepted besides 0.03. We choose the comparatively ideal features set ['0', '1', '11', '12'] as our target features set Final-X or F1 for all concerned, those columns in $D-PCA1$ whose column names are not in F1 are eliminated, then we can get the data set D-Final1. D-Final1 and F1 are used as the input of the Bayesian network structure learning algorithm. In this paper SA and K2 algorithms are chosen to learn Bayesian network structure B1. The detailed parameters of SA in experiment are recorded in Table 4.

Table 4. Some detailed parameters of SA.

Parameters	Values
Minimum temperature	1
Initial temperature	2000
Loop count in each isothermal temperature	7000
Attenuation parameter	0.99

4.2 Experiments 2

In second experiment, the first module of PAM is skipped, we only use PCA to process the original 122 dimensional training set TrainS directly and then use MIC to select features set most relevant to the class label allcase as target set F2 from the low dimensional space generated by PCA. As what is done in part A, data pre-processing including zero-mean and z-score should be performed on TrainS before PCA is used to reduce dimension. Here we still set the minimum threshold of support min_support to be 90%. We find that j is 74 when the first j PCs arrives at a cumulative contribution rate of 90% when we figure out the eigenvalues and eigenvectors of the covariance matrix, these 74 features form the features set PX.

A features set PX and a 74 dimensional data set D-PCA2 are obtained after PCA. But not all features in PX are strong related to the classification result and the class label allcase, hence the third and last module in PAM is introduced where MIC is used to select features relevant to allcase from PX. We still select three data sets here with different number of samples to compute the MIC values between these 74 features and allcase, training set of 10% of D-PCA2 (DataRatio = 0.1), training set of 50% of D-PCA2 (DataRatio = 0.5) and training set of 100% of D-PCA2 (DataRatio = 1). Limited by space, here we only present the orders of the top 10 features whose MIC values with allcase are in top 10 on three different number of data sets in Table 5. The orders of the top 8 features on three different data sets are found stable and not changed according to the experiment result, and the ninth feature begins to change on the different data sets.

Table 5. The top ten features selected according to MIC values.

The number of training samples	Several features sorted by the descending order of corresponding MIC values
DataRatio = 1	'2', '15', '1', '18', '6', '12', '57', '14', '33', '68'
DataRatio = 0.5	'2', '15', '1', '18', '6', '12', '57', '14', '68', '33'
DataRatio = 0.1	'2', '15', '1', '18', '6', '12', '57', '14', '33', '68'

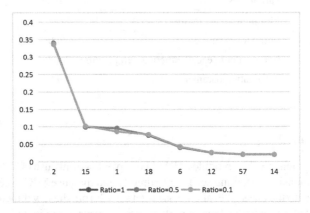

Fig. 3. MIC values between 9 features and allcase in 3 different data sets in 2nd experiment.

We plot these stable eight points on Fig. 3, the curve of '12' and the latter features is basically stable with no obvious fluctuation. The difference between the MIC value of '6' and the value of '12' is large and the MIC values of '12' and the latter features are less than 0.05, which can be regard as irrelevant to the class label allcase here. So here we only take ['2', '15', '1', '18', '6'] as our ideal features set Final-X2 and F2. Then we select those columns in $D-PCA2$ whose column names are in F2 are selected to form the data set D-Final2. D-Final2 and F2 are inputted to the SA and K2 algorithms to learn the Bayesian network structure B1. The detailed parameters of SA in this experiment are showed in Table 6.

Table 6. Some detailed parameters of SA in the second experiment.

Parameters	Values
Minimum temperature	1
Initial temperature	2000
Loop count in each isothermal temperature	7000
Attenuation parameter	0.99

4.3 Experiments 3

In this experiment, at first, we directly build a Naive Bayes network B3 using the original 122 features of TrainS, and at the same time we use the features set ['0', '1', '11', '12'] which is generated by PAM on TrainS in the first experiment to build the Naive Bayes network B4 as a comparative experiment. In Naive Bayes network, attribute variable nodes are assumed to be independent from others, so these nodes are not connected in Naive Bayes structure. Each attribute variable node only is connected to the class label node in the structure. Given limited space, the structure of B3 and B4 are not presented here. The Bayesian network structures of B1 and B2 are shown in Fig. 4 and Fig. 5 respectively.

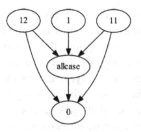

Fig. 4. B1 DAG.

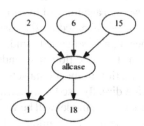

Fig. 5. B2 DAG.

We compare the accuracy of Bayesian networks B1, B2, B3 and B4 on the three same test sets TestS1, TestS2, TestS3, the results are shown in tabular form in Table 7. Considering that PCA is used to process the training set TrainS in both first and second experiments, so here we multiply the test sets of these two experiments by the projection matrix of the corresponding training sets of first and second experiments respectively. Test sets having identical distribution with their corresponding training sets are obtained to test B1 and B2. In the third experiment, the training set of B3 is not projected, so the test sets of B3 are not processed either. But the training set of B4 are the same training set of B1, which is processed by PAM. So the three test sets of B4 also should be same as the test sets of B1. The accuracy of these four Bayesian networks are shown in Table 7.

Table 7. Results of experiments.

Bayesian network	Test 1	Test 2	Test 3	Average accuracy
B1	0.98505	0.98471	0.98566	0.98514
B2	0.88512	0.88613	0.88863	0.88662
B3	0.77864	0.77369	0.77552	0.77595
B4	0.97392	0.97667	0.97535	0.97531

We can find that B1 is the most accurate model from Table 7, which proves our analysis above that F1 is an ideal features set and the PAM algorithm is feasible. The accuracy of B2 is comparatively low in the second experiment, we directly use PCA to process the original TrainS with 122 features and then use MIC to select 5 features from the low space generated by PCA to learn B2. In other words, B2 is learned from features processed by PAM with first step skipped, which is a common practice in many former methods using PCA to reduce dimension. We analyze the phenomenon carefully and get an explanation for the low accuracy of B2. There are a large number of attributes irrelevant to classification results or having little to do with breast cancer in the original attributes set. If the original attributes set is directly inputted into PCA instead of being processed and selected by the first step of PAM or other feature selection methods, then most of the features generated by PCA for building the Bayesian network may also be irrelevant to classification results, which will result in the low accuracy of B2. From this aspect, the low accuracy of B2 also proves that the feature selection step in PAM, which uses MIC algorithm and Apriori algorithm to select features, are important before the PCA dimension reduction step.

The average accuracy of the original Naive Bayesian network B3 is only about 77%. Naive Bayesian is based on the assumption that feature nodes must be independent of each other, this is so-called condition independence assumption. But in most reality applications the condition independence assumption is not met. There is a lot of redundant information in most data sets used to construct Naive Bayesian network, if these features are directly used to construct Naive Bayesian model without being processed, the accuracy of the model will be very low. On the other hand, the low accuracy of B3 also proves that there is a fair amount of redundant information in our original data set. The accuracy of the comparative Naive Bayes model B4 is around 97%, which proves that there does exist a large amount of redundant information in our data and the redundancy between attribute variables is greatly reduced by PAM. To a large extent, the condition independence assumption of Naive Bayes can be meet by PAM, which improves the performance of Naive Bayes. The original 122 features are reduced into 4 features by PAM, and at the same the high accuracy of the model can still be promised.

5   Conclusion

From several experiments, we prove that PAM can significantly reduce the dimension of the nodes set and at the same time the original classification characteristics is also well maintained. There are three main novel contributions of PAM, (i). Instead of adopting only a single method to reduce dimension, PAM is a hybrid dimension reduction method which absorbs respective advantages of Feature selection methods and Feature extraction methods. (ii). Features pre-selection step is introduced before PCA, in which features irrelevant to classification result are eliminated to improve the quality of the Principal Components (PCs) found by PCA. (iii). Features pre-selection step consists of two algorithms from two different angle, and two features set relevant to classification result are found by the two different algorithms respectively, we adopt the union of the two sets to make complementary advantages. We hope that this proposed hybrid approach, PAM, will gives readers more inspiration.

Acknowledgment. This work was supported in part by the National Key Research and Development Program of China under Grant No. 2016YFC0901303.

References

1. Rodrigues, P.P., Santos, D.F., Leite, L.: Obstructive sleep Apnea diagnosis: the Bayesian network model revisited. In: Proceedings of the IEEE Symposium on Computer-Based Medical Systems, pp. 115–120 (2015)
2. Wang, X., Sontag, D., Wang, F.: Unsupervised learning of disease progression models (2014)
3. LóPez, M., RamíRez, J., GóRriz, J.M., et al.: Automatic tool for Alzheimer's disease diagnosis using PCA and Bayesian classification rules. Electron. Lett. **45**(8), 389 (2009)
4. Exarchos, K.P., Exarchos, T.P., Bourantas, C.V., et al.: Prediction of coronary atherosclerosis progression using dynamic Bayesian networks. In: Conference Proceedings: Annual International Conference of the IEEE Engineering in Medicine and Biology Society, pp. 3889–3892 (2013)
5. Marshall, A.H., Hill, L.A., Kee, F.: Continuous Dynamic Bayesian networks for predicting survival of ischaemic heart disease patients. In: IEEE 23rd International Symposium on Computer-Based Medical Systems (CBMS 2010), Perth, Australia, 12–15 October 2010. IEEE (2010)
6. Karabatak, M.: A new classifier for Breast Cancer detection based on Naïve Bayesian. Measurement **72**, 32–36 (2015)
7. Desantis, C., Ma, J., Bryan, L., et al.: Breast Cancer statistics, 2013. CA Cancer J. Clin. **64**(1), 52–62 (2014)
8. Fan, L., Strasser-Weippl, K., Li, J.J., et al.: Breast Cancer in China. Lancet Oncol. **15**(7), e279–e289 (2014)
9. Tyrer, J., Duffy, S.W., Cuzick, J.: A Breast Cancer prediction model incorporating familial and personal risk factors. Stat. Med. **23**(7), 1111–1130 (2004)
10. Huang, S., Li, J., Ye, J., et al.: A sparse structure learning algorithm for Gaussian Bayesian network identification from high-dimensional data. IEEE Trans. Pattern Anal. Mach. Intell. **35**(6), 1328–1342 (2013)
11. Li, S., Zhang, J., Sun, B., et al.: An incremental structure learning approach for Bayesian network. In: Control & Decision Conference. IEEE (2014)
12. Li, X.-L.: Bayesian network classification text algorithm based on EM- PCA. Electron. Qual. (2015)
13. Adedigba, S.A., Khan, F., Yang, M.: Dynamic failure analysis of process systems using principal component analysis and Bayesian network. Ind. Eng. Chem. Res. **56**(8), 2094–2106 (2017)
14. Reshef, D.N., Reshef, Y.A., Finucane, H.K., et al.: Detecting novel associations in large data sets. Science **334**(6062), 1518–1524 (2011)
15. Zhang, Y.-h., Hu, Q.-p., Zhang, W.-s., et al.: A novel Bayesian network structure learning algorithm based on maximal information coefficient. In: Proceedings of the 5th IEEE International Conference on Advanced Computational Intelligence, pp. 862–867. IEEE Press (2012)
16. Agrawal, R., Srikant, R.: Fast algorithms for mining association rules. In: Proceedings of the 20th VLDB Conference, pp. 487–499 (1994)
17. Zhao, H.Y., Cai, L.C., Li, X.-J.: Overview of association rules Apriori mining algorithm. J. Sichuan Univ. Sci. Eng. **24**(1), 66–70 (2011)
18. Camacho, J., Picó, J., Ferrer, A.: Data understanding with PCA: structural and variance information plots. Chemometr. Intell. Lab. Syst. **100**(1), 48–56 (2010)
19. Jiang, Q., Yan, X., Huang, B.: Performance-driven distributed PCA process monitoring based on fault-relevant variable selection and Bayesian inference. IEEE Trans. Industr. Electron. **63**(1), 377–386 (2015)

20. Wu, X., Kumar, V., Quinlan, J.R., et al.: Top 10 algorithms in data mining. Knowl. Inf. Syst. **14**(1), 1–37 (2008)
21. Song, Z.: Research on the algorithm Apriori of mining association rules. J. S. Cent. Coll. Natl. (Nat. Sci.) (2003)
22. Fan, J., Sun, Q., Zhou, W.X., et al.: Principal component analysis for big data (2018)
23. Xingjia, T., Xiufang, Z., Science, S.O., et al.: Multi-dimensional Bayesian network classifiers based on ICA. Electron. Sci. Technol., 501–511 (2014)

Indoor Activity Recognition by Using Recurrent Neural Networks

Yu Zhao[1], Qingjuan Li[1], Fadi Farha[1], Tao Zhu[2], Liming Chen[3],
and Huansheng Ning[1(✉)]

[1] University of Science and Technology Beijing, Beijing 100083, China
ninghuansheng@ustb.edu.cn
[2] Nanhua University, Beijing 421001, China
[3] Ulster University, 601, Belfast BT48 7JL, UK

Abstract. Because of the development of the ageing population, most countries are facing an increasingly serious pension resources problem. With the development of Internet of Things, the integration of smart home and smart retirement provides a new solution for the new smart home for the elderly, to achieve the elderly to intelligently support the elderly. This paper is based on the development of this background, mainly to solve the problem of indoor activity recognition of the elderly, so as to prepare for the construction of smart medical care. The specific research process is to process the sensor data collected from the smart environment, identify different activities using RNN, LSTM and GRU models with strong ability to process time series data, realize the target of activity recognition.

Keywords: Activity recognition · Smart environment · Recurrent Neural Network

1 Introduction

Along with the development of the ageing population, we are facing an increasingly serious supporting resource problem for the elderly. Among the many problems, the problem that the old-age resources obviously do not keep up with the expansion rate of the aging population is particularly prominent, because it makes the quality of life of most elderly people declined. This problem is manifested in the following: as the aging population continues to increase, more and more elderly people will have to live alone and cannot get timely help assistance in the event of an accident. Now it is a very prominent problem how to let the elderly live more comfortably in their later years [1]. Smart home provides a new solution to relieve this stress. As a sensor-based system, smart homes are designed to create an intelligent, safe and comfortable environment for the elderly and the disabled [2]. In traditional smart house, cameras are used to get the information of the residents to offer appropriate service. But the video includes too much personal information and if the video is leaked, there may be great trouble in the future. For protecting personal information, we use the various sensors to collect the information about the activity of the residents and the environment. Because the

© Springer Nature Singapore Pte Ltd. 2019
H. Ning (Ed.): CyberDI 2019/CyberLife 2019, CCIS 1138, pp. 205–215, 2019.
https://doi.org/10.1007/978-981-15-1925-3_15

information of a single sensor is limited, the data collected from the sensor side only has the time and the name of the specific triggered sensor, which does not contain too much personal information. With the improvement of the speed of computer operation, processing data and feedbacking to personal side have been easier, which makes the data analysis faster and greatly improves the user experience. Now the various sensors have slowly replaced the cameras in recording the activity information.

Nowadays, the integration of smart home and smart retirement provides a new solution for the new smart home for the elderly [3]. At the same time, the remarkable progress on Machine Learning methods has opened up a wide variety of applications through combining different types of sensors and machine learning techniques for data acquisition and processing, respectively [4].

It has become more and more popular to combine the machine learning and sensor technology to settle the problem on human activity recognition (HAR), because the machine learning can achieve the goal that computers can learn to settle the problem like person using numerous training data. To achieve this goal, we will deploy simple and various sensors to build a smart home to collect the activity information of the residents for training an activity-recognized model. Because in the most time the same activity has the same features, we can use large amounts of data to train a model which can automatically recognize activity by analyzing the data from the online sensors. If we want to build a such model, we first need to determine whether activities will be recognized and use various sensors to build a smart environment to get a rich dataset for training the model to learn the features on these specific activities. Washington University (WSU) prepared a laboratory to study human living activities and behaviors continually to recognize human activities in throughout a whole day [5]. We refer their datasets and environment setting to build a smart environment to collect data.

In the process of building a dataset, we not only need to deploy simple and various sensors in a smart home, but also need to record what activities happen when the activity happens, if not we cannot match the sensor sequences and the labels. Only when we carefully record the activity can we get a less error label and a high accurate model. In the process of the training model, several methods and related parameters will be used to improve the accuracy of model. After training the model well, the model can deal with the data from the smart environment and analyze what activities have happened.

In this paper, we have investigated the accuracy of Recurrent Neural Network (RNN) and the improved models of RNN in detecting and recognizing the indoor activities of the residents such as drinking, eating, bathing, etc.

The paper is organized as follows. Section 2 presents an overview of activity recognition models and related work in machine learning techniques and introduces Recurrent Neural Network model (RNN) [6], Long Short-Term Memory model (LSTM) [7] and Gated Recurrent Unit model (GRU) [8]. Section 2.1 describes the datasets that were used and explains the results in comparison to different models. Section 2.3 discusses the outcomes of the experiments. Finally, Sect. 3 gives suggestions for future work.

2 Related Work

In previous research, activity recognition models have been classified into data-driven and knowledge-driven models. The data-driven models are able to handle uncertainties and temporal information but require large datasets for training the model [9]. In this process, in face of the large datasets, what we need to do is to preprocess the data and make label in advance. Then we can train the model to learn the features in the datasets to mine the features. The major challenge of the data-driven model is to how to get large datasets. If the size of the datasets is too small, the learning ability of the model is limited and it is very likely that overfitting has occurred, which will lead to the wrong recognition results [10]. The knowledge-driven approach is to acquire sufficient knowledge of personal preferences in advance by applying knowledge engineering and management technology methods [4]. But the knowledge-driven model needs enough known regulation, which is more difficult to get than data-driven model in the most time. In this paper, we use large datasets to train the model to learn how to recognize the activity automatically, which is a data-driven approach.

Using sensors to collect data for activity recognition has been proposed in many years ago. In 1988, Moser has developed a system to control the basic residential comfort systems [11]. Later more and more researches on HAR using sensors have been published. In terms of sensor placement, we deploy multiple types of sensors on the specific objects. Multiple sensors have different triggering sequences, and different triggering sequences may represent different activities. So, we can observe what kind of activity each sequence represents, sum up the rules, and label the training data [12].

Many methods have been used for activity recognition, such as Decision Trees [13], Nearest Neighbor (NN) [14], Support Vector Machines (SVM) [15] and different boosting techniques.

Recurrent Neural Network have a recursive structure between their neurons, making it well able to process information with contextual inputs [8]. This unique structure makes it have the memory function, which can retain the information of the previous time, and is suitable for dealing with time series problems, such as speech recognition, natural language processing and other context-related situations [16]. Some activities only require one or two sensors to be triggered to determine the action. However, some complex activities or simple activities involving multiple items may involve multiple sensors, and also have a certain relationship with the order in which the sensors are triggered. Different trigger sequences may symbolize different activities. It is necessary to consider the triggering of the sensor and the order in which the sensor is triggered.

Because of the special feature of RNN, many researchers have used this method to resolve activity recognition [4, 17, 18]. In [17], the datasets have different representation forms. The paper analyzes which data form will get a better result and determine the last-fired sensor data which are the data received from the sensor that was fired last will significantly improve the precision of the model.

In [4], the paper addresses window size problem which is the length of the sensors that should be considered to identify an activity and recognize the interrupted sensor data which occur in the same location when other activity is happening.

In [17], the paper survey the different Machine Learning techniques for Human Activity Recognition, which compares three different methods. In this paper, the survey result shows that CT-PCA reduces error rate by 15%, CNN Architecture reduces computational complexity by 16% and LSTM increases human activity recognition accuracy by 38%.

Compared with these methods, we segment the data in the process of training model according to the activity recording and use the sliding windows approach to analyze the data after finishing model training. We choose the RNN model and the improved model of the RNN-LSTN and GRU models to train model and improve the accuracy of the three models using different neuron numbers and learning rates.

2.1 Dataset

To design a proper experimental scenario for getting more activity information, we observe the datasets of the WSU which has been opened on the Internet. In this dataset, they describe the place of the sensors in detail. After summarizing the regulation in the dataset, we design several most representative activities to detect the basic situations of the residents, which can reflect the activity track of the residents. The representative activities include drinking, washing, eating, opening the refrigerator, turning on the light, opening the door, using computer, watching TV and cooking.

To detect the occurrence of the drinking, we deploy the touch sensor, tilt sensor and height sensor on a cup. When someone take the cup and hold it to drink, the sensor will be triggered. Even if different people have different drinking water habits, the important action features can be captured by these sensors and then we can recognize the activity through the sensor sequence. According to similar idea, we design different activity environment. When we wash hands, we will touch the faucet, so the touch sensors deployed on the faucet will detect the occurrence of the washing. When we eat, the weight of the bowl and the orientation of the bowl will be changed, and we can use the touch sensor, tilt sensor, height sensor and weight sensor to collect the information of the bowl. The reed switch can detect the object getting far away, which will be triggered when the refrigerator is opened, so we choose the reed sensor to detect whether the refrigerator is opened. We can use the similar method to detect whether the door is opened. To detect the occurrence of the turning on the light, using computer and watching TV, we can deploy a touch sensor on the switch of the light, computer and TV. When we cook, we will turn on the switch of the gas tank, so we can use the touch sensor to detect whether we turn on the switch and the infrared sensor to detect the position of the resident. The correspondence between sensors and the activities is in the Table 1.

After setting up the experimental scene, we recruit volunteers to do these activities in the experimental scene according to their favor and their own habits, so the experiment data will be greatly different between different volunteers, which is meaningful for the model scalability. The model which is trained by these data will have better scalability, because these data contains the features of these activities, but the sensor sequences of the specific activity will be a little different among them, which will help the model recognize more activity information of different people. Figure 1 is the partial dataset.

2019/4/29,16:22:52,cupM,ON,
2019/4/29,16:22:52,cupI,ON,
2019/4/29,16:22:54,cupM,ON,
2019/4/29,16:22:55,cupT,ON,
2019/4/29,16:22:56,cupI,ON,
2019/4/29,16:22:57,cupM,ON,
2019/4/29,16:22:58,cupT,ON,
2019/4/29,16:22:58,cupI,ON,
2019/4/29,16:23:14,bowlT,ON,

Fig. 1. Figure 1 is the partial dataset

What's more, what is more important is that we need to record the current label for later labeling data. We use supervised learning methods to train model, so the labeling is one of the most important parts in this process, because if the label don't match the data, then the model will learn the wrong relation between the training data and the label.

Table 1. The correspondence between sensors and the activities

Activity	Sensors
Drinking	Touch sensor, tilt sensor, height sensor
Washing	Touch sensor
Eating	Touch sensor, tilt sensor, height sensor, weight sensor
Opening the refrigerator	Reed switch
Turning on the light	Touch sensor
Opening the door	Reed switch
Using computer	Touch sensor
Watching TV	Touch sensor
Cooking	Touch sensor, infrared sensor

In the set experimental scenario, we need to take enough experimental data, and observe the data to summarize the sensor sequence corresponding to different activities and label it. In the process of data collection, we not only need to collect the data, but also need to record what occur for later label.

2.2 Data Preprocessing

The experimental data is unprocessed and raw, which contains a lot of interference infor-mation. For example, at some point, the sensors deployed in the living for is triggered, but at the same time the sensor used to detect the washing activity in the bathroom is likely to be invalid information and needs to be manually deleted, so such interfering data in the training data needs preprocessing and after the model is trained well, we can predict the maximum probability activity according to the probability value of each activities. Each activity has different trigger sequence, and the number of sensors that can be triggered by the same activity is different, so it is necessary to preprocess the collected data.

First, because possibly sensors suddenly have something wrong or environment have some impersonal interference factors, there may be some interference data in the dataset. For such data, we need to manually delete such sensor data to avoid the model learning the relationship between the wrong data and label and bring to the wrong recognition results.

Second, we need to observe all the sensor sequences to figure out the length of the longest sensor sequence. Because each sensor sequence represents an activity and each activity has different sensor sequence, the various sensor sequences all have their own features which the model will learn. To train the model to learn the feature of each activity, each sensor sequence must be selected according to the recording when the volunteers do experiments.

In addition, in the process of model training, the data length of each input model is the shortest dependent length, which is determined by the structure of the RNN model. This shortest dependent length is the length of the longest sequence in the dataset. In the actual training process, we need make sure that the length of each sequence is the same and extend the short sequences to long sequences, otherwise the model cannot start training.

After data segmentation, the data collected from the experimental scene should be encapsulated in a prescribed format. The specific data format depends on the selection of the model. In this paper, we choose RNN, LSTM and GRU models and use Keras to build the model, which is a high-level neural network API written in Python. The format of the input data is a three-dimensional list in the following format: batch_input_shape = (batch_size, time_step, input_size). Batch_input represents how many sets of data are trained at a time. Time_step indicates the length of the sensor sequence. In this experimental scenario, sometimes four sensors are required to determine an activity, and sometimes three sensors can also determine an activity, so the time_step of this activity is four. The time_step of the model is the longest sequence of all the activity sequence. Input_size represents the size of each input in each time_step. In this experiment, we used 30 sensors. With a binary representation, each input is five digits. Figure 2 shows the input data format of the RNN model and how to put the data into the model.

The last part of data preprocessing is to label the data. In this process, each sequence owns a label. The reason why we don't use the sliding window methods [18, 19] in the process of training model is that if we use the sliding window methods, the size of the dataset will be great bigger than the size of dataset we use and the model will learn the relationship between the different activities which is useless for the model. Because the process of training model is to teach the model to learn how to recognize various activities like human using the sensor data, the complexity and scale of the training set directly determines the ability of model to solve the real problem. If the sliding window methods are used to read training set, the model not only need to learn the sensor sequence of each activity, but also need to learn the relationship between different activities. Without

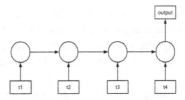

Fig. 2. The picture shows the input data format of the RNN model and how to put the data into the model.

huge training dataset, the model cannot learn the sensor sequence of each activity and the sequence of different activities, which is a complex and arduous task for training model.

After the data and model framework are prepared, iterative training can be repeated through repeated iterations. The training model is a long process, which requires a large amount of experimental data to be repeated and trained to ensure that the final accuracy and loss are satisfactory, and the final predicted output of the model is more accurate.

When predicting activity with the trained model, the data processed by the model is not pre-processed and is not segmented. To make sure that all the data can be read by the model, we use the sliding window method to read the sensor sequence one by one. The size of the sliding window is the same with the time_step. In the process of reading data each time, the overlapping part of the sliding window is two sensor data rather than one or three sensor data.

The output of the model is the probability of each activity. We will summarize the model result and the label of the test data to set proper threshold, and only when the probability of one activity is higher than the threshold, the model will make sure the occurrence of the activity. In the real world, the orders are various, and the model don't know the next sensor which is triggered, so it must combine the sensors triggered before and after. And the data format of the test dataset must be the same with the training set.

2.3   Result

Once the data is ready, we can start model training. In this experiment, three models were tried - RNN, LSTM and GRU models. In this process, we need to compare different models and find the optimal solution. Different models have a large impact on the accuracy of the experimental results. However, there are many factors affecting the accuracy of the same model, such as the number of hidden layer units, the learning rate, and the size of the packaged data. We not only compare different models, but also compare the effects of other parameters of each model on the results.

The number of neurons in the hidden unit will also largely affect the accuracy of the model. If the number of neurons is too small, the ability of the model to learn the hidden feature is poor and cannot handle complex problems. If the number of neurons is too large, it will lead to over-fitting problems, that is, the model learning ability is too strong, and it is only suitable for processing training set data. It cannot process the data of the test set except the training set or other real data which brings about that the model expansion ability is weak, and the hiding should be appropriately reduced. The number of units and the number of neurons must match the size of the problem. Too much or too little neurons all will cause problems.

Learning step size is also a key factor. Although we have the Adam optimization function, which is an optimization algorithm for automatic adaptation of learning rate, the initial value of the learning rate still plays a decisive role. Therefore, in the course of the experiment, the influence of the asynchronous length on the training precision is also compared.

At the beginning of the training, the error was generally too large, and the model was greatly adjusted accordingly. However, during the process of model adjustment, there may be oscillations at this time. The reason is that after many iterations, the error is very small and is very close to the lowest point of loss. At this time, if the learning rate η is too large, the model is likely to be over-adjusted. Assume that the error value is at the left of the lowest point at this time. After one iteration, skip the lowest point and jump to the right of the lowest point. After reaching the right side, for the same reason, it will return to the left side with greater probability, and the left and right sides will continually jump and oscillate. This will cause the model to fail to reach the desired accuracy.

In order to explore the impact of different parameters on the accuracy of the model, we conducted several experiments to find the most accurate situation. The prediction accuracy of the three models under different neuron numbers and learning rates is as follows.

Table 2. Results of RNN model

Learning rate\Neurons	4	6	8	10
0.1	Concussion	Concussion	Concussion	Concussion
0.01	94.44%	94.89%	95.74%	95.79%
0.001	94.54%	95.03%	95.87%	95.96%
0.0001	94.56%	95.04%	95.89%	96.96%

Table 2 shows the accuracy of Recurrent Neural Network model (RNN) under different neuron numbers and learning rates. If the learning rate is 0.1, the model will certainly oscillate regardless of the number of neurons. And when the learning rate is 0.01, 0.001 or 0.0001, the accuracy of RNN model under different neuron is almost the same. The number of the neuron will improve the accuracy of RNN model, but the accuracy will not always increase with more neurons, because this is likely to cause overfitting, which will cause that ability of the model to deal with actual problems is weak.

Table 3 shows the accuracy of Long Short-Term Memory model (LSTM) under different neuron numbers and learning rates. The structure of the LSTM model is more complicated than the RNN model, which greatly improves the accuracy of the model to a certain extent. In addition, the law of change in the accuracy of LSTM model under different neuron numbers and learning rates is similar with the RNN model.

Table 4 shows the accuracy of Gated Recurrent Unit model (GRU) under different neuron numbers and learning rates. Because the LSTM model adds a three-layer gating structure to the original RNN model, which greatly increases the calculation speed of the model, the GRU model simplifies the three-layer gating structure into a two-layer gating

Table 3. Results of LSTM model

Learning rate\Neurons	4	6	8	10
0.1	Concussion	Concussion	Concussion	Concussion
0.01	96.32%	96.83%	97.43%	97.48%
0.001	96.40%	96.92%	97.64%	97.81%
0.0001	96.41%	96.92%	97.65%	97.84%

Table 4. Results of GRU model

Learning rate\Neurons	4	6	8	10
0.1	Concussion	Concussion	Concussion	Concussion
0.01	96.31%	96.85%	97.44%	97.50%
0.001	96.43%	96.94%	97.64%	97.71%
0.0001	96.45%	96.94%	97.65%	97.75%

structure, which reduces the parameters that need to be trained and greatly increases the calculation speed of the model. In addition, the law of change in the accuracy of LSTM model under different neuron numbers and learning rates is similar with the RNN model and the LSTM model.

In the end, according to the three tables, it can be obviously seen that no matter which model, when the learning speed is 0.1, the model cannot converge to the global minimum point of error due to the excessive adjustment range of each matrix to be trained. Therefore, when the learning rate is 0.1, no matter which model, the oscillation occurs. This shows that at the beginning, if the learning rate of the model is set to a large extent, even if there is an optimization function to make appropriate adjustments to the learning rate during each training, the ability to adjust is very limited, and the model oscillation cannot be completely solved.

The number of the neurons will improve the accuracy of the model, but this attribute should be in keeping with the scale of the problem, because the scale of the problem decides the number of the neurons which is best suitable for the model. If the number of the model is too small, the learning ability of the model will be too weak to learn enough features, which will cause that the model cannot recognize the activities that are hidden in the sensor sequence. However, if the number of the model is too much, the learning ability of the model will almost recognize all the activities in the training data, but cannot ideally recognize the activities in the test data. The reason for this phenomenon is that the model learns some features which is not necessary and even harmful.

To get a better model, we should not only consider the advantages and disadvantages of different models, but also consider the influence of different parameters of the same model.

3   Conclusion

In this paper, we consider that sensor data has strong timing characteristics and use the Recurrent Neural Network, Long Short-Term Memory and Gated Recurrent Unit to settle the problem of activity recognition. To improve the accuracy of the model, we carry the experiment to research the accuracy of the three models under different neuron numbers and learning rates. The results of the experiments presented in this paper show that the LSMT model and GRU model have higher precision than RNN model, and GRU performs better than LSTM, because the parameters that need to be trained is smaller and own faster computing. Specific neuron numbers and learning rates of the model need to be combined with the scale of the problem and the size of the dataset.

Our future work will focus on reducing the variance on our model. Because the GRU model simplifies the hidden layer of LSTM, which decreases model calculation speed, and doesn't greatly affect the model accuracy, we will try different models and improve existing models in the future.

References

1. Udofia, E.A., Aheto, J.M., Mensah, G., Biritwum, R., Yawson, A.E.: Prevalence and risk factors associated with non-traffic related injury in the older population in Ghana: wave 2 of the WHO Study on Global AGEing and adult health (SAGE). Prev. Med. Rep. **15**, 100934 (2019)
2. Badlani, A., Bhanot, S.: Smart home system design based on artificial neural networks. In: The World Congress on Engineering and Computer Science (WCECS), vol. I.9 0 (2011)
3. Mehr, H.D., Polat, H., Cetin, A.: Resident activity recognition in smart homes by using artificial neural networks. In: 2016 4th International Istanbul Smart Grid Congress and Fair (ICSG), pp. 1–5 (2016)
4. Asghari, P., Soelimani, E., Nazerfard, E.: Online human activity recognition employing hierarchical hidden Markov models. arXiv preprint arXiv:1903.04820 (2019)
5. Tapia, E.M., Intille, S.S., Larson, K.: Activity recognition in the home using simple and ubiquitous sensors. In: Ferscha, A., Mattern, F. (eds.) Pervasive 2004. LNCS, vol. 3001, pp. 158–175. Springer, Heidelberg (2004). https://doi.org/10.1007/978-3-540-24646-6_10
6. Dmitriev, A.V., et al.: Solar Activity Forecasting on 1999-2000 by Mean of Artificial Neural Network. EGS XXIV General Assembly, The Hague (1999)
7. Hochreiter, S., Schmidhuber, J.: Long short-term memory. Neural Comput. **9**(8), 1735–1780 (1997)
8. Chung, J., Gulcehre, C., Cho, K.H., et al.: Empirical evaluation of gated recurrent neural networks on sequence modeling. arXiv preprint arXiv:1412.3555 (2014)
9. Yuen, J., Torralba, A.: A data-driven approach for event prediction. In: Daniilidis, K., Maragos, P., Paragios, N. (eds.) ECCV 2010. LNCS, vol. 6312, pp. 707–720. Springer, Heidelberg (2010). https://doi.org/10.1007/978-3-642-15552-9_51
10. Kampars, J., Grabis, J.: Near real-time big-data processing for data driven applications. In: 2017 International Conference on Big Data Innovations and Applications (Innovate-Data), Prague, pp. 35–42 (2017)
11. Mozer, M.C.: The neural network house: an environment hat adapts to its inhabitants. In: Proceedings of the AAAI Spring Symposium on Intelligent Environments, vol. 58 (1998)
12. Chowdhury, N., Kashem, M.A.: A comparative analysis of Feed-forward neural network & Recurrent Neural network to detect intrusion. In: International Conference on Electrical and Computer Engineering, Dhaka, pp. 488–492 (2008)

13. Fan, L., Wang, Z., Wang, H.: Human activity recognition model based on decision tree. In: International Conference on Advanced Cloud and Big Data, Nanjing, pp. 64–68 (2013)
14. Wu, W., Dasgupta, S., Ramirez, E.E., Peterson, C., Norman, G.J.: Classification accuracies of physical activities using smartphone motion sensors. J. Med. Internet Res. **14**, e130 (2012)
15. Patil, C.M., Jagadeesh, B., Meghana, M.N.: An approach of understanding human activity recognition and detection for video surveillance using HOG descriptor and SVM classifier. In: 2017 International Conference on Current Trends in Computer, Electrical, Electronics and Communication (CTCEEC), Mysore, pp. 481–485 (2017)
16. Islam, A.B.M.A.A., Sabrina, T.: Detection of various denial of service and Distributed Denial of Service attacks using RNN ensemble. In: 2009 12th International Conference on Computers and Information Technology, Dhaka, pp. 603–608 (2009)
17. Singh, D., Merdivan, E., Psychoula, I., et al.: Human activity recognition using recurrent neural networks (2017)
18. Noor, M.H.M., Salcic, Z., Wang, K.I.: Dynamic sliding window method for physical activity recognition using a single tri-axial accelerometer. In: 2015 IEEE 10th Conference on Industrial Electronics and Applications (ICIEA), Auckland, pp. 102–107 (2015)
19. Kim, H.G.: A structure for sliding window equijoins in data stream processing. In: 2013 IEEE 16th International Conference on Computational Science and Engineering, Sydney, NSW, pp. 100–103 (2013)

Petal-Image Based Flower Classification via GLCM and RBF-SVM

Zhihai Lu[1] and Siyuan Lu[1,2(✉)]

[1] School of Education Science, Nanjing Normal University, Nanjing, Jiangsu 210023, China
sl672@le.ac.uk
[2] School of Informatics, University of Leicester, Leicester LE1 7RH, UK

Abstract. Flower identification is a difficult problem in practice. Because there are over 250,000 different kinds of species worldwide so far. Even an experienced flower expert needs reference book to categorize a flower because of the high intra-class variation and inter-class similarity. In this study, an automatic flower recognition method was proposed based on digital image processing and artificial intelligence for petal image. Gray level co-occurrence matrix was employed as the image feature and a support vector machine was trained as the classifier. Three different kernel functions were tested and radial basis function performed best. Experimental results revealed that our approach can achieve state-of-the-art classification performance.

Keywords: Flower classification · Gray level co-occurrence matrix · Support vector machine · Radial basis function · Pattern recognition

1 Introduction

Our environment is decorated with flowers. They are beautiful and fragrant which makes people relaxed and happy. We like flowers but unfortunately, we rarely know their names. The high intra-class variation and inter-class similarity make flower classification a difficult problem. Recently, the development of computer vision technology and machine learning algorithms have achieved good performance on image classification. Digital images can be easily obtained, stored, and transmitted nowadays with the popularization of smart phones. Inspired by this, researchers started to propose flower identification methods based on digital flower images.

Saitoh, Aokiy [1] proposed a blooming flower recognition approach. Firstly, a novel flower extraction method was used to segment flower boundaries from the image background. Then, shape feature and color feature were extracted. Finally, a piecewise linear discriminant function was selected as the classifier. Their system yielded boundary detection accuracy of 97% and recognition accuracy of 90%. Nilsback and Zisserman [2] suggested to create a number of vocabularies to describe the shape, color, and texture of flower images. They also combined these features in a flexible way in order to further improve classification performance. Then, Nilsback and Zisserman [3] built a dataset of 103 flower species. They firstly segmented flower from image background by Markov

© Springer Nature Singapore Pte Ltd. 2019
H. Ning (Ed.): CyberDI 2019/CyberLife 2019, CCIS 1138, pp. 216–227, 2019.
https://doi.org/10.1007/978-981-15-1925-3_16

random field cost function and iteration. Four different kinds of features were extracted: shape, boundary, petal distribution and color. For classifier, they employed support vector machine (SVM) with multiple kernels. The kernels were optimized in weighted linear manner. Guru, Sharath [4] put forward a flower classification algorithm using texture feature and k nearest neighbors (kNN). The flower images were segmented by threshold algorithm and the gray level co-occurrence matrix (GLCM) were calculated as the features. They also compared the classification performance of kNN with three distance measurements: city block, Euclidean and cosine. Guru, Sharath Kumar [5] proposed to combine three different texture features and trained a probabilistic neural network for identification. In experiment, a dataset of 35 categories with 50 samples for each class was collected. They tested the performance of using single feature and their combinations. The proposed method achieved the best accuracy of 79% with combination of all three features. Sari and Suciati [6] proposed a combined a*b*color (ABC) approach to classify flowers. Sarkate and Khanale [7] mapped the RGB flower image into hue, saturation and value (HSV) space for segmentation. The three HSV matrixes were fed into artificial neural network (ANN) for classification. Sari and Suciati [8] employed pillbox filtering and OTSU's thresholding to remove the background. Texture feature and color features were extracted by color space transfer and segmentation-based fractal texture analysis. The kNN with cosine distance was chosen as the classification method. Experimental results showed that the accuracy of their method was 73.63%. Vasudevan, Joshi [9] adopted skeleton pruning method for flower image segmentation and modified symbolic classifier for recognition. Gurnani, Mavani [10] introduced deep neural network to flower categorization, and tested the performance of AlexNet and GoogleNet. Tao and Shih [11] employed particle swarm optimization (PSO) approach to classify angiosperm images.

In this study, a new flower recognition algorithm was put forward which combined GLCM and SVM with radial basis function kernel. The proposed method achieved state-of-the-art classification performance in terms of accuracy. The rest of this paper is organized as follows: Sect. 2 is about the flower image dataset, Sect. 3 presents our methodology including GLCM, SVM, radial basis function and K-fold cross validation, Sect. 4 gives experiment results and discussion, and Sect. 5 concludes the paper.

2 Dataset

We created our own flower petal image dataset using a 3-CCD digital camera. The dataset contains three different classes of petals: peach blossom, phalaenopsis, and hibiscus, which are all common species. There 157 samples in total with 47, 58 and 52 for peach blossom, phalaenopsis, and hibiscus, respectively. The images are resized into 400 × 400 pixel and the background is removed manually. Figure 1 presents several image samples.

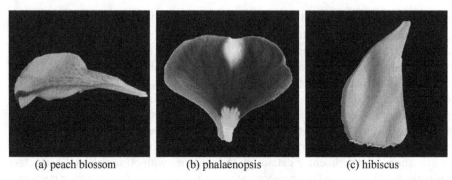

(a) peach blossom   (b) phalaenopsis   (c) hibiscus

Fig. 1. Petal samples

3 Methodology

Usually, an image classification system contains two parts: feature extraction and classification algorithm. In this work, we employed GLCM to extract feature from images and SVM for classification. The kernel for SVM was selected as radial basis function. K-fold cross validation was used to evaluate the proposed method and prevent overfitting.

There are recently deep learning methods [12–15] which can learn features instead of manual extracting features. Nevertheless, we do not choose deep learning for this task, because our dataset is too small to train that large number of parameters.

3.1 Gray Level Co-occurrence Matrix

Gray level co-occurrence matrix (GLCM) was proposed by Haralick, Shanmugam [16] to describe texture information of images. Texture is formed with the recurring of gray value in spatial position. GLCM can be seen as the joint distribution of the gray values of two pixels that suffice certain spatial location relationship [17, 18]. Given an image $I(x, y)$ and an offset (a, b), the GLCM can be calculated by:

$$GLCM(i, j) = N(I(x, y) = i, I(x + a, y + b) = j), \ \forall (x, y) \in I \qquad (1)$$

Where N denotes the sum of the pixel pairs whose gray values equal to (i, j). Figures 2 and 3 present a simple example to generate GLCM.

0	1	2	3	0	1	2
1	2	3	0	1	2	3
2	3	0	1	2	3	0
3	0	1	2	3	0	1
0	1	2	3	0	1	2
1	2	3	0	1	2	3
2	3	0	1	2	3	0

Fig. 2. Input image

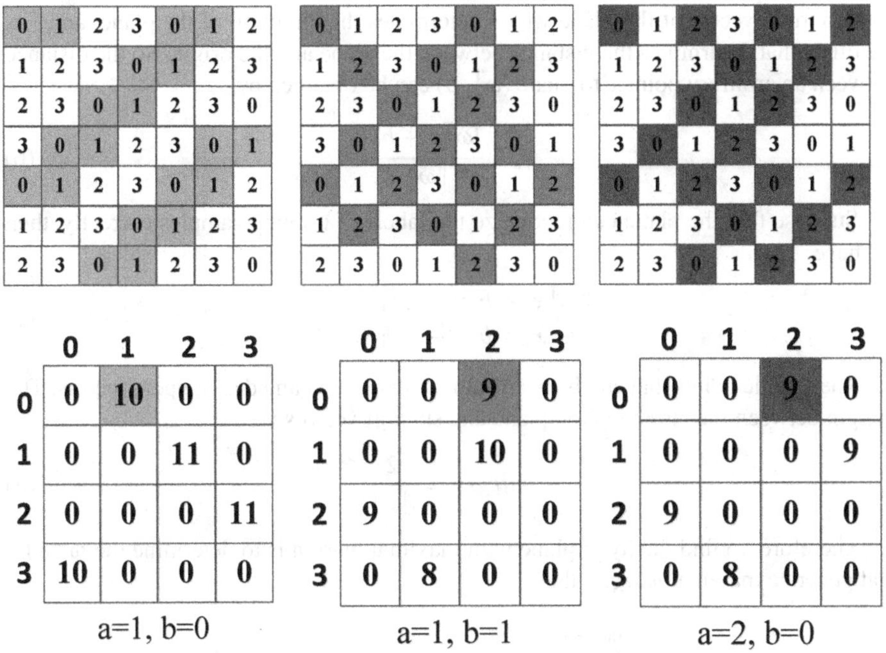

0	1	2	3	0	1	2
1	2	3	0	1	2	3
2	3	0	1	2	3	0
3	0	1	2	3	0	1
0	1	2	3	0	1	2
1	2	3	0	1	2	3
2	3	0	1	2	3	0

0	1	2	3	0	1	2
1	2	3	0	1	2	3
2	3	0	1	2	3	0
3	0	1	2	3	0	1
0	1	2	3	0	1	2
1	2	3	0	1	2	3
2	3	0	1	2	3	0

0	1	2	3	0	1	2
1	2	3	0	1	2	3
2	3	0	1	2	3	0
3	0	1	2	3	0	1
0	1	2	3	0	1	2
1	2	3	0	1	2	3
2	3	0	1	2	3	0

	0	1	2	3
0	0	10	0	0
1	0	0	11	0
2	0	0	0	11
3	10	0	0	0

a=1, b=0

	0	1	2	3
0	0	0	9	0
1	0	0	10	0
2	9	0	0	0
3	0	8	0	0

a=1, b=1

	0	1	2	3
0	0	0	9	0
1	0	0	0	9
2	9	0	0	0
3	0	8	0	0

a=2, b=0

Fig. 3. GLCM of different offsets

3.2 Support Vector Machine

Support vector machine (SVM) belongs to a classification algorithm, created by CORTES and VAPNIK [19]. SVM is based on statistical learning and it is effective for small datasets. The idea of SVM is to find the best hyperplane in feature space that maximizes the margins between the two classes [20–23]. Given a training set S_N,

$$S_N = \{(x_i, y_i)|x_i \in R^n, y_i \in \{1, -1\}\} \tag{2}$$

where x_i denotes the feature and y_i denotes the sample label, the classification is to find a mapping from feature space to label space

$$f : R^n \rightarrow \{1, -1\} \tag{3}$$

which can classify testing samples correctly. A hyperplane in space can be expressed as

$$\omega^T x + b = 0 \tag{4}$$

Where

$$\omega^T = (\omega_1; \omega_2; \omega_3; \ldots; \omega_d) \tag{5}$$

is the normal vector of the plane which determines the direction of the plane, and b is the offset that determines the distance between the plane and the origin. So the distance between an arbitrary point x to plane $(\boldsymbol{\omega}^T, b)$ can be obtained by

$$r = \frac{|\boldsymbol{\omega}^T x + b|}{\|\boldsymbol{\omega}\|} \tag{6}$$

Suppose that the plane can recognize the labels of training samples correctly, then we have

$$\begin{cases} \boldsymbol{\omega}^T x_i + b \geq +1, \, y_i = +1 \\ \boldsymbol{\omega}^T x_i + b \leq -1, \, y_i = -1 \end{cases} \tag{7}$$

The sample points that are the nearest to the plane are named as support vectors. The margin between the classification plane and support vectors is

$$margin = \frac{2}{\|\boldsymbol{\omega}\|} \tag{8}$$

Therefore, to find the hyperplane with maximal margin is to determine the $\boldsymbol{\omega}$ and b that generate maximal $margin$, that

$$\begin{aligned} &\max_{\omega,b} \frac{2}{\|\boldsymbol{\omega}\|} \\ &\text{s.t.} \, y_i(\boldsymbol{\omega}^T x_i + b) \geq 1, \, i = 1, 2, 3 \ldots N \end{aligned} \tag{9}$$

The equivalent version is

$$\begin{aligned} &\min_{\omega,b} \|\boldsymbol{\omega}\|^2 \\ &\text{s.t.} \, y_i(\boldsymbol{\omega}^T x_i + b) \geq 1, \, i = 1, 2, 3 \ldots N \end{aligned} \tag{10}$$

That's basic model of SVM, which can be obtained by convex quadratic programming. Due to its outstanding performance and efficiency, SVM is now widely applied in machine learning tasks [24–26]. In this study, we utilized SVM for flower identification.

3.3 Radial Basis Function

The basic SVM is effective to solve linear separable problems. Unfortunately, most classification problems are linear inseparable [27–30]. Kernel function was introduced to handle this problem. Kernel function maps the input from feature space to higher dimension space so that the original linear inseparable problem can be converted into linear separable one. Radial basis function (RBF) is often the first choice [31] because it can provide nonlinear mapping with less parameters. Moreover, the RBF kernel has less numerical difficulty [32–34]. The expression of radial basis function is

$$\kappa(x_i, x_j) = exp\left(-\frac{\|x_i - x_j\|^2}{2\sigma^2}\right) \tag{11}$$

where $\sigma > 0$ denotes the width of the kernel. Support vector machine with radial basis function kernel was employed as the classification algorithm.

3.4   K-Fold Cross-Validation

Cross validation (CV) is an evaluation method. CV is often applied in small datasets to evaluate the out of sample performance of the model because it enables the model to learn from the dataset sufficiently. Figure 4 illustrated the diagram of 5-fold CV.

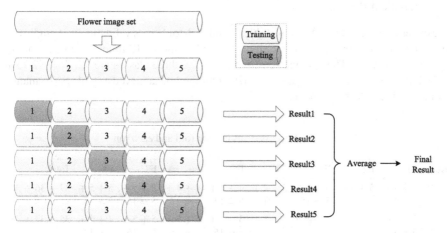

Fig. 4. 5-fold cross validation

Suppose the confusion matrix is

$$CM = \begin{bmatrix} c_{11} & c_{12} & c_{13} \\ c_{21} & c_{22} & c_{23} \\ c_{31} & c_{32} & c_{33} \end{bmatrix} \tag{12}$$

The sensitivity of all three classes is defined as:

$$S_1 = \frac{c_{11}}{c_{11} + c_{12} + c_{13}} \tag{13}$$

$$S_2 = \frac{c_{22}}{c_{21} + c_{22} + c_{23}} \tag{14}$$

$$S_3 = \frac{c_{33}}{c_{31} + c_{32} + c_{33}} \tag{15}$$

Besides, the accuracy is defined as:

$$Acc = \frac{c_{11} + c_{22} + c_{33}}{\sum_{j=1}^{3} \sum_{i=1}^{3} c_{ij}} \tag{16}$$

In this study, 8-fold cross validation was employed, so each fold contains 5 peach images, 5 phalaenopsis images, and 5 hibiscus images, respectively. We run it 10 times and compute the average and standard deviation results.

4   Experiment Results and Discussions

The algorithm was implemented on MATLAB 2018a and our experiment was done on a laptop with i5 8250U and 8 GB RAM.

4.1   Statistical Results

The experiment results are listed in Tables 1 and 2. 8-fold cross validation was employed to evaluate the performance of our algorithm and it run for 10 times. There are totally 1112 success cases. The average sensitivities for the three classes are 92.75%, 92.75% and 92.50%. The overall accuracy is 92.67%. Our method achieved a stable performance for all the three classes of petals.

Table 1. Success cases in each fold and each run

Run	F1	F2	F3	F4	F5	F6	F7	F8	Total
1	$4+4+$ $4=12$	$5+5+$ $5=15$	$5+5+$ $5=15$	$5+4+$ $5=14$	$4+4+$ $5=13$	$4+4+$ $4=12$	$5+5+$ $5=15$	$5+5+$ $5=15$	$37+36+38$ $=111$
2	$5+4+$ $5=14$	$4+5+$ $5=14$	$3+4+$ $4=11$	$4+5+$ $4=13$	$5+5+$ $5=15$	$5+5+$ $5=15$	$5+5+$ $5=15$	$4+4+$ $5=13$	$35+37+38$ $=110$
3	$5+5+$ $4=14$	$5+5+$ $5=15$	$5+5+$ $4=14$	$5+4+$ $5=14$	$3+4+$ $5=12$	$5+4+$ $5=14$	$5+5+$ $4=14$	$4+4+$ $5=13$	$37+36+37$ $=110$
4	$5+5+$ $4=14$	$5+4+$ $5=14$	$4+5+$ $5=14$	$5+4+$ $5=14$	$4+4+$ $5=13$	$4+5+$ $5=14$	$5+5+$ $4=14$	$4+5+$ $4=13$	$36+37+37$ $=110$
5	$4+5+$ $5=14$	$4+5+$ $5=14$	$5+5+$ $5=15$	$5+4+$ $3=12$	$4+5+$ $5=14$	$5+4+$ $5=14$	$5+5+$ $5=15$	$5+5+$ $4=14$	$37+38+37$ $=112$
6	$4+4+$ $4=12$	$5+5+$ $5=15$	$5+5+$ $5=15$	$5+4+$ $5=14$	$5+4+$ $4=13$	$5+4+$ $5=14$	$5+5+$ $5=15$	$4+4+$ $5=13$	$38+35+38$ $=111$
7	$4+5+$ $5=14$	$5+4+$ $5=14$	$5+5+$ $4=14$	$5+5+$ $5=15$	$5+5+$ $4=14$	$5+5+$ $5=15$	$5+5+$ $5=15$	$4+5+$ $4=13$	$38+39+37$ $=114$
8	$5+4+$ $5=14$	$5+5+$ $4=14$	$5+5+$ $3=13$	$5+5+$ $5=15$	$5+5+$ $5=15$	$4+5+$ $4=13$	$5+5+$ $5=15$	$4+4+$ $4=12$	$38+38+35$ $=111$
9	$4+5+$ $5=14$	$4+4+$ $5=13$	$4+5+$ $5=14$	$5+4+$ $5=14$	$5+5+$ $5=15$	$5+5+$ $5=15$	$5+5+$ $4=14$	$5+5+$ $4=14$	$37+38+38$ $=113$
10	$5+5+$ $5=15$	$5+5+$ $5=15$	$5+5+$ $5=15$	$3+4+$ $3=10$	$5+5+$ $5=15$	$5+4+$ $4=13$	$5+5+$ $4=14$	$5+4+$ $4=13$	$38+37+35$ $=110$
Total									$371+371+$ $370=1112$

Table 2. Performance of our method

Run	S1	S2	S3	Acc
1	92.50	90.00	95.00	92.50
2	87.50	92.50	95.00	91.67
3	92.50	90.00	92.50	91.67
4	90.00	92.50	92.50	91.67
5	92.50	95.00	92.50	93.33
6	95.00	87.50	95.00	92.50
7	95.00	97.50	92.50	95.00
8	95.00	95.00	87.50	92.50
9	92.50	95.00	95.00	94.17
10	95.00	92.50	87.50	91.67
Mean + SD	92.75 ± 2.49	92.75 ± 2.99	92.50 ± 2.89	92.67 ± 1.17

4.2 Kernel Comparison

We compared using RBF kernel with linear kernel and polynomial kernels. The comparison results are shown below in Table 4. Linear kernel performed the worst with accuracy of 85.42%, while polynomial kernel achieved accuracy of 88.08%. RBF performed the best among the three with accuracy of 92.67%. RBF uses a set of basis to implement nonlinear mapping. RBF is activated only if the input is near to one of the basis. So, it is a better option than linear and polynomial functions as the kernel (Table 3).

Table 3. Performance of using other kernel techniques

Linear kernel	S1	S2	S3	Acc
1	82.50	85.00	87.50	85.00
2	85.00	85.00	82.50	84.17
3	82.50	92.50	82.50	85.83
4	87.50	82.50	87.50	85.83
5	90.00	87.50	80.00	85.83
6	82.50	90.00	87.50	86.67
7	87.50	85.00	85.00	85.83
8	87.50	82.50	85.00	85.00
9	77.50	87.50	90.00	85.00
10	87.50	82.50	85.00	85.00

(*continued*)

Table 3. (*continued*)

Linear kernel	S1	S2	S3	Acc
Mean + SD	85.00 ± 3.73	86.00 ± 3.37	85.25 ± 2.99	85.42 ± 0.71
Polynomial kernel	S1	S2	S3	Acc
1	87.50	90.00	82.50	86.67
2	87.50	85.00	90.00	87.50
3	87.50	92.50	82.50	87.50
4	87.50	85.00	95.00	89.17
5	87.50	90.00	87.50	88.33
6	90.00	87.50	87.50	88.33
7	87.50	92.50	87.50	89.17
8	92.50	85.00	87.50	88.33
9	87.50	90.00	85.00	87.50
10	90.00	82.50	92.50	88.33
Mean + SD	88.50 ± 1.75	88.00 ± 3.50	87.75 ± 3.99	88.08 ± 0.79

Table 4. Comparison among different kernel methods

Kernel	S1	S2	S3	Acc
Linear	85.00 ± 3.73	86.00 ± 3.37	85.25 ± 2.99	85.42 ± 0.71
Polynomial	88.50 ± 1.75	88.00 ± 3.50	87.75 ± 3.99	88.08 ± 0.79
RBF (Ours)	92.75 ± 2.49	92.75 ± 2.99	92.50 ± 2.89	92.67 ± 1.17

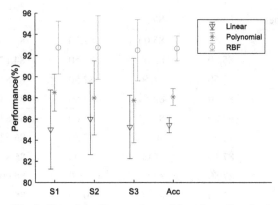

Fig. 5. Error bar of comparison of three kernel methods

Figure 5 gave a direct illustration of the errors of the three kernels. It is revealed that RBF kernel performed the best and linear kernel performed the worst in terms of average statistics. However, linear kernel sometimes can outperform polynomial kernel and polynomial kernel can reach the performance of RBF kernel in terms of S1, S2 and S3. For overall accuracy, the stability of linear kernel was the best while RBF was the worst. That's because the basis in RBF was randomly chosen, which had a bad effect on the robustness.

4.3 Comparison to State-of-the-Art Approaches

We compared our GLCM-RBF-SVM approach with traditional methods: ABC [6] and PSO [11]. We can see that GLCM-RBF-SVM yielded higher classification performance than ABC and PSO, and the robustness of our method is better than PSO. Parameter optimization is one of our future research directions. We shall try some more advanced methods like bat algorithm [35] (Table 5).

Table 5. Performance comparison

Method	Acc (%)
ABC [6]	73.63
PSO [11]	86.00 ± 1.83
GLCM-RBF-SVM (Ours)	92.67 ± 1.17

5 Conclusions

In this paper, a novel petal classification method was proposed based on GLCM and SVM. GLCM was employed for feature extraction and SVM served as the classifier. We also carried out experiments to select the best kernel function for SVM and found out RBF was better than linear and polynomial kernels. Our approach achieved an accuracy of 92.67% which outperformed two state-of-the-arts.

For future research directions, we shall continue to collect petal images and build larger datasets with more classes of petals. We will also try to employ deep learning models for petal recognition such as VGG and ResNet and introduce transfer learning to improve generalization ability. We will employ more advanced parameter optimization algorithms to further boost the classification performance.

References

1. Saitoh, T., Aokiy, K., Kaneko, T.: Automatic recognition of blooming flowers. In: Proceedings of 17th International Conference on Pattern Recognition, vol. 1, pp. 27–30 (2004)
2. Nilsback, M.-E., Zisserman, A.: A visual vocabulary for flower classification. Proceedings of the IEEE Conference on Computer Vision and Pattern Recognition, vol. 2, pp. 1447–1454 (2006)
3. Nilsback, M.-E., Zisserman: Automated flower classification over a large number of classes. In: Proceedings of the Indian Conference on Computer Vision, Graphics and Image Processing, pp. 722–729 (2008)
4. Guru, D.S., Sharath, Y.H., Manjunath, S.: Texture features and KNN in classification of flower images. IJCA Spec. Issue Recent Trends Image Process. Pattern Recognit. **37**(1), 21–29 (2010)
5. Guru, D.S., Sharath Kumar, Y.H., Manjunath, S.: Textural features in flower classification. Math. Comput. Modell. **54**(3–4), 1030–1036 (2011)
6. Sari, Y.A., Suciati, N.: Flower classification using combined a*b*color and fractal-based texture feature. Int. J. Hybrid Inform. Technol. **7**(2), 357–368 (2014)
7. Sarkate, R., Khanale, P.B.: Domain specific knowledge based machine learning for flower classification using soft computing. Int. J. Comput. Appl. **104**(1), 14–17 (2014)
8. Sari, Y.A., Suciati, N.: Flower classification using combined a* b* color and fractal-based texture feature. Int. J. Hybrid Inform. Technol. **7**(2), 357–368 (2014)
9. Vasudevan, H., et al.: Delaunay triangulation on skeleton of flowers for classification. Procedia Comput. Sci. **45**, 226–235 (2015)
10. Gurnani, A., et al.: Flower Categorization using Deep Convolutional Neural Networks, pp. 4321–4324 (2017)
11. Tao, Y., Shih, M.-L.: Classification of angiosperms by gray-level cooccurrence matrix and combination of feedforward neural network with particle swarm optimization. In: 23rd International Conference on Digital Signal Processing (DSP), Shanghai, China. IEEE (2018)
12. Govindaraj, V.V.: High performance multiple sclerosis classification by data augmentation and AlexNet transfer learning model. J. Med. Imaging Health Inform. **9**(9), 2012–2021 (2019)
13. Jiang, X.: Chinese sign language fingerspelling recognition via six-layer convolutional neural network with leaky rectified linear units for therapy and rehabilitation. J. Med. Imaging Health Inform. **9**(9), 2031–2038 (2019)
14. Tang, C.: Cerebral micro-bleeding detection based on densely connected neural network. Front. Neurosci. **13**, 422 (2019)
15. Xie, S.: Alcoholism identification based on an AlexNet transfer learning model. Front. Psychiatry **10**, 205 (2019)
16. Haralick, R.M., Shanmugam, K., Dinstein, I.H.: Textural features for image classification. IEEE Trans. Syst. Man Cybern. **3**(6), 610–621 (1973)
17. Li, W.: A gingivitis identification method based on contrast-limited adaptive histogram equalization, gray-level co-occurrence matrix, and extreme learning machine. Int. J. Imaging Syst. Technol. **29**(1), 77–82 (2019)
18. Khaldi, B., Aiadi, O., Kherfi, M.L.: Combining colour and grey-level co-occurrence matrix features: a comparative study. IET Image Proc. **13**(9), 1401–1410 (2019)
19. Cortes, C., Vapnik, V.: Support-vector networks. Mach. Learn. **20**(20), 273–297 (1995)
20. Wu, L.: An MR brain images classifier via principal component analysis and kernel support vector machine. Prog. Electromagn. Res. **130**, 369–388 (2012)
21. Wu, L.: Classification of fruits using computer vision and a multiclass support vector machine. Sensors **12**(9), 12489–12505 (2012)
22. Dong, Z.: Classification of Alzheimer disease based on structural magnetic resonance imaging by kernel support vector machine decision tree. Prog. Electromagn. Res. **144**, 171–184 (2014)

23. Wang, S., et al.: Identification of green, oolong and black teas in china via wavelet packet entropy and fuzzy support vector machine. Entropy **17**(10), 6663–6682 (2015)
24. Jayachandran, A., Dhanasekaran, R.: Brain tumor detection and classification of MRI using texture feature and fuzzy SVM classifiers. Res. J. Appl. Sci. Engg. Tech. **6**, 2264–2269 (2013)
25. Chen, Q., et al.: Feasibility study on identification of green, black and Oolong teas using near-infrared reflectance spectroscopy based on support vector machine (SVM). Spectrochim Acta A Mol. Biomol. Spectrosc. **66**(3), 568–574 (2007)
26. Ortiz, A., et al.: LVQ-SVM based CAD tool applied to structural MRI for the diagnosis of the Alzheimer's disease. Pattern Recogn. Lett. **34**(14), 1725–1733 (2013)
27. Chen, S., Yang, J.-F., Phillips, P.: Magnetic resonance brain image classification based on weighted-type fractional Fourier transform and nonparallel support vector machine. Int. J. Imaging Syst. Technol. **25**(4), 317–327 (2015)
28. Liu, G.: Pathological brain detection in MRI scanning by wavelet packet Tsallis entropy and fuzzy support vector machine. SpringerPlus **4**(1), 716 (2015)
29. Chen, M.: Morphological analysis of dendrites and spines by hybridization of ridge detection with twin support vector machine. PeerJ. **4**, e2207 (2016)
30. Chen, P.: Computer-aided detection of left and right sensorineural hearing loss by wavelet packet decomposition and least-square support vector machine. J. Am. Geriatr. Soc. **64**(S2) (2016)
31. Rashidinia, J., Khasi, M.: Stable Gaussian radial basis function method for solving Helmholtz equations. Comput. Methods Differ. Equations **7**(1), 138–151 (2019)
32. Gorriz, J.M., Ramírez, J.: Wavelet entropy and directed acyclic graph support vector machine for detection of patients with unilateral hearing loss in MRI scanning. Front. Comput. Neurosci. **10**, 106 (2016)
33. Lu, H.M.: Facial emotion recognition based on biorthogonal wavelet entropy, fuzzy support vector machine, and stratified cross validation. IEEE Access. **4**, 8375–8385 (2016)
34. Li, Y.: Detection of dendritic spines using wavelet packet entropy and fuzzy support vector machine. CNS Neurol. Disord.-Drug Targets **16**(2), 116–121 (2017)
35. Yang, X.-S.: A new metaheuristic bat-inspired algorithm. In: Nature Inspired Cooperative Strategies for Optimization (NICSO 2010), vol. 284, pp. 65–74 (2010)

A Convolutional Neural Network-Based Semantic Clustering Method for ALS Point Clouds

Zezhou Li, Tianran Tan, Yizhe Yuan, and Changqing Yin[✉]

School of Software Engineering, Tongji University, Shanghai, China
yin_cq@qq.com

Abstract. Point clouds semantic clustering is an important technique for extracting information from point clouds. A large number of point clouds datasets now can be obtained using airborne laser scanning (ALS). Existing clustering methods of point clouds are mostly focus on regular objects and restrictions, these may lead to errors when the scenario is complex or they are segmentation not semantic clustering. In recent years, CNN based methods for point clouds data have achieved good results. So this paper proposed an new method for ALS point clouds semantic clustering using convolutional neural network (CNN). In the approach, firstly the feature of every point with its adjacent points are extracted and transformed into corresponding pixel, the whole point cloud are transformed into single image. Then the image is used as input of a model that based on CNN and superpixel segmentation to achieve semantic clustering goal. We evaluate our method on the public datasets provided by the 2019 IEEE GRSS Data Fusion Contest and compared with common methods for point cloud clustering. The method performs good, shows the potential of deep-learning-based methods in semantic clustering of point clouds.

Keywords: Convolutional neural network · Semantic clustering · ALS

1 Introduction

The appearance of large-scale 3D point clouds scanning technologies such as ALS has produced many massive point cloud datasets, the datasets may include large objects like entire village or region. Using massive point clouds is an effective method for 3D spatial analyzing of objects in a variety of fields, such as digital terrain model, urban mapping, 3D modelling. The ALS can offer a description of surface on object with high density and accuracy. With the massive point clouds data, one of the challenges in leveraging the information carried by such large streams is need for efficient methods that extract the valuable information from the data. However, manually labelling such big numbers of data for supervised learning tasks is difficult, such as semantic segmentation problems because the most of real-environment point clouds have billions of points. Therefore, it is very meaningful to develop a method that extract information from point clouds with no label data. Semantic Clustering can meet this point.

H. Ning (Ed.): CyberDI 2019/CyberLife 2019, CCIS 1138, pp. 228–240, 2019.
https://doi.org/10.1007/978-981-15-1925-3_17

The current algorithms of point cloud clustering can be considered as three categories according to their characteristics, namely edge-based clustering, model-based clustering and region-based clustering. When evaluate unknown objects, such as roads, buildings and civil engineering fields, The clustering of region-based and edge-base are often applied [1]. For known objects like the standardized objects that are fit to the model-based clustering, of course it's unfit for modeling unknown objects such as composite objects and natural environment [2]. So the existing common methods cannot handle large entities such as entire regions or cities from ALS point clouds.

The convolutional neural network (CNN) [3] are enlightened by biological vision systems; these CNN have shown their superiority to extract high-level representations in recent researches, by compositions of low-level features [4]. Following the success of CNN in many 2D image unsupervised learning tasks, their results show that CNN can extract the key feature of the input image, the CNNs have great potential for extracting specific information from image, it is very helpful for unsupervised learning about image segmentation. So we want convert the 3D point clouds to 2D image so that use unsupervised image segmentation method to solve the problem of point clouds semantic clustering.

At first, we consider to convert point cloud into a single image. A point is represented by a pixel in the method we proposed instead of an image. In addition, the channel values of image are generated from information of the pixel related point and the neighborhood point. Finally we combined the superpixel segmentation and the CNN to tackle unsupervised image segmentation. The clustering model is alternately iterate CNN prediction and label redistribution by superpixel refinement to meet the specified conditions.

This rest of paper is organized as follows: Sect. 2 introduce the related work. Section 3 describe the method we proposed. Section 4 we present the experimental result and comparison. And we concludes our study and propose ideas for improvement in Sect. 5.

2 Related Work

2.1 The CNNs on Point Clouds

Deep Convolutional neural network exhibit great performance on common structured data representations such as time series and images which are all ordered vector sets. But the point clouds are unordered vector sets, it became a series of problems that challenges for deep learning. Although there are some researches of deep learning for unordered sets [5, 6] have already applied to point clouds [7], they did not consider the spatial structure of point clouds.

For solving this problem, researchers try use point-to-image [18, 19] and multi-view approaches [8], in these methods 3D point cloud are projected into 2D image so that fit the 2D CNN. Despite there are some problems of using those method, they still have shown strong performance in classification. Hu and Yuan [18] proposed to produce the digital terrain model (DTM) form the ALS point clouds by using CNN. Their methods use the point-to-image framework for every point with spatial context of point itself and the feature of its neighboring points, all of them are extracted and transformed into an image. The problem of classification of a point is equivalent to the classification of an image. They use those image to train CNN model, this method performs better than typical algorithm for DTM extraction, which is a binary classification problem. Yang and Jiang [20] improved Hu's method in image-generation and CNN model that they solve multiple-classification problem well.

More recent approaches, such as PointNet [9] and its improved version PointNet++ [21], they both using the original cloud data directly, learn from 3D spatial information. The PointNet sort a single feature vector that express the global context from a flexible number of points by max-pooling operation. PointNet++ utilizes local context by stacking multiple PointNet layers. The Dynamic Graph CNNs (DGCNNs) [10] has made further improvements, it use the graph convolution to edges of the k-nearest neighbor graph of the point clouds.

2.2 Unsupervised Deep Learning

The deep learning has already shows the ability to tackle unsupervised learning tasks. In which the train labels are produced by the data itself [11–13]. Unsupervised learning can significantly decrease the amount of labeled data for training that required for good performance in different challenges [14], so unsupervised learning is meaningful to point cloud semantic information extraction,

The GANs [15] and Autoencoders [16] have made the amazing performance on the 2D image, some results of research on point clouds using unsupervised learning have already utilized these researches. However, GANs for point clouds has some shortcomings [17]. Autoencoder-based methods for point clouds depend on similarity metrics such as the pseudo distance when operating on raw point clouds data.

3   Methods

The work flow of our method is shown in Fig. 1. We transform the point cloud semantic clustering problem into an unsupervised image segmentation problem. the image is the point clouds' corresponding feature image. The Main steps include the feature extraction for each point and its the neighboring points, the transformation of the feature into an pixel, and using the generated image into the unsupervised CNN model, finally, we project the clustering result from pixel to point.

Fig. 1. Workflow of the proposed approach.

3.1   Point Cloud to Feature Image Conversion

For cooperation with CNN for unsupervised image segmentation, we designed a Points-to-Image method. The previous method by Hu [18] converts one point into one image and extract the height relationship of points within a certain range, then train their

CNN model by these image. The method based grids works well in the DTM-extraction task or binary classification. However, only height relationships are not enough to take multi-classification problem and their method need a lot computational resource whether point-to-image conversion or training.

In order to solve the above problem, firstly, the multiple classes clustering task, we extract features not only height relations but also the intensity and return number. Secondly, for the computational cost, we do not convert one point into one image but converts the whole point cloud into only one image. In this way, point is represented by pixel instead of image.

In summary, our approach is to generate RGB images using the point's spatial information, intensity, return number and the height difference relations of its neighboring.

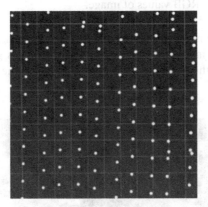

Fig. 2. The grid and points.

First of all, we using a grid which is two dimensional window to achieve the projection of point-to-pixel as shown in Fig. 2. For our data, the grid is consists of 512×512 grids, a pixel size set to 1×1 m based on the point cloud density. One grid corresponds to one pixel, If there is more than one point within a pixel, we calculate the mean value of those points, and consider they as one point to extract feature, they will be mapped back from the corresponding pixel of the result. Moreover, if there are no points in the pixel, the pixel value are calculated from the neighboring points, and the label will not be given for that pixel in the result.

To the points of pixel, we extract four features based on the point and its neighborhoods, they are the z-coordinate, intensity, return number and the standard deviation of z coordinate of neighboring points:

(1) All the features are calculated from the point $P_0 = (x_0, y_0, z_0)$ in the pixel and its neighboring points $P_i = (x_i, y_i, z_i)$ that within a sphere of radius r.
(2) The z coordinate, intensity and return number are derived from the original data. The z-coordinate are using to divide the ground and non-ground points. The intensity and return number are clearly different for objects that have different physical properties.

(3) The standard deviation of z coordinate of neighboring points shows good performance of distinguish flat objects in Chehata's research [22], and it is helpful for determining the edge between objects. We it as follow Eq. (1), the n is the number of points in sphere, the $\bar{z}$ is mean of the z coordinate of points in sphere.

$$\sigma_o = \sqrt{\frac{1}{n-1} \sum_{i=1}^{n} (z_i - \bar{z})} \tag{1}$$

(4) We normalize the four features from 0 to 1, for following computing needs, we adopt the principal component analysis (PCA) [34] method to reduce the four dimensional features to three dimensions. And Finally we convert three feature into three integers from 0 to 255 as the RGB values of image.

An example of the conversion of point clouds to feature image can be seen in Fig. 3. The left one is the original point cloud data, which includes ground, building, high vegetation, water and others (as one class), the right one is the corresponding feature image.

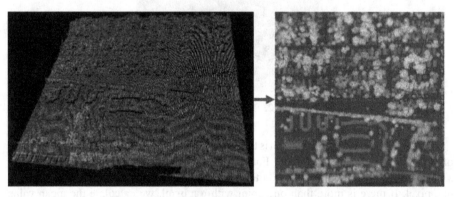

Fig. 3. The example of feature image generation.

3.2 Unsupervised CNN

For unsupervised CNN, Kanezaki [23] proposed a way of thinking, which combined the traditional machine learning and CNN. We learn from the approach and designed out own network structure.

The main steps in this method as followings:

(1) Pre-cluster image before work with CNN get the result of pre-cluster.
(2) Using the network to classify every pixel get the result of predict.
(3) Counting the result of step (2) in every clustering of step (1), set the most number of classes in one cluster as labels for all pixels of that cluster.

(4) Using the result of step (3) as ground truth to calculate the loss with result of step (2)

(5) Iterating step (2) (3) (4) until the termination condition is met.

For step (1), it's expected that spatial consecutive pixels will be give the same label. So first we take a pre-clustering method to gather pixels with similar feature which called superpixel segmentation. The result of pre-clustering will get the spatial information of some adjacent pixels with similar characteristics, and every epoch of training will use it to generate target image. In order to make the final result more accurate, in the pre-clustering phase, a sufficiently fine-grained classification is needed, Separating enough areas that ensure the performance of following classification. We choose felzenszwalb [24] to do superpixel extraction which performance is shown in Fig. 4. The felz algorithm compared with SLIC that commonly used method in superpixel extraction, it cost less computational time but hit more correct boundaries with fewer regions, and the boundaries of felz are more accurate.

Fig. 4. The example of superpixel extraction by felzenszwalb method.

For step (2), the different pixels with similar features should have the same label, we use the CNN to complete the task of classification to let the different pixels with similar feature get same label. Considered the demands of our research, we use the experience of ResNet [27] and FCN [28] for reference, and designed the network structure which shows in Fig. 5 according to the detail of our method.

For step (3), we want the amount of pixels belongs to one cluster as large as possible. So we according the result of step (1) to refine labels in superpixel. And use the result of superpixel refinement as the ground truth to calculate loss with the prediction of CNN. Finally we backpropagate loss information to optimize the CNN parameters.

These steps are somewhat antipathic so that it will never be perfectly satisfied. But after the iteration, it get gradual optimization, we can be close to the better result as possible. The termination condition is the maximum number of iterations or the minimum number of label categories.

As shown in Fig. 5, the feature image as input to neural network, and then step (2), step (3) using the superpixel segmentation result of step (1) to refine the output of neural network. It reassign the labels of all pixels in each superpixel to the category with the most occurrences in that superpixel. The result of superpixel refinement as the truth of this iteration to calculate the loss.

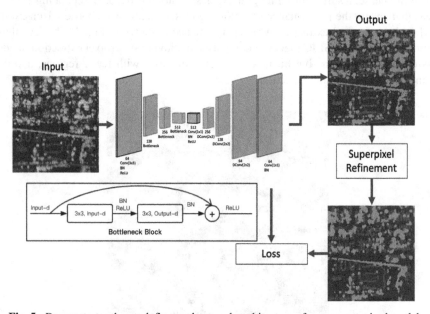

Fig. 5. Demonstrates the work flow and network architecture of our unsupervised model

In the method we use stochastic gradient descent update the parameters, we also found we can use RMSprops as the optimizer, which is an unpublished, adaptive learning rate method proposed by Geoff Hinton [29], it can reduce the number of iterations significantly, but it also reduces accuracy, in this paper, we use SGD as optimizer. We initialized the parameters by Xavier initialization [25], which samples values from the uniform distribution normalized according to the input and output layer size.

We can see the example of the clustering result calculated by our proposed method in Fig. 6.

Fig. 6. The example of the semantic clustering result.

3.3 Accuracy Evaluation

We use the result of prediction to label the ALS point clouds via the mapping relations of point-pixel conversion. And then We choose the Fowlke-Mallows index (FMI) [26] to assess the performance of our result. The FMI is defined as the geometric mean between of the precision and recall:

$$FMI = \frac{TP}{\sqrt{(TP + FP) * (TP + FN)}} \tag{2}$$

The TP is the number of true positive, the number of pair of points that belongs to the same clusters in ground truth and clustering result, FP is the number of false positive, the number of pair of points that belongs to the same clusters in ground truth but not in clustering result, FN is the number of false negative, the number of pair of points that belongs to the same clusters not in ground truth but in clustering result [30].

The indicator ranges from 0 to 1. The higher value means more similarity between two clusters.

4 Experimental Results

4.1 Experimental Data

We evaluate the proposed method on the open datasets Urban Semantic 3D (US3D) offered by 2019 IEEE GRSS Data Fusion Contest [31]. The dataset includes ALS point cloud and its semantic labels of urban city. We use part of airborne lidar data, the data include {x, y, z, intensity, return number} for every point. The aggregate nominal pulse spacing (ANPS) is approximately 80 cm.

The Fig. 7 shows the example of experimental data. The left one is the origin data with labels, the right one is the result of our method.

Fig. 7. The example of experimental data.

4.2 Results and Comparisons

We choose five samples from the datasets to evaluate the performance of our approaches by FMI. We trained the unsupervised CNN with 50 iterations for each feature image. There are two main stages in our experiment as described in the Sect. 2, the first stage is points-to-image conversion, the z-coordinate standard deviation among the four features is affected by the radius r of sphere. For the second stage, we consider the influence of the superpixel extraction to the final result, there are two key parameters in Felz method, they are *scale* and *min_size*, for *scale,* higher means larger clusters, *min_size* means the minimum component size.

We first set the to *scale* to 16, *min_size* to 32 *to* discuss the influence of the neighboring spherical radius:

- R_n means a radius of n m;
- $Sample_m$ mean the sample number m;

Table 1. The FMI value for different neighborhood radius and samples

Sample	R_2	R_4	R_6	R_8	R_{10}
Sample 1	0.673	0.682	**0.720**	0.678	0.654
Sample 2	0.622	0.644	**0.679**	0.652	0.631
Sample 3	0.703	**0.711**	0.699	0.690	0.678
Sample 4	0.711	0.714	**0.725**	0.716	0.710
Sample 5	0.680	**0.745**	0.742	0.737	0.720
Average	0.678	0.699	**0.713**	0.695	0.679

The bold number show the highest values within the different radius. According to the Table 1. We set the neighborhood radius to 6 m, and then we discuss the influence of the *scale* and *min_size*.

- minSize-j means set *min_size* to j;
- Scale-k mean set *scale* to k;

Table 2. The FMI value for different *scale* and *min_size*

	minSize-8	minSize-16	minSize-32	minSize-64	minSize-128
Scale-8	0.712	0.710	0.712	0.701	0.662
Scale-16	0.704	**0.716**	0.712	0.701	0.661
Scale-32	0.705	0.714	0.712	0.699	0.660
Scale-64	0.696	0.705	0.708	0.696	0.672
Scale-128	0.676	0.685	0.692	0.669	0.659

The bold number show the highest values, according to the Table 2, We determine the *min_size* to 16, the *scale* to 16. And the example of result can be seen in the right part of Fig. 7.

For comparison, we choose to use k-means clustering and the Gaussian Mixture model (GMM). With both of them, we use the three-dimensional features made by PCA algorithm that were discussed in Sect. 2.1 as the attributes. The comparison of FMI is show in Table 3.

Table 3. The FMI value for different methods

	K-means	GMM	Proposed
Sample 1	0.524	0.486	0.719
Sample 2	0.440	0.490	0.680
Sample 3	0.548	0.532	0.702
Sample 4	0.561	0.513	0.725
Sample 5	0.580	0.530	0.742
Average	0.531	0.510	0.714

Through the comparison of Fig. 8, we find that there are more misclassifications of K-means and GMM, but the method we proposed also ignores some details. The main reason for the shortcomings of our method is the converted feature image has lost some details.

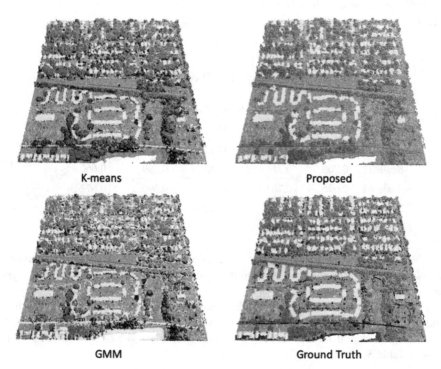

Fig. 8. Comparison of we proposed method and other methods on the US3D dataset

5   Conclusions

In recent research, the CNN was used to classify feature images or directly classify from raw point cloud, they are all supervised learning that rely on a large amount of labeled data. The motivation for our work is to propose a effective method for extracting ALS point cloud semantic information from unlabeled data.

At first, we convert all points into one image which is more efficient than transform one point to a image, then a method of superpixel segmentation is used, and finally we combine the result of superpixel segmentation and the CNN model we is used to do self-training. They jointly assigned cluster labels to pixel and updated the convolutional network for better separation of cluster. The comparison between our method and other clustering method prove the feasibility and effectivity of the method.

For future work, the approach we proposed still lack of robustness, it need subjective selection of features that transform point cloud to image. Recently, many CNN architectures like PointNet++ [21] that allow the network to learn directly from point cloud without manual feature extraction. Even though they are supervised learning, but we can learn the pre-processing and feature-extraction from them to avoid point-to-image conversion.

Acknowledgements. This work was supported by Shanghai Science and Technology Commission's Scientific Research Program (#17DZ1204903). The authors would like to thank the Johns Hopkins University Applied Physics Laboratory and IARPA for providing the data used in this study, and the IEEE GRSS Image Analysis and Data Fusion Technical Committee for organizing the Data Fusion Contest.

References

1. Tsai, A., Hsu, C.F., Hong, I., et al.: Plane and boundary extraction from lidar data using clustering and convex hull projection. Int. Arch. Photogrammetry Remote Sens. Spatial Inf. Sci. **38**, 175–179 (2010)
2. Boyko, A., Funkhouser, T.: Extracting roads from dense point clouds in large scale urban environment. ISPRS J. Photogrammetry Remote Sens. **66**(6), S2–S12 (2011)
3. Meng, L.: Application of neural network in cartographic pattern recognition. In: Proccedings 16th International Cartographic Conference, Cologne, pp. 192–202 (1993)
4. LeCun, Y., Boser, B., Denker, J.S., et al.: Backpropagation applied to handwritten zip code recognition. Neural Comput. **1**(4), 541–551 (1989)
5. Vinyals, O., Bengio, S., Kudlur, M.: Order matters: sequence to sequence for sets. arXiv preprint arXiv:1511.06391 (2015)
6. Zaheer, M., Kottur, S., Ravanbakhsh, S., et al.: Deep sets. In: Advances in Neural Information Processing Systems, pp. 3391–3401 (2017)
7. Ravanbakhsh, S., Schneider, J., Poczos, B.: Deep learning with sets and point clouds. arXiv preprint arXiv:1611.04500 (2016)
8. Su, H., Maji, S., Kalogerakis, E., Learned-Miller, E.: Multi-view convolutional neural networks for 3D shape recognition. In: Proceedings of the IEEE International Conference on Computer Vision, pp. 945–953 (2015)
9. Qi, C.R., Su, H., Mo, K., et al.: PointNet: deep learning on point sets for 3d classification and segmentation. In: Proceedings of the IEEE Conference on Computer Vision and Pattern Recognition, pp. 652–660 (2017)
10. Wang, Y., Sun, Y., Liu, Z., et al.: Dynamic graph CNN for learning on point clouds. arXiv preprint arXiv:1801.07829 (2018)
11. Lee, H., Grosse, R., Ranganath, R., et al.: Convolutional deep belief networks for scalable unsupervised learning of hierarchical representations. In: Proceedings of the 26th Annual International Conference on Machine Learning, pp. 609–616. ACM (2009)
12. Doersch, C., Gupta, A., Efros, A.A.: Unsupervised visual representation learning by context prediction. In: Proceedings of the IEEE International Conference on Computer Vision, pp. 1422–1430 (2015)
13. Gidaris, S., Singh, P., Komodakis, N.: Unsupervised representation learning by predicting image rotations. arXiv preprint arXiv:1803.07728 (2018)
14. Yang, Y., Feng, C., Shen, Y., et al.: FoldingNet: point cloud auto-encoder via deep grid deformation. In: Proceedings of the IEEE Conference on Computer Vision and Pattern Recognition, pp. 206–215 (2018)
15. Goodfellow, I., Pouget-Abadie, J., Mirza, M., et al.: Generative adversarial nets. In: Advances in Neural Information Processing Systems, pp. 2672–2680 (2014)

16. Hinton, G.E., Salakhutdinov, R.R.: Reducing the dimensionality of data with neural networks. Science **313**(5786), 504–507 (2006)
17. Han, Z., Shang, M., Liu, Y.S., et al.: View inter-prediction GAN: unsupervised representation learning for 3D shapes by learning global shape memories to support local view predictions. In: Proceedings of the AAAI Conference on Artificial Intelligence, vol. 33, pp. 8376–8384 (2019)
18. Hu, X., Yuan, Y.: Deep-learning-based classification for DTM extraction from ALS point cloud. Remote Sens. **8**, 730 (2016)
19. Yang, Z., Jiang, W., Xu, B., et al.: A convolutional neural network-based 3D semantic labeling method for ALS point clouds. Remote Sens. **9**(9), 936 (2017)
20. Wu, Z., Song, S., Khosla, A., et al.: 3D ShapeNets: a deep representation for volumetric shapes. In: Proceedings of the IEEE Conference on Computer Vision and Pattern Recognition, pp. 1912–1920 (2015)
21. Qi, C.R., Yi, L., Su, H., et al.: PointNet++: deep hierarchical feature learning on point sets in a metric space. In: Advances in Neural Information Processing Systems, pp. 5099–5108 (2017)
22. Chehata, N., Guo, L., Mallet, C.: Airborne lidar feature selection for urban classification using random forests. Int. Arch. Photogrammetry Remote Sens. Spatial Inf. Sci. **38**(Part 3), W8 (2009)
23. Kanezaki, A.: Unsupervised image segmentation by backpropagation. In: 2018 IEEE International Conference on Acoustics, Speech and Signal Processing (ICASSP), pp. 1543–1547. IEEE (2018)
24. Felzenszwalb, P.F., Huttenlocher, D.P.: Efficient graph-based image segmentation. Int. J. Comput. Vis. **59**(2), 167–181 (2004)
25. Glorot, X., Bengio, Y.: Understanding the difficulty of training deep feedforward neural networks. In: AIS-TATS (2010)
26. Fowlkes, E.B., Mallows, C.L.: A method for comparing two hierarchical clusterings. J. Am. Stat. Assoc. **78**(383), 553–569 (1983)
27. He, K., Zhang, X., Ren, S., et al.: Deep residual learning for image recognition. In: Proceedings of the IEEE Conference on Computer Vision and Pattern Recognition, pp. 770–778 (2016)
28. Long, J., Shelhamer, E., Darrell, T.: Fully convolutional networks for semantic segmentation. In: Proceedings of the IEEE Conference on Computer Vision and Pattern Recognition, pp. 3431–3440 (2015)
29. RMSprops. http://www.cs.toronto.edu/~tijmen/csc321/slides/lecture_slides_lec6.pdf
30. Fowlkes_mallow_score. https://scikit-learn.org/stable/modules/generated/sklearn.metrics.fowlkes_mallows_score.html
31. IEEE GRSS Data Fusion Contest. http://www.grss-ieee.org/community/technical-committees/data-fusion
32. Bosch, M., Foster, G., Christie, G., Wang, S., Hager, G.D., Brown, M.: Semantic stereo for incidental satellite images. In: Proceedings of Winter Conference on Applications of Computer Vision (2019)
33. Le Saux, B., Yokoya, N., Hänsch, R., Brown, M., Hager, G.D., Kim, H.: 2019 IEEE GRSS data fusion contest: semantic 3D reconstruction [technical committees]. IEEE Geosci. Remote Sens. Mag. (2019)
34. Tipping, M.E., Bishop, C.M.: Probabilistic principal component analysis. J. R. Stat. Soc. Ser. B (Stat. Methodol.) **61**(3), 611–622 (1999)

Aligning Point Clouds with an Effective Local Feature Descriptor

Xialing Feng, Tianran Tan, Yizhe Yuan, and Changqing Yin[✉]

School of Software Engineering, Tongji University, Shanghai, China
yin_cq@qq.com

Abstract. Point cloud registration is a crucial step and gaining more importance in many challenging 3D computer vision tasks including 3D reconstruction, autonomous navigation, 3D object recognition and remote sensing. In this work, we proposed a highly discriminative local feature descriptor named Local Point Feature Histogram (LPFH) for 3D point cloud registration. LPFH formulates a simple and comprehensive histogram for surface representation, which encompassed a 3D descriptor. Based on the proposed LPFH, we use Random Sample Consensus (RANSAC) algorithm to form our coarse registration stage, followed by an Iterative Closest Point (ICP) fine registration stage, these two steps form our registration algorithm. Validations and comparisons with other point cloud registration algorithms showed that LPFH is low-dimension, efficient, effective and easy to compute.

Keywords: Point cloud · Registration · Local feature descriptor · RANSAC · ICP

1 Introduction

3D point cloud data is often obtained from optical devices, including laser scanner and depth camera like Microsoft Kinect. However, there is no way to get all the surface point clouds at the same viewpoint due to the limitations of devices. Therefore, developing a method which integrate all the parts of the point clouds obtain from different viewpoint into one single model automatically is necessary. Registration is the process to align more than one point clouds by estimating the relative 3DOF transformations between overlapping point clouds, and put all scans into a common coordinate frame, so that assemble all these point clouds into a complete model. Point cloud registration is an important part of many computer vision tasks like 3D localization and mapping, object pose estimation, 3D reconstruction, reverse engineering and remote sensing.

The most widely used registration algorithms is the Iterative Closest Point (ICP) algorithm proposed by Besl and McKay [1]. ICP correct the transformation between point clouds which is needed to minimize an error metric iteratively, usually a distance between two point clouds, such as Euclidean distance. ICP algorithm received significant attention because of it's accuracy and simplicity, but it also has many drawbacks. Due to ICP's greedy nature, it's possible that ICP would converge to a local minimum, to

© Springer Nature Singapore Pte Ltd. 2019
H. Ning (Ed.): CyberDI 2019/CyberLife 2019, CCIS 1138, pp. 241–255, 2019.
https://doi.org/10.1007/978-981-15-1925-3_18

use ICP algorithm effectively and accurately, a good initial guess of the alignment is required.

Since ICP performs very well if started within a good initial alignment, to overcome the limitations of ICP algorithm, an initial registration stage is necessary. In general, point cloud registration methods that involve ICP algorithm consist of two steps: coarse and fine registration. The goal of coarse registration step is to estimate a rough alignment between point clouds, it is often done by matching features and estimating correspondences between two point clouds. As we can see that the coarse registration consists of two important steps: feature extraction and feature matching. Feature extraction plays an important role in coarse registration, it aims to extract features of point cloud surface, so that feature matching step can be done correctly.

Generally, point cloud coarse registration methods can be classified into two categories: global feature-based coarse registration and local feature-base coarse registration. The difference between global feature-based and local feature-based registration is the scope of extracted feature. Global feature-based method describes the entire point cloud model shape, the extracted features are global feature descriptors, whereas local feature-base methods encode the relative geometric feature of point cloud, the extracted information is local feature descriptors. Global feature-base methods often suffer from limited discriminative power to describe partially overlapped data, in most of the point cloud registration application scenarios, local feature-base methods are more suitable.

The considerably crucial steps involved in point cloud local feature-based coarse registration is local feature descriptors, which have a great effect on the overall performance of descriptive result. A powerful and discriminative local feature descriptor should be able to capture the geometric structure and invariant to translation, scaling and rotation at the same time. Therefore, how to extract such a good 3D local feature descriptor from 3D point cloud, especially in environment with occlusion and clutter, it is still a very hot and challenging research area so far.

Numerous local feature descriptors have been proposed by researchers, such as Spin Image [2], Point Feature Histogram (PFH) [3], Fast Point Feature Histogram (FPFH) [4], Rotational Projection Statistics (RoPS) [5], and Signature of Histogram of Orientations (SHOT) [6], etc. Most of the LRF-based local feature descriptor such as Spin Image, RoPS, Intrinsic Shape Signatures (ISS) and MeshHOG provide a descriptive description of point cloud local surface. However, these local feature descriptors encode the local spatial feature in a fair detail way, as a result, these local feature descriptors have high-dimensional feature vectors and cost too much computational time, thus can not ensure the efficiency of the overall registration process.

1.1 Related Work

Point cloud registration has been an active research topic more than two decades. Among all the point cloud registration algorithms in the literatures, ICP algorithm received significant attention since proposed in 1992. The basic concept of ICP is to minimize the Euclidean distances between neighboring points iteratively.

The traditional ICP algorithm align two point clouds precisely when there is a good initial alignment. But in practice, it is difficult to obtain without a good coarse registration process. Under the circumstances, the initial position between two point clouds is too far away from the ground truth, the traditional ICP algorithm would converge to local minimum so that the registration cannot be done. Besides, traditional ICP algorithm has the weakness of large computation. Therefore, many algorithms have been proposed to overcome the drawbacks of traditional ICP. Chen proposed a method by using the distance of the point-to-plane instead of point-to-point distance, to be the ICP algorithm's error metric [7]. This method improved traditional ICP algorithm's computation complexity slightly. However, when the point cloud surface curvature of the target point cloud changes significantly, this method may have bad performances. Zhang improve ICP algorithm's robustness by adding a outlier rejection step in the correspondence selection of the ICP algorithm [8]. Traditional ICP algorithm need to search the nearest point of each point in the source point cloud in the target cloud in every iteration, if the size of point cloud is too large, the search process would cost too much computation time. Gelfand [9] and Wu [10] improved the efficiency of iteration by down-sampling the original point clouds. Many other researchers used tree search data structure like KD-tree [11], projection [12], and point cloud invariant feature [13] to effectively find corresponding point pairs between source cloud and target cloud, thus accelerate the search phase and improve the precision of correspondence.

Due to these limitations of ICP algorithm we discuss above, before apply ICP algorithm to our point clouds, we need to provide a good initial alignment for ICP fine registration first, this pre-step is so-called coarse registration. An essential step of coarse registration is feature extraction. One of the most common way for feature extraction is to create feature descriptor. 3D point cloud feature descriptors can be classified into two groups: local feature descriptors, global feature descriptors.

For local feature descriptor, these descriptors have been introduced to encode the information of local geometric like surface normal and curvature. Johnson et al. proposed a local feature descriptor on 3D point clouds called Spin Image. The Spin Image first build a own local coordinate for each point and encode its neighboring points' feature as a pair of distances (α, β). Finally, the Spin Image is generated by accumulating the neighbors of feature points in discrete 2D bins. This descriptor is robust to occlusion and clutter [15], but the presence of high-level noise will lead to degradation of performance. Rusu et al. proposed Point Feature Histogram (PFH) [3, 16], PFH constructed Darboux frame first and use the relationship between point pairs in the support region and estimated surface normal to represent the geometric properties. PFH describe local point feature in detail, but it compares every point pair in k neighbors, so its computational complexity is $O(nk^2)$, which makes PFH inappropriate to be used in the real-time application. Rusu et al. introduced Fast Point Feature Histogram (FPFH) [4] to simplify the PFH descriptor but keep the PFH descriptive in an acceptable way. Signature of Histogram of Orientation descriptor, presented by Tombari et al. can be considered as a combination of Signatures and Histograms, it computes a repeatable Local Reference Frame (LRF) with disambiguation and uniqueness. In order to improve the accuracy of feature matching, Tombari et al. incorporated texture information to extend the SHOT to form the color version [17], i.e. SHOTCOLOR or CSHOT. The Viewpoint Feature Histogram (VFH) [57, 58] extends the idea and properties of FPFH by including additional viewpoint variance, it is a global descriptor composed of a viewpoint direction component and a surface shape component. The Clustered Viewpoint Feature Histogram [20] descriptor can be considered as an extension of VFH, which takes into account the advantage from stable object region obtained by applying a region growing algorithm after removing the point with high curvature. CVFH can produce promising result, but lacking of notion of an aligned Euclidean space causes missing a proper spatial description.

All the feature descriptors we mention above have contributed to 3D point cloud data processing a lot. Nevertheless, few of these feature descriptors can balance computation complexity between time efficiency fairly.

1.2 Contributions

Figure 1 shows the whole process of our proposed registration algorithm. The input data of our point cloud registration process is two overlapping clouds, namely the source cloud and the target cloud. Due to the limitation of sensors, it's inevitable that the data we obtained would have noise data, therefore, we first eliminate noises from source cloud and target cloud. Besides, the size of input cloud could be very large and our registration algorithm would be very time-consuming, for this consideration, we down sample our input clouds to reduce computation. Then we present our feature descriptor LPFH for point cloud surface local description. Our local feature descriptor LPFH encodes local geometric feature from different aspects, such as point density, curvature and point-to-point deviation angles. By extracting and combining these features we can obtain LPFH descriptor, the detail of feature extraction we will discuss later in this paper. LPFH describe point cloud local surface information efficiently, as will be shown in Sect. 2. Next, correspondences between source cloud and target cloud will be compute via feature matching. In this stage, we use Random Sample Consensus (RANSAC) algorithm to match correspondence feature between two cloud and produce an initial rigid transformation. At last, the initial alignment is further refined by ICP algorithm.

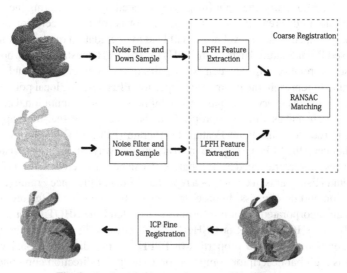

Fig. 1. Registration process proposed in this paper

The remainder of this paper is organized as follows: Sect. 2 describe our LPFH feature descriptor in detail. Section 3 presents the detail of our whole registration process. Section 4 will validate the correctness and effectiveness of our algorithm, and demonstrate our experimental results in different datasets. Conclusions and future work are discussed in Sect. 5.

2 Local Point Feature Histogram

Point cloud registration method using local feature descriptor is often done by finding correspondences between source point cloud and target point cloud. Correspondences point pairs is often generated by matching local feature descriptor. Hence, for the purpose of offering sufficient correct corresponding point pairs, a robust and descriptive local feature descriptor is necessary. In this section, the proposed point cloud local feature descriptor LPFH is introduced in detail.

2.1 Definition

In this work, we compute our LPFH descriptor by encoding the point cloud local surface feature from the following three aspects: point density, deviation angles between normal and point curvature.

For the point cloud P that contains N points $\{p_1, p_2, p_3, \ldots, p_N\}$. For each query point p_i in point cloud P, a radius r is given, and forms a sphere centered at p_i. Points within p_i's sphere excluding p_i consist of p_i's neighboring points, which are represented by $p_n^i = \left\{ p_i^j \mid j = 1, 2, 3, \ldots, k \right\}$, where k is the number of points p_i's neighborhood points. This estimation process is often performed by using Principal Component Analysis. Given a point p_i, and the neighboring points of p_i which are p_n^i. Assume that $\overline{p_i}$ is the centroid of p_n^i. So that $Cov(p_i)$ can be represent as:

$$Cov(p_i) = \begin{bmatrix} p_i^1 - \overline{p_i} \\ \ldots \\ p_i^k - \overline{p_i} \end{bmatrix}^T \cdot \begin{bmatrix} p_i^1 - \overline{p_i} \\ \ldots \\ p_i^k - \overline{p_i} \end{bmatrix} \tag{1}$$

the normal of p_i can be represented by the equation below:

$$n_i = V_0, \lambda_0 \leq \lambda_1 \leq \lambda_2 \tag{2}$$

where n_i is the normal vector of point p_i, V_0 is the eigenvector corresponding to the smallest eigenvalue λ_0.

The first local feature we considered is points density. This sub feature can be obtained by projecting 3D points onto a 2D plane. This idea is inspired by Guo [5], Johnson [2] and Yang [21]. First, we project point p_i and all it's neighborhood points to a sphere's tangent plane where perpendicular to n_i. In this case, the projective point of p_i we denoted by p_i', and projections of all the neighborhood points of p_i denoted by $p_n'^i =$

$\left\{ p_i^{\prime j} \mid j = 1, 2, 3, \ldots, k \right\}$. d^j represent the Euclidean distance between projected point and projected center point, which can be computed by the following equation:

$$d^j = \sqrt{\left\| p_i - p_i^j \right\|^2 - \left(n_i \cdot \left(p_i - p_i^j \right) \right)^2} \tag{3}$$

where $\|\cdot\|$ denoted the Euclidean distance. As we can see that $0 < d^j \leq r$. Then, we normalize the value of d^j by divide r, we define this value by ρ^j, which can be denoted by the following equation:

$$\rho^j = \frac{d^j}{r} \tag{4}$$

where the range of ρ^j is [0, 1]. Finally, we form a 2D histogram to describe the distribution of ρ^j by divide the range of ρ^j which is [0, 1] with 15 bin, each bin's value is computed by counting the number of points that it's ρ^j fall into correspondent bin. Therefore, we can obtain our first local feature by generating a sub histogram to describe point density within each sphere of point p_i. The illustration of compute the first local feature can be seen in Fig. 2.

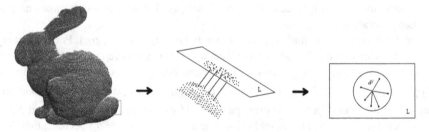

Fig. 2. The illustration of point density feature

The second feature we used in our local feature descriptor is the deviation angle between point p_i and its neighboring points. Local feature descriptors in the literature like PFH and FPFH have demonstrated that feature based on deviation angle between points has great discriminative power, [21] directly use deviation angle θ_j between point p_i and its neighboring point p_n^j as a local feature. The angle θ_j can be compute using the equation below:

$$\theta_j = arccos\left(n_i \cdot n_i^j \right) \tag{5}$$

For the consideration of simplicity and convenience, we use cos value to represent the deviation angle between point, which can be compute by the equation below:

$$cos\left(p_i, p_i^j \right) = n_i \cdot n_i^j \tag{6}$$

Compared to using deviation angle θ_j directly, our method reduces a computation step, thus improve the efficiency of feature descriptor computation. Local feature descriptor in literature compute deviation angle between every point pair in sphere, thus the time complexity is $O(k^2)$, this could have bad effect on the efficiency of algorithm. Thus, we do not calculate every point-to-point deviation angle between normal but only calculating the deviation angles between p_i and its neighboring points, so our computational complexity is reduced to $O(k)$. The range of $cos\langle p_i, p_i^j \rangle$ is $[-1, 1]$, we divide this range into 25 bins and form the second sub-histogram by counting the number of $cos\langle p_i, p_i^j \rangle$ value fall into correspondence bin. The illustration of computing second local feature is shown in Fig. 3.

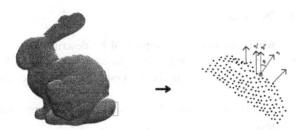

Fig. 3. Illustration of the second local feature

The third feature we use is point curvature. Point cloud surface curvature is a fundamental feature of point cloud feature extraction. In particular, curvatures like Gaussian curvature, mean curvature and principal curvature is often considered due to their descriptiveness and invariant to translation and rotation. In this work, we use mean curvature as our third local feature. The mean curvature is equal to half the sum of the principal curvatures, which is denoted by the following equation:

$$H = \frac{k_1 + k_2}{2} \tag{7}$$

where H is the mean curvature, k_1 and k_2 are principal curvatures, which can be computed by decomposing the singular value of Weingarten matrix [22].

As we presented above, the three local sub-features can be encoded as three sub-histograms. There are some Local feature descriptors like PFH generate a three-dimension histogram, each sub-histogram occupied one dimension, thus the data structure of feature descriptor is very large. Inspired by FPFH, in this work, we just concatenate the three sub-histograms into one histogram, to form our LPFH descriptor.

2.2 Analyzation

In the previous sub-section, we describe the definition and theory of our LPFH descriptor in detail. The major characteristics of LPFH are summarized as follows:

- The three sub features we use in LPFH descriptor are low-dimensional and easy to compute. Besides, for point p_i in point cloud P, we only compute the relative local feature between p_i and it's neighboring points p_n^i, but not all the point pairs within p_i's k-neighbors. This can effectively reduce the complexity of algorithm to $O(k)$.

- The LPFH descriptor describe point cloud surface from several different aspects, like point density, normal deviation and point curvature. Compared with the descriptor that only use one single aspect, LPFH descriptor is supposed to be more descriptive.
- The three sub local feature we use in LPFH descriptor are unique and all invariant to rotation and transformation. Besides, 3D projection is robust to noise according to the previous research which done by other researchers [24], the deviation between normal [3, 4] and point curvature are also proved to be descriptive [22].

3   Registration Scheme

In this section, the whole registration process will be described. First, for the input source cloud and target cloud, we need to eliminate the noise points from our input clouds. Besides, the down sample step is also necessary before enter coarse registration process.

After the above preprocess, we enter the coarse registration stage, and followed by a fine registration stage.

3.1   Preprocess

Due to the limitation, randomness error and fault of the optical devices, the point cloud data we obtained from these devices are inevitably contain noise. These noisy points would affect the precision of some basic point cloud feature estimation such as normal estimation and curvature estimation. And further affect the registration and other point cloud process stages. Therefore, in the preprocess step, we need to wipe out noise data. In this work, we use Statistical Outlier Removal [24] algorithm to solve this problem. Statistical Outlier Removal solved this problem by conducting a statistical analysis on point's neighboring points, and trimming those noisy point which do not satisfy a certain condition. This method assumes that the spatial distribution between each point and it's neighboring point can form a Gaussian distribution, and compute the mean distance and standard deviation of the distribution. For the point whose mean distance are outside the interval which is the global mean distance, would be considered as noisy point and eliminated from point cloud. Experiment shows that Statistical Outlier Removal is very effective.

Usually, huge point cloud data collected by sensors are very dense, after noise points elimination, we need to decrease the point number of input cloud, that is down sample the point cloud, for the consideration of registration efficiency. In our experiment, we used a voxel grid approach, call Voxel Grid filter. The Voxel Grid filter creates a 3D voxel grid over the input point cloud. After that, the points in each voxel would be replace with their centroid. Experiment shows that this method can approximated the point cloud surface and reduce the size of the data effectively.

3.2   Coarse Registration

As we discuss in the first section, the purpose of coarse registration is to provide an initial alignment for ICP fine registration stage.

In our coarse registration stage, the proposed LPFH descriptor of every points in source cloud and target cloud is computed, then follows the conjugate point set computation step. We applied Random Sample Consensus (RANSAC) [25] algorithm to complete the remaining of our coarse registration steps. The RANSAC algorithm repeats the following three steps, until reach a user-define maximum iteration time or convergence:

1. Select s points from source cloud S. Especially, the distance between points in S must be grater than a user-defined threshold.
2. For each point s_i in s, compare the feature descriptor similarity with every point in target cloud T, and form a list of candidate correspondence within a similarity threshold. After that, random pick one candidate as s_i's correspondence.
3. According to the correspondences we obtained in step 2, we compute the rigid transformation between source cloud and target cloud, and further compute the error metric such as distance between two point cloud and the quality of transformation.

In step 2 of the RANSAC, we use Kullback-Leibler distance (divergence) to measure the similarity between histograms. KL divergence can be defined as the following equation:

$$KL \, divergence = \sum_{i=1}^{n} \left(p_i^f - \mu_i \right) \cdot \ln \left(\frac{p_i^f}{\mu_i} \right) \tag{8}$$

where p_i^f denoted the histogram at bin i, μ_i represent mean histogram of the entire dataset at bin i.

In the third step, the error metric is evaluated using a Huber Penalty measure L_h:

$$L_h(e_i) = \begin{cases} \frac{1}{2}e_i^2 & \|e_i\| \le t_e \\ \frac{1}{2}t_e(2\|e_i\| - t_e) & \|e_i\| > t_e \end{cases} \tag{9}$$

The above steps are repeated until the error is less than the threshold or reach the maximum iteration time. During the iteration, the transformation which hold the smallest error metric would be storage and use to coarse align two point clouds.

Finally, a non-linear optimization algorithm called Levenberg-Marquardt [26] is applied. The process of preprocess and coarse registration is shown in Fig. 4.

3.3 Fine Registration

In this stage, we use the famous ICP algorithm to compute our final transformation. The ICP algorithm aims to minimize the distance between source point set S and target point set T by iteratively determining the translation vector T and rotation R parameters of 3D rigid transformation. Assuming that $S = \{S_i | S_i \in R^3, i = 1, 2, \ldots N_S\}$, T $= \{T_i | T_i \in R^3, i = 1, 2, \ldots N_M\}$. The rigid body transformation between S and T is computed, in each iteration the transformation is update, until the error metric between S and T is not greater than the user-defined threshold τ. Generally, ICP consist of the following steps:

1. Calculate the corresponding point S_i in the source cloud, so that the distance between points $\|T_i - S_i\|$ reach the minimum.
2. Compute the rigid body transformation parameters, including rotation matrix r^k and translation vector t^k, where k is the number of iterations, so that $\sum_{i=1}^{N} \|r^k S_i + t^k - T_i\|^2$ reach the minimum.
3. Transform the source cloud S with the grid body transformation parameters calculated in step 2.
4. Calculate the distance $d^k = \sum_{i=1}^{N} \|S_i - T_i\|^2$.
5. If d^k is greater than threshold τ, repeat the above five steps until $d^k < \tau$ or iteration number k reach the maximum number of iteration that user-defined.

The fine registration result of our experiment data is shown in Fig. 5.

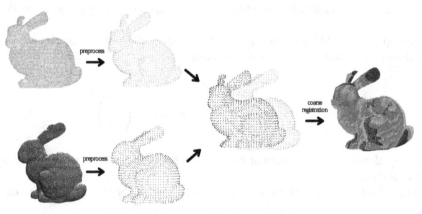

Fig. 4. Preprocess and coarse registration

Fig. 5. Coarse to fine registration

4 Experiment Result

In this section, we conduct the proposed coarse-to-fine registration scheme in different dataset. Besides, the comparison with the state-of-the-art algorithm is present. The experimental dataset we used in this section is The Stanford 3D Scanning Repository. Our algorithm is implemented with C++ 11, Point Cloud Library (PCL). All the experiments are conducted on a 3.5 GHz Intel(R) Core 3 processor with 8 GB RAM.

We use Stanford Bunny, Happy Buddha, Dragon as our validation datasets. They are shown in Fig. 6.

Fig. 6. Experiment dataset, from left to right: Stanford Bunny, Happy Buddha, Dragon

Our partially experiment result are shown in Fig. 7.

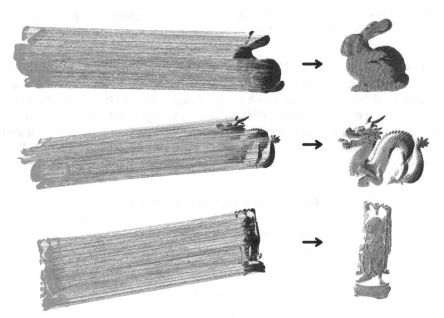

Fig. 7. Partially experiment result, left: correspondences, right: registration result

The registration results are visualization and evaluation by human eyes at first, and the root mean square error (RMSE) [27] between correspondence of source and target.

The RMSE are defined as following equation:

$$RMSE = \sqrt{\frac{\sum_{i=1}^{N_c} \left\| R_{GT} \cdot c_S^i + t_{GT} - c_T^i \right\|^2}{N_C}} \tag{10}$$

where c_S^i and c_T^i are corresponding point pair belong to source cloud and target cloud respectively. N_C is the size of the correspondence set. We can see that RMSE describe the distance between correspondence.

Figure 8 shows the RMSE tendency during our experiment and the RMSE compare between LPFH descriptor and another state-of-the-art surface representation.

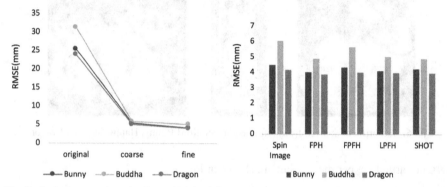

Fig. 8. Left: proposed method's RMSE in different dataset. Right: RMSE comparison between proposed method and state-of-the-art descriptor.

As we can see that our LPFH descriptor describes point cloud surface correctly and it is suitable for point cloud feature extraction. We compared LPFH with some state-of-the-art local feature descriptors such as Spin Image, FPH, FPFH and SHOT with the three different datasets, the finally RMSE is shown in Fig. 8. All the search radius of local feature descriptor is 13%. As we can see that LPFH perform quite well compared with the state-of-the-art descriptors. Meanwhile, we can see from Table 1 that our LPFH is low dimension compare to the others.

Table 1. Dimensionalities of descriptors

Descriptors	Dimensionality	Length
Spin Image	15 * 15	225
PFH	5 * 5 * 5	125
FPFH	11 + 11 + 11	33
SHOT	8 * 2 * 2 * 10	320
LPFH	15 + 25 + 20	60

Table 2 demonstrate the timing statistics of the whole registration process.

Table 2. Registration timing statistics

Dataset	Source size	Target size	T_{coarse}(s)	T_{fine}(s)	Sum(s)
Stanford Bunny	57499	59544	4.332	0.820	5.152
Happy Buddha	244860	244200	10.348	3.489	13.837
Dragon	126546	132000	7.825	2.435	10.26

5   Conclusion and Future Work

In our paper, a novel descriptive descriptor LPFH was proposed. LPFH combine a set of low-dimension geometric feature such as projection point density, deviation angle between normal and point curvature. The LPFH feature descriptor are rotation and transformation invariant and have discriminative power, which makes it a great candidate for feature extraction. Besides, based on LPFH descriptor, we implemented a coarse-to-fine registration scheme. The overall registration process is described as follow: Before the input source cloud and target cloud go into registration process, we need to do some preprocess work including noisy data elimination and point cloud down sampling. After the preprocess stage, we compute LPFH descriptor for each point of the source and target. Based on the computed LPFH descriptor, we use RANSAC algorithm to match corresponding points and compute initial alignment of the source and the target. That is our coarse registration stage. Coarse registration method based on LPFH we mention above provide a good initial alignment for fine registration. Finally, we use ICP to refine the alignment we got in the coarse registration stage and obtain final result. Experiment shows that our registration method has good performance in real world datasets.

Our feature plan is to study the robustness of our feature descriptor for multiple point cloud resolutions. This problem very challenging in the field of local feature extraction and matching. Another direction of future study is to improve the efficiency of ICP.

Acknowledgement. In this work, we like to acknowledge the Stanford 3D Scanning Repository for making their datasets available to us. And all the developers of Point Cloud Library for developing some general algorithms for us. This work is supported by Shanghai Science and Technology Commission's Scientific Research Program (#17DZ1204903).

References

1. Besl, P.J., McKay, N.D.: Method for registration of 3D shapes. In: Sensor Fusion IV: Control Paradigms and Data Structures, vol. 1611. International Society for Optics and Photonics (1992)
2. Johnson, A.E., Martial, H.: Surface matching for object recognition in complex three-dimensional scenes. Image Vis. Comput. **16**(9–10), 635–651 (1998)
3. Rusu, R.B., et al.: Aligning point cloud views using persistent feature histograms. In: IEEE/RSJ International Conference on Intelligent Robots and Systems. IEEE (2008)

4. Rusu, R.B., Nico, B., Michael, B.: Fast point feature histograms (FPFH) for 3D registration. In: IEEE International Conference on Robotics and Automation. IEEE (2009)
5. Guo, Y., et al.: Rotational projection statistics for 3D local surface description and object recognition. Int. J. Comput. Vis. **105**(1), 63–86 (2013)
6. Tombari, F., Salti, S., Di Stefano, L.: Unique signatures of histograms for local surface description. In: Daniilidis, K., Maragos, P., Paragios, N. (eds.) ECCV 2010. LNCS, vol. 6313, pp. 356–369. Springer, Heidelberg (2010). https://doi.org/10.1007/978-3-642-15558-1_26
7. Chen, Y., Medioni, G.: Object modelling by registration of multiple range images. Image Vis. Comput. **10**(3), 145–155 (1992)
8. Zhang, Z.: Iterative point matching for registration of free-form curves. IRA Rapports de Recherche, Programme 4: Robotique, Image et Vision, no. 1658 (1992)
9. Gelfand, N.: Feature analysis and registration of scanned surfaces. Ph.D. thesis. Stanford University, Stanford, CA, USA (2007)
10. Wu, Y.F., et al.: A new method for registration of 3D point sets with low overlapping ratios. Procedia CIRP **27**, 202–206 (2015)
11. Simon, D.A.: Fast and Accurate Shape-Based Registration. Carnegie Mellon University, Pittsburgh (1996)
12. Rusinkiewicz, S., Levoy, M.: Proceedings of the Third International Conference on 3D Digital Imaging and Modeling, pp. 145–152 (2001)
13. Sharp, G.C., Lee, S.W., Wehe, D.K.: ICP registration using invariant features. IEEE Trans. Pattern Anal. Mach. Intell. **24**(1), 90–102 (2002)
14. Bae, K.-H.: Evaluation of the convergence region of an automated registration method for 3D laser scanner point clouds. Sensors **9**(1), 355–375 (2009)
15. Johnson, A.E., Hebert, M.: Using spin images for efficient object recognition in cluttered 3D scenes. IEEE Trans. Pattern Anal. Mach. Intell. **21**(5), 433–449 (1999)
16. Rusu, R.B., et al.: Persistent point feature histograms for 3D point clouds. In: Proceedings of the 10th International Conference on Intel Autonomous System (IAS-2010), Baden-Baden, Germany (2008)
17. Tombari, F., Salti, S., Di Stefano, L.: A combined texture-shape descriptor for enhanced 3D feature matching. In: 18th IEEE International Conference on Image Processing. IEEE (2011)
18. Muja, M., et al.: Rein-a fast, robust, scalable recognition infrastructure. In: IEEE International Conference on Robotics and Automation. IEEE (2011)
19. Rusu, R.B., et al.: Fast 3D recognition and pose using the viewpoint feature histogram. In: IEEE/RSJ International Conference on Intelligent Robots and Systems. IEEE (2010)
20. Aldoma, A., et al.: CAD-model recognition and 6DOF pose estimation using 3D cues. In: IEEE International Conference on Computer Vision Workshops (ICCV Workshops). IEEE (2011)
21. Yang, J., Cao, Z., Zhang, Q.: A fast and robust local descriptor for 3D point cloud registration. Inf. Sci. **346**, 163–179 (2016)
22. Crosilla, F., Visintini, D., Sepic, F.: Reliable automatic classification and segmentation of laser point clouds by statistical analysis of surface curvature values. Appl. Geomatics **1**(1–2), 17–30 (2009)
23. Malassiotis, S., Strintzis, M.G.: Snapshots: a novel local surface descriptor and matching algorithm for robust 3D surface alignment. IEEE Trans. Pattern Anal. Mach. Intell. **29**(7), 1285–1290 (2007)
24. Rusu, R.B., et al.: Towards 3D point cloud based object maps for household environments. Robot. Auton. Syst. **56**(11), 927–941 (2008)
25. Fischler, M.A., Bolles, R.C.: A paradigm for model fitting with applications to image analysis and automated cartography. Commun. ACM **24**(6), 381–395 (1981). Reprinted in readings in computer vision, Fischler, M.A. (ed.)

26. Fitzgibbon, A.W.: Robust registration of 2D and 3D point sets. Image Vis. Comput. **21**(13-14), 1145–1153 (2003)
27. Rusinkiewicz, S., Marc L.: Efficient variants of the ICP algorithm. In: 3 DIM, vol. 1 (2001)
28. Zhang, J.: Multi-source remote sensing data fusion: status and trends. Int. J. Image Data Fusion **1**(1), 5–24 (2010)
29. Min, Z., Wang, J., Meng, M.Q.H.: Robust generalized point cloud registration using hybrid mixture model. In: IEEE International Conference on Robotics and Automation (ICRA), pp. 4812–4818. IEEE (2018)
30. Tong, L., Ying, X.: 3D point cloud initial registration using surface curvature and SURF matching. 3D Res. **9**(3), 41 (2010)

A Tutorial and Survey on Fault Knowledge Graph

XiuQing Wang[1] and ShunKun Yang[2(✉)]

[1] School of Computer and Communication Engineering, University of Science and Technology Beijing, Beijing 100083, China
[2] School of Reliability and Systems Engineering, Beihang University, Beijing 100083, China
ysk@buaa.edu.cn

Abstract. Knowledge Graph (KG) is a graph-based data structure that can display the relationship between a large number of semi-structured and unstructured data, and can efficiently and intelligently search for information that users need. KG has been widely used for many fields including finance, medical care, biological, education, journalism, smart search and other industries. With the increase in the application of Knowledge Graphs (KGs) in the field of failure, such as mechanical engineering, trains, power grids, equipment failures, etc. However, the summary of the system of fault KGs is relatively small. Therefore, this article provides a comprehensive tutorial and survey about the recent advances toward the construction of fault KG. Specifically, it will provide an overview of the fault KG and summarize the key techniques for building a KG to guide the construction of the KG in the fault domain. What's more, it introduces some of the open source tools that can be used to build a KG process, enabling researchers and practitioners to quickly get started in this field. In addition, the article discusses the application of fault KG and the difficulties and challenges in constructing fault KG. Finally, the article looks forward to the future development of KG.

Keywords: Fault Knowledge Graph · Key technologies · Tools · Applications

1 Introduction

The Fault Knowledge Graph (KFG) is based on the KG in the field of faults. Now it has been applied to finance [1, 2], medical [3–5], biological [6, 7], agriculture [8, 9], journalism, education, question answering [10–12], and other industries. Because KG can display the relationship between various data types and efficient and intelligent search information. However, fault areas such as mechanical engineering [13], trains [14], power grids [15], power equipment [16], etc. All of this raised the need for building KG, but there are not many KGs that actually build success. Each domain has the same place and different places in building KGs. Therefore, this paper investigates the architecture and key technologies of KGs in other fields, and also investigates the construction process and key technologies of KFG. At the same time, KGs in other fields can be used as a guide to complete the construction of KFG. The appearance of the FKG can well analyze the relationship between various faults, achieve prediction, and promote development.

© Springer Nature Singapore Pte Ltd. 2019
H. Ning (Ed.): CyberDI 2019/CyberLife 2019, CCIS 1138, pp. 256–271, 2019.
https://doi.org/10.1007/978-981-15-1925-3_19

This paper provides an overview of the FKG, summarizes its build process and key technologies, and summarizes the open source tools and applications that may be used.

The structure of this paper is as follows:

- Section 2 outlines the concept of the FKG and the reasons for its development, some key technologies, as well as the application of faults;
- Section 3 provides the architecture of the general KG and some key technologies, as well as open source tools that may be used to guide the construction of the FKG. These include steps such as data acquisition, information extraction, knowledge fusion, knowledge processing, and knowledge storage;
- Section 4 analyzes the application of the FKG and the challenges and problems to be solved in the establishment process;
- Section 5 summarizes the FKG, and predicts the future development of FKG.

2   Overview

In this section, we will describe the origins of the concept of the FKG and some of the reasons that motivated its development. We will also briefly present scope of some of its application fault devices.

The KG [17] was originally proposed by the Google knowledge graph project to enhance the google search engine and enhance the user experience. Later, the KG was applied to many fields with its advantages. However, the application of the KG to the field of failure is still relatively few. Still, there is still a certain demand for constructing KG in the field of failure. For example, [13–16] all addressed the needs of the field of failure. Therefore, in 2107, Yuan-cheng et al. provided a definition of a formal Fault Knowledge Graph [13] based on the collection of a large number of mechanical engineering fault data. The specific applications and methods used are listed below in tabular form (Table 1).

Table 1. Fault domain application and construction technology

Fault area	Fault application	Technology
Engineering machinery	Fault knowledge question-answering and troubleshooting assistance	Data-driven iterative
Train	Anomaly detection	Building consistency matrices
Power grids	Achieve automatic/semi-automatic disposal of faults	"hidden Markov model" + "punctuation-based segmentation"
Power equipment	Fault diagnosis	Entity relationship extraction + RDF

As can be seen from the above chart, the types of faults includes engineering machinery, train, power grids, power equipment and so on. In [13], Yuan-cheng LU et al. provided construction process of FKG and proposed a data-driven KG iterative automatic construction method. In [16], the author used entity extraction and relation extraction techniques and combined the relevant data of multi-source heterogeneous power equipment to construct a power equipment KG, and to improve the efficiency of power equipment management. In [15], the author used information extraction.

3   Key Technologies

The public construction process of the knowledge graph includes Data Acquisition, Knowledge Acquisition, Knowledge Fusion and Knowledge Processing [18], as shown in below (Fig. 1).

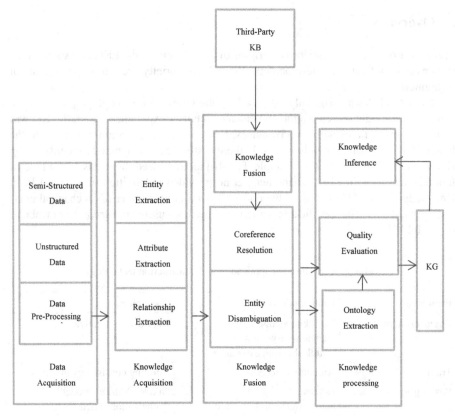

Fig. 1. The knowledge graph construction process

3.1 Data Acquisition

The first problem facing the construction of a KG is the source of the data. There are three types of data sources, which are structured, semi-structured, and unstructured. Structured data is stored in the database and tables [19]. Semi-structured data has a specific structure but is not very strict, such as xml data. Finally, unstructured data has no predefined data model information, such as publications, web pages or social media [20]. Most data sources are obtained by crawling unstructured data (such as an encyclopedia) as a data source.

As mentioned earlier, data sources can get data from web pages or extract them from databases. Then the technology needed to get data from the Internet, namely web crawler technology [21]. It described Mercator, a web crawler that is fully written in Java and can be extended. In [22], the author pointed out that web crawlers are a recursive process, and the user has to add restrictions, such as specifying the maximum number of tags or documents to retrieve, and the time limit. To improve the quality of the data, users can limit the crawl to a specific domain or file format and apply a blacklist of unwanted URLs/domains.

In [23], it provided the technology of Micro-blog Website, including depth control, breadth control and URL controller. URL controller is important. Because we have to consider whether this URL is suitable for crawling is very important for the results obtained, such as too many nodes or poor data quality, etc.

3.2 Information Extraction

KG consists of nodes and their relationships. The nodes include entities, concepts and literals. Entities are real-world individuals. Concepts represent a set of individuals with the same characteristics. Literals are strings indicated specific values of some relations. Information extraction is a kind of automatic entity extraction from semi-structured data and unstructured data [24]. Information extraction includes entity extraction, relationship extraction and attribute extraction [25].

(1) Named Entity Recognition

Named Entity Recognition was presented at the 6th MUC Conference (MUC-6, the Sixth Message Understanding Conferences) in November 1995. The core idea is to identify and classify the proper nouns needed for a given text. NER's method research is divided into three main categories: firstly, dictionary-based and rule-based methods; secondly, traditional machine learning methods, the main conditional random field (CRF) and support vector machine (SVM), Long Short Term Memory Network (LSTM), Bi-directional LSTM (BILSTM), Part-Aware LSTM (PLSTM); thirdly, Deep learning methods such as Convolutional Neural Network (CNN) [73] and Recurrent Neural Network (RNN) [74].

A rule- based approach is to manually build rules and then match the strings of those rules in the text. The most representative of these is the DLCoTrain method proposed by Collins [35]. They proposed two algorithms. One is boosting-like framework. Another described an algorithm that directly optimizes this function. Similar to this, there was also a way to automatically generate rules through Bootstrapping [36].

The machine learning based approach is implemented by a categorical approach. Mainly through the method of serialization annotation, its main models include SVM (Support Vector Machine) [37], ME (Maximum Entropy) [38], HMM (Hidden Markov Model) [39, 40], CRF (Conditional Random Field) [41].

In [85], Thanh Hai Dang et al. proposed a novel named entity model. The model uses conditional random fields and bidirectional long-term short-term memory, and the experiment works well. In [86], Hui Chen et al. proposed a simple but effective CNN-based network, the Gate Correlation Network (GRN). This network is better at capturing context information than CNN. At the same time, Parallel Recurrent Neural Networks and CNN-RNN have promoted the development of NER

(2) Relationship Extraction

The entity relationship extraction task was first proposed at the 7th Message Understanding Conference (MUC) in 1998 [60]. Many research methods for entity relationship extraction include pattern matching [61], dictionary-driven [62] and machine learning methods. At present, the study of entity relationship extraction mainly uses machine learning or deep learning. The machine learning algorithm considers the relationship extraction as a classification problem, constructs the classifier when training the corpus, and finally applies it to the category judgment of the entity relationship.

Machine learning-based entity relationship extraction methods include supervised methods, unsupervised methods, weakly supervised methods, remotely supervised methods, and open domain-oriented methods. There was a supervised method in [63] that could obtain better performance by determining the relationship of entities in a sentence by a given entity and a sentence containing a pair of entities. The disadvantage of this method is that it requires a lot of manual labeling when training data. The unsupervised learning [64] method did not require manual labeling of corpus, but the performance of relational extraction was poor. The method of weak supervised learning was bootstrapping at the earliest [65]. This algorithm is simple and easy to operate, but it uses a lot of statistical assumptions, so it is assumed whether the accuracy of sampling is established. Later, remote supervision appeared, using the alignment of the open knowledge base to automatically mark the corpus, reducing the dependence on manual annotation data, and enhancing cross-domain adaptability. However, its shortcoming is that it will bring a lot of noise to the corpus, so solving the noise has become a problem that scholars are concerned about. If there is a relationship between the two entities in the knowledge base, then the sentence containing the two entities has more or less the relationship. Open-domain-based relationship extraction is not limited to relationship categories and text classification, and does not need to be labeled corpus, which is suitable for processing large-scale network data, but the extracted results require a lot of processing, and there is no objective evaluation standard.

The above method requires the use of the NLP tool, but the NLP itself will have system errors, so when the algorithm is used later, the performance of the algorithm will be degraded. The deep learning method is applied to the relationship extraction, and the advantages of feature extraction and automatic learning are taken. SemEval-2010 task 8 [66] was used as the test standard. Deep learning based algorithms include recurrent neural networks and convolutional neural networks. The RNN model [67] set vectors and matrices, learned the meanings of propositional logic and natural language operators, but relied on traditional NLP tools to focus on semantic learning, thus reintroducing tool noise. Therefore, Zeng et al. [68] used convolutional neural networks to extract the hierarchical features of sentences and vocabulary for relational extraction, reducing the pre-marking processing of input materials. Then Nguyen et al. [69] added convolution kernels of different sizes to the convolutional layer as filters to extract more features and add position vectors. Lin et al. [70] introduced PCNN (piecewise CNN), which pooled the feature map into three segments for the two physical locations of the pooled layer of the traditional convolutional neural network, while using the attention mechanism to establish the sentence level. Selective attention to the neural model mitigated the problem of mislabeling.

Remote supervisory relationship extraction extends relation extraction to a very large corpus. Neural relation extraction has made great progress in modeling sentences in low-dimensional space, but lacks consideration of simulated entities. Firstly, the context coding on the dependency tree is embedded as a tree-GRU-based sentence entity, and then the relationship between the sentence and the sentence is used to obtain the entity embedding of all sentence sets. Finally, the sentence embedding and the entity embedding are combined to classify the relationship. Better performance extraction [78]. In 2019, XIAOYU GUO proposed a combination of CNN and RNN for relationship classification based on single attention. In 2019, PLSTM-CNN [81] based on remote supervision was used to implement relationship extraction. At the same time, Multi-Gram CNN-Based Self-Attention [82] is also based on remote supervision for relationship classification. In the same year, attention-based Att-RCNN [83] was also used to achieve relationship classification. For the classification of semantic relations, SHEN Y proposed a neural network based ED-LSTM [84] algorithm.

However, the above entity relationship extraction is to extract the entity and the relationship separately, which will generate redundant information, so the joint extraction is proposed. Extract the entities and their relationship types in the same model, realize parameter sharing and synchronization optimization, and reduce the possibility of extracting errors before extraction. Zheng [71] et al. used the underlying expression of shared neural network for joint learning. Li et al. [72] proposed a joint structured extraction method for incremental cluster search algorithm and a constraint method using global features.

3.3 Knowledge Fusion

Information extraction obtains attribute information of entities, relationships, and entities from raw unstructured and semi-structured data. But this information contains a lot of redundancy and error messages. Therefore, it needs to clean up the extracted knowledge items. Knowledge fusion aims to solve errors in network data and errors in knowledge extraction. Knowledge fusion mainly includes entity links and knowledge integration [26].

(1) Entity Linking

Entity link is an operation that connects an entity object extracted from text to the corresponding item in external databases or knowledge bases. Wikipedia is often used as one of the knowledge bases for entity linking. The entity link was to assign the reference item to the correct entity object through disambiguation processing and co-finding digestion [27]. Tomoaki Urata proposed a disambiguation method for Wikipedia's Weibo entity links [28]. Entity disambiguation has four steps. The first step is to obtain the candidate entity. This requires obtaining a Wikipedia article for the candidate entity from the Wikipedia disambiguation page and preparing for the next entity disambiguation. The second step is to compare the nearest entity in the target entity with the Wikipedia article for each candidate entity and match the entity in Wikipedia. The third step is to calculate the similarity. They use word2vec to get the relevant entity of the nearest entity. The final step is to multiply the results of the second and third steps and extract the Wikipedia article with the highest score as the correct entity.

However, there is a lack of relationship between some entities and entities, so link prediction is required. In some algorithms, Daniel Neil proposed a new model, Graph Convolutional Neural Networks (GCNNs) [77], which improves performance on clear datasets and accommodates noise in KGs pervasive issue in real world applications. What's more, this model is more interpretable. Because it allowed measurement the effect of a particular edge on prediction by adjusting the link weight or completely removing the edge. In 2019, Binling Nie combined latent feature models and graph feature models to propose a model of Text-enhanced KG Embedding (TKGE), which can perform inference over entities, relations and text [80].

(2) Knowledge Integration

Knowledge Integration is to combine knowledge from multiple, distributed, heterogeneous knowledge sources [18]. In the previous entity link, it linked data extracted from semi-structured or unstructured data, but there were structured data packages for external or relational databases. The process of knowledge integration involves concept matching, entity matching, the evaluation of knowledge and the resolution of conflicts [29].

3.4 Knowledge Processing

(1) Ontology Construction

Ontology is a formal language used to construct ontology. It is a descriptive language and a metaphysical form of framework language. Ontologies include individuals, classes, attributes, function terms, constraints, rules, axioms, events, and so on. In the structure of ontology, it contains the relationships between things, the nature of things, the constraints of things and so on. In [30], the author proposed an automatic ontology extension method based on supervised learning and text clustering. This method used the k-means algorithm to separate domain knowledge and guided the creation of the Naïve Bayes classifier training set. First, the candidate set words will be added to the target ontology. At the same time, the noise words will be added to the stop word dictionary, which can automatically expand the ontology. The experimental results show that the expansion effect is very good.

(2) Knowledge Inference

After completing the ontology construction step, the relationship between many nodes and nodes of the KG is still vacant. Therefore, we need to fill in these relationships through knowledge reasoning to better improve the KG. In [31], Lucas Fonseca Navarro et al. proposed the Graph Rule Learner (GRL), a method for extracting inference rules from the ontology knowledge base graphed to graphs, and explored the combination of link prediction indicators. The input to GRL is a ontology graph, and the output is a list of induced inference rules. GRL uses the link prediction metric of the extra neighbor [32] to rank the possible rules, and it's scalable, using a structure called Graph DB-Tree [33] to store the graphical representation on disk.

In [34], the author proposed a unified ontology reasoning framework to incorporate co-occurrence and subclass relationships, and this framework can also automatically build ontology. The article pointed out that subclass relationships could be obtained from WordNet. It covers almost all common nouns and can find the relationship between words and words. Then get the co-occurrence relationship from the training set. This framework effectively combines the concept of reasoning with superior performance and other methods.

In 2013, BORDES proposed to encode the triplet into a low-dimensional distributed vector, the TransE model. TransE [42] is a translation model, which considers the relationship between the three heads (head, relation, and tail) as a translation from head to tail. By constantly adjusting the relationship between head, relation, and tail, h + r is equal to t as much as possible. But TransE is dealing with the properties of some relational graphing, such as reflexivity. So, in 2014, ZhenWang1, Jianwen Zhang and others proposed TransH [43] model. TransH models the relationship as a relational hyperplane along with the translation operations on it, which can preserve some graphing properties such as reflexivity, one-to-many, many-to-one, many-to-many, etc. when sneaking, and in efficiency It is almost the same as TransE. TransH has a predictive accuracy comparable to that of TransE. Both TransE and TransH assume that entities and relationships are in the same space, but in fact one entity has multiple attributes. Different relationships focus on different attributes of the entity. In other words, some similar entities should be close to each other in space, not similar in space. The middle should be away from each other. So in 2015, Yankai Lin et al. proposed the TransR model [44], Embedding entities and relationships into different spaces and implementing translations in the corresponding spatial relationships. The disadvantage of the TransR model is that the parameters are too many and too complicated. So in the same year Guoliang Ji, et al. proposed the TransD model [45], which not only considers the diversity of relationships, but also considers the entity. The main idea is to use two vectors to represent entities and relationships, one (h, r, t) for entities or relationships, and one for dynamically constructing graphing matrices. The advantage is that there are fewer parameters and no multiplication of matrix vectors, and it is applied to large-scale graphics. The previous representations oversimplified the loss metric and did not have enough power to simulate complex entities and relationships in the knowledge base. Han Xiao proposed the TransA model [46], which replaced the metric function, used the metrics to learn the sneak method, and treated each dimension in the vector differently, improving the representation ability. In 2016, in order to solve the problem of heterogeneity and imbalance, the connection relationship of the entity is complicated and simple, and the number of head and tail of many connection relationships is not equal. Guoliang Ji proposed the TranSpare model [47]. The core idea is that the transfer matrix is replaced by an adaptive matrix, and the sparsity of the adaptive matrix is determined by the number of pairs of entities, thus preventing under-fitting of complex relationships or over-simplification of simple relationships. Then, in order to solve the problem of multi-relational semantics, the TransG model emerged, and a Bayesian nonparametric Gaussian mixture model was used to generate multiple translation parts for a relationship. This discovered the potential semantic relationship and embeds a triple with the mixture-specific component vector [48]. Finally extended to the KG2E model [49]. The new idea of using the covariance of multidimensional Gaussian distributions to represent the uncertainty of entities and relationships can properly represent its certainty. The above models are all extensions to the TransE model.

3.5 Knowledge Storage

After obtaining the relationship between the entity and the entity, the data is used to form a KG. Usually, the ontology acts as a carrier for KG. Ontology language based on Web ontology includes Resource Description Framework (RDF) [57] and Web Ontology Language (OWL) [58]. RDF can be used to describe resources on the network and their relationship to each other, including resources, attributes, and relationships, in the form of triples. Later, RDFS appeared which mainly added some vocabulary to expand the ability of RDF. OWL is still an extension of RDF's vocabulary. It has strong knowledge representation and knowledge reasoning ability, and adds representations of classes, vocabulary and relationships. So OWL is more expressive and more powerful than RDF.

The other is storage based on a graph database, such as Neo4j [59]. The Neo4j graphics database is based on attribute graphs and focuses on queries and searches. The disadvantage of Neo4j is that it does not support distributed. Although OrientDB and JanusGraph (formerly Titan) support distributed, they are immature. So choosing Neo4j to store KG is better. And it separates the relationship between nodes and nodes. The last set of triplet row vectors is stored in csv format and then imported into the Neo4j database (graphics database), which is also convenient for future queries and optimizations.

4 Tools and Platforms

For the processing of data, in addition to some algorithms, it is also possible to implement information extraction through some open source natural language tool processing packages. Below we introduce the open source toolkits written by Python and Java, and analyze their functions. And the language that is suitable for processing, Chinese or English.

Gensim [50] is an open source Python toolkit that handles raw unstructured text and unsupervised learning of topic vector representations of text. It supports a variety of topic model algorithms including TF-IDF, LSA, LDA, and word2vec, supports streaming training, and provides API interfaces for common tasks such as similarity calculation and information retrieval.

Stanford CoreNLP [51] is an open source toolkit for relation extraction based on Java. Its methods such as supervising, remote monitoring, and neural network can realize the analysis of natural language texts, including part of speech restoration, part-of-speech tagging, named entity tagging, co-finger decomposition, syntax analysis and dependency analysis.

NLTK [52] is a third-party toolkit based on Python. NLTK is suitable for processing English, but there are some restrictions on handling Chinese. NLTK does not have a Chinese stop word due to the lack of a Chinese corpus. And the stop words of Chinese text cannot be filtered, so Chinese text cannot be segmented. Therefore, NLTK is not suitable for processing Chinese text.

FudanNLP [53] is a Java-based Chinese open source toolkit that contains machine learning algorithms and data sets. It enables information retrieval such as text categorization and news clustering. For Chinese processing, it includes Chinese word segmentation, part of speech tagging, entity name recognition, keyword extraction, dependency syntax analysis and time phrase recognition.

Stanford University has developed a toolkit that supports Chinese processing, namely deepdive [54]. It is used for knowledge extraction, the extraction of triples. It extracts structured relational data from unstructured text through weakly supervised learning. This project has modified the model package for natural language processing to support Chinese and provide Chinese tutorial.

SOFIE [55] is an automated ontology extension system developed by the max planck institute. It can extract ontology-based events from text, implement ontology links, and perform logical reasoning for disambiguation.

OpenCCG [56] is a Java-based open source natural language processing library that implements text grammar based on Mark Steedman's combination of grammatical forms, including syntax analysis and dependency analysis.

OLLIE is a three-tuple extraction component of KnowItAll, a knowledge base developed by the University of Washington. It enables the extraction of relationships based on grammar-dependent trees and can extract relationships over long distances. Reverb is also an open ternary extraction tool developed by the University of Washington. It can extract triples of entity relationships from English sentences. The advantage is that it does not need to specify relationships in advance, and supports information extraction on a network-wide scale. This reduces a lot of manual intervention.

ICTCLAS (NLPIR) is a Chinese language open source tool based on multiple languages, such as Java, C++, C, C#. It is mainly used to deal with Chinese, such as Chinese word segmentation, named entity recognition, part-of-speech tagging, custom user dictionary, new word micro-blog word segmentation, new word discovery and keyword extraction. Visual interface operation and API call provide users with Great convenience.

In summary, most open source toolkits for Chinese natural language processing are implemented in the Java language, while most of the self-language toolkits that handle English are written in the Python language. NLTK is suitable for English processing. FudanNLP, deepdive, ICTCLAS are suitable for Chinese processing. Gensim processes text based on an unsupervised learning algorithm. Stanford CoreNLP processes text based on remote monitoring and neural networks.

5 Conclusion and Futures Challenges

KG is an important branch of artificial intelligence. It simulates the way people think, and carries out efficient knowledge management and knowledge acquisition on data.

This article aims to use crawling technology to crawl some fault data and then generate KG through natural language processing. This allows a clearer understanding of the relationship between faults and faults, and analysis and prediction to prevent failures. Natural language processing mainly uses the entity relationship extraction technology. In the previous introduction, there are many algorithms and tools. The algorithms generally have unsupervised methods, supervised methods, remote monitoring methods and bootstrapping methods. Entity and relationship extraction have first entity extraction and relationship extraction and joint extraction of entities and relationships. Nowadays, although there are more separate extraction algorithms, its extraction is limited by entity extraction. Joint extraction is better than separate extraction, but there are no particularly mature algorithms. The specific algorithms of entity relationship extraction mainly include entity relationship extraction based on deep convolutional neural network, circular convolutional neural Venaero algorithm, CRF-based named entity, semi-supervised learning method combined with word rules and SVM model and open Chinese. A major problem in the extraction of entity relationships is the need for a large number of manual annotations. However, many Chinese entities can also be used to extract relational tools such as NTLK, Sanford CoreNLP, OLLIE, SOFIE, and so on. However, NTLK is not suitable for the processing of Chinese texts. It lacks Chinese corpus and stop words. Most of the tools developed are not very suitable for processing Chinese texts.

The important point is that most of the projects related to KG involve English. However, Chinese is a different language. It is not feasible to convert English KGs into Chinese. Thus, the construction of Chinese KG is very significant. The main difficulty lies in the following three points: (I) quality of data sources, (II) taxonomy derivation and (III) knowledge harvesting [75].

The construction of previous KG was constructed with node-relationship-nodes, but we still lack the unified definition and standard expression of KG. Therefore, in order to make the KG clearer, Yucong Duan clarified the structure of the KG from the aspects of data, information, knowledge and wisdom, and suggest to specify the KG in a gradual manner, including data graphs, infographics, KG and wisdom graph [76].

References

1. Fu, X., Ren, X., Mengshoel, O.J., et al.: Stochastic optimization for market return prediction using financial knowledge graph. In: 2018 IEEE International Conference on Big Knowledge (ICBK). IEEE Computer Society (2018)
2. Liu, Y., Zeng, Q., Yang, H., Carrio, A.: Stock price movement prediction from financial news with deep learning and knowledge graph embedding. In: Yoshida, K., Lee, M. (eds.) PKAW 2018. LNCS (LNAI), vol. 11016, pp. 102–113. Springer, Cham (2018). https://doi.org/10.1007/978-3-319-97289-3_8
3. Shen, Y., Yuan, K., Dai, J., et al.: KGDDS: a System for Drug-Drug Similarity Measure in therapeutic substitution based on knowledge graph curation. J. Med. Syst. **43**(4), 43 (2019)
4. Shengtian, S., Zhihao, Y., Lei, W., et al.: SemaTyP: a knowledge graph based literature mining method for drug discovery. BMC Bioinform. **19**(1), 193 (2018)
5. Sang, S., Yang, Z., Liu, X., et al.: GrEDeL: a knowledge graph embedding based method for drug discovery from biomedical literature. IEEE Access **7**, 8404–8415 (2018)
6. Ali, M., Hoyt, C.T., Domingo-Fernandez, D., et al.: BioKEEN: a library for learning and evaluating biological knowledge graph embeddings. BioRxiv, 475202 (2018)

7. Alshahrani, M., Khan, M.A., Maddouri, O., et al.: Neuro-symbolic representation learning on biological knowledge graphs. Bioinformatics **33**(17), 2723–2730 (2017)
8. Xiaoxue, L., Xuesong, B., Longhe, W., et al.: Review and trend analysis of knowledge graphs for crop pest and diseases. IEEE Access **7**, 62251–62264 (2019)
9. Chenglin, Q., Qing, S., Pengzhou, Z., et al.: Cn-makg: China meteorology and agriculture knowledge graph construction based on semi-structured data. In: Proceedings of the 2018 IEEE/ACIS 17th International Conference on Computer and Information Science (ICIS), F, 2018. IEEE (2018)
10. Sawant, U., Garg, S., Chakrabarti, S., et al.: Neural architecture for question answering using a knowledge graph and web corpus. Inf. Retrieval J. **22**(3–4), 324–349 (2019)
11. Shin, S., Jin, X., Jung, J., et al.: Predicate constraints based question answering over knowledge graph. Inf. Process. Manage. **56**(3), 445–462 (2019)
12. Zheng, W., Cheng, H., Yu, J.X., et al.: Interactive natural language question answering over knowledge graphs. Inf. Sci. **481**, 141–159 (2019)
13. Lu, Y.-C., Wen, Y.-J., Xuan, L., et al.: Exploration of the construction and application of knowledge graph in equipment failure. DEStech Transactions on Computer Science and Engineering, (smce) (2017)
14. Qin, Z., Cen, C., Jie, W., et al.: Knowledge-graph based multi-target deep-learning models for train anomaly detection. In: Proceedings of the 2018 International Conference on Intelligent Rail Transportation (ICIRT). IEEE (2018)
15. Shan, X., Zhu, B., Wang, B., et al.: Research on deep learning based dispatching fault disposal robot technology. In: Proceedings of the 2018 2nd IEEE Conference on Energy Internet and Energy System Integration (EI2). IEEE (2018)
16. Tang, Y., Liu, T., Liu, G., et al.: Enhancement of power equipment management using knowledge graph. arXiv preprint arXiv:190412242 (2019)
17. Steiner, T., Verborgh, R., Troncy, R., et al.: Adding realtime coverage to the google knowledge graph. In: Proceedings of the 11th International Semantic Web Conference (ISWC 2012). Citeseer (2012)
18. Zheng, M., Ma, Y., Zheng, A., et al.: Constructing method of public opinion knowledge graph with online news comments. In: Proceedings of the 2018 International Conference on Robots & Intelligent System (ICRIS). IEEE (2018)
19. Choudhury, S., Agarwal, K., Purohit, S., et al.: Nous: construction and querying of dynamic knowledge graphs. In: Proceedings of the 2017 IEEE 33rd International Conference on Data Engineering (ICDE). IEEE (2017)
20. Zheng, M., Ma, Y., Zheng, A., et al.: Constructing method of public opinion knowledge graph with online news comments. In: Proceedings of the 2018 International Conference on Robots & Intelligent System (ICRIS). IEEE (2018)
21. Heydon, A., Najork, M.: Mercator: a scalable, extensible web crawler. World Wide Web **2**(4), 219–229 (1999)
22. De Groc, C.: Babouk: focused web crawling for corpus compilation and automatic terminology extraction. In: Proceedings of the 2011 IEEE/WIC/ACM International Conferences on Web Intelligence and Intelligent Agent Technology. IEEE (2011)
23. Xia, J., Wan, W., Liu, R., et al.: Distributed web crawling: a framework for crawling of micro-blog data (2015)
24. Cowie, J., Wilks, Y.: Information extraction. Handbook Nat. Lang. Process. **56**, 57 (2000)
25. Lian, H., Qin, Z., He, T., et al.: Knowledge graph construction based on judicial data with social media. In: Proceedings of the 2017 14th Web Information Systems and Applications Conference (WISA). IEEE (2017)
26. Wang, X., Ma, C., Liu, P., et al.: A potential solution for intelligent energy management-knowledge graph. In: Proceedings of the 2018 IEEE International Conference on Energy Internet (ICEI). IEEE (2018)

27. Li, Y., Wang, C., Han, F., et al. Mining evidences for named entity disambiguation. In: Proceedings of the 19th ACM SIGKDD International Conference on Knowledge Discovery and Data Mining. ACM (2013)

28. Urata, T., Maeda, A.: An entity disambiguation approach based on wikipedia for entity linking in microblogs. In: Proceedings of the 2017 6th IIAI International Congress on Advanced Applied Informatics (IIAI-AAI). IEEE (2017)

29. Wang, X., Ma, C., Liu, P., et al.: A potential solution for intelligent energy management-knowledge graph. In: Proceedings of the 2018 IEEE International Conference on Energy Internet (ICEI). IEEE (2018)

30. Song, Q., Liu, J., Wang, X., et al.: A novel automatic ontology construction method based on web data. In: Proceedings of the 2014 Tenth International Conference on Intelligent Information Hiding and Multimedia Signal Processing. IEEE (2014)

31. Navarro, L.F., Hruschka, E.R., Appel, A.P.: Finding inference rules using graph mining in ontological knowledge bases. In: Proceedings of the 2016 5th Brazilian Conference on Intelligent Systems (BRACIS). IEEE (2016)

32. Appel, A.P., Junior, E.R.H.: Prophet–a link-predictor to learn new rules on NELL. In: Proceedings of the 2011 IEEE 11th International Conference on Data Mining Workshops. IEEE (2011)

33. Navarro, L.F., Appel, A.P., Junior, E.R.H.: GraphDB – storing large graphs on secondary memory. In: Catania, B., et al. (eds.) New Trends in Databases and Information Systems. AISC, vol. 241, pp. 177–186. Springer, Cham (2014). https://doi.org/10.1007/978-3-319-01863-8_20

34. Tsai, S.-F., Tang, H., Tang, F., et al.: Ontological inference framework with joint ontology construction and learning for image understanding. In: Proceedings of the 2012 IEEE International Conference on Multimedia and Expo. IEEE (2012)

35. Collins, M., Singer, Y.: Unsupervised models for named entity classification. In: Proceedings of the Joint SIGDAT Conference on Empirical Methods in Natural Language Processing and Very Large Corpora, pp. 100–110 (1999)

36. Cucerzan, S., Yarowsky, D.: Language independent named entity recognition combining morphological and contextual evidence. In: Proceedings of the 1999 Joint SIGDAT Conference on EMNLP and VLC, pp. 90–99 (1999)

37. Isozaki, H., Kazawa, H.:[Association for Computational Linguistics the 19th international conference - Taipei, Taiwan (2002.08.24–2002.09.01)] Proceedings of the 19th international conference on Computational linguistics, - - Efficient support vector classifiers for named entity recognition[In: Proceedings of the 19th International Conference on Computational Linguistics, vol. 1, pp. 1–7 (2002)

38. Borthwick, A.E.: A Maximum Entropy Approach to Named Entity Recognition. New York University, New York (1999)

39. Bikel, D.M., Miller, S., Schwartz, R., et al.: Nymble: a High-Performance Learning Name-finder. Anlp 94–201 (1998)

40. Bikel, D.M.: An algorithm that learns what's in a name. Machine Learning 34 (1999)

41. Mccallum, A., Li, W. [Association for Computational Linguistics the seventh conference - Edmonton, Canada (2003.05.31-.)] Proceedings of the Seventh Conference on Natural Language Learning at HLT-NAACL 2003, - - Early Results for Named Entity Recognition with Conditional Random Fields, Feature Induction and Web-Enhanced Lexicons, vol. 4, pp. 188–191 (2003)

42. Bordes, A., Usunier, N., Garcia-Duran, A., et al.: Translating embeddings for modeling multi-relational data. In: Proceedings of the Advances in Neural Information Processing Systems (2013)

43. Wang, Z., Zhang, J., Feng, J., et al.: Knowledge graph embedding by translating on hyperplanes. In: Proceedings of the Twenty-Eighth AAAI Conference on Artificial Intelligence (2014)
44. Lin, Y., Liu, Z., Sun, M., et al.: Learning entity and relation embeddings for knowledge graph completion. In: Proceedings of the Twenty-Ninth AAAI Conference on Artificial Intelligence (2015)
45. Ji, G., He, S., Xu, L., et al.: Knowledge graph embedding via dynamic mapping matrix. In: Proceedings of the 53rd Annual Meeting of the Association for Computational Linguistics and the 7th International Joint Conference on Natural Language Processing (Volume 1: Long Papers) (2015)
46. Xiao, H., Huang, M., Hao, Y., et al.: TransA: An adaptive approach for knowledge graph embedding. arXiv preprint arXiv:150905490 (2015)
47. Ji, G., Liu, K., He S., et al.: Knowledge graph completion with adaptive sparse transfer matrix. In: Proceedings of the Thirtieth AAAI Conference on Artificial Intelligence (2016)
48. He, S., Liu, K., Ji, G., et al.: Learning to represent knowledge graphs with gaussian embedding. In: Proceedings of the 24th ACM International on Conference on Information and Knowledge Management. ACM (2015)
49. Xiao, H., Huang, M., Hao, Y., et al.: TransG: a generative mixture model for knowledge graph embedding. arXiv preprint arXiv:150905488 (2015)
50. Rehurek, R., Sojka, P.: Software framework for topic modelling with large corpora. In: Proceedings of the LREC 2010 Workshop on New Challenges for NLP Frameworks. Citeseer (2010)
51. Manning, C., Surdeanu, M., Bauer, J., et al.: The Stanford CoreNLP natural language processing toolkit. In: Proceedings of 52nd Annual Meeting of the Association for Computational Linguistics: System Demonstrations (2014)
52. Bird, S., Klein, E., Loper, E.: Natural Language Processing with Python: Analyzing Text with the Natural Language Toolkit. O'Reilly Media Inc., Beijing (2009)
53. Qiu, X., Zhang, Q., Huang, X.: Fudannlp: a toolkit for chinese natural language processing. In: Proceedings of the 51st Annual Meeting of the Association for Computational Linguistics: System Demonstrations (2013)
54. Zhang, C.: DeepDive: A Data Management System for Automatic Knowledge Base Construction. University of Wisconsin-Madison, Madison (2015)
55. Suchanek, F.M., Sozio, M., Weikum, G.: SOFIE: a self-organizing framework for information extraction. In: Proceedings of the 18th International Conference on World wide web. ACM (2009)
56. Baldridge, J., Chatterjee, S., Palmer, A., et al.: DotCCG and VisCCG: Wiki and programming paradigms for improved grammar engineering with OpenCCG; proceedings of the CSLI Studies in Computational Linguistics Online. Citeseer (2007)
57. Miller, E.: An Introduction to the Resource Description Framework. Bull. Am. Soc. Inf. Sci. Technol. 25(1), 15–19 (1998)
58. Bechhofer, S.: OWL: web ontology language. Encyclopedia Inf. Sci. Technol. Second Ed. 63(45), 990–996 (2004)
59. Partner, J., Vukotic, A., Watt, N.: Neo4j in Action. Pearson Schweiz Ag (2014)
60. Chinchor, N., Marsh, E.: Muc-7 information extraction task definition. In: Proceeding of the Seventh Message Understanding Conference (MUC-7), Appendices (1998)
61. Vilain, M., Burger, J., Aberdeen, J.: Proceedings of the 6th Conference on Message Understanding (MUC-6) (1995)
62. Brants, T.: Proceedings of the Sixth Conference on Applied Natural Language Processing (2000)
63. Kambhatla, N.: Proceedings of the ACL 2004 on Interactive Poster and Demonstration Sessions (2004)

64. Gonzalez, E., Turmo, J.: Unsupervised relation extraction by massive clustering. In: Proceedings of the 2009 Ninth IEEE International Conference on Data Mining. IEEE (2009)

65. Liu, X., Yu, N.: Multi-type web relation extraction based on bootstrapping. In: proceedings of the 2010 WASE International Conference on Information Engineering. IEEE (2010)

66. Hendrickx, I., Kim, S.N., Kozareva, Z., et al.: Semeval-2010 task 8: multi-way classification of semantic relations between pairs of nominals. In: Proceedings of the Workshop on Semantic Evaluations: Recent Achievements and Future Directions. Association for Computational Linguistics (2009)

67. Socher, R., Huval, B., Manning, C.D., et al.: Semantic compositionality through recursive matrix-vector spaces. In: Proceedings of the 2012 Joint Conference on Empirical Methods in Natural Language Processing and Computational Natural Language Learning. Association for Computational Linguistics (2012)

68. Zeng, D., Liu, K., Lai, S., et al.: Relation classification via convolutional deep neural network (2014)

69. Nguyen, T.H., Grishman, R.: Relation extraction: perspective from convolutional neural networks. In: Proceedings of the 1st Workshop on Vector Space Modeling for Natural Language Processing (2015)

70. Lin, Y., Shen, S., Liu, Z., et al.: Neural relation extraction with selective attention over instances. In: Proceedings of the 54th Annual Meeting of the Association for Computational Linguistics (Volume 1: Long Papers) (2016)

71. Zheng, S., Hao, Y., Lu, D., et al.: Joint entity and relation extraction based on a hybrid neural network. Neurocomputing **257**, 59–66 (2017)

72. Li, Q., Ji, H.: Incremental joint extraction of entity mentions and relations. In: Proceedings of the 52nd Annual Meeting of the Association for Computational Linguistics, vol. 1: Long Papers (2014)

73. Krizhevsky, A., Sutskever, I., Hinton, G.E.: Imagenet classification with deep convolutional neural networks. In: proceedings of the Advances in Neural Information Processing Systems (2012)

74. Goller, C., Kuchler, A.: Learning task-dependent distributed representations by backpropagation through structure. In: Proceedings of International Conference on Neural Networks (ICNN 1996). IEEE (1996)

75. Wang, C., Gao, M., He, X., et al.: Challenges in chinese knowledge graph construction. In: Proceedings of the 2015 31st IEEE International Conference on Data Engineering Workshops. IEEE (2015)

76. Duan, Y., Shao, L., Hu, G., et al.: Specifying architecture of knowledge graph with data graph, information graph, knowledge graph and wisdom graph. In: Proceedings of the 2017 IEEE 15th International Conference on Software Engineering Research, Management and Applications (SERA). IEEE (2017)

77. Neil, D., Briody, J., Lacoste, A., et al.: Interpretable graph convolutional neural networks for inference on noisy knowledge graphs. arXiv preprint arXiv:181200279 (2018)

78. He, Z., Chen, W., Li, Z., et al.: SEE: syntax-aware entity embedding for neural relation extraction. In: Proceedings of the Thirty-Second AAAI Conference on Artificial Intelligence (2018)

79. Guo, X., Zhang, H., Yang, H., et al.: A single attention-based combination of CNN and RNN for relation classification. IEEE Access **7**, 12467–12475 (2019)

80. Nie, B., Sun, S.: Knowledge graph embedding via reasoning over entities, relations, and text. Future Gener. Comput. Syst. **91**, 426–433 (2019)

81. Yan, D., Hu, B.: Shared representation generator for relation extraction with Piecewise-LSTM convolutional neural networks. IEEE Access **7**, 31672–31680 (2019)

82. Zhang, C., Cui, C., Gao, S., et al.: Multi-gram CNN-based self-attention model for relation classification. IEEE Access **7**, 5343–5357 (2019)

83. Guo, X., Zhang, H., Yang, H., et al.: A single attention-based combination of CNN and RNN for relation classification. IEEE Access **7**, 12467–12475 (2019)
84. Shen, Y., Sun, J, Jia, P., et al.: Entity-dependent long-short time memory network for semantic relation extraction. In: Proceedings of the 2018 5th IEEE International Conference on Cloud Computing and Intelligence Systems (CCIS). IEEE (2019)
85. Le, H.Q., Nguyen, T.M., Vu, S.T., et al.: D3NER: biomedical named entity recognition using CRF-biLSTM improved with fine-tuned embeddings of various linguistic information. Bioinformatics (2018)
86. Chen, H., Lin, Z., Ding, G., et al.: GRN: gated relation network to enhance convolutional neural network for named entity recognition (2019)

An Attention-Based User Profiling Model by Leveraging Multi-modal Social Media Contents

Zhimin Li, Bin Guo[✉], Yueqi Sun, Zhu Wang, Liang Wang, and Zhiwen Yu

Northwestern Polytechnical University, Xi'an 710100, China
guob@nwpu.edu.cn

Abstract. With the popularization of social media, inferring user profiles from the user-generated content has aroused wide attention for its applications in marketing, advertising, recruiting, etc. Most existing works focus on using data from single modality (such as texts and profile photos) and fail to notice that the combination of multi-modal data can supplement with each other and can therefore improve the prediction accuracy. In this paper, we propose AMUP model, namely the Attention-based Multi-modal User Profiling model, which uses different tailored neural networks to extract and fuse semantic information from three modalities, i.e., texts, avatar, and relation network. We propose a dual attention mechanism. The word-level attention network selects informative words from the noisy and prolix texts and the modality-level attention network addresses the problem of imbalanced contribution among different modalities. Experimental results on more than 1.5K users' real-world data extracted from a popular Q&A social platform show that our proposed model outperforms the single-modality methods and achieves better accuracy when compared with existing approaches that utilize multi-modal data.

Keywords: User profile inferring · Multi-modal learning · Deep neural networks · Dual attention mechanism

1 Introduction

Since the beginning of the 21st century, netizens have generated more and more content on social media platforms. User-generated content (UGC), as a kind of crowd-sourced data, contains data of various modalities, including their texts, uploaded photos, comments, etc. Users' behavior and generated content online can finely indicate their personality and profile, which include their gender, age, occupation, etc. [1–4]. For example, in social media platforms, we can easily recognize the user's gender and age from the user's selfies, uploaded photos or avatars. The user's occupation can also be inferred by the topics he/she discusses with his friends online. For the Internet industries, knowing user profile is quite valuable for precision marketing, expanding market, and targeted adverting [5].

In recent years, a large number of user profiling studies using online digital traces have emerged in the field of data mining and natural language processing. One of the most

© Springer Nature Singapore Pte Ltd. 2019
H. Ning (Ed.): CyberDI 2019/CyberLife 2019, CCIS 1138, pp. 272–284, 2019.
https://doi.org/10.1007/978-981-15-1925-3_20

representative works is that the IBM Watson Personality Insights project [6] leverages users' texts in the social media to predict user personality, proving that UGC can be an effective tool for portrait prediction. Besides textual information, netizens have also generated considerable digital traces of other modalities, which can also reflect their attributes. Fong et al. [7] have found users choose their avatar for certain intentions of how they want other people to perceive them, and they explore the relationship between avatar and personality. A research by Staiano [8] reveals that the social network and the friends never lie and can reflect the user's true personality. According to the above studies, we can see that not only texts but also data of other modality is useful to user profiling. Different modalities can supplement to each other to form an integrated profile prediction of the user. However, most of current studies are based on single modality, and they fail to capture other modalities which can make the result more complete. There're a few studies that use multi-modal content for user profile inferring [9, 10]. To the best of our knowledge, these studies mainly use voting or feature fusion to predict the final results. However, different modalities may weigh differently for different users. For example, the relation network structure of users who have few online friends may provide little useful information for portrait prediction. The texts for the people who're not willing to talk on the Internet may not be as informative as others. Therefore, among different modalities, there should be a trade-off mechanism.

Our research is motivated by the following findings. First, users' digital trace in the social media contains multi-modal information, such as texts, avatars, relation network relationship. Those data contain implicit user profile information. For example, photographers like to use their selfies with beautiful sceneries as avatars, programmers are used to discussing new technologies with other programmers in related fields. With the above information, even if the user does not label their information in the profile page directly, we can easily figure out their profiles through their comprehensive data in various forms. Second, among all the modalities, the textual information is quite different and needs special processing. Online users can generate massive and various texts online, but not all of them are useful for profile inferring because they are too long and noisy. Like the sentence "How can I evaluate my photography work and how can it be improved" is a real-world sentence we saw on the social media. Even though it's a long one, the only part that can be useful for user profiling is the word "photography". Those informative words should be given higher weight in the user's corpus. And on the contrary, the less informative words should be less weighed. This is conducive to better learning the semantic representation of the text. Third, although users' portrait can be embodied in multi-modal content, different modalities may take on different importance for different users. The information we can get from the social network structure is different for the user who has numerous friends and the user who have few friends.

Based on above characteristics of information in social media, we propose an Attention-based Multi-modal User Profile (AMUP) model. This model leverages multi-modal information from social media platform using dual attention networks to predict user profile. First, we use different methods to embed the multimodal user information to get an implicit representation of the content. For example, for text in social media, we first use the GRU structure to extract contextual semantic representation. In order to select the informative words of the user's massive texts, we use the word-level attention network and give the informative words higher weights. For user avatars, we use CNN ResNet-50 [25] for feature extraction. In the process of learning the structure of the social network, we use weighed-Node2Vec to learn the structural representation of the network and the contextual information of the neighborhood. Next, we use a modality-level attention network to give the more informative modality a higher weight. The multi-modal content features are fused according to the weight. At last, we use the Softmax layer to complete the final user profile classification.

There are three main contributions of this article:

(1) We propose AMUP model to predict user profile using multi-modal information including textual information, avatar, and social network structure. The combination of different modalities learns users' portraits embodied in different aspects and performs well when compared with single-modality learning.

(2) A dual attention mechanism is proposed in this paper. For one thing, unlike existing user portrait related works, which simply concatenate the text together, we use the word-level attention network to find informative information from the noisy and long texts to build effective text representations. Moreover, due to the imbalance among different modal information of users, we use a modality-level attention network to integrate these information according to their weight, and then we get the user's representation.

(3) Experiments on gender, age, and occupation prediction have been conducted. The results show that our proposed model outperforms existing studies both on single modality data and on multi-modal modality data.

The rest of this paper is organized as follows: Sect. 2 presents the related work. Section 3 describes the framework and Implementation of AMUP model. Section 4 presents the evaluation and results. We conclude the paper in Sect. 5.

2 Related Work

Inferring user profile through users' digital traces in social media is a research issue that has received wide attention in recent years [11–13]. This research topic covers a wide range of modalities, such as blogs, tweets, text messages, photos, and online behaviors. From linear models to machine learning and deep learning approaches, researchers are taking increasingly deeper exploration.

Text is the most commonly used tool to explore user profile, Schwartz et al. [14] use the 700 million Facebook messages of 75,000 volunteers to extract words, phrases and topics that are related to demographic information in their social media data. They discover the language the netizens use have significant relations to their age, personality, and gender. Garcia et al. [15] learn the semantic information of 304 Facebook users' updates using Latent Semantic Analysis, and use it to predict personality traits such as neuroticism, mental illness, and narcissism. Others used text splicing features for the Twitter dataset, using Naive Bayes, K-Nearest Neighbors and Support Vector Machine to determine the user's Big Five personality. Qiu et al. [17] apply the concatenated feature of texts from Twitter to Naïve Bayes, K-Nearest Neighbors, and Support Vector Machine to predict the users' Big Five personality. With users' original microblog messages, Zhang et al. [16] use LSTM to predict the demographic information including gender and age. They also put retweeted messages, self-written comments and comments from others into consideration to learn the semantic information from texts in different aspects.

In addition to the text, researchers also take pictures and social relationships of users in social media as an important reference for predicting user portraits. Ferwerda et al. [18] try to infer users' personality traits from the filters used to the uploaded photos in Instagram. They extracted the characteristics of the picture including hue, brightness, saturation, etc., and successfully define a relationship between users' personality traits and the way they feel like making their pictures look. Face++ and EmoVu are two open tools for image feature exaction, which can extract the colors, aesthetics, facial expressions and emotions of the users' pictures. Liu et al. [19] employed the tools for feature extraction of users' avatars and found that they have a certain correlation with users' Big Five Personality test results. For example, extravert people tend to use profile photos that are colorful and with multiple people, and people with Agreeableness personality tend to use colorful, blurry, and bright photos as their avatars. As for the relation network, Youyou et al. [20] found that the "voteup mode" of user social network has a direct correlation with people's personality. Based on 70,000 users, they built a matrix for the voteup between people and used linear regression to make an analysis between the voteup mode and the big five personality. 90% of the users are used as training data and the relationship is verified with the remaining 10%. Using the model to predict the Big Five personality, the results indicated that the machine prediction is more accurate than humans did.

The works above have demonstrated a good basis for our hypothesis that the information of user-generated content of different modalities can all indicate the user's profile to a certain extent. However, most of the existing studies just used single-modal data to predict user characteristics and have not considered combining multiple modes. What's more, in the studies of textual contribution to user profile prediction, most of the researches have concatenated the feature of the texts and directly send them into the classifier or neural network for training, which cannot distinguish the useful and informative texts from the noisy expressions.

In the past three years, there've been a few studies on user profile prediction based on multi-modal data. Wei et al. used multi-modal information, i.e., self-language, avatar, emotion, and responsive patterns, and extract features from different models through deep learning. Then they use the stacked generalization-based ensemble method to output the final prediction. Another work is finished by Farnadi et al. [8], who use a power-set combination to combine shared and non-shared representations among data of different modalities and integrates them both at the feature level and the decision level. Although the above studies have considered the complementarity between multimodalities, they all simply combine the modalities in decision fusion or feature fusion. They didn't consider there may be distinctions among modalities for different users because of their personal preferences and habits. This encourages us to further explore the contribution of different modalities to user portraits in our work.

3 Our Approach

In this section, we present our Attention-based Multi-modal User Profile (AMUP) model in detail. Given user i's multi-modal contents from his/her social media, the aim of our model is to classify the user i into a profile category set $C_i = \{c_{ig}, c_{ia}, c_{io}\}$. To be specific, user i's contents include T_i, A_i, R_i. T_i represents the texts user i has written by himself/herself. A_i is the avatar of user i. R_i means the relation network user i is in. In this paper, we denote the social network as an undirected graph $G_i = <V_i, E_i>$. The vertex set N_i represents the people who follow or is being followed by user i. The edge E_i is formed whenever there's a follower-following relationship between the two users. There's a weight of each edge, which is the amount of the posts they have interacted with in common. As for the profile category, it is composed of three dimensions, including gender, age (5 age groups), and occupation (8 kinds of occupations).

3.1 System Overview

Our AMUP model is composed of three modules, namely feature extraction, user encoder, and user profile classification. First, the feature extraction module extracts the feature of users' texts, avatars, and relation network. The text features are extracted by a word-level attention-based Bi-GRU network. The attention network, by selecting informative words, well addresses the noisiness and the prolixity of texts. For the feature of avatars and social network, we use ResNet-50 and Weighed-Node2Vec to learn their representations respectively. Next, in the user encoder module, we leverage a modality-level attention network to deal with the imbalance among modalities and then generate user's representation. At last, the profile classification module classifies each user into different profile categories according to the hidden representation. The overall system framework of our AMUP model is shown in Fig. 1.

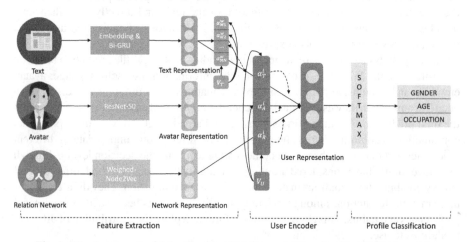

Fig. 1. The framework of Attention-based Multi-modal User Profile (AMUP) model

3.2 Feature Extraction

We use different methodologies to extract semantic representations from users' texts, avatars, and social networks.

Texts. Users' texts have been considered as an effective indicator of their profiles. Therefore, users' self-written texts should be an indispensable modal in this task. On the whole, the text feature extraction is composed of three layers, namely word embedding, GRU, and attention network. The structure of text features extraction module is illustrated in Fig. 2.

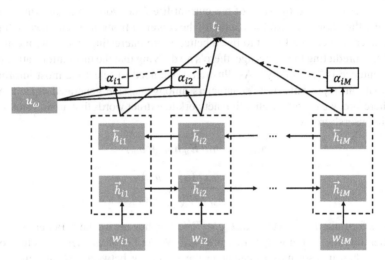

Fig. 2. The structure of text extraction module

The first layer is the word embedding layer, which uses words' pre-trained embedding vector dictionary $E \in \mathbb{R}^{V \times D}$ to map the word $w_{ij} \in Ti$ to low-dimension dense representation $w_{ij} \in \mathbb{R}^{D}$. Note that V is the vocabulary size, D is the dimension of the word vector, and w_{ij} represents the jth word of the user i.

The second layer is bidirectional GRU [21] layer. It's used to learn the contextual semantic representation of words by summarizing information of words from both directions. The forward GRU $\vec{f}$ reads texts from the users' first word w_{il} to the last word w_{iM}, and the backward GRU $\overleftarrow{f}$ reads texts from the last word w_{iM} to the first word w_{il}:

$$x_{im} = E\omega_{im}, m \in [1, M] \tag{1}$$

$$\vec{h}_{im} = \overrightarrow{GRP}(x_{im}) \in [1, M] \tag{2}$$

$$\overleftarrow{h}_{im} = \overleftarrow{GRP}(x_{im}) \in [M, 1] \tag{3}$$

$\vec{h}_{im}$ and $\overleftarrow{h}_{im}$ represent the hidden state of forward GRU and backward GRU respectively. We get the hidden state of a word by concatenating them together.

$$h_{im} = \left[\vec{h}_{im}, \overleftarrow{h}_{im} \right] \qquad (4)$$

At the third layer, attention network is used to select the most informative words among all the texts generated by the user. The semantic information extracted by the second layer, Bi-GRU layer, is actually hard to be leveraged for user profile inferring. It's because in most occasions the texts are massive and not all the words are equally useful. The main parts we actually focus on when predicting user profile are the things that the user is interested in and the state of her current life. The words embodies their profile are far less than noisy words. For example, the sentence from the social media "I study in my university because I want to live a better, more interesting, and more meaningful life." When predicting the user's age, there's no denying that the most informative words in this sentence is "university". As illustrated, we need to select the most informative words that can represents users' characters among massive generated content of the user. Therefore, we introduce attention network to extract words that are informative for predicting users' profile.

$$u_{im} = tanh(E_w h_{im} + b_w) \qquad (5)$$

$$\alpha_{im} = \frac{exp\left(u_{im}^T \cdot u_w\right)}{\sum_t exp\left(u_{im}^T \cdot u_w\right)} \qquad (6)$$

In this formula, $E_w \in \mathbb{R}^{V \times D} a, b_w \in \mathbb{R}^D$, and they are all the parameters of word-level attention. α_{im} is the weight of word w_{im}. We set the parameter u_w as the context vector,and then use softmax to calculate the similarity between w_{im} and the context vector.

At the third layer, the representation of user i's texts is calculated by the sum of all the words weighed vectors.

$$T_i = \sum_m \alpha_{im} h_{im} \qquad (7)$$

Avatars. Avatar is the visualization of the profile of the user. It embodies the user's portrait profile photo of the user in a direct-viewing way. Learning its semantic is essential to let the machine understand what the avatar is about and what hidden information in it implies. CNN model has been popular because its good performance in deep semantic feature extraction and visual classification. In order to efficiently and clearly extract visual features, we use ResNet-50 model [21]. Before the result output layer, we add a fully connected layer to extract as well as adjust the dimension of visual feature of the avatar. We also fine-tuned and pre-trained the ResNet-50 on our dataset. The feature of the avatar A_i can be calculated using the follow equation.

$$f A_i = \sigma(W_v \cdot f_R), \qquad (8)$$

where σ is the ReLU activation function, W_v is the weight matrix of the fully connected layer in the ResNet-50, $f A_i$ is the final representation of user I's avatar, and f_R is the representation from the pre-trained Resnet-50. $f A_i$, $f_R \in \mathbb{R}^D$ because in this paper we set all of the representations among different modalities to the same dimension.

Relation Network. The user's following and followers' profile, the number of user's friends in the social media, and the interaction mode of the user and his/her friends all help to reflect and describe the profile of the user to a certain extent. Therefore, it is important to learn the structural information of users in their social networks. In this subsection, we introduce the weighed-Node2Vec, which is an improved model of Node2Vec [22] and can capture the semantic representation of the user in his/her neighborhoods as well as structural information of the social network.

First of all, we model the relation network of user i as a weighed graph $G_i = \langle V, E \rangle$. V represents the set of user is' followings and followers, and E represents the set of the links between them. An edge is formed when there's a fellowship between two nodes. The weight of the edges is the number of the common contents they have interacted with.

After building the graph, the next step is to extract the structural information of the graph based on the weight of the edges. We adopt a modified method of DeepWalk [23], which transfers graph to sequences composed of the nodes. Here we introduce two parameters, the return parameter p and in-out parameter q. p determines the probability of revisiting the node. If the value of p is bigger than 1, it's guaranteed that the probability of sampling the visited node in two steps is relatively low. If $q > 1$, the model chooses to go to the node near the current node. Suppose the current node is v, the probability of the next node to be chosen is illustrated in Fig. 3. In order to value the intimacy among the nodes, we add weights to the forums.

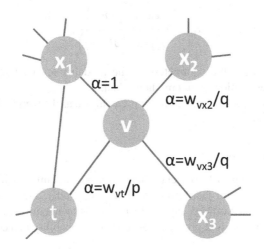

Fig. 3. A diagram for the principle of choosing the next node

Let c_i be the i-th node of the walk, and the node c_i obey the following distribution,

$$P(c_i = x | c_{i-1} = v) = \begin{cases} \frac{\pi_{vx}}{Z} & if (v, x) \in E \\ 0 & otherwise \end{cases}, \quad (8)$$

where

$$\pi_{vx} = \alpha_{pq}(t, x) \begin{cases} \frac{w}{p} \ if \ d_{tx} = 0 \\ w \ if \ d_{tx} = 1, \\ \frac{w}{q} \ if \ d_{tx} = 2 \end{cases} \tag{9}$$

After transferring the graph into the sequence made up of nodes, we adopt the skip-gram [24] model to calculate the node vectors. The loss function is as

$$J(\theta) = -\frac{1}{T} log L(\theta) = -\frac{1}{T} \sum_{t=1}^{T} \sum_{\substack{-m \le i \le m \\ j \ne 0}} log P\left(w^{(t+j)} | w^{(t)}\right) \tag{10}$$

3.3 User Encoder

In real world, not every person's habit on the social media is the same. For example, some people would like to read texts without interacting with the author, and some people are not willing to use unchangeable avatar. The different habits and personalities of people make the weight among modalities different. Thus, simply assuming that all modalities are of the same weight does not work in some occasions. Therefore, we propose the modality-level attention network to capture such differences among various modalities.

The Feature Extraction module described in Sect. 3.2, we have acquired the feature set of user i's feature $\{Ti, Ai, Ri\}$. The attention weight of each modality is calculated as follows:

$$v_{ip} = tanh\left(V_p^T \times F_i + b_p\right) \tag{11}$$

$$\alpha_{ip} = \frac{exp\left(v_p^T \cdot v_{ip}\right)}{\sum_i exp\left(v_p^T \cdot v_j\right)}, \tag{12}$$

where V_p, v_p, and b_p are the parameters of the modality-level attention network. F_i represents the representation of any modality data.

The final hidden representation of user i is calculated as the weighed sum of the three modality representations.

3.4 Profile Classification

We transfer the profiling task to a classification task, the aim of which is to classify the user into the category of their profile such as age, gender, and occupation. The last layer of our model is a SoftMax layer which intends to maximize the probability of categorizing the user into right profilelasses. User i's probability in a certain category is formulated as:

$$P_i = softmax\left(W^T \times u + b_u\right), \tag{13}$$

where W and b_u are the parameters to be learned. Here, note that the dimension of W is $\mathbb{R}^{C \times D}$, C is the number of categories. We use cross entropy as the loss function. The category with the highest probability is used as the inferred user profile.

4 Experiments

4.1 Dataset and Experimental Settings

We build a dataset of 1577 random users from Zhihu, a social Knowledge question-answer sharing platform. There're 1092 males and 485 females in the dataset. All of them are from 8 kinds of occupations: art creation, IT, educator, financial analysts, catering service, law service, medical service, journalists. Note that their age profile is not available for the public so we labelled them manually and deleted those whose age are hard to figure out from the dataset. The users are categorized into 5 age groups, under 18, 18–30, 31–45, 45–60, over 60. We crawled the users' avatars, following and followers, and all activities of 8 categories, including creating article, creating question, creating answer, following column, following question, following topic, liking answers, and liking article. In average, each user contributes 664 activities to the dataset. In our experiments, the word vector dictionary were pre-trained using the word2vec tool on a corpus acquired from the Zhihu platform. The embedding dimensions for texts, avatars, and social network are all set to 256.

4.2 Performance Evaluation

We evaluate the performance of our AMUP model by comparing the Accuracy and Fscore (macro-averaged Fscore) with several baselines.

SVM: Support Vector Machine, which is widely used in user profile prediction [9].

TextCNN: A kind of CNN network intended for text classification [28].

KNN: A supervised learning algorithm, K-Nearest Neighbor algorithm based on distance calculation.

MLP: Multi Layer Perceptron, a three-layer feedforward neural network.

Following Dong et al's work [26], the features used in SVM are word unigrams. The features used in KNN and MLP are SURF [27].

As is shown in Table 1, our proposed method outperforms other baselines and the adopted feature extraction methods are the most effective ones among all the baselines. For users' text, attention-based Bi-GRU method achieves the best result. It's because the features used for SVM is handcraft and cannot capture semantic and contextual information of texts. Moreover, while TextCNN is one of the state-of-the-art methods for text classification, which can well capture the contextual information, it only performs well with short texts. However, TextCNN only performs well with short texts. Compared with TextCN, Bi-GRU can learn the sequential information well and can extract good representations in forward and backward directions. Also, the word-level attention can select the words with abundant and useful information, therefore it can extract the most efficient features. For avatars, compared to traditional machine learning method—KNN and MLP—and the SURF feature, ResNet-50 can achieve better results with deep representation. For social network, weighed-Node2Vec works better than the unweighted one because when we add weights to the edges, the network can depict the social network in a more realistic way. The intimate degree between friends can be well reflected and thus the weighed graph achieves better results. Among all the methods, our AMUP model achieves the best result. For one thing, it can leverage the information among multi-modal data, which can supplement each other. For another, it efficiently uses the features among different modalities.

Table 1. The result of experiment with baselines

		Age		Gender		Occupation	
		Accuracy	Fscore	Accuracy	Fscore	Accuracy	Fscore
Text	SVM	0.4925	0.5874	0.6237	0.6028	0.4035	0.5176
	TextCNN	0.5127	0.6095	0.7935	0.6316	0.4429	0.5428
	Bi-GRU	0.5283	0.6111	0.7862	0.6487	0.4903	0.5749
Avatar	KNN	0.4092	0.5384	0.6472	0.4562	0.3264	0.5369
	MLP	0.4686	0.5549	0.6528	0.5438	0.4284	0.5466
	ResNet-50	0.5124	0.6042	0.6791	0.5860	0.4018	0.5765
Relationship	Node2Vec	0.6382	0.6031	0.8457	0.8337	0.5981	0.5762
	Weighed-Node2Vec	0.6531	0.6102	0.8519	0.8587	0.6272	0.5815
AMUP	No Attention	0.6637	0.6190	0.8771	0.8625	0.6374	0.5928
	Attention	0.7245	0.6337	0.8527	0.8890	0.6758	0.6037

5   Conclusion and Future Work

In this paper, we study the problem of online user profile inferring, which is of great business value. We propose AMUP, i.e., an Attention-based Multi-modal User Profile. Our model leverages features from users' texts, avatars, and relation network. To fully explore useful information in the multi-modal data, we use different neural network-based methods to extract features from each modality, namely word-level attention-based Bi-GRU for text, ResNet-50 for avatars, and Node2Vec for relation network. We also build an modality-level attention network to tackle with the imbalanced contribution of different modalities. Experimental results based on the real-world dataset from the Zhihu platform demonstrate our proposed model outperforms the baselines.

There're two main directions for the future work of AMUP. First, because users' interests are changing with time, integrating other user data sources, such as time stamps and geography location information, can help the module learn the dynamic evaluation of the users. In addition, learning the confidence level of data is a complex but necessary task that can be taken into account due to the massive noise in the data. Designing a method to learn the confidence level of the data using deep neural networks is an open issue to explore.

Acknowledgement. This work was partially supported by the National Key R&D Program of China (2017YFB1001800), the National Science Foundation of China (No. 61772428, 61725205), and the University-Enterprise Cooperation of Northwestern Polytechnical University (No. G2019KY04302).

References

1. Farnadi, G., et al.: Computational personality recognition in social media. User Model. User-Adap. Inter. **26**(2–3), 109–142 (2016)
2. Segalin, C., Cheng, D.S., Cristani, M.: Social profiling through image understanding: personality inference using convolutional neural networks. Comput. Vis. Image Underst. **156**, 34–50 (2017)
3. Liu, Y., et al.: CrowdOS: a ubiquitous operating system for crowdsourcing and mobile crowd sensing. arXiv preprint arXiv:1909.00805 (2019)
4. Guo, B., et al.: Mobile crowd sensing and computing: the review of an emerging human-powered sensing paradigm. ACM Comput. Surv. (CSUR) **48**(1), 7 (2015)
5. Nowson, S., Oberlander, J.: The identity of bloggers: openness and gender in personal weblogs. In Proceedings of AAAI Spring Symposium: Computational Approaches to Analyzing Weblogs, pp. 163–167 (2006)
6. Gou, L., Zhou, M.X., Yang, H.: Knowmeandshareme: understanding automatically discovered personality traits from social media and user sharing preferences. In: Proceedings of the 32nd Annual ACM Conference on Human Factors in Computing Systems, pp. 955–964. ACM (2014)
7. Fong, K., Mar, R.A.: What does my avatar say about me? Inferring personality from avatars. Pers. Soc. Psychol. Bull. **41**(2), 237–249 (2015)
8. Staiano, J., et al.: Friends don't lie: inferring personality traits from social network structure. In: Proceedings of the 2012 ACM Conference on Ubiquitous Computing. ACM (2012)
9. Wei, H., et al.: Beyond the words: predicting user personality from heterogeneous information. In: Proceedings of the Tenth ACM International Conference on Web Search and Data Mining. ACM (2017)
10. Farnadi, G., et al.: User profiling through deep multimodal fusion. In: Proceedings of the Eleventh ACM International Conference on Web Search and Data Mining. ACM (2018)
11. Mischel, W., Shoda, Y., Smith, R.E., Mischel, F.W.: Introduction to Personality. University of Phoenix: A John Wiley & Sons, Ltd., Publication (2004)
12. Tausczik, Y.R., Pennebaker, J.W.: The psychological meaning of words: LIWC and computerized text analysis methods. J. Lang. Soc. Psychol. **29**(1), 24–54 (2010)
13. Yarkoni, T.: Personality in 100,000 words: a large-scale analysis of personality and word use among bloggers. J. Res. Pers. **44**(3), 363–373 (2010)
14. Schwartz, H.A., et al.: Personality, gender, and age in the language of social media: the open-vocabulary approach. PLoS ONE **8**(9), e73791 (2013)
15. Garcia, D., Sikström, S.: The dark side of Facebook: Semantic representations of status updates predict the Dark Triad of personality. Personality Individ. Differ. **67**, 92–96 (2014)
16. Zhang, D., et al.: User classification with multiple textual perspectives. In: Proceedings of COLING 2016, the 26th International Conference on Computational Linguistics: Technical Papers (2016)
17. Qiu, L., et al.: You are what you tweet: Personality expression and perception on Twitter. J. Res. Personality **46**(6), 710–718 (2012)
18. Ferwerda, B., Schedl, M., Tkalcic, M.: Predicting personality traits with Instagram pictures. In: Proceedings of the 3rd Workshop on Emotions and Personality in Personalized Systems 2015. ACM (2015)
19. Liu, L., et al.: Analyzing personality through social media profile picture choice. In: Tenth International AAAI Conference on Web and Social Media (2016)
20. Youyou, W., Kosinski, M., Stillwell, D.: Computer-based personality judgments are more accurate than those made by humans. Proc. Natl. Acad. Sci. **112**(4), 1036–1040 (2015)

21. Bahdanau, D., Cho, K., Bengio, Y.: Neural machine translation by jointly learning to align and translate. Comput. Sci. (2014)
22. Grover, A., Leskovec, J.: node2vec: scalable feature learning for networks. In: ACM SIGKDD International Conference on Knowledge Discovery & Data Mining (2016)
23. Perozzi, B., Al-Rfou, R., Skiena, S.: DeepWalk: online learning of social representations. In: ACM SIGKDD International Conference on Knowledge Discovery & Data Mining (2014)
24. Mikolov, T., et al.: Efficient estimation of word representations in vector space. arXiv preprint arXiv:1301.3781 (2013)
25. Akiba, T., Suzuki, S., Fukuda, K.: Extremely large minibatchsgd: training resnet-50 on ImageNet in 15 minutes. arXiv preprint arXiv:1711.04325 (2017)
26. Nguyen, D., et al.: How old do you think I am? A study of language and age in Twitter. In: Seventh International AAAI Conference on Weblogs and Social Media (2013)
27. Bay, H., Tuytelaars, T., Van Gool, L.: SURF: speeded up robust features. In: Leonardis, A., Bischof, H., Pinz, A. (eds.) ECCV 2006. LNCS, vol. 3951, pp. 404–417. Springer, Heidelberg (2006). https://doi.org/10.1007/11744023_32
28. Kim, Y.: Convolutional neural networks for sentence classification. arXiv preprint arXiv:1408.5882 (2014)

Face Anti-spoofing Algorithm Based on Depth Feature Fusion

Jingying Sun and Zhiguo Shi[✉]

University of Science and Technology Beijing, Beijing 100083, China
szg@ustb.edu.cn

Abstract. With the development of face recognition system towards automation and unsupervised, illegal intruders have become a serious threat to face authentication system by disguising face authentication system, how to ensure the security of the face recognition system has become an urgent problem in face recognition technology. Therefore, living face detection has become an important issue that must be solved in the face authentication system. By deeply studying the importance of facial image color feature information for human face detection, a deep feature fusion network structure is constructed by deep convolutional neural networks ResNet and SENet to effectively train the involved face anti-spoof data. The feature with large amount of information, while suppressing the features with low usefulness, the experimental results are greatly improved compared with the traditional methods, and have higher recognition effect and accuracy.

Keywords: Face spoofing detection · Convolutional neural network · Color texture analysis

1 Introduction

Face recognition technology has been widely used in current biometric technology because of its low cost, convenient use and high accuracy. However, As biometric information, especially face information, is easily acquired and forged, illegal certifiers can spoof the face of a legitimate certifier by stealing and recording a video of a legitimate certifier [1]. If the face recognition system does not have the detection function of these fake attacks, this will bring threat to the security of the face recognition system, and it will also hinder the promotion and application of face recognition technology.

The face biopsy has been studied for decades. The academic research institutions for face biopsy mainly include the Biometrics group led by the Li Ziqing team of the Institute of Automation of the Chinese Academy of Sciences, the senior researcher of the IDAIP laboratory in Switzerland, Sebastien Marcel, and the University of Southampton, UK. The Department of Visual Learning and Control, part of the Department of Visual Learning and Control, and the international biometrics expert Anil K. Jain, the Michigan State University Biometrics Research Group. In recent years, high-quality articles on living body detection by the above-mentioned organizations have been published in some top journals such as IEEE TIFS/TIP, and Springer published "Handbook of

© Springer Nature Singapore Pte Ltd. 2019
H. Ning (Ed.): CyberDI 2019/CyberLife 2019, CCIS 1138, pp. 285–300, 2019.
https://doi.org/10.1007/978-981-15-1925-3_21

Biometric Anti-Spoofing" by Sebastien Marcel in 2014. In-depth introduction of fingerprint, face, sound, iris, gait and other biometrics anti-spoofing methods, as well as performance evaluation indicators, international standards, legal aspects, ethical issues, etc. for biometric identification The further development of deception technology makes an important contribution.

Researchers have proposed a large number of anti-spoofing face detection methods by analyzing the essential differences between living faces and remake faces, such as texture features, spectral features, light reflection differences, and motion information [2], living face detection methods are mainly divided into the following categories, the living detection method based on the micro-texture feature of the face: the method mainly distinguishes the true and false face according to the difference between the forged face and the real face in the texture detail, Since the forged face is mainly acquired by secondary acquisition or multiple acquisitions, it has the difference between the shape and the local high-light fine micro-texture compared with the real face. Based on these differences, Li et al. [3] used Fourier spectral analysis to analyze the high-frequency components of face images to identify active faces. Since the size of the fake face is generally smaller than the real face, and the fake face is mainly planar, the high-frequency component in the frequency domain is less than the real face, and the fake face does not have local motion in time, so the frequency domain change is very small in the time domain. Using the spatial information of the image, the frequency domain features can be transformed into spatial domain features to distinguish. Maatta et al. proposed using the LBP operator to measure the micro-texture details of the image, extracting the LBP features of the face [5], Gabor wavelet features [6] and the directional gradient histogram, and using the support vector machine (SVM) to perform live detection on faces, The micro-texture-based methods can effectively distinguish true and false faces to a certain extent, but micro-texture-based methods are susceptible to illumination, image resolution, etc., especially in the case of video forgery.

Based on the dynamic feature detection of facial features, in addition to the static feature differences such as face micro-texture, there are dynamic feature differences between real face and fake face. The live face detection method based on dynamic features is currently applied on the mobile side. Widely, most of them use random dynamic instructions to interact. Singh et al. [7] proposed using Haar cascade classifier to analyze lip and eye movement. Schwartz [8] proposed video recognition gray-scale co-occurrence matrix, HOG (histogram of oriented gradient), LBP (local binary pattern) and other features to distinguish video authenticity. However, this method faces a serious challenge, that is, the attacker hollows out the eye area and the mouth area of the face of the legitimate user and makes corresponding action instructions behind the photo, and the method requires high user interaction, and The cost of fraud is low.

In recent years, face recognition algorithms have made great progress. One of the key points of this success is these existing data sets, which help to stimulate interest and progress in anti-spoofing research. However, compared with large-scale image classification and face recognition data sets, public face anti-spoofing data sets are far from meeting the requirements of practical applications, which is also a big problem hindering the development of new technologies. With the needs of human face detection research, more and more research institutions are devoted to the construction of human face detection data sets. This paper mainly uses the NUAA [9] face anti-spoof data set to study and propose a face anti-spoofing algorithm based on depth feature fusion.

2 Related Work

2.1 Image Color Space Feature Analysis

The RGB color space is the closest to our naked eye. The various colors are different combinations of red (R), green (G), and blue (B). They are also the most used color space. Most of the acquisition devices on the market collect images through the RGB color space, while the image output devices also display images in the RGB color space. The RGB color space uses three primary colors of R (red), G (green), and B (blue) as the three colors. The component, in which R, G, and B have a value range of 0 to 255, other colors can be represented by a combination of different values of R, G, and B components, for a total of 256 * 256 * 256 = 16777216 colors. The RGB color space can be represented by a three-dimensional unit cube. As shown in Fig. 1, the three axes of the three-dimensional space represent the three primary colors of red, green, and blue, respectively. The origin (0, 0, 0) represents black, and the vertex that is furthest from the origin. (1, 1, 1) represents white, the point on the R axis (1, 0, 0) represents red, the point on the G axis (0, 1, 0) represents green, and the point on the B axis (0, 0, 1) represents Blue, other colors are located in this unit cube. In the RGB color space, the color information and the brightness information are mixed. When the color is used to detect a specific object in the computer, the bright-ness information of the image is interference information, which is counterproductive to the detection effect, so the color detection effect using the RGB color space is not good enough.

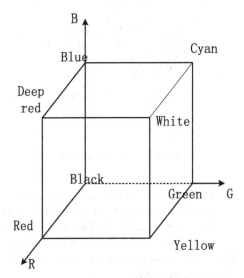

Fig. 1. RGB color space cube model

The HSV color space is visually uniform and has good consistency with human color vision. The representation based on HSV color space has important applications in image segmentation and feature extraction. The HSV color space consists of three components: Hues, Saturation, and Value. Hue (H) refers to the color we usually say, such as red, blue, etc., the value range of H is between (0–360), the different values represent different colors. For example, when H = 60, it means red, H = 120 represents yellow. Saturation (S) refers to the purity of the hue. The value of S ranges from (0 to 100%). The higher the purity, the more vivid the color. For example, when S = 100%, the color has the purest color, and the lower the purity, the color. The closer to gray, for example, when S = 0, it means gray, the brightness (V) represents the brightness of the color, the value of V is in the range of (0–100%), the higher the brightness, the brighter the color, such as V = 100%. The color has the brightest color. The lower the brightness, the darker the color. For example, when V = 0, it represents black. The HSV color model can be represented by a hexahedral model, as shown in Fig. 2, where the S axis represents saturation (S), the V axis represents brightness (V), and the chromaticity (H) is rotated from saturation (S).

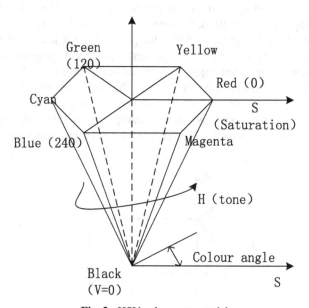

Fig. 2. HSV color space model

YCbCr color space is the color space used by many video compression coding. It is often used in video products such as video cameras and digital TV. For example, YCbCr color space is widely used in standards such as MPEG and JPEG [11]. The YCbCr color space is derived from the YUV color model, where Y refers to brightness, while Cb and Cr are U and V adjusted a small amount, and the Cr and Cb components represent the chromaticity of red and blue, respectively.

The R, G, and B components in the RGB color space contain luminance information, and there is a certain correlation between them. The color space has better performance and is relatively simple. The YCbCr color space can be directly obtained by linearly converting the RGB color space, which can separate the luminance information from the color component information and is less affected by the brightness variation. The HSV color space can separate the luminance V from the saturation S and the chrominance H. Therefore, the advantages of various color space models can be fully utilized, and the color space features can be effectively utilized.

2.2 Analysis of Human Face Detection Based on Image Color Information

In general, forged face images may contain local color changes due to media production. By using a printed or video image to display facial faces to perform facial spoofing attacks, it is conventionally possible to detect low-surface texture quality by analyzing the texture and quality of images. Attack attempts. Currently, most of the traditional research work on facial biopsy is based on analyzing the brightness (i.e., grayscale) information of the face image. If the image resolution (quality) is good enough, the details of the observed face can be captured. The texture analysis of the grayscale facial image can provide sufficient means to reveal the false face [12]. However, if we look closely at the face image of the real face and the corresponding dummy face, it is basically impossible to clearly see any texture differences between them, because the input image resolution is not high enough, so only the webcam quality image is used. The brightness information to detect higher quality false faces is more difficult or almost impossible to detect.

At present, the research based on image color information mainly focuses on how image color information is used for facial anti-spoofing. It is found that color information can be an important visual cues to distinguish false faces from real faces. The human eye does have more brightness than chromaticity. Sensitive, so when the same facial image is displayed in color, the fake face still looks very similar. However, if only the corresponding chrominance components are considered, some feature differences can be noted. Although gamut mapping and other artifacts are not clearly visible in grayscale or color images, they are very unique in chroma channels. Therefore, color texture analysis of chrominance images can be used to detect these gamut mappings and other (color) rendering factors.

In the field of face image color information research, Boulkenafet et al. [13, 14] introduced a novel method, using color texture analysis and proving that chrominance components are very useful in distinguishing between true and false faces, in order to understand which colors Space is most suitable for distinguishing between true and false faces. In this paper, three color spaces are considered, namely RGB, HSV and YCbCr, and several descriptors are used, namely color local binary mode (CLBP) [15], local phase quantization (LPQ).), adjacent local binary mode (CoALBP), binarized statistical image feature (BSIF), and scale variable descriptor (SID) [16] study facial color texture content, these texture features have been demonstrated in gray-scale texture-based facial living body Effective in testing, these features are used to analyze color textures by extracting facial descriptions from different color bands. The paper calculates low-level features from different descriptors extract-ed from different color spaces, using joint color texture information of luminance and chrominance channels. Experiments have been conducted to investigate the extent to which different color spaces and descriptors can be used to describe the inherent differences in color texture between real faces and false faces. At the same time, the study also performs fusion analysis to analyze the complementarity of different descriptors and different color spaces. The experiment used a common evaluation standard to conduct extensive experimental analysis of the dataset, and compared several of the most advanced face detection techniques. The results show that the proposed method can achieve stable performance in the benchmark data set and shows better. Generalization ability.

Through the above analysis, it can be found that all research methods based on image color information basically use hand-made features such as LBP, HoG and GLCM [9]. These methods need to perform texture feature analysis based on images and manually design features, and extract features. More cumbersome and less efficient. In recent years, with the rapid development of deep learning [17], convolutional neural networks have a place in the field of computer vision with extremely high performance [18]. Deep learning brings the main advantages to the learning model, and by designing a specific deep convolutional neural network (CNN) [19], the high-level features generated by the different layers of the model improve the classification results, through larger training data and A more complex, deeper learning model achieves better classification than traditional methods. Based on the above analysis, this paper analyzes the traditional manual design features, uses deep convolutional neural networks for feature extraction and classification, combines the idea of color space and feature fusion, designs a deep fusion network, and uses multiple streams with three sub-networks. The architecture performs fusion of complementary information between the three images to take full advantage of the features between the different modes of the image.

3   Face Anti-spoofing Algorithm Based on Depth Feature Fusion

3.1   The Overall Framework

In recent years, face recognition algorithms have made great progress, and the security of face recognition has been paid more and more attention. The problem of face detection is the classification of real faces and fake faces. Because these two types of images have high similarity, usually difficult to distinguish. At present, face recognition may face a variety of spoofing attacks. Usually, a person's face photo and face video are easily acquired. Currently, the most common face spoofing attacks are as follows: (1) Face photo spoofing attacks, such as flat photos, curved photos, cropping, bending, etc. camouflage faces by wearing real face masks and cutting facial features in different ways to perform face deception. (2) Face video spoofing attack, the spoofing attacker will first record a video containing a legal real face, the video can show a very similar effect to the real face, and the face video has more dynamic than the face photo Features such as blinking, shaking his head and other regular movements. The following figure shows several common face spoofing attacks (Fig. 3).

It is easy for humans to distinguish between real faces and fake faces, usually because humans can recognize the physical characteristics of many human faces, such as the details of face textures and the color details of face images. Based on such characteristics, the main idea of this paper is Aiming at the method of deep learning, combined with the idea of color space and feature fusion, this paper proposes a feature fusion human

Fig. 3. Common face spoofing attacks

face detection network structure. Through the model training for a large number of face images, experiments show that the network structure of this paper has better detection effect on false face images than traditional methods.

3.2 Face Feature Fusion Network Structure

It has been found through analysis that our human eyes are generally more sensitive to luminance than to chrominance, so when the same facial image is displayed in color, the false face still looks very similar. However, if only the corresponding chrominance components are considered, some feature differences can be noted. In order to gain a deeper understanding of which color space is more distinguishable between real faces and false faces, this paper considers three color spaces, RGB, HSV and YCbCr, which are the most common color spaces for sensing, representing and displaying color images. However, due to the high correlation between the three color components (red, green and blue) and the incomplete separation of luminance and chrominance information, its application in image analysis is very limited. Both the HSV and YCbCr color spaces are based on the separation of luminance and chrominance components. In the HSV color space, the hue and saturation dimensions define the chrominance of the image, while the dimension values correspond to the luminance. The YCbCr space separates the RGB components into luminance (Y), chroma blue (Cb), and chrominance red (Cr). The representations of the chrominance components in the HSV and YCbCr spaces are different. Different color channels are used to detect fake faces. provides higher contrast for different visual cues from natural skin tones. So they can provide a complementary facial color texture description for live face detection, and you can find more details about these color spaces (Fig. 4):

Since the face images of different color spaces have different characteristics, this paper mainly studies how to design a network structure to fuse complementary information between these three modes. For three different color space image data sets, we mainly use three sub-networks. A multi-stream architecture in which RGB, HSV, and YCbCr images are learned by each stream structure, and then a shared layer is added midway to learn joint representation and collaborative decision making. In this way, features from different modes can be merged. In order to improve the classification effect, in order to make full use of the features between different modes of images, this paper uses the squeeze and excitation fusion method, uses the SENet [20] structure branch to

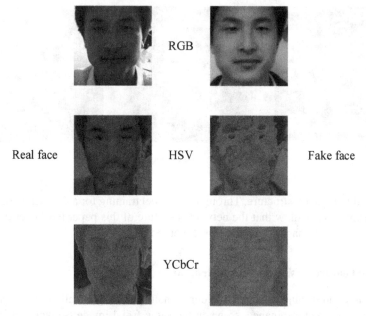

Fig. 4. Real face (left) and Fake face in different color spaces (right)

enhance the representation of different mode features by explicitly modeling the correlation between different mode feature convolution channels. then connect the shared layer structure and learn more distinguishing features from the fused feature structure. The basic network structure of this paper is shown in the figure below (Fig. 5).

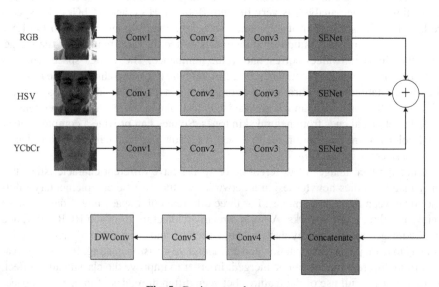

Fig. 5. Basic network structure

In this paper, the face image training data set is divided into three color space images as the input image data, and the input face images are all the same face. The basic model selects the basis of ResNet34 [21] with better classification effect. On the basis of the basic network structure ResNet34 has five convolution blocks (Conv1, Conv2, Conv3, Conv4, Conv5), of which Conv1, Conv2 and Conv3 convolutional blocks are proprietary for each layer structure, used to extract different modes The characteristics, and embedding the SENet structure after the Conv3 convolution constitutes the SE-ResNet structure (Fig. 6).

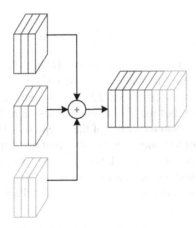

Fig. 6. Depth feature fusion structure

First, the convolution process before each layer of feature fusion structure SE-ResNet is the feature extraction operation of each type of image. The definition of input and output is as follows:

$$F_{tr} : X \rightarrow U, X \in R^{W' \times H' \times C'}, U \in R^{W \times H \times C} \tag{1}$$

$$u_c = v_c * X = \sum_{s=1}^{C'} v_c^s * x^s \tag{2}$$

$$v_c = \left[v_c^1, v_c^2, \ldots, v_c^{c'} \right], X = \left[x^1, x^2, \ldots, x^{c'} \right] \tag{3}$$

The F_{tr} convolution output feature U is a C feature map of size H * W, expressed as $U = [U_1, U_2, \ldots, U_C,]$, V_c represents the cth convolution kernel, X^s represents the sth input, and Uc Represents the cth two-dimensional matrix in U, and the subscript c represents the channel.

The feature compression is then performed along the spatial dimension, and each two-dimensional feature channel is transformed into a real number, which represents the global distribution of the response on the feature channel.

$$z_c = F_{sq}(u_c) = \frac{1}{H \times W} \sum_{i=1}^{H} \sum_{j=1}^{W} u_c(i, j) \tag{4}$$

Next, a full capture of the channel dependencies is performed, and a weight is generated for each feature channel by the parameter W, wherein the parameter W is learned to explicitly model the correlation between the feature channels.

$$s = F_{ex}(z, W) = \sigma(g(z, W)) = \sigma(W_2\delta(W_1 z)) \tag{5}$$

Finally, the weight s of the output is regarded as the importance of each feature channel after feature selection, and then weighted to the previous feature u by multiplication by channel, and the re-calibration of the original feature in the channel dimension is completed.

$$\tilde{x}_c = F_{scale}(u_c, s_c) = s_c \cdot u_c \tag{6}$$

Concat connects three features into the Conv4 and Conv5 shared layers to learn more distinguishing features. For face-related recognition tasks, as shown in Fig. 7, although the feature map is in the corner cell and the FMap-end central unit theory. The sensing domains have the same size, but the central unit carries more facial information than the corner cells, and the pixels in the center of the receptive field have a greater impact on the output. Therefore, for the importance of different units, in the last convolutional layer After Conv5, the deep convolution (DWConv) layer is used instead of the global average pooling (GAP) to down-sample its output. At the same time, the fully connected layer is removed, the network parameters are reduced, and the network is more concise.

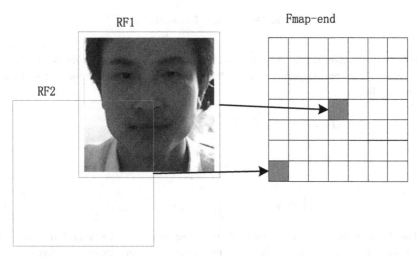

Fig. 7. Receptive field area of the middle part and the edge part of the feature map

By using the deep convolution (DWConv) layer to down-sample the output, the feature is directly flattened in-to a one-dimensional feature vector. The calculation process is expressed as follows:

$$FV_{n(y,x,m)} = \sum_{i,j} K_{i,j,m} \cdot F_{IN_y(i),IN_x(j),m} \tag{7}$$

Where is the flattened feature vector, (H, W, and C represent the height, width, and channel of the DWConv layer output feature map, respectively), representing the FV of the (y, x) unit in the first channel corresponding to the DWConv layer output feature map. The first element.

$$n(y, x, m) = m \times H' \times W' + y \times H' + x \tag{8}$$

On the right side of the equation, K is the convolution kernel in the opposite direction, and F is the Fmap end of the size h × w × c (h, w, and c represent the height, width, and channel of the feature map, respectively). m denotes the channel index, i, j denotes the spatial position in the k core, and denotes the corresponding position in the F core, and the calculation formula is as follows:

$$IN_y(i) = y \times S_0 + i \tag{9}$$

$$IN_x(j) = x \times S_1 + j \tag{10}$$

S0 is the vertical span and S1 is the horizontal span. After flattening the feature map, do not add a fully connected layer as this increases the risk of more parameters and overfitting.

4 Experiments

4.1 Data Collection and Environment Configuration

In order to evaluate the performance of the algorithm in live face detection, this paper selects the public data set for training. The experiment uses Pytorch framework as the experimental platform. The test platform hardware configuration is processor Intel Core i7-6700 K@4.00 GHz*8; graphics card GeForce GTX 1080; 4 GB of memory and 8 GB of video memory. The software uses the 64-bit version of Ubuntu 16.04.

The NUAA dataset is a public facial anti-spoofing dataset that is captured using a webcam. Face photos contain various appearance changes that facial recognition systems typically encounter (gender, lighting, with/without glasses), all Face photos are cropped and only the face area is preserved, for fake face samples, high definition photos taken by a typical Canon camera, the face area occupies at least 2/3 of the entire area of the photo, and fake face photos are made in two ways, The first method is to print them on photographic paper using a conventional method, and their ordinary sizes are 6.8 cm × 10.2 cm (small) and 8.9 cm × 12.7 cm (large), respectively. On the other hand, each photo was printed on 70 g of A4 paper using a conventional color HP printer. At the same time, five types of light attacks were simulated before the webcam: (1) horizontal, vertical, backward and forward photos; (2) rotating photos along the vertical axis; (3) same as (2) but along the horizontal axis (4) Bending the photo inward and outward along the vertical axis; (5) Same as (4) but along the horizontal axis. The data set sample and distribution are as follows (Fig. 8 and Table 1):

Fig. 8. Face image sample

Table 1. Data set distribution.

Type	Training set	Test set
Real face	3300	1700
Fake face	5700	1700
All face	9000	3400

4.2 Model Training Process

1. Image preprocessing: The training samples are adjusted according to the requirements of the model input size. The training first uses the RGB images in the dataset. The HSV and YCbCr color space images are obtained by color space conversion of RGB images; There is a sample imbalance phenomenon, and data enhancement processing is performed on samples with insufficient sample numbers, such as random scaling, random clipping, random offset, random rotation, etc. and finally standardized.

2. Training process: The initial learning rate of model training starts from 0.001, decays by 0.1 after every 60 cycles, and the momentum is set to 0.9. The Focal loss function weight parameter is initialized to $\alpha = 0.25$ and $\gamma = 3$. The training process of the model is monitored by the callback function, and the result with the least loss on the test set is saved as the final weight file of the model, and the training is ended when the model training reaches the set training round or the test set loss is not reduced.

4.3 Experimental Results and Analysis

In the experiment, accuracy rate (ACC), false positive rate (ERR) and half error rate (HTER) were used as indicators to evaluate the classification results of the living face detection algorithm. HTER is defined as:

$$HTER = \frac{FRR + FAR}{2} \tag{11}$$

In the above formula, FRR refers to the probability that a real face is mistakenly judged as a fake face, and FAR refers to the probability that a fake face is mistakenly judged as a real face.

In order to verify the effectiveness of the multi-mode color space fusion method in this paper, the spatial characteristics of RGB, HSV and YCbCr are verified experimentally on the NUAA dataset, and different fusion modes between different color space features are carried out. Various combinations of experimental analysis, The experimental comparison results are shown in Fig. 9 and Table 2 below.

In order to verify the validity and reliability of the proposed algorithm, the proposed algorithm and some classical convolutional neural networks such as ResNet and some classical face detection methods M-DOG (Multiple-DOG) algorithm [22], LTP The (Local Ternary Pattern) algorithm [23] and other comparative experiments were carried

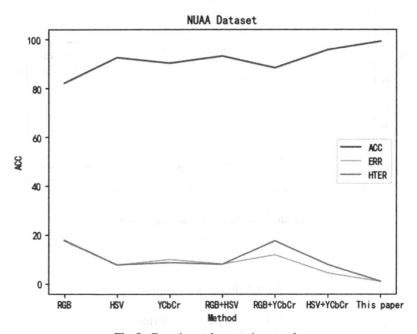

Fig. 9. Experimental comparison results

Table 2. Experimental results

Model	ACC	ERR	HTER
RGB	82.1	17.9	17.5
HSV	92.5	7.5	7.6
YCbCr	90.2	9.8	8.5
RGB+HSV	93.1	7.9	7.8
RGB+YCbCr	88.3	11.7	17.5
HSV+YCbCr	95.7	4.3	7.7
This paper	99.2	0.8	0.9

out to analyze the detection effect of different network structures on living faces under the same hardware conditions and preprocessing methods. The experimental comparison results are shown in Fig. 10 and Table 3 below.

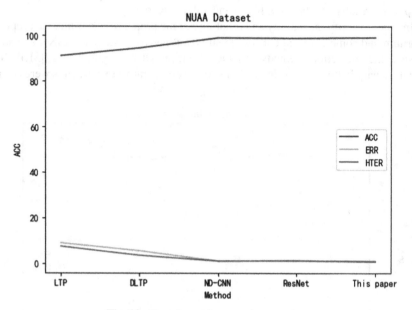

Fig. 10. Experimental comparison results

Table 3. Experimental results

Model	ACC	ERR	HTER
LTP	91.1	8.9	7.4
DLTP	94.5	5.5	3.5
ND-CNN	99.0	1.0	0.98
ResNet	98.9	1.1	1.2
This paper	99.2	0.8	0.9

The above table shows the effects of different color space features and different color space feature combinations on the detection rate of the algorithm. It can be seen that for a single color space feature, the HSV color space feature has the highest recognition accuracy on the NUAA dataset, and the recognition accuracy (ACC) on the NUAA dataset is 96.4%, and the false positive rate (ERR) is 3.6%. The half error rate (HTER) is 4.2%. The data are superior to the other two color space features, indicating that the HSV color space feature is the most important in the detection of human face. For different color space feature combination experiments, It can be seen that the three color space combinations of RGB+HSV+YCbCr are better than any other two combinations. The recognition accuracy (ACC) of the three color space combinations on the NUAA dataset is 96.4%, and the false positive rate (ERR) The 1.4% and half error rate (HTER) is 4.2%, which shows that the three color space combination methods can extract the texture information of the feature map more accurately and improve the overall performance of the live face detection algorithm.

From the experimental results of different detection algorithms in the table on the NUAA dataset, it can be seen that in the biometric detection method based on manual extraction of facial features, LTP (Local Ternary Pattern) algorithm and DLTP (Dynamic Local Ternary Pattern) The algorithm has poor detection performance on living faces. The detection performance of deep learning-based detection algorithm DB-CNN [24] and classical convolutional neural network ResNet is better than traditional methods. The accuracy of this algorithm is 96.4% and the false positive rate (ACC) on the NUAA dataset. ERR) is 1.4%, and the half-error rate (HTER) is 4.2%, which indicates that the live face algorithm has a good defense effect on photo attacks.

5   Conclusions

This paper mainly uses a deep feature fusion structure to detect the living face. Through the training experiment analysis on the face anti-spoofing data, the experimental results show that the training model has achieved good results and has large scalability. however, there is still a lot of work to be done in the direction of live face detection. The data is very important for the research based on deep learning algorithms. For now, the data set of live face detection is still very limited. There is also a need to further collect data, update and optimize the model.

In summary, the detection of live face is a challenging research field. At the same time, it is also a research direction full of opportunities. Through this research, we hope to inject new ideas into the research in this field, and the future will also attract more researchers to develop more advanced theoretical algorithms.

Acknowledgement. This study was supported by the National Natural Science Foundation of China (No. 61977005). The work was conducted at University of Science and Technology Beijing.

References

1. Galbally, J., Marcel, S., Fierrez, J.: Biometric anti-spoofing methods: a survey in face recognition. IEEE Access **2**, 1530–1552 (2017)
2. Xie, X., Lam, K.M.: Face recognition under varying illumination based on a 2D face shape model. Pattern Recogn. **38**, 221–230 (2005)
3. Li, J., Wang, Y.: Live face detection based on the analysis of Fourier spectra. Proc Spie **5404**, 296–303 (2004)
4. Maatta, J., Hadid, A., Pietikainen, M.: Face spoofing detection from single images using micro-texture analysis. In: 2011 International Joint Conference on Biometrics, pp. 1–7. IEEE (2011)

5. Tiago, D.F.P., Komulainen, J., Anjos, A.: Face liveness detection using dynamic texture. EURASIP J. Image Video Process. **1**, 1–15 (2014)
6. Yang, J., Lei, Z., Liao, S., Li, S.Z.: Face liveness detection with component dependent descriptor. ICB **1**, 1–6 (2013)
7. Singh, A.K., Joshi, P., Nandi, G.C.: Face recognition with liveness detection using eye and mouth movement. In: International Conference on Signal Propagation and Computer Technology, pp. 592–597 (2014)
8. Pinto, A., Schwartz, W.R., Pedrini, H.: Using visual rhythms for detecting video-based facial spoof attacks. IEEE Trans. Inf. Forensics Secur. **10**, 1025–1038 (2017)
9. Tan, X., Li, Y., Liu, J., Jiang, L.: Face liveness detection from a single image with sparse low rank bilinear discriminative model. In: Daniilidis, K., Maragos, P., Paragios, N. (eds.) ECCV 2010. LNCS, vol. 6316, pp. 504–517. Springer, Heidelberg (2010). https://doi.org/10.1007/978-3-642-15567-3_37
10. Konstantinos, N., Plataniotis, R.: Color Image Processing: Methods and Applications, vol. 8. CRC, New York (2007)
11. Malacaed, D.: Color vision and colorimetry: theory and applications. Color Res. Appl. **28**, 77–78 (2012)
12. Wen, D., Han, H., Jain, A.K.: Face spoof detection with image distortion analysis. IEEE Trans. Inf. Forensics Secur. **10**, 746–761 (2015)
13. Boulkenafet, Z., Komulainen, J., Hadid, A.: Face anti-spoofing using speeded-up robust features and fisher vector encoding. IEEE Signal Process. Lett. **24**, 141–145 (2017)
14. Boulkenafet, Z., Komulainen, J., Hadid, A.: Face spoofing detection using colour texture analysis. IEEE Trans. Inf. Forensics Secur. **11**, 1818–1830 (2017)
15. Choi, J.Y., Plataniotis, K., Ro, Y.M.: Using colour local binary pattern features for face recognition. In: Proceedings of the IEEE International Conference Image Process (ICIP), pp. 4541–4544 (2010)
16. Gragnaniello, D., Poggi, G., Sansone, C., Verdoliva, L.: An investigation of local descriptors for biometric spoofing detection. IEEE Trans. Inf. Forensics Secur. **10**(4), 849–863 (2015)
17. LeCun, Y., Bengio, Y., Hinton, G.: Deep learning. Nature **521**(7553), 436–444 (2015)
18. Szeliski, R.: Computer Vision: Algorithms and Applications. Springer, London (2010). https://doi.org/10.1007/978-1-84882-935-0
19. Hinton, G., Osindero, S.: A fast learning algorithm for deep belief nets. Neural Comput. **18**(7), 1527–1554 (2006)
20. Hu, J., Shen, L., Sun, G.: Squeeze-and-excitation networks. In: CVPR (2018)
21. He, K., Zhang, X., Ren, S., Sun, J.: Deep residual learning for image recognition. In: CVPR, pp. 770–778 (2016)
22. Zhang, Z., Yan, J., Liu, S.: A face anti-spoofing database with diverse attacks. In: Proceedings of the 5th IAPR International Conference on Biometrics, New Delhi, pp. 26–31 (2012)
23. Parveen, S., Ahmad, S.M.S., Abbas, N.H.: Face liveness detection using dynamic local ternary pattern (DLTP). Computer **5**(2), 10 (2016)
24. Alotaibi, A., Mahmood, A.: Deep face liveness detection based on nonlinear diffusion using convolution neural network. SIViP **11**(4), 713–720 (2017)

Facial Micro-expression Recognition Using Enhanced Temporal Feature-Wise Model

Ruicong Zhi[1,2]([✉]), Mengyi Liu[1,2], Hairui Xu[1,2], and Ming Wan[1,2]

[1] School of Computer and Communication Engineering, University of Science and Technology Beijing, Beijing 100083, People's Republic of China
zhirc_research@126.com
[2] Beijing Key Laboratory of Knowledge Engineering for Materials Science, Beijing 100083, People's Republic of China

Abstract. Automatic facial micro-expression recognition is challenging for the subtlety and transience in facial motion, and limited databases. Most researches focus on handcrafted techniques for facial micro-expression analysis on two-dimensional images. However, spatiotemporal facial feature representation is a critical issue for facial micro-expression recognition due to its short duration and subtle facial movement. To deeply extract the appearance characteristics and facial changes effectively from facial image sequences, a feature-wise deep learning model was proposed by applying temporal Convolutional Neural Network (3D-CNN) and Long Short-Term Memory (LSTM) to enhance temporal feature learning. There are two stages involved: (1) The CNN was extended to convolute along spatio and temporal simultaneously, to better represent the facial texture and motion. (2) The feature vector obtained by 3D-CNN was fed into LSTM for temporal enrichment. It was demonstrated that the proposed model achieved promising good performance on CASME II and SMIC databases on person-independent and cross-database experiments.

Keywords: Facial micro-expression · 3D Convolutional Neural Networks · Long Short Term Memory · Spatiotemporal features

1 Introduction

Facial expression is a vital nonverbal behavior in which human beings express their own emotional information. According to the formula of emotion expression defined by Mehrabian et al. [1], facial expression accounts for 55% of the emotional expression. It can be seen that the emotional information contained in the facial expression plays a prominent role in communication between people.

In 1960, Haggard and Isaacs [2] discovered a special emotion which was considered as an expression of repressed emotions related to self-defense mechanisms. However, it was not until 1969 that Ekman and Friesen discovered this particular emotion when invited to watch a video of a patient with depression. They found that the patient in the video suddenly showed a very fast and painful expression and disappeared in an instant. Ekman and Friesen [3] defined this special facial expression as a micro-expression.

© Springer Nature Singapore Pte Ltd. 2019
H. Ning (Ed.): CyberDI 2019/CyberLife 2019, CCIS 1138, pp. 301–311, 2019.
https://doi.org/10.1007/978-981-15-1925-3_22

Facial micro-expression can reveal the true feelings that people try to hide, and it has many great values in diverse fields, such as, clinical diagnosis, criminal detection, and security. Facial micro-expression recognition has posed a huge challenge to researchers, as a standard micro-expression lasts between 1/5 to 1/25 of a second and usually occurs in only specific parts of the face [4]. Nevertheless, facial micro-expression has received increasing attention in recent years.

Automatic facial micro-expression recognition is challenging for the subtlety and transience in facial motion, and limited databases. To deeply extract the appearance characteristics and facial changes effectively from facial image sequences, a feature-wise deep learning model was proposed by applying temporal Convolutional Neural Network (3D-CNN) and Long Short-Term Memory (LSTM) to enhance temporal feature learning. There are two stages involved: (1) The CNN was extended to convolute along spatio and temporal simultaneously, to better represent the facial texture and motion. (2) The feature vector obtained by 3D-CNN was fed into LSTM for temporal enrichment.

1.1 Related Work

Compared with facial macro-expression, the facial micro-expression has short duration and small intensity, leading to more challenge in extracting facial micro-expression features. Through the in-depth study of literature related to facial micro-expression recognition, it is found that the Local Binary Patterns (LBP) is a commonly applied feature extraction method for facial micro-expression recognition. LBP was put forward by Ojala [5], and it is a powerful texture descriptor to represent the features of facial expression due to its ability to derive local statistical patterns. However, the traditional LBP operator does not work well on dynamic texture extraction. Therefore, the ordinary LBP is extended to LBP from three orthogonal planes (LBP-TOP) [6], which is able to encode the spatiotemporal and motion features of image sequences.

Pfister et al. [7] utilized LBP-TOP for analyzing spontaneous facial micro-expression, and the temporal interpolation model (TIM) together with Multiple Kernel Learning (MKL) and Random Forest (RF) classifiers were used on spontaneous micro-expression corpus (SMIC) database. Li et al. [8] used LBP-TOP to extract dynamic features from facial micro-expression image sequences and SVM to perform classification and obtained the recognition accuracy of 48.78% on the SMIC. Wang et al. [9] proposed a lightweight descriptor based on the LBP-TOP, called Local Binary Patterns with Six Intersection Points (LBP-SIP), which reduced the redundancy in LBP-TOP patterns and leaded to more efficient computational complexity.

Motion information is also considered by optical flow methods. Shreve et al. [10] proposed a derivative of optical flow called an optical strain for facial micro-expression spotting and recognition. Some invariances of optical flow method were proposed such as Bi-Weighted Oriented Optical Flow (Bi-WOOF) [11] and Facial Dynamics Map [12]. These hand crafted features have achieved some success in facial macro- and micro-expression recognition, but there is still space to improve the performances, as the quality of the features depends heavily on the experience of the researchers.

Recently, deep learning has demonstrated its outstanding performance in image analysis. Deep learning is evolved from the study of neural networks which is composed of multiple processing layers and has powerful ability to learn hierarchical feature representations automatically. The Convolutional Neural Network (CNN) were one of the most popular deep learning models, which is successfully applied to image processing and pattern recognition tasks.

More recently, the deep learning methods have been applied to facial macro-expression recognition. And temporal facial features are utilized to describe the facial motion characteristics. Kahou et al. [13] presented a hybrid CNN-RNN architecture for facial expression analysis that outperformed a previously applied CNN approach using temporal averaging for aggregation. Byeon et al. [14] designed a 3D-CNN structure for video-based facial expression recognition. Hasani et al. [15] proposed a 3D-CNN for facial expression recognition which consisted of 3D Inception-ResNet layers followed by an LSTM unit to extract the spatial relations within facial images as well as temporal relations between different frames in the videos. Fan et al. [16] presented a video-based hybrid emotion recognition system which combined recurrent neural network (RNN) and 3D Convolutional networks (C3D) in a late-fusion fashion. Obviously, 3D-CNN and RNN/LSTM are superior to learn spatiotemporal features for video analysis. However, rare researches reported the performance of deep learning methods on facial micro-expression recognition problem, which may mainly due to the limited facial samples in available databases.

2 Proposed Method

In this work, we propose a feature-wise deep learning model for automatic facial micro-expression recognition. The proposed model aims to deeply extract the appearance characteristics and facial changes effectively from facial image sequences, and enhance the temporal characteristics for facial micro-expression. There are three stages involved: (1) The CNN was extended to convolute along spatio and temporal simultaneously, to better represent the facial texture and motion. (2) The feature vector obtained by 3D-CNN was fed into LSTM for temporal enrichment. The overview of the proposed architecture is illustrated in Fig. 1.

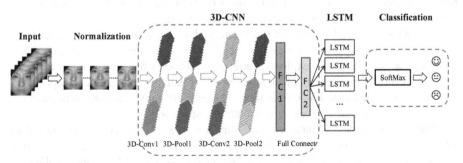

Fig. 1. Overview of proposed method.

2.1 Short-Term Spatiotemporal Features Learning

The CNN is deep feed-forward artificial neural networks using a variation of multilayer perceptron (MLP). The convolution operations are applied on the 2D static images from the spatial dimensions only. Compared to 2D-CNN, 3D-CNN has the ability to model temporal information better owing to 3D convolution and 3D polling operations. Recent researches proved that 3D-CNN is well-suited for spatiotemporal feature learning on video data [17, 18].

The 2D convolution is applied on an image and the output is an image, and even though the 2D convolution is applied on multiple images, the output is still an image. For 3D-CNN, the 3D convolution is conducted on image sequences and results in an output volume [19]. Hence 3D convolution preserves both the texture information and temporal information of the input signals. The same phenomena are applicable for 2D and 3D pooling.

Formally, the value at position (x, y, z) in the j th feature map in the i th layer is calculated as follows:

$$v_{ij}^{xyz} = \sigma\left(\sum_m \sum_{p=0}^{P_i-1} \sum_{q=0}^{Q_i-1} \sum_{r=0}^{R_i-1} w_{ijm}^{pqr} v_{(i-1)m}^{(x+p)(y+q)(z+r)} + b_{ij}\right) \tag{1}$$

where w_{ijm}^{pqr} is the (p, q, r) th value of the filter connected to the m th feature map in the previous layer. P_i and Q_i are the height and width of the 3D kernels, respectively. R_i is the depth of the 3D kernels along the temporal dimension, and the nonlinear activation function $\sigma(x) = \max(0, x)$ is utilized in our model, named Rectified Linear Units (ReLU).

2.2 High-Level and Long-Term Spatiotemporal Features

Recurrent neural network is a special structure of recursive neural network, which is a recursive neural network with linear chain structure expanded in time dimension [20]. Compared with the traditional feedforward neural network, the recurrent neural network has neuron feedback connection. This unique way of connection enables recurrent neural network to store data information in recent time periods.

One of the improved architectures of the normal recurrent neural network is the long short-term memory (LSTM), which has the ability to learn long-term dynamic while avoiding the vanishing and exploding gradients problems by introducing a memory cell [21]. LSTM unit consists of three gates: input gate, forget gate and output gate, which control how much new information is entered, how much history information is forgotten, and how much information is output, respectively.

The input sequence of data with arbitrary length $x = (x_1, x_2, \ldots, x_T)$ is fed to LSTM architecture. The output sequence $o = (o_1, o_2, \ldots, o_T)$ is estimated in an iterative manner in the recurrent hidden layer. The key operations of the LSTM are as follows:

forget gate:

$$f_t = \sigma(W^{(f)}x_t + U^{(f)}h_{t-1} + b^{(f)}) \tag{2}$$

input gate:

$$i_t = \sigma(W^{(i)}x_t + U^{(i)}h_{t-1} + b^{(i)}) \tag{3}$$

current input:

$$\tilde{c}_t = \tanh\left(W^{(c)}x_t + U^{(c)}h_{t-1} + b^{(c)}\right) \tag{4}$$

update of current memory information:

$$c_t = f_t c_{t-1} + i_t \tilde{c}_t \tag{5}$$

output gate:

$$o_t = \sigma(W^{(o)}x_t + U^{(o)}h_{t-1} + b^{(o)}) \tag{6}$$

The output of LSTM is determined by the output gate and the current memory information:

$$h_t = o_t \tanh(c_t) \tag{7}$$

2.3 Hybrid Network for Spatiotemporal Features Integration

The 3D convolution kernel only extracts spatiotemporal features from a small number of continuous images, while LSTM takes the whole image sequence as input. To enhance the 3D-CNNs' ability of temporal feature learning, a structure that combines 3D-CNNs with LSTM is proposed in this paper (Fig. 1).

The input data is a continuous 10-frame sequence of images. The 3D convolutional layers filters the input image sequence with 32 kernels of $3 \times 5 \times 5$ with stride of $1 \times 1 \times 1$, where 3 is the temporal depth and 5×5 is the size of the receptive field spatially. Then the data will be adjusted to enter the LSTM component, which has 10 LSTM units. At last, Softmax is adopted to classify the data for the final classification.

3   Results and Discussion

3.1   Datasets Description

CASME II: CASME II database [22] is the most extensive spontaneous facial micro-expression dataset to date, and it includes 247 facial micro-expression video clips recorded by a 200 fps camera with spatial resolution of 280×340 pixel from 26 participants. There are five classes of the facial micro-expressions in this dataset: happiness, surprise, disgust, regression and others.

SMIC: The SMIC-HS database [8] is used in the experiment, which includes 16 subjects with 164 spontaneous facial micro-expression recorded in a controlled scenario by 100 fps camera with resolution of 640×480. These video clips are classified into three main categories: positive (happiness), negative (sad, fear, disgust) and surprise.

In our experiments, the cropped images were converted to 32×32 pixel with gray scale. To normalize the length of facial micro-expression sequence, the spline interpolation was utilized to down-sample the clips along time axis and each facial micro-expression image sequence was normalized to 10 frames. The result of pretreatment is shown in Fig. 2.

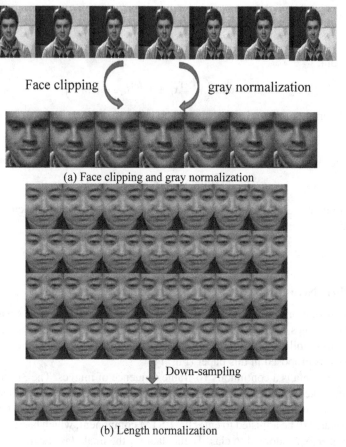

(a) Face clipping and gray normalization

Down-sampling

(b) Length normalization

Fig. 2. Example of data pretreatment.

3.2 Subject-Independent Experiments

Subject-independent facial micro-expression recognition task is more challenge and practical than subject-dependent task. In this subsection, two facial micro-expression databases were utilized, and the proposed algorithm was explored and compared to state-of-art for facial micro-expression analysis evaluation.

Implementation: There are 26 subjects in CASME II database, and the leave-one-subject-out (LOSO) cross validation was carried out to evaluate the proposed method. The experimental setup in details on CASME II was: the batch size was set to 50, the initial learning rate was set to 0.015, and the decay rate was 0.95. For SMIC-HS database, the 16 participants were involved in the experiments using LOSO cross validation. The experimental setup in details on SMIC-HS was: the batch size was set to 150, the initial learning rate was set to 0.015, and the decay rate was 0.95.

Comparison to State-of-Art

CASME II. The proposed 3D-CNN plus LSTM model was conducted for five-class facial micro-expression recognition on CASME II, and the accuracy was compared to some recent approaches (Table 1). The confusion matrix was shown in Fig. 3(a). It was shown that TIM+DCNN [25] obtained the best recognition accuracy of 64.9% under leave-one-video-out protocol, and the other methods were evaluated with leave-one-subject-out protocol. It demonstrated that the person-independent experiments were stricter than person-dependent experiments. Methods of [12, 23, 24] and [25] were traditional hand-craft methods, while the rest were deep learning based methods. It was observed from Table 1 the result of the deep learning methods was superior to the traditional methods despite of the limited training samples in the database. These results demonstrated that our method achieved better performance than most of the researches.

Table 1. Comparison of facial micro-expression accuracy on CASME II database.

Algorithms	Accuracy
Local LBP-TOP [23]	43.9%
FDM [12]	45.9%
Sparse sampling [24]	49.0%
HIGO [25]	57.1%
TIM+DCNN [26]	64.9%
CNN+LSTM [27]	61.0%
3D-CNN+LSTM	**62.5%**

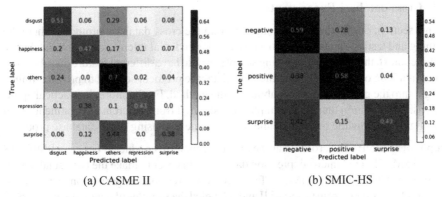

(a) CASME II   (b) SMIC-HS

Fig. 3. Confusion matrix for CASME II and SMIC-HS database.

SMIC-HS. Similarly, our method was applied for three-class facial micro-expression recognition on SMIC-HS, and the comparative results were illustrated in Table 2. The confusion matrix was shown in Fig. 3(b). We could see that the TIM+DCNN [26] achieved the highest recognition accuracy. However, TIM+DCNN [26] was the only one using leave-one-video-out cross validation protocols among the listed methods in Table 2. From comparative results, we could see that 3DCNN-LSTM obtained the promising and competitive performance.

Table 2. Comparison of facial micro-expression accuracy on SMIC-HS database.

Algorithms	Accuracy
AdaBoost + STM [28]	44.3%
LBP-TOP [6]	48.8%
FDM [12]	54.9%
OSW-LBP-TOP [29]	53.1%
CNN+SFS [30]	53.6%
TIM+DCNN [26]	65.9%
3D-CNN+LSTM	**56.6%**

In summary, we obtained the recognition accuracy of 62.5% and 56/6% on CASME II and SMIC-HS, respectively. We think the experimental results were reasonable. Firstly, the facial micro-expression sequences are not easy to identify and label, and it is difficult to train a good model using the insufficient samples. The 3D-CNN plus LSTM model is very simple implemented and the databases are utilized without data argument. Secondly, in terms of cross validation protocols, subject independent protocol is stricter than other protocol, i.e., leave-one-video-out. When samples of the same person appear both in the training set and testing set, it is more probably to classify correctly.

3.3 Cross-database Experiments

In this section, our method was evaluated using the cross-database protocol, where the training and testing samples were from two different facial micro-expression databases. In this context, the training and testing samples would have different feature distributions. The evaluation of a method using the intra-database protocol seemed to provide limited insight into the generalization capability of the method. Furthermore, in practical applications, it is common that facial samples are acquired from different environments, so it was highly worthwhile to evaluate the proposed method using cross-database protocol.

Implementation: The cross-database experiments were conducted between SMIC-HS and CASME II facial micro-expression databases. In order to make the two facial micro-expression databases share the same facial micro-expression categorization, we removed the samples of others from CASME II and relabeled the samples of happiness category as positive, and the samples of disgust and repression were categorized into negative. The labels of surprise samples from CASME II were unchanged. We referred the processed CASME II as New CASME II, which included the same three classes of the facial micro-expression as the SMIC-HS. The same network was applied for this experiment. The batch size was set to 150, the initial learning rate was set to 0.015, and the decay rate was 0.95.

Results and Analysis: Table 3 showed the recognition accuracy achieved in the cross-database experiments. It could be found that the recognition accuracy was higher in the case when the SMIC-HS was served as the training set than the case where the New CASME II was used as the training set. This was probably because the samples of the three classes of the facial micro-expressions in SMIC-HS were relatively balanced. In the literature of [31], the authors proposed a Target Sample Re-Generator (TSRG) method for facial micro-expression recognition, and achieved a rate of 50.61% when CASME II was served as the training set, and the accuracy was 64.19% when training with SMIC-HS database. It could be seen that the results of our method was much higher than the results of [31].

Table 3. Recognition rates in cross-database experiments.

	Train	Test	TSRG [31]	3D-CNN+LSTM
Case 1	New CASME II	SMIC-HS	50.61%	**54.0%**
Case 2	SMIC-HS	New CASME II	64.19%	**67.3%**

4 Conclusions

In this paper, we proposed a simple and effective model for facial micro-expression recognition based on three-dimensional Convolutional Neural Networks (3D-CNNs) and Long Short Term Memory (LSTM) networks, which firstly learned the short-term spatiotemporal features from the facial image sequences using 3D-CNNs and then learned the higher-level and long-term spatiotemporal features using LSTM. Two extensive experiments: subject-independent and cross-database between CASME II and SMIC database were conducted to evaluate the generalization of our proposed method. Experimental results showed that our method could achieve promising results and outperformed the recent state-of-the-art. Our future work will focus on detecting facial micro-expression in the wild environment and more evaluation of our method is to be conducted.

Acknowledgments. This work was supported by the National Research and Development Major Project (2017YFD0400100), the National Natural Science Foundation of China (61673052), the Fundamental Research Fund for the Central Universities of China (2302018FRF-TP-18-014A2), and the grant from Chinese Scholarship Council (CSC).

References

1. Mehrabian, A.: Communication without words. Communication Theory, pp. 193-200 (2008)
2. Haggard, E.A., Isaacs, K.S.: Micromomentary facial expressions as indicators of ego mechanisms in psychotherapy. In: Methods of Research in Psychotherapy, pp. 154–165. Springer US (1966). http://doi.org/10.1007/978-1-4684-6045-2_14

3. Ekman, P., Friesen, W.V.: Nonverbal leakage and clues to deception. Psychiatry **32**(1), 88–106 (1969)
4. Frank, M.G., Herbasz, M., Sinuk, K., et al.: I see how you feel: training laypeople and professionals to recognize fleeting emotions. In: The Annual Meeting of the International Communication Association. Sheraton, New York, New York City (2009)
5. Ojala, T., Harwood I.: A comparative study of texture measures with classification based on feature distributions. Pattern Recogn. **29**(1), 51–59 (1996)
6. Zhao, G., Pietikainen, M.: Dynamic texture recognition using local binary patterns with an application to facial expressions. IEEE Trans. Pattern Anal. Mach. Intell. **29**(6), 915–928 (2007)
7. Pfister, T., Zhao, G., Pietikainen, M.: Recognising spontaneous facial micro-expressions. In: International Conference on Computer Vision, pp.1449–1456. IEEE Computer Society (2011)
8. Li, X., Pfister, T., Huang, X., et al.: A spontaneous micro-expression database: inducement, collection and baseline. In: IEEE International Conference and Workshops on Automatic Face and Gesture Recognition, pp. 1–6. IEEE Computer Society (2013)
9. Wang, Y., See, J., Phan, R.C.-W., Oh, Y.-H.: LBP with six intersection points: reducing redundant information in LBP-TOP for micro-expression recognition. In: Cremers, D., Reid, I., Saito, H., Yang, M.-H. (eds.) ACCV 2014. LNCS, vol. 9003, pp. 525–537. Springer, Cham (2015). https://doi.org/10.1007/978-3-319-16865-4_34
10. Shreve, M., Godavarthy, S., Goldgof, D., et al.: Macro- and micro-expression spotting in long videos using spotio-temporal strain. In: The Ninth IEEE International Conference on Automatic Face and Gesture Recognition (2011)
11. Liong, S., See, J., Wong, K., Phan, R.C.-W.: Less is more: Micro-expression recognition from video using apex frame. Sig. Process Image Commun. **62**, 82–92 (2018)
12. Xu, F., Zhang, J., Wang, J.Z.: Microexpression identification and categorization using a facial dynamics map. IEEE Trans. Affect. Comput. **8**(2), 1-1 (2016)
13. Kahou, S.E., Michalski, V., Konda, K., et al.: Recurrent neural networks for emotion recognition in video. In: Acm International Conference on Multimodal Interaction, pp.467-474. ACM (2015)
14. Byeon, Y.H., Kwak, K.C.: Facial expression recognition using 3D convolutional neural network. Int. J. Adv. Comput. Sci. Appl. **5**(12) (2014)
15. Hasani, B., Mahoor, M.H.: Facial expression recognition using enhanced deep 3D convolutional neural networks. arXiv preprint arXiv:1705.07871 (2017)
16. Fan, Y., Lu, X., Li, D., Liu, Y.: Video-based emotion recognition using cnn-rnn and c3d hybrid networks. In: Proceedings of the 18th ACM International Conference on Multimodal Interaction, pp. 445–450. ACM (2016)
17. Zhu, G., Zhang, L., Shen, P., et al.: Multimodal gesture recognition using 3D convolution and convolutional LSTM. IEEE Access PP(99), 1–1 (2017)
18. Tran, D., Bourdev, L., Fergus, R., et al.: Learning spatiotemporal features with 3D convolutional networks. In: IEEE International Conference on Computer Vision, pp. 4489–4497 (2014)
19. Ji, S., Xu, W., Yang, M., et al.: 3D convolutional neural networks for human action recognition. IEEE Trans. Pattern Anal. Mach. Intell. **35**(1), 221-231 (2013)
20. Nair, V., Hinton, G.E.: Rectified linear units improve restricted Boltzmann machines. In: International Conference on Machine Learning, DBLP, pp. 807–814 (2010)
21. Irsoy, O., Cardie, C.: Deep recursive neural networks for compositionality in language. In: Advances in neural information processing systems, pp. 2096–2104 (2014)
22. Yan, W.J., Li, X., Wang, S.J, et al.: CASME II: an improved spontaneous micro-expression database and the baseline evaluation. Plos One 9(1), e86041 (2014)

23. Zhang, S., Feng, B., Chen, Z., Huang, X.: Micro-expression recognition by aggregating local spatio-temporal patterns. In: Amsaleg, L., Guðmundsson, G.Þ., Gurrin, C., Jónsson, B.Þ., Satoh, S. (eds.) MMM 2017. LNCS, vol. 10132, pp. 638–648. Springer, Cham (2017). https://doi.org/10.1007/978-3-319-51811-4_52

24. Le Ngo, A.C, See, J., Phan, C.W.R.: Sparsity in Dynamics of Spontaneous Subtle Emotion: Analysis & Application. IEEE Trans. Affective Comput. **8**(3), 396–411 (2017)

25. Li, X., et al.: Towards reading hidden emotions: a comparative study of spontaneous micro-expression spotting and recognition methods. IEEE Trans. Affect. Comput. **9**(4), 563–577 (2018)

26. Mayya, V., Pai, R.M., Pai, M.M.M.: Combining temporal interpolation and DCNN for faster recognition of micro-expressions in video sequences. In: 2016 International Conference on. IEEE Advances in Computing, Communications and Informatics (ICACCI), pp. 699–703 (2016)

27. Kim, D.H., Baddar, W.J., Ro, Y.M.: Micro-expression recognition with expression-state constrained spatio-temporal feature representations. In: Proceedings of the 2016 ACM on Multimedia Conference, pp. 382–386. ACM (2016)

28. Le Ngo, A.C., Phan, R.C.-W., See, J.: Spontaneous subtle expression recognition: imbalanced databases and solutions. In: Cremers, D., Reid, I., Saito, H., Yang, M.-H. (eds.) ACCV 2014. LNCS, vol. 9006, pp. 33–48. Springer, Cham (2015). https://doi.org/10.1007/978-3-319-16817-3_3

29. Liong, S.-T., See, J., Phan, R.C.-W., Le Ngo, A.C., Oh, Y.-H., Wong, K.: Subtle expression recognition using optical strain weighted features. In: Jawahar, C.V., Shan, S. (eds.) ACCV 2014. LNCS, vol. 9009, pp. 644–657. Springer, Cham (2015). https://doi.org/10.1007/978-3-319-16631-5_47

30. Patel, D., Hong, X., Zhao, G.: Selective deep features for micro-expression recognition. In: 2016 23rd International Conference on Pattern Recognition (ICPR), pp. 2258–2263. IEEE (2016).

31. Zong, Y., Huang, X., Zheng, W., et al.: Learning a target sample re-generator for cross-database micro-expression recognition. arXiv preprint arXiv:1707.08645 (2017)

Dynamic Facial Feature Learning by Deep Evolutionary Neural Networks

Ruicong Zhi[1,2(✉)], Caixia Zhou[1,2], and Tingting Li[1,2]

[1] School of Computer and Communication Engineering, University of Science and Technology Beijing, Beijing 100083, People's Republic of China
`Zhirc_research@126.com`
[2] Beijing Key Laboratory of Knowledge Engineering for Materials Science, Beijing 100083, People's Republic of China

Abstract. Facial Action Coding System is a comprehensive and anatomical system which could encode various facial movements by the combination of basic AUs (Action Units), and makes the emotion categories much wider. Recently, deep learning has been shown its superiority on recognition tasks. Despite the powerful feature learning ability of deep learning, there are still several problems remained. Firstly, a large amount of training data is needed to fully extract features and avoid overfitting. Secondly, the parameters optimization of deep neural network is complex, and the direct guidance of the results is insufficient. In this paper, a spatiotemporal self-learning method is designed by evolutional deep neural network model, and spatial augmentation is utilized to deal with the two problems facing in practical application. The proposed method is conducted on AUs analysis task which is important for emotion identification. The 3D convolutional neural network which could learn dynamic facial features from AUs image sequences is optimized automatically for the topology and hyper-parameters by evolutional scheme. Extensive experiments demonstrated the effectiveness of EVONET (Deep Evolutionary Neural Networks) on the facial databases over alternative methods, including 3DCNNs (3D Convolutional Neural Networks), and several convolutional neural network based models.

Keywords: Deep Evolutionary Neural Networks · Convolutional Neural Networks · AUs intensity estimation · Data augmentation · Dynamic facial features

1 Introduction

Facial expression analysis is a rapidly growing field of research, due to the constantly increasing interest in applications for automatic human behavior analysis and novel technologies for human-machine communication and multimedia retrieval [1]. Presently, discrete facial expressions categories are popular in affective analysis, and the widely used example is the six primary facial expressions, including anger, fear, sadness, disgust, surprise, and happiness, based on the underlying assumption that humans express the primary emotions universally regardless of cultures. Another popular facial behavior

© Springer Nature Singapore Pte Ltd. 2019
H. Ning (Ed.): CyberDI 2019/CyberLife 2019, CCIS 1138, pp. 312–327, 2019.
https://doi.org/10.1007/978-981-15-1925-3_23

measurement is the Facial Action Coding System (FACS) which is a sign judgment method. Facial expression categories describe facial behaviors in a global way, while facial action units depict the local variations on face. FACS is a comprehensive and anatomical system which could encode various facial movements by the combination of basic AUs (Action Units), and makes the emotion categories much wider [2].

Recently, deep learning has been shown its superiority on recognition tasks. Despite the powerful feature learning ability of deep learning, there are still several problems remained. Firstly, a large amount of training data is needed to fully extract features and avoid overfitting. Secondly, the parameters optimization of deep neural network is complex, and the direct guidance of the results is insufficient. To address the problem of insufficient expression data, data augmentation [3] and transfer learning [4–6] are two commonly used means. Moreover, the construction of neural networks is designed deeper and deeper to improve the ability of tackling big-data problems, which leading to an increasing number of parameters. To implement DNN optimization automatically, a number of neural network evolution algorithms have been developed over the last decade. With the popularization of deep learning, NeuroEvolution which refers to methodologies that aim at the automatic search and optimization of the NNs' parameters, has been vastly used.

Generally, the automatic generation of deep neural networks can be grouped into three categories: evolution of the network parameters, evolution of the network topology, and evolution of both the topology and parameters. The parameters optimization of networks could be automatically implemented by Evolutionary Computation (EC), such as CoSyNE [7], Gravitational Swarm and Particle Swarm Optimization applied to OCR [8], training of deep neural networks [9]. The topology of the neural network is fixed during evolution when EC is utilized for parameters optimization. Besides, approaches tackling the automatic evolution of the topology have also been proposed by using off-the-shelf learning methodology for finding adequate weights. Both the weights and the topology of neural networks could be evolved simultaneously regarding network representation by structured grammar. Methodologies using direct encodings and indirect encodings for neural networks representation are applied for deep learning evolution. For example, Topology-optimization Evolutionary Neural Network [10] and NeuroEvolution of Augmenting Topologies [11] utilized direct representation for genotype of the deep neural networks. Cellular Encoding [12] utilized indirect encoding scheme by transforming the genotype into an interpretable network.

Grammatical Evolution (GE) is a genotypic representation for neural networks, and several researches applied it to NeuroEvolution for the evolution of both the parameters and the topology of neural networks. Tsoulos et al. [13] conducted GE to evolve the topology and weights of one-hidden-layered neural networks. In another research, GE was applied to topology evolution, and Genetic Algorithm (GA) was utilized for parameters evolution [14]. The GE is not suited for real value tuning, so that GA was utilized to optimize the real values such as weights and bias. The vast majority of NE works target the evolution of small networks for very specific tasks. Therefore, evolving deep neural networks often resort to layer-based encodings [15, 16]. Assuncao et al. [17] proposed a Dynamic Structured Grammatical Evolution (DSGE) to generate a new genotypic representation for neural networks, so that the GE could deal with evolving networks with more than one hidden-layer. Denser was proposed in [18] encoding the structure of network by GA and the parameters associated to a layer by DSGE, and the scheme was utilized to automatic design of CNNs (Convolutional Neural Networks).

In this paper, a spatiotemporal self-learning method is designed by evolutional deep neural network model, and spatial augmentation is utilized to deal with the two problems facing in practical application. The proposed method is conducted on AUs analysis task which is important for emotion identification. The 3D convolutional neural network which could learn dynamic facial features from AUs image sequences is optimized automatically for the topology and hyper-parameters by evolutional scheme. AUs intensity estimation is a challenge task and fine facial features are required to represent facial changes. Automatic optimized deep learning model with large scale of samples could benefit the AUs analysis task effectively.

2 Data Augmentation

Large scale of samples is an important condition for excellent performance of deep learning. It is similar as the more books you read, the better you will write. AUs annotation is expensive and the number of labeled samples is limited. Therefore, it is necessary for data expansion to obtain more annotated facial samples.

Deep neural networks require a large amount of training data, and the existing facial expression databases are not sufficient to train the well-known neural network with deep architecture that achieved the most promising results in facial expression recognition tasks.

In this paper, a variance of Generative Adversarial Networks which converges more stably and is easy to train is utilized, that is, Boundary Equilibrium Generative Adversarial Networks. For BEGAN, the Wasserstein distance $W(\mu_1, \mu_2)$ is introduced to express the distance between the two distributions of auto-encoder losses, i.e. μ_1 is the distribution of the loss $\mathcal{L}(x)$, μ_2 is the distribution of the loss $\mathcal{L}(G(z))$. x and $G(z)$ denote real samples and generated samples individually.

The principle of BEGAN is maximizing discriminator function while minimizing the generator function, which could be expressed as

$$\begin{cases} \mathcal{L}_D = \mathcal{L}(x) - k_t \mathcal{L}(G(z_D)) & for\, \theta_D \\ \mathcal{L}_G = \mathcal{L}(G(z_D)) & for\, \theta_G \\ k_{t+1} = k_t + \lambda_k(\gamma \mathcal{L}(x)) - \mathcal{L}(G(z_D)) & for\, each\, training\, step\, t \end{cases} \tag{1}$$

The parameters θ_D and θ_G are updated by minimizing the losses $\mathcal{L}_D$ and $\mathcal{L}_G$. In practice the balance between the generator and discriminator losses is important to maintain. The Proportional Control Theory is utilized to maintain the equilibrium $\mathbb{E}[\mathcal{L}(G(z))] = \gamma \mathbb{E}[\mathcal{L}(x)]$. The variable $k_t \in [0, 1]$ is used to control how much emphasis is put on $\mathcal{L}(G(z_D))$ during gradient descent, and λ_k is the proportional gain for k, i.e. learning rate. In contrast to traditional GANs which require alternating training D and G, the BEGAN algorithm requires neither to train stably.

The BEGAN is trained on BP4D facial dataset to generate facial samples, and the new facial samples are obtained on step 40,000. Adam is used during training with the default hyper parameters, with an initial learning rate of 0.0001, with resolution of 32 × 32. In the structure of BEGAN, the 3 × 3 convolutions with exponential linear units (ELUs) are applied at their outputs. Each layer is repeated at twice. To observe the quality of facial samples, the generated results are output on every 500 steps, and the images are shown in Fig. 1. Moreover, the difference between real image and generated image is measured by the histogram of two images (Fig. 2). The Hamming distance is utilized to calculate the similarity and the similarity degree is 0.625. Data argumentation is conducted for AUs facial images and the dataset scale is expanded to be balanced for each AU type.

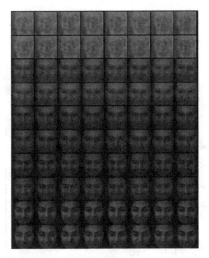

Fig. 1. The generating procedure of BEGAN.

3   Evolution of 3DCNNs

3.1   Three-Dimensional Convolutional Neural Networks (3DCNNs)

The deep neural network utilized in this paper is derived from the classic LeNet-5 [19], while the convolution layer is conducted by 3D convolutional kernel, which could effectively learn spatiotemporal features from image sequences. The 3D convolution operations are performed by convolving the 3D filters on the frame cube and sharing the weights along the time axis [20]. The specific formula for weight calculation is as follows:

$$v_{ij}^{xyz} = \sigma\left(\sum_m \sum_{p=0}^{P_i-1} \sum_{q=0}^{Q_i-1} \sum_{r=0}^{R_i-1} w_{ijm}^{pqr} v_{(i-1)m}^{(x+p)(y+q)(z+r)} + b_{ij}\right) \tag{2}$$

where v_{ij}^{xyz} represents the value at point (x, y, z) in the j-th feature map of the i-th layer in the network. w_{ijm}^{pqr} is the value of the point at (p, q, r) on the m-th feature map, which is the output of the upper layer connected by the same convolution kernel. P_i and Q_i are the height and width of the 3D filter respectively, R_i is the depth of the 3D filter along the temporal dimension. b_{ij} denotes the bias and $\sigma(\cdot)$ is the activation function.

There are three layers of convolution and two pooling layers, followed by two full connection layers, so that to map the high dimensional features to low dimensional space, and to get more effective features for classification task. There are several parameters that can be adjusted and optimized in the neural network, such as learning rate, regularized penalty coefficient, dropout rate during training, and the number of training iterations. Moreover, the size of input images and the size of convolutional kernel also have an impact on the performance.

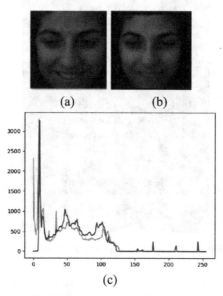

Fig. 2. Facial sample generated by BEGAN. (a) original image (b) generated image (c) histogram of the two images, where the red line represents the original image and the blue line represents the generated image. (Color figure online)

3.2 Evolution of Multi-layered Neural Networks

The Deep Neural Network (DNN) optimization is implemented based on evolutionary algorithm. The evolutionary deep neural network represents the structure of network by an ordered sequence of feedforward layers, and the parameters associated to the layers. Genetic Algorithms (GAs) and Dynamic Structured Grammatical Evolution (GSGE) are combined to encode the neural network structure at two different levels (i.e. DENSER in [18]), corresponding to layer sequences design and parameters learning. The 3DCNNs are optimized follow the DENSER evolution framework. The main steps in deep neural network evolution are illustrated in Fig. 3. Firstly, encode the macro structure of the neural network and their respective parameters through a user-defined Context-Free-Grammar

(CFG). A population is randomly initialized where the individuals in the population are the coding of neural network. Secondly, a fitness function is defined according to the target task, and each individual is evaluated by the fitness function. Thirdly, several variation operators are conducted to generate offspring from the parent through crossover and mutation, including layer lever and parameters lever. Finally, the optimal solution is selected when the evaluation of individual is satisfied, and the solution is evolutionary deep neural network.

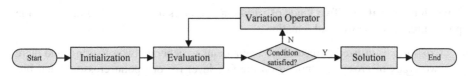

Fig. 3. Flow chart of the Evolutionary DNN algorithm.

The main component of evolutionary procedure is detailed as follows:

Representation: The deep neural network contains several feedforward layers with different function and several parameters need to be set to the layers. An innovation encoding manner is applied in neural network definition, including GA level for layer structure and DSGE level for parameter setting. The DSGE is a form of Genetic Programming, which consists of a list of genes, one for each non-terminal symbol, and a variable length representation is used. The grammar is defined as [18]: a CFG is a tuple $G = (N, T, S, P)$, where N is a non-empty set of non-terminal symbols, T is a non-empty set of terminal symbols. S is the starting symbol, and P is the set of production rules of the form $A ::= \alpha$, with $A \in N$ and $\alpha \in (N \cup T)^*$. N and T is disjoint. Language $L(G)$ is defined as $L(G) = \left\{ w : S \overset{*}{\Rightarrow} w, w \in T^* \right\}$, which composed by all sequences of terminal symbols that can be derived from the starting symbol.

The coding form of each individual is composed of network structure definition, and network parameters definition. For example, the valid sequence of layers can be specified as [(feature, 1,10),(classification, 1,2), (softmax, 1,1),(learning, 1,1)], where each tuple indicates the valid starting symbols, and the minimum and maximum number of times they can be used. The parameters of each layer are encoded by valid values or ranges. For example, the parameters of pooling layer include kernel size, stride, and padding, and the value of padding can be set to SAME or VALID, while stride can be set as [int, 1,1,3].

Fitness Evaluation: The candidate solutions of evolutionary deep neural networks (EDNN) need to be evaluated during the evolution by fitness function. As the EDNN model is applied for AUs analysis which is a typical classification task, the RMSE is utilized to evaluate the performance of candidate solutions. The fitness function is defined as

$$F = \prod_{c=1}^{n} \exp\left(\sqrt{\sum_{i=1}^{n_c} (y_i - o_i)^2 / n_c} \right) \tag{3}$$

where n is the number of pattern classes, n_c is the sample number in class c, y_i is the actual indicator of sample i, and o_i is the output indicator of sample i predicted by EDNN.

Fitness function is used to compare the performance of individual after multiple variation operators, and it is the main clue for offspring selection during evolution. The evolution process of the neural network is terminated when the fitness function reaches the goal set by user, or the number of iteration steps achieves the upper limit which is also set by user.

Variation Operators: Two kinds of variation operators are involved in the evolution procedure, i.e. crossover and mutation.

Crossover is utilized to recombine two parents to generate two offspring, and there are two types of crossover applied on the GA level for one-point crossover and bit-mask crossover. The one-point crossover is conducted to change layers within the same module, while bit-mask crossover is utilized to exchange modules between individuals. The module means a set of layers belongs to the same network structure, for example classification module.

- For one-point crossover, the cutting point is selected at the layer level randomly, and the offspring is obtained by exchanging genetic material delimited by the cutting point. The cutting point in one-point crossover takes consideration of the size variation of the same module for different individuals.
- For bit-mask crossover, a user defined bit-mask needs to be created with the size of the number of modules. The offspring is generated by exchanging modules between two individuals, and the selection rule is copying a module from the first individual if the bit is 1, while copying a module from the second individual if the bit is 0.

Mutation is utilized to change the genetic material upon two evolution levels respectively, i.e. GA level and DSGE level. The GA level mutation deal with the structure of neural network, and the layer sequences are updated by mutation operator implemented on layers, including add layer, replicate layer, and remove layer. The DSGE level mutation treats the content of each layer, and the parameters respective to layers are changed within valid value scope.

- GA level mutation: The layer sequence design could be conducted through three operators: add layer (generates a new layer randomly while not violate the maximum number of layers), replicate layer (copies an existing layer to another position of the module), and remove layer (delete a layer from a module randomly while not violate the minimum number of layers).
- DSGE level mutation: The parameters of layers respective to each module are updated in two manners: valid value mutation (the parameter value is replaced by another valid value), or numeric mutation (integer or float value is replaced by a new generated value). For example, the parameter padding mutated from SAME to VALID, the kernel size changed from 3×3 to 4×4, and float value can be generated randomly with Gaussian perturbation.

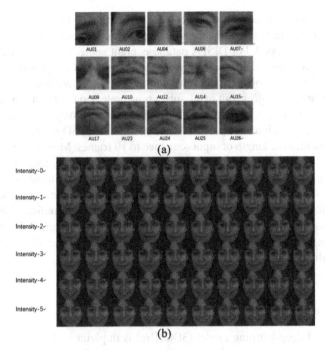

Fig. 4. Examples of facial images. (a) action units (b) AU sequences with different intensity.

4 Experiments

4.1 Datasets and Settings

Datasets: The proposed method is evaluated on two widely used datasets for facial AUs intensity estimation, that is, DISFA and BP4D. DISFA [21] consists of 27 adults with different ethnicity, with 12 women and 15 men. Each video frame is manually coded for the presence of facial action units, also for intensities of action units on a five ordinal scale. The number of events and frames for each intensity level of each AU are reported. BP4D [22] is a dynamic 3D video database of spontaneous facial expressions of 41 subjects, with 23 females and 18 males. Facial action units are annotated by frame-level, and the onset and offset of 27 action units are coded.

There are 12 action units coded in DISFA database, i.e. AU1, AU2, AU4, AU5, AU6, AU9, AU12, AU15, AU17, AU20, AU25, and AU26. And the BP4D database annotated 12 action units for AU1, AU2, AU4, AU6, AU7, AU10, AU12, AU14, AU15, AU17, AU23, and AU24. Figure 4 shows some examples of the AUs facial image.

Implementation: In our experiments, the dataset is divided into three subsets, i.e. training set, verification set and testing set. The training set and verification set are utilized for deep neural network optimization, and the performance of the deep learning model is verified on testing set.

To evaluate the generalization ability and robustness of the learning model, the person-independent scheme is utilized in the experiments, namely, the subjects of each subsets are not overlapped. For DISFA database, eight subjects are selected randomly, and their facial samples are utilized to form training set, facial samples of four subjects are used to form verification set, and the remaining 15 subjects are conducted as testing. For BP4D dataset, 13 subjects out of 41 subjects are selected to obtain training set, and the verification set and testing set contains 10 subjects and 18 subjects respectively. The length of facial expression sequence is different. Therefore, 3D spline interpolation is applied to normalize the length of input sequence to 10 frames. Moreover, the facial size is normalized to 32×32 by taking consideration of both accuracy and efficiency.

The AUs intensities are varied from A to E, indicating weak to strong. Due to the sample limitation of intensity 4 and 5, we utilize the facial samples with intensity A-C for classification. The performance of AUs intensity estimation is indicated by accuracy and ICC (Intra-class Correlation Coefficient), which is a powerful reliability measure for multi-class identification problems as it takes consideration of the difference between large and small errors between judges [23].

4.2 Evolutionary Topology of 3DCNNs

The evolution of deep learning model (3DCNNs) is implemented based on Keras, i.e. the grammatical derivation of defined grammar is fed to a Keras model running on top of TensorFlow. Keras is specified with its modularity which could combines modules freely with low cost.

The basic genes of Keras are available to form various combinations to get better network structure. Fitness is utilized to evaluate the superiority of the individual, and the bigger the fitness of individual is, the better the neural network model is. High accuracy is expected for the AUs intensity estimation task, therefore, the classification accuracy is chosen as the fitness function to assess the generalization and scalability ability of the neural networks.

The GA structure is set as: [(feature, 1, 30), (classification, 1, 10), (softmax, 1, 1)]. The 3DCNNs is trained with batches of 150, and varying learning rate policy is employed, i.e. the initial learning rate is 0.001, and it is varied along with decay steps. The structure of 3DCNNs for AUs intensity estimation is optimized by evolutionary scheme and the best neural network which achieved the best accuracy on test set is obtained as illustrated in Fig. 5. The figure depicted the best topology of the 3DCNNs, which is called as EVONET (Deep Evolutionary Neural Networks).

Conv3D is the three-dimensional convolution layer, and it learns spatio-temporal facial features with both texture information and facial changes along with time. Activation is a layer for mapping, and MaxPool3D is the pooling layer to eliminate irrelevant and unnecessary information out from facial features. The feature dimension can be reduced and the computational complexity is reduced too. Flatten layer turns the input multidimensional data to one dimension. Dropout refers to increasing elimination rate

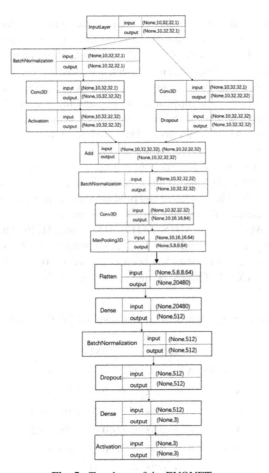

Fig. 5. Topology of the EVONET.

in training procedure to prevent overfitting, and improve generalization ability of the model. The deep neural network performs better when these layers combines in certain manner.

4.3 AUs Intensity Estimation

Experiments are implemented to evaluate the performance of optimized network in AUs intensity estimation. The original 3DCNNs and the optimized network by evolution (EVONET) are trained on DISFA database and BP4D database individually. The experimental results on the testing set are compared with accuracy (Table 1) and ICC (Fig. 6).

It can be seen from the illustrations that the proposed model performs well for most of the AUs with intensity B and C. As shown in the table, the optimized EVONET leads to higher accuracies than the original 3DCNNs network. High intensity of facial action units means great strength of the facial movement, and more obvious facial features could be extracted to represent the facial changes during AUs occurrence. ICC is an indicator representing the reliability of identification results, and the ICC of AU6, AU12, AU15, AU17, AU25 are high, which means high reliability for these AUs, while the ICC of AU1, AU2, AU4, and AU26 is not good. The possible reason is some of the AUs are similar and easy to be confused (e.g. AU1 and AU2).

Moreover, to further investigate the effectiveness of our proposed method, the ICC is compared with several CNN based algorithms (Table 2), i.e. SCNN [24], CCNN-IT, CCNN, and OCNN [25]. The bracketed and bold numbers indicate the best performance, and bold numbers indicate the second best. The ICC of AUs achieved by EVONET is the best for AU1, AU2, AU12, AU15, AU17, AU20, and AU26. There is an average enhancement of 6% (accuracy) and 0.21 (ICC) comparing to 3DCNNs. The comparison indicates that the average AUs ICC has been improved by 0.32, 0.27, 0.21, and 0.18 compared to OCNN, CCNN, CCNN-IT, and SCNN, respectively.

Table 1. Accuracies comparison between 3DCNNs and EVONET on DISFA database.

	3DCNNs			EVONET		
Intensity	A	B	C	A	B	C
AU1	0.36	0.62	0.45	0.46	0.59	0.51
AU2	0.11	0.64	0.41	0.20	0.61	0.45
AU4	0.15	0.58	0.62	0.64	0.58	0.62
AU6	0.82	0.45	0.29	0.78	0.69	0.64
AU9	0.63	0.75	0.71	0.21	0.54	0.67
AU12	0.87	0.75	0.71	0.86	0.82	0.79
AU15	0.89	0.75	0.42	0.87	0.65	0.73
AU17	0.67	0.67	0.89	0.71	0.69	0.87
AU20	0.25	0.35	0.53	0.34	0.47	0.56
AU25	0.63	0.33	0.57	0.75	0.52	0.64
AU26	0.85	0.74	0.43	0.86	0.81	0.67
Average	0.57	0.60	0.54	0.61	0.63	0.65

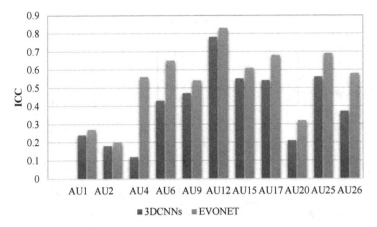

Fig. 6. ICC comparison between 3DCNNs and EVONET on DISFA database.

Table 2. Comparison of different methods on DISFA database.

	CCNN-IT	SCNN	CCNN	OCNN	3DCNNs	EVONET
AU1	0.18	0.16	0.14	0.04	**0.24**	**[0.27]**
AU2	0.15	0.12	0.14	0.04	**0.18**	**[0.20]**
AU4	**[0.61]**	0.43	0.37	0.41	0.12	**0.56**
AU6	**[0.65]**	0.62	0.46	0.35	0.43	**0.65**
AU9	**[0.55]**	**0.54**	0.44	0.19	0.47	**0.54**
AU12	**0.82**	**0.82**	0.64	0.72	0.78	**[0.83]**
AU15	0.44	0.43	0.25	0.23	**0.55**	**[0.61]**
AU17	0.37	0.37	0.37	0.45	**0.54**	**[0.68]**
AU20	**0.28**	**0.28**	0.09	0.06	0.21	**[0.32]**
AU25	**[0.77]**	**[0.77]**	0.58	0,53	0.56	**0.69**
AU26	**0.54**	0.53	0.31	0.44	0.37	**[0.58]**

The BP4D database is also utilized to evaluate the performance of proposed EVONET. There are only five AUs annotated with intensity levels in BP4D, therefore, AU6, AU10, AU12, AU14 and AU17 are involved in the experiments. Firstly, the accuracy of different AUs intensity estimation obtained by 3DCNNs and EVONET are compared in Fig. 7. Secondly, the performance of proposed model is compared with other researches as shown in Table 3.

According to the results in Fig. 2, it could also be observed that higher accuracy is achieved with higher intensity, and AUs with intensity C obtained the highest accuracy. The accuracies obtained by AU6, AU10, and AU12 exceed 80%, mainly due to the large difference between these three AUs, and the relatively large scale of facial samples. The ICC of EVONET is 0.71, while it is 0.68 for 3DCNNs.

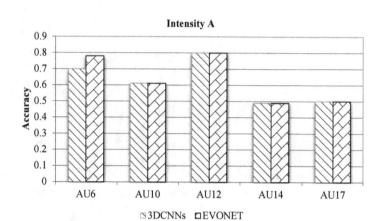

(a)

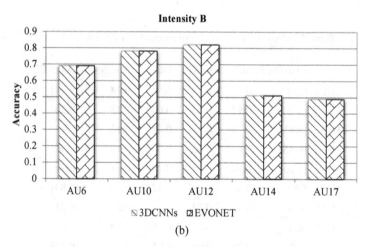

(b)

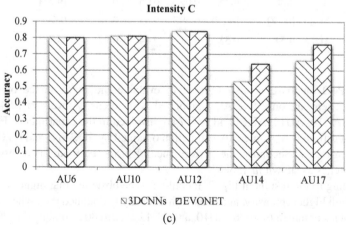

(c)

Fig. 7. Comparison of accuracy between 3DCNNs and EVONET on BP4D database.

Table 3. Comparison of different methods on BP4D database.

	AU6	AU10	AU12	AU14	AU17
Single Heatmap	**0.78**	**0.79**	0.68	**0.54**	0.49
ResNet	0.71	0.76	0.84	0.36	0.44
2DC	0.76	0.71	0.84	0.43	**0.53**
VGP-AE	0.75	0.66	**0.88**	0.47	0.49
3DCNNs	0.75	0.78	0.83	**[0.55]**	0.51
EVONET	**[0.79]**	**[0.80]**	**[0.89]**	0.52	**[0.54]**

The results are also compared with a number of algorithms for AUs analysis on BP4D database (Table 3), including Single Heatmap [26], ResNet [27], 2DC, and VCP-AE [2858]. Single Heatmap is proposed by Du et al. [26], it trained a network model for each AU, and each network model returned a heatmap. It can be seen from the table that the Single Heatmap method achieved the second best results for AUs intensity estimation. Besides, we also compared out model with two hidden layer model (2DC) and Deep structured network (VGP-AE). The average ICC of EVONET is 0.71, and it is 0.05 higher than Single Heatmap, and 0.08 higher than ResNet.

5   Conclusion

To address the two practical problems of deep learning method (facial samples limitation and neural network optimization), the spatial augmentation and evolutional deep neural network model is conducted in this paper. The deep evolutional neural network is automatically designed and applied to AUs analysis task. The proposed algorithm is evaluated on the well applied facial database, i.e. DISFA and BP4D, and the indicators of accuracy and ICC are used for evaluation. Extensive experiments demonstrated the effectiveness of EVONET on the facial databases over alternative methods, including 3DCNNs, and several convolutional neural network based models. Further work will focus on exploring the relationship between AUs and emotions, so that to develop a comprehensive automatic facial expression system, which could learn facial features automatically from optimized deep learning model.

Acknowledgments. This work was supported by the National Research and Development Major Project (2017YFD0400100), the National Natural Science Foundation of China (61673052), the Fundamental Research Fund for the Central Universities of China (2302018FRF-TP-18-014A2), and the grant from Chinese Scholarship Council (CSC).

References

1. Valstar, M.F., et al.: Second facial expression recognition and analysis challenge. In: 11th IEEE International Conference and Workshops on Automatic Face and Gesture Recognition, FERA 2015, pp. 1–8 (2015)

2. Zhi, R., Liu, M., Zhang, D.: A comprehensive survey on automatic facial action unit analysis. Vis. Comput. 1–27 (2019)
3. Lopes, A.T., Aguiar, E.D., Souza, A.F.D., Oliveira-Santos, T.: Facial expression recognition with convolutional neural networks: coping with few data and the training sample order. Pattern Recogn. **61**, 610–628 (2017)
4. Ruiz-Garcia, A., Elshaw, M., Altahhan, A., Palade, V.: Stacked deep convolutional auto-encodes for emotion recognition from facial expressions. In: International Joint Conference on Neural Networks, pp. 1586–1593 (2017)
5. He, K., Zhang, X., Ren, S., Sun, J.: Deep residual learning for image recognition. In: 2016 IEEE Conference on Computer Vision and Pattern Recognition (CVPR), pp. 770–778 (2016)
6. Zhi, R., Xu, H., Wan, M., Li, T.: Combining 3D convolutional neural networks with transfer learning by supervised pre-training for facial micro-expression recognition. IEICE Trans. Inf. Syst. **E102-D**(5), 1054–1064 (2019)
7. Gomez, F., Schmidhuber, J., Miikkulainen, R.: Accelerated neural evolution through cooperatively coevolved synapses. J. Mach. Learn. Res. **9**, 937–965 (2008)
8. Fedorovici, L.O., Precup, R.E., Dragan, F., Purcaru, C.: Evolutionary optimization-based training of convolutional neural networks for OCR applications. In: 2013 17th International Conference on System Theory, Control and Computing (ICSTCC), pp. 207–212 (2013)
9. David, O.E., Greental, I.: Genetic algorithms for evolving deep neural networks. In: Proceedings of the 2014 Conference on Companion on Genetic and Evolutionary Computation Companion, pp. 1451–1452. ACM (2014)
10. Rocha, M., Cortez, P., Neves, J.: Evolution of neural networks for classification and regression. Neurocomputing **70**(16), 2809–2816 (2007)
11. Stanley, K.O., Miikkulainen, R.: Evolving neural networks through augmenting topologies. Evol. Comput. **10**(2), 99–127 (2002)
12. Gruau, F.: Genetic synthesis of Boolean neural networks with a cell rewriting developmental process. In: International Workshop on Combinations of Genetic Algorithms and Neural Networks, pp. 55–74. IEEE (1992)
13. Tsoulos, I., Gavrilis, D., Glavas, E.: Neural network construction and training using grammatical evolution. Neurocomputing **72**(1), 269–277 (2008)
14. Ahmadizar, F., Soltanian, K., AkhlaghianTab, F., Tsoulos, I.: Artificial neural network development by means of a novel combination of grammatical evolution and genetic algorithm. Eng. Appl. Artif. Intell. **39**, 1–13 (2015)
15. Suganuma, M., Shirakawa, S., Nagao, T.: A genetic programming approach to designing convolutional neural network architectures. In: Proceedings of the Genetic and Evolutionary Computation Conference, pp. 497–504 (2017)
16. Miikkulainen, R., et al.: Evolving deep neural networks. arXiv preprint arXiv:1703.00548 (2017)
17. Assuncao, F., Lourenco, N., Machado, P., Ribeiro, B.: Towards the evolution of multi-layered neural networks: A dynamic structured grammatical evolution approach. In: Proceedings of the Genetic and Evolutionary Computation Conference, pp. 393–400 (2017)
18. Assuncao, F., Lourenco, N., Machado, P., Ribeiro, B.: DENSER: deep evolutionary network structured representation. Genet. Program Evolvable Mach. **20**(1), 5–35 (2019)
19. Lecun, Y., Bottou, L., Bengio, Y., Haffner, P.: Gradient-based learning applied to document recognition. Proc. IEEE **86**(11), 2278–2324 (1998)
20. Ji, S., Xu, W., Yang, M., Yu, K.: 3D convolutional neural networks for human action recognition. IEEE Trans. Pattern Anal. Mach. Intell. **35**(1), 221–231 (2013)
21. Mavadati, S.M., Mahoor, M.H., Bartlett, K., Trinh, P., Cohn, J.F.: DISFA: a spontaneous facial action intensity database. IEEE Trans. Affect. Comput. **4**(2), 151–160 (2013)
22. Zhanag, X., et al.: A 3D spontaneous dynamic facial expression database. In: IEEE International Conference on Automatic Face and Gesture Recognition, pp. 395–406 (2013)

23. Mohammadi, M.R., Fatemizadeh, E., Mahoor, M.H.: Intensity estimation of spontaneous facial action units based on their sparsity properties. IEEE Trans. Cybern. **46**(3), 817–826 (2016)

24. Walecki, R., Rudovic, O., Pavlovic, V., Schuller, B., Pantic, M.: Deep structured learning for facial action unit intensity estimation. In: IEEE Conference on Computer Vision and Pattern Recognition, pp. 5709–5718 (2017)

25. Lin, G., Shen, C., van den Hengel, A., Reid, I.: Efficient piecewise training of deep structured models for semantic segmentation. In: IEEE Conference on Computer Vision and Pattern Recognition (CVPR), pp. 1–8 (2016)

26. Du, S., Tao, Y., Martinez, A.M.: Compound facial expressions of emotion. Proc. Natl. Acad. Sci. **111**(15), E1454 (2014)

27. elKaliouby R.: Mind-reading machines: the automated inference of complex mental states from video. Ph.D. Thesis, University of Cambridge (2005)

28. Zhao, K., Chu, W., Zhang, H.: Deep region and multi-label learning for facial action unit detection. In: IEEE Conference on Computer Vision and Pattern Recognition, pp. 3391–3399 (2016)

Baseball Pitch Type Recognition Based on Broadcast Videos

Reed Chen[1], Dylan Siegler[2], Michael Fasko Jr.[3], Shunkun Yang[4], Xiong Luo[5], and Wenbing Zhao[3(✉)]

[1] Duke University, Durham, NC 27708, USA
[2] Georgia Institute of Technology, North Avenue, Atlanta, GA 30332, USA
[3] Department of Electrical Engineering and Computer Science,
Cleveland State University, Cleveland, OH 44115, USA
wenbing@ieee.org
[4] School of Reliability and Systems Engineering, Beihang University,
37 Xueyuan Road, Beijing 100191, China
[5] School of Computer and Communication Engineering,
University of Science and Technology Beijing, Beijing 100083, China

Abstract. In this paper, we report our work on baseball pitch type recognition based on broadcast videos using two-stream inflated 3D convolutional neural network (I3D). To improve the state-of-the-art of research, we developed our own high-quality dataset, trained and tuned the I3D model extensively, primarily combating the problem of overfitting while still trying to improve final validation accuracy. In the end, we are able to achieve an accuracy of $53.43\% \pm 3.04\%$ when oversampling and $57.10\% \pm 2.99\%$ when not oversampling, which is a significant improvement over the published best result of an accuracy of 36.4% on the same six pitch type classes.

Keywords: Two-stream inflated 3D convolutional neural network · Baseball pitch type recognition · Overfitting · Support vector machine · Regularization

1 Introduction

Computer-vision-based human activity recognition has been under intense study due to its huge implications, such as video surveillance, patient health assessment and intervention, ambient assisted living, human machine interaction, and self-driving vehicles [1–6]. In recent years, professional sports teams started to rely on data science to better understand athlete performance in games where video-based analysis of player activities is an important part [7]. Player activities are highly sophisticated in any professional sport games. In this work, we aim to identify pitch type automatically based on the video recordings of broadcast games of the US major league baseball. We chose to study this problem as the starting point because this a well defined task and it is relatively easy to establish the ground truth with the availability of the recorded games and the PitchF/X

© Springer Nature Singapore Pte Ltd. 2019
H. Ning (Ed.): CyberDI 2019/CyberLife 2019, CCIS 1138, pp. 328–344, 2019.
https://doi.org/10.1007/978-981-15-1925-3_24

data. However, it is actually quite hard to classify the pitch type because the pitcher always tries to hide which pitch type he is throwing to the batter, which makes the task challenging and interesting.

The research on fine-grained activity recognition in professional sports, despite its apparent significance to sports teams, has just started, presumably due to its high complexity. The earliest effort appears to be the publication of the Sports-1M dataset in 2014 [8]. However, the work has been predominantly coarse-grained activity recognition (such as the type of sports event). For fine-grained activity, we were only able to find a single research article related to baseball [7]. We intentionally decided to take the same approach as that in this article to learn from these pioneers and subsequently make improvements. Indeed, we were able to significantly improve the baseball pitch type recognition accuracy compared to this work. For the six pitch types used in both studies, namely, fastball, sinker, curveball, changeup, slider, and knuckle-curve, we managed to achieve an average accuracy of $53.43\% \pm 3.04\%$ when oversampling and $57.10\% \pm 2.99\%$ when not oversampling, which is a significant improvement over the result in [7], with an average accuracy of 36.4% on the same 6 pitch type classes.

We attribute the improvements to the following factors:

- Dataset size. Our dataset was about double the size of the previous one [9]. Some research suggests that the classification performance increases logarithmically based on the volume of the training data [10]. Hence, doubling the dataset size should increase the performance by $\log_{10}(2)$, or about 30%.
- Dataset focus. Our dataset is cropped to just focus on the pitcher, whereas the original dataset includes the whole TV broadcast. We think this focus should reduce noise and thus improve the network. We also did not overlay OpenPose [11] heatmaps onto the video. Instead, we relied on the raw Open-Pose figure overlay. While this removes confidence data, it may provide more usable features when training our neural network.
- Small training details. The work in [7] and related literature [12–14] all point out a few training details, some of which they mention are critically important. As is well-known, machine learning research results are often not reproducible, and this is probably one of the reasons why. These deep networks can be incredibly fickle during training. In this paper, we used a temporal stride of 1 with an FPS of 5, whereas the authors in [7] used a temporal stride of 8 with an FPS of 24, making an effective FPS of 3 in the Flow network but 24 in the other networks. These are also many small things that we changed and tuned.
- Optimizer. We used Adam optimizer which had fast convergence speed but quickly led to overfitting. We tested the SGD optimizer, but it seemed to be much less stable. Literature suggests that in large networks the optimizer should not really matter that much in the end, though our (fairly quick) tests did seem to contradict this.
- Fine tuning. We spent a considerable amount of time fine tuning our parameters. We tested multiple learning rate schedules including manually setting

the learning rate after specific epochs and experimenting with exponential and logistic functions. We also experimented with L1 regularization and noise layers, but most of these layers had a detrimental effect on our results.

2 Related Work

Video-based activity recognition has been under intense study and has been reviewed multiple times [3]. Zhang et al. provided an excellent summary of state-of-the-art deep learning based approaches [15]. Additionally, Kong and Fu [3] provided an extensive overview of approaches before I3D, though looks primarily at network structure and not results.

2.1 CNN + LSTM

The first deep learning method on video activity classification took a fairly simple form. First, a convolutional neural network (CNN) is applied to every frame of a video, possibly with a pre-trained CNN like Inception [16], then the last layer (*i.e.*, the layer before the softmax and predictions) is taken and fed into a long short-term memory network (LSTM) [17]. Though this may seem intuitive (*i.e.*, CNN extracts spatial features, and LSTM extracts temporal features), in practice this model is hard to train, takes a lot of data, and produces relatively poor results [18,19].

2.2 3D CNN

First introduced by the C3D model [20], the 3D CNN took the already existing idea of a 3D convolutional layer and applied this to video activity classification. Making no distinction between the two spatial dimensions and the temporal dimension, these networks again required a lot of data, took up a lot of memory, were hard and slow to train, and very poor at picking up temporal features. Some researchers proposed attention mechanisms, 2+1D convolutions, 1+2D convolutions, and many other mechanisms but none of these could boost the 3D CNN's performance significantly.

2.3 Inflated Two-Stream Convolutional Neural Networks

The inflated convolutional neural network model is best understood by first understanding the basic two-stream convolutional neural network. A number of researchers had been looking into ways of effectively capturing spatial and temporal features in one model. However, in this seminal paper [5], Simonyan and Zisserman went down a relatively unexplored and largely rejected path: having completely separate models for spatial and temporal features. They prove that a two stream model handily beats any previous state-of-the-art in many tasks. A two stream model is a bit unintuitive at first. The model consists of two completely separate but (almost) identical in structure CNNs. These two separate streams are called RGB and Flow.

The RGB stream looks a small group of consecutive frames from a video. Just like Inception tries to perform image recognition, the RGB network is doing more or less the same thing. Obviously, by looking at a small number of frames it is impossible to classify the action throughout a whole video, but a good CNN can get pretty good at it. This CNN, obviously, gets very good at detecting spatial features, but has no sense of temporal features because it is not performing convolutions along the temporal axis.

The Flow stream is identical in structure to the RGB stream. It too could use the same base structure as InceptionV1 [21]. In [5], Simonyan and Zisserman do change the Flow stream slightly just to reduce on memory size. However, the key difference between the Flow and RGB streams is the input. The flow stream takes the dense optical flow as input. Dense optical flow is an image processing algorithm that attempts to generate a vector for every pixel that describes where that pixel has moved between two consecutive frames. The Flow network takes, for example, 10 frames worth of flow data as input. If the input image was of size 224, the Flow input size would then be $10 \times 224 \times 224 \times 2$ (*i.e.*, 10 frames, 224 width, 224 height, the optical flow vectors have an x and a y component, so that is where the 2 is from). Thus, after training, the Flow network has a very strong understanding of temporal features.

To get a final prediction, fusion is performed. The two networks independently make their predictions on a video (*e.g.*, looking at random selections of RGB and 10 flow frames respectively), and the softmax layers are then either averaged, or fed into a pre-trained support vector classifier (SVC). Many authors report slightly better results with averaging the two softmaxes, but we have found that the SVC is more generalizable. Thus, by combining these two streams, we get a near state-of-the-art video activity predictor.

Finally, Carreira and Zisserman [22] take the two stream model one step further. The 2D convolutional layers found in InceptionV1 was inflated into 3D convolutional layers, bootstrapping on a third dimension.

Obviously with all of these there are an incredible number of implementation details. Probably the most important note is that the RGB stream is pre-trained on Imagenet and the Kinetics-600 dataset, and when we trained it we were really just performing fine-tuning transfer learning. The Flow stream can also be pre-trained, but it seems to provide almost no boost in performance.

2.4   Fine-Grained Activity Recognition in Baseball Videos

The work by Piergiovanni and Ryoo [23] is by far the most related work to ours. The primary task of the paper is to classify actions throughout a whole baseball broadcast, such as whether the clip is in play, ball, strike, no activity, etc. The authors describe extensively their experiments in that task. However, they only briefly describe their experiments into pitch type classification (as well as pitch speed regression). They use OpenPose heatmaps, which are similar to the skeletons we used, as input features to networks based on InceptionV3 [16] with no transfer learning. As the paper notes, pitches are hard to classify as the pitcher is trying to hide which pitch type he is throwing to the batter.

3   Deep Learning for Pitch Type Classification

We use the two-stream inflated 3D convolutional neural network (I3D) first intro-duced in [22] and was later used in [7] for baseball action recognition. In this section we will elaborate on some of the details that are critically important to the training process. Modern deep networks can be very sensitive to seemingly unimportant small details, such as preprocessing or slightly modified training procedures.

3.1   Data Normalization

We want our input data to be as close to following a normal distribution as possible (*i.e.*, zero mean, one standard deviation). This is accomplished with the RGB data by scaling the pixel values to be between −1 and 1, which means that it comes very close to fitting this normal distribution. For the Flow data, we did not explicitly normalize. As shown in Fig. 1, the Flow data was zero centered, though it was not perfectly normally distributed. We decided that the distribution was sufficient to keep it as it.

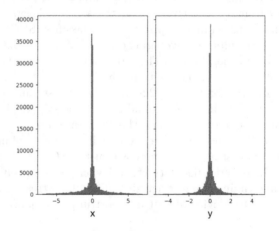

Fig. 1. The distribution of the Flow data.

3.2   Oversampling

Because we clipped the pitching actions from multiple entire games to build our dataset, the pitch type distribution basically follows that in actual games where some types are much more popular than others. Hence, the dataset has very imbalanced classes. In our dataset, fastballs account for 55.65% of our samples, while none of the other five classes account for more than 15%, with knuckle-curve accounting for a mere 1.93%. This much imbalance can be exceptionally detrimental for training. The model has to overcome the very tempting strategy of, guess fastball every time, which would get an astonishing 55.65% accuracy. Of course, this is a rather terrible strategy by most metrics. Thus, for training, the dataset that the network uses must be made close to balanced. In order to achieve this, we used a fairly naive version of oversampling. We simply copied random samples of underrepresented classes until they were equally represented.

We are unsure if evaluation should be done on the oversampled or non-oversampled dataset. The general consensus seems to suggest that evaluation should be done on non-oversampled data. However, we could find no paper that explicitly states really anything about oversampling or working with their imbalanced datasets. The MLB-Youtube dataset has a similar class distribution to ours, yet the authors of [23] did not mention how they handled this. Furthermore, if the network was better at predicting a common class, it would get an unfair advantage if non-oversampled data were used to generate final accuracy numbers. Thus, in our analysis, we looked at metrics on both the oversampled and non-oversampled datasets and made our code versatile, so we could switch between these two easily.

3.3 Learning Rate Schedule

We spent an exceptional amount of time tuning our learning rate schedules. For the Flow stream, we used a fixed learning rate schedule, shown in Fig. 2, whereas for the RGB stream, we used a logistic decay curve on the learning rate schedule, shown in Fig. 3, because this combination achieved the best result for us.

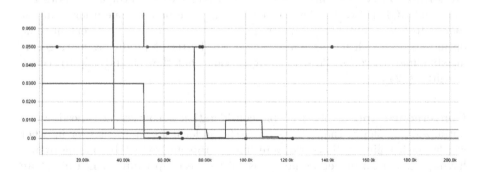

Fig. 2. This is an example of the many learning rate fixed learning rate schedules we used for training the Flow stream. The learning rate is jumped down at certain fixed step points.

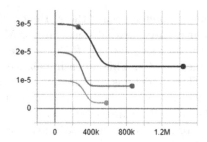

Fig. 3. This is an example of the logistic decay learning rates that we used for training the RGB stream. The learning rate is reduced smoothly from an initial value to an end value over a period of time.

3.4   Optimizer

Often the gradient descent algorithm is presented as the way to train models in machine learning. In short, gradient descent tries to minimize a loss function (it is also often called a cost function, it is just a measure of error) by taking its derivative with respect to each layer of the network, propagating back from the output layer, and adjusting each layer in the direction of the negative gradient, which is the direction of steepest descent. This simple algorithm can work, but in practice, there are a lot more considerations and alternatives. All optimizer algorithms operate on the same principal of gradient descent - back-propagating the derivative of loss and step weights in the direction of the negative gradient - but each one does it in a bit of a different way that has certain benefits and drawbacks. The following are the main considerations when picking an optimizer in the order of importance for our task.

Generalizability. We encountered a lot of issues with overfitting in both the Flow and RGB network. Thus, we believe that the number one priority in our task is to find a suitable optimizer. Due to the stochasticity inherent in some optimizers, research has shown that they are better at producing more generalizable final networks. This, in theory, means that they should help to reduce overfitting [24].

Convergence Speed. Another big consideration is convergence speed. Some optimizers are slower than others to reach the local loss minimum, or in other words finish training. Since we were limited to some extent by our hardware, having a faster optimizer would be useful to speed up training. Learning rate has a large influence on convergence speed. A small learning rate results in longer training time, but a large learning rate can result in the gradient descent overshooting the minimum or diverging. However, this can be useful in reducing overfitting as the model converges to wider and more general minima. In addition, a high learning rate is associated with issues such as dead neurons.

Memory. We were also heavily restricted by the amount of VRAM we had. Thus, we could not use memory-intensive optimizers. Luckily, almost every optimizer works in batches. This means that it does not look at the whole data set at the same time, but just a small subsample of it (n samples, where n is the batch size). Therefore, we can simply decrease the batch size to use up less VRAM.

Stability. Finally, we do have to consider whether an optimizer is stable. This means that it produces generally the same results from run to run, and does not include any weird effects like spiking loss randomly or causing wild swings in the weight values even when the learning rate it low. Every modern optimizer is stable if used and tuned properly, but this range of stability (*i.e.*, the amount of variability that occurs while still being regarded as stable) does vary slightly from optimizer to optimizer [25].

Our Optimizer Tests. We tested two optimizers in our network, Adam and Stochastic Gradient Descent.

Adam. This optimizer has an adaptive learning rate for every parameter. It keeps track of the change of each parameter over time, and adjusts the learning rate for that parameter to try to keep the change fairly constant, decaying only a bit over time. This was the first optimizer we tried, and it has performed well.

SGD with Momentum. Stochastic gradient descent is basically just gradient descent, but performed only on a random subsample of the training set (*i.e.*, a batch). Momentum is added to help accelerate the convergence speed. Most two-stream papers mention using SGD with momentum as their optimizer. It is also generally held that SGD is better at generalizing than Adam. We also tested a slightly different form of momentum called Nesterov momentum [26]. We found the same results.

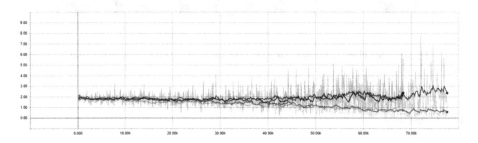

Fig. 4. This is a comparison between two training runs. The green and pink lines represent the train and test loss respectively for the Adam run, and the blue and dark orange lines represent the train and test loss respectively for the SGD run. Exponential smoothing (set to 0.9) was applied to make this graph more readable. (Color figure online)

We tried using SGD with momentum set to 0.9, as is recommended by the literature (such as [5]). We found almost no difference between SGD and Adam in the Flow network, as shown in Fig. 4. In the RGB network, we encountered the previously documented problem of fluctuation [27]. Due to the high stochasticity (randomness) of SGD, occasionally an update can cause loss to spike abnormally high, as depicted in Fig. 1 in [27]. We used Adam in both the Flow and RGB network due to its fast convergence rate and stability. Testing was done to determine the best optimizer for the RGB network. Due to issues of overfitting, we tried fine tuning our model with the SGD optimizer. However, while SGD did decrease overfitting, even after 100 epochs with SGD, validation loss was still higher than our Adam model trained over only 3 epochs. After 3 epochs, validation loss began increasing due to overfitting. While validation loss and accuracy approached the training loss and accuracy when using SGD, the training accuracy was only around 50%, and the loss was also quite high compared to when using the Adam optimizer.

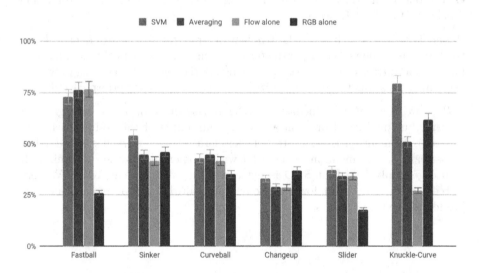

Fig. 5. Per class accuracy for our best model on the oversampled data.

4   Results

We report the results of our study using accuracy (per class and average) and confusion matrices. Because we have a multi-class task, confusion matrix is a better visualization tool to demonstrate the results than evaluation metrics such as F-1 score and the receiver operating characteristics curve (ROC), which are best used for binary classification problems.

We compared the results with four different combinations: (1) Flow alone, *i.e.*, using only the Flow stream; (2) RGB alone, *i.e.*, using only the RGB stream;

(3) two-stream using SVM as the classifier; and (4) two-stream using average to classify. For accuracy, we choose to use 95% confidence intervals for each of the four methods.

We used 90% of the dataset to train and the remaining data for evaluation. For the two-stream with SVM configuration, we used 70% of the evaluation data to train the SVM classifier and the remaining 30% for testing. Hence there are much fewer data shown in the corresponding confusion matrix.

The per class (*i.e.*, pitch type) accuracy based on the oversampled dataset is shown in Fig. 5. We choose not to show the results based on the non-oversampled dataset because they are similar. The average accuracy across all six classes are $53.43\% \pm 3.04\%$ with oversampling and $57.10\% \pm 2.99\%$ when not oversampling, which is a significant improvement over the result of [23] with an accuracy of 36.4% on the same 6 pitch type classes. No per class accuracy was reported in [23]. The confusion matrices for the four different methods are provided in Figs. 6, 7, 8, and 9.

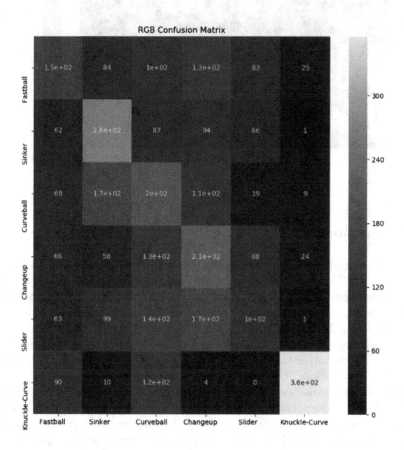

Fig. 6. Confusion matrix for RGB only.

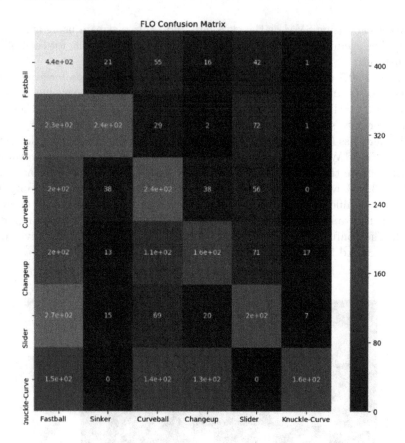

Fig. 7. Confusion matrix for Flow only.

5   Discussion

One of the biggest challenges we faced in this project was combatting overfitting in our model. This is first and foremost evidenced by the numerous training graphs that look like Fig. 10. For accuracy, training accuracy approaches 100% while validation accuracy increases then levels off. For loss, training loss decreases while validation loss decreases a bit then increases slowly. This trend is seen over and over again. It does not seem to look like the typical overfitting graph, but rather much flatter (and noisier, but this is to be expected with very deep networks). Nonetheless, it is a classic case of overfitting.

One fairly intuitive way to combat overfitting is by simply adding noise at various places in the network. This, in theory, forces the network to generalize, rather than memorizing specific examples, as is common in overfit networks. The most common way to add noise is to add Gaussian noise to the input. We did this in our network, experimenting with different standard deviations by dividing our data standard deviation by different constants (to preserve some consistency

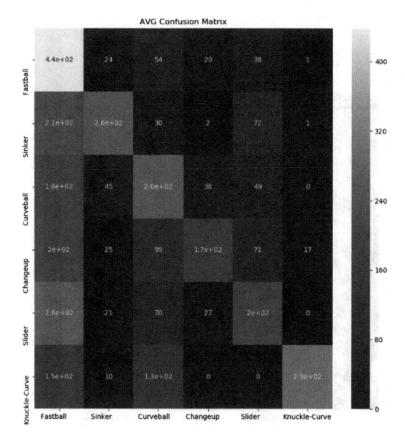

Fig. 8. Confusion matrix for two-stream with average.

between Flow and RGB). However, noise can be added to the activations, gradients, weights, and outputs of the network as well [28]. We used a noise layer as the input layer, which decreased overfitting by a small amount. The Gaussian noise was also added to convolutional layers. However, this did not appear to have a positive impact, so we removed it from the convolutional layers. Further testing may find a more balanced solution that results in an increased reduction of overfitting.

Another idea is to apply regularization. The idea comes from looking at the decision boundaries generated by neural networks. Overfit networks tend to have very complex and detailed decision boundaries. It was observed that if the magnitude of weights are decreased, decision boundaries tend to become less complex. Thus, regularization methods were introduced to try to encourage smaller weights.

This regularization comes in many sometimes mathematically complex forms, but the three simplest and most common regularization methods are L1 [29], L2 [29], and dropout [30].

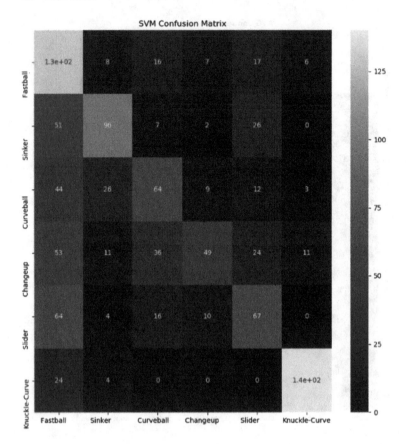

Fig. 9. Confusion matrix for two-stream with support vector machine as classifier.

Both L1 and L2 regularization add a penalty for large weights in the loss function. L1 normalization makes this penalty a normalized constant times the L1 norm of the weights, whereas L2 makes this penalty a normalized constant times the L2 norm of the weights. If the batch normalization is only applied to some layers, then only those layers will be added in this loss penalty. Thus, the network via gradient descent tries to minimize weight magnitudes. L1 and L2 normalization can be applied to just about any layer, whether it is convolutional or fully connected. It should be noted that L2 normalization does not have a regularizing effect when used in conjunction with batch normalization [31]. We have not experimented with L2 normalization much as it is not recommended that it is added on top of batch normalization, which we already had. We experimented with L1 regularization instead, and found that it did appear to have a regularizing effect resulting in decreased overfitting. While L1 regularization did result in validation loss decreasing, both training and validation accuracy suffered greatly. Not only did they start at a lower value, they also improved very slowly. As a result, we decided not to use L1 or L2 regularization.

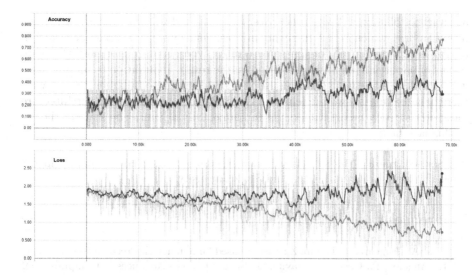

Fig. 10. This is one of the many training runs that exhibit typical overfitting. 0.9 exponential smoothing was applied for readability. The blue line is for training, the dark orange is for validation. The top figure is for training (blue) and testing (dark orange) accuracy, and the bottom figure is for loss. (Color figure online)

Dropout can be applied to any layer type, but is most commonly applied to fully connected layers. During each backpropagation, each neuron's weight and bias has a (1 - p)% chance of being set to a random value. For example, with dropout on a fully connected set to 0.8 keep probability, each neuron has a 20% chance of being set to a random value each back pass. The keep probability in dropout can get as low as 0.5. Dropout, in theory, forces the network to come up with a highly robust classifier, so if it loses a few neurons it can still perform effectively. This also acts as a regularizer, since neurons are reset back to values close to 0 every so often, so the network is discouraged from trying to make the weights large. We experimented with dropout on our network. But found that it made little difference as there is only 1 fully connected layer in our network, *i.e.*, the last classification layer. We did not use dropout with the convolutional layers as combining dropout and batch normalization can have a negative effect as discussed in [32].

6   Conclusion

We used the pioneering work on baseball fine grained activity recognition [7] as the starting point for our project. However, we made numerous changes, both creating our own much bigger dataset, and exploring many ways of combating the overfitting problem, which is rampant in deep learning research. Not surprisingly, we were able to achieve much better pitch type recognition accuracy. For the six pitch types used in both studies, we managed to achieve an average accuracy of $53.43\% \pm 3.04\%$ when oversampling and $57.10\% \pm 2.99\%$ when not oversampling, which is a significant improvement over the average accuracy of 36.4% as reported in [7] for the same 6 pitch type classes. That said, there are plenty of opportunities to further improve the result. In addition to finding more ways of addressing the overfitting problem, the creation of a larger dataset with more balanced classes would help, and the design of a better network architecture for sports activity recognition could fundamentally advance the state-of-the-art in this new field.

Acknowledgment. This work is partially supported by the Undergraduate Summer Research Award program at Cleveland State University.

References

1. Bux, A., Angelov, P., Habib, Z.: Vision based human activity recognition: a review. In: Angelov, P., Gegov, A., Jayne, C., Shen, Q. (eds.) Advances in Computational Intelligence Systems. AISC, vol. 513, pp. 341–371. Springer, Cham (2017). https://doi.org/10.1007/978-3-319-46562-3_23
2. Chen, M., Li, Y., Luo, X., Wang, W., Wang, L., Zhao, W.: A novel human activity recognition scheme for smart health using multilayer extreme learning machine. IEEE Internet Things J. **6**(2), 1410–1418 (2018)
3. Kong, Y., Fu, Y.: Human action recognition and prediction: a survey. arXiv preprint arXiv:1806.11230 (2018)
4. Lun, R., Zhao, W.: A survey of applications and human motion recognition with Microsoft Kinect. Int. J. Pattern Recognit. Artif. Intell. **29**(5), 1555008 (2015)
5. Simonyan, K., Zisserman, A.: Two-stream convolutional networks for action recognition in videos. In: Advances in Neural Information Processing Systems, pp. 568–576 (2014)
6. Zhao, W.: A concise tutorial on human motion tracking and recognition with Microsoft Kinect. Sci. China Inf. Sci. **59**(9), 93101 (2016)
7. Piergiovanni, A., Fan, C., Ryoo, M.S.: Learning latent subevents in activity videos using temporal attention filters. In: Thirty-First AAAI Conference on Artificial Intelligence (2017)
8. Karpathy, A., Toderici, G., Shetty, S., Leung, T., Sukthankar, R., Fei-Fei, L.: Large-scale video classification with convolutional neural networks. In: Proceedings of the IEEE conference on Computer Vision and Pattern Recognition, pp. 1725–1732 (2014)
9. Siegler, D., Chen, R., Fasko Jr., M., Yang, S., Luo, X., Zhao, W.: Semi-automated development of a dataset for baseball pitch type recognition. In: Ning, H. (ed.) CyberDI 2019/CyberLife 2019. CCIS, vol. 1138, pp. 345–359. Springer, Singapore (2019)
10. Sun, C., Shrivastava, A., Singh, S., Gupta, A.: Revisiting unreasonable effectiveness of data in deep learning era. In: Proceedings of the IEEE International Conference on Computer Vision, pp. 843–852 (2017)
11. Cao, Z., Simon, T., Wei, S.E., Sheikh, Y.: Realtime multi-person 2D pose estimation using part affinity fields. In: Proceedings of the IEEE Conference on Computer Vision and Pattern Recognition, pp. 7291–7299 (2017)

12. Aghdam, H.H., Heravi, E.J., Puig, D.: Analyzing the stability of convolutional neural networks against image degradation. In: VISIGRAPP (4: VISAPP), pp. 370–382 (2016)
13. Laermann, J., Samek, W., Strodthoff, N.: Achieving generalizable robustness of deep neural networks by stability training. arXiv preprint arXiv:1906.00735 (2019)
14. Xu, Z., Yu, F., Chen, X.: DoPa: a comprehensive CNN detection methodology against physical adversarial attacks (2019)
15. Zhang, H.B., et al.: A comprehensive survey of vision-based human action recognition methods. Sensors **19**(5), 1005 (2019)
16. Szegedy, C., Vanhoucke, V., Ioffe, S., Shlens, J., Wojna, Z.: Rethinking the inception architecture for computer vision. In: Proceedings of the IEEE Conference on Computer Vision and Pattern Recognition, pp. 2818–2826 (2016)
17. Gers, F., Schmidhuber, J., Cummins, F.: Learning to forget: continual prediction with LSTM. In: IET Conference Proceedings, pp. 850–855(5). https://digital-library.theiet.org/content/conferences/10.1049/cp_19991218
18. Luo, X., et al.: Short-term wind speed forecasting via stacked extreme learning machine with generalized correntropy. IEEE Trans. Ind. Inform. **14**(11), 4963–4971 (2018)
19. Luo, X., et al.: Towards enhancing stacked extreme learning machine with sparse autoencoder by correntropy. J. Franklin Inst. **355**(4), 1945–1966 (2018)
20. Tran, D., Bourdev, L., Fergus, R., Torresani, L., Paluri, M.: Learning spatiotemporal features with 3D convolutional networks. In: Proceedings of the IEEE International Conference on Computer Vision, pp. 4489–4497 (2015)
21. Szegedy, C., et al.: Going deeper with convolutions. In: Proceedings of the IEEE Conference on Computer Vision and Pattern Recognition, pp. 1–9 (2015)
22. Carreira, J., Zisserman, A.: Quo vadis, action recognition? A new model and the kinetics dataset. In: Proceedings of the IEEE Conference on Computer Vision and Pattern Recognition, pp. 6299–6308 (2017)
23. Piergiovanni, A., Ryoo, M.S.: Fine-grained activity recognition in baseball videos. In: Proceedings of the IEEE Conference on Computer Vision and Pattern Recognition Workshops, pp. 1740–1748 (2018)
24. Luo, X., Jiang, C., Wang, W., Xu, Y., Wang, J.H., Zhao, W.: User behavior prediction in social networks using weighted extreme learning machine with distribution optimization. Future Gener. Comput. Syst. **93**, 1023–1035 (2019)
25. Hardt, M., Recht, B., Singer, Y.: Train faster, generalize better: stability of stochastic gradient descent. arXiv preprint arXiv:1509.01240 (2015)
26. Sutskever, I., Martens, J., Dahl, G., Hinton, G.: On the importance of initialization and momentum in deep learning. In: International Conference on Machine Learning, pp. 1139–1147 (2013)
27. Ruder, S.: An overview of gradient descent optimization algorithms. arXiv preprint arXiv:1609.04747 (2016)
28. Reed, S., Lee, H., Anguelov, D., Szegedy, C., Erhan, D., Rabinovich, A.: Training deep neural networks on noisy labels with bootstrapping. arXiv preprint arXiv:1412.6596 (2014)
29. Ng, A.Y.: Feature selection, L 1 vs. L 2 regularization, and rotational invariance. In: Proceedings of the Twenty-First International Conference on Machine Learning, p. 78. ACM (2004)
30. Wager, S., Wang, S., Liang, P.S.: Dropout training as adaptive regularization. In: Advances in Neural Information Processing Systems, pp. 351–359 (2013)
31. van Laarhoven, T.: L2 regularization versus batch and weight normalization. arXiv preprint arXiv:1706.05350 (2017)

32. Li, X., Chen, S., Hu, X., Yang, J.: Understanding the disharmony between dropout and batch normalization by variance shift. In: Proceedings of the IEEE Conference on Computer Vision and Pattern Recognition, pp. 2682–2690 (2019)

Semi-automated Development of a Dataset for Baseball Pitch Type Recognition

Dylan Siegler[1], Reed Chen[2], Michael Fasko Jr.[3], Shunkun Yang[4], Xiong Luo[5], and Wenbing Zhao[3(✉)]

[1] Georgia Institute of Technology, North Avenue, Atlanta, GA 30332, USA
[2] Duke University, Durham, NC 27708, USA
[3] Department of Electrical Engineering and Computer Science, Cleveland State University, Cleveland, OH 44115, USA
wenbing@ieee.org
[4] School of Reliability and Systems Engineering, Beihang University, 37 Xueyuan Road, Beijing 100191, China
[5] School of Computer and Communication Engineering, University of Science and Technology Beijing, Beijing 100083, China

Abstract. In this paper, we report our work on developing a new dataset for baseball pitch type recognition based on youtube videos of the US Major League Baseball games. The core innovation is a largely automated procedure to extract relevant clips from the full game, and automatically label the clips by aligning the infographic information included in the broadcast and the PitchF/X data. We adopted the Needleman-Wunsch algorithm to address the challenges imposed by the aligning the two streams of data based on pitch speed, *i.e.*, minimize gaps and mismatches between the two streams. Manual inspection is used only to select games that include infographic information for clip extraction and to remove erroneous clips for improve the quality of the dataset.

Keywords: Pitch type · Video-based human activity recognition · Dataset · PitchF/X · Needleman-Wunsch algorithm

1 Introduction

Datasets are instrumental in machine-learning based research such as video-based human activity recognition [1–6]. On the one hand, the quality of datasets would directly impact the recognition accuracy. On the other hand, it may be a labor-intensive task to manually label the data for supervised learning. In this paper, we present a semi-automated procedure to build a dataset for pitch type recognition based on the broadcast games of the US major league baseball posted on youtube.com. There are only two steps that require manual intervention. First, we rely on manual selection of games such that the broadcast of the games contain the speed of each pitch because we use the included pitch speed

© Springer Nature Singapore Pte Ltd. 2019
H. Ning (Ed.): CyberDI 2019/CyberLife 2019, CCIS 1138, pp. 345–359, 2019.
https://doi.org/10.1007/978-981-15-1925-3_25

in the broadcast to associate a clip with the PitchF/X data [7] of the game for automated labeling of the clip. The second place that relies on manual intervention is to remove erroneous clips of pitching. Even though only a tiny fraction of the clips automatically generated off the games, they may significantly impact the training and recognition accuracy [2,8]. For both manual tasks, we have built tools to ease the work.

This paper makes the following research contributions:

- The semi-automated procedure presented in this paper could enable other researchers who are interested in doing video-based fine-grained activity recognition to develop their own datasets as needed. While there are numerous datasets for video-based activity recognition, few provided sufficient details on how the dataset was created, and even less so on creating a procedure to build the dataset in a semi-automated manner.
- We introduce a novel mechanism for automated video clipping based on jump cuts in the broadcast and automatically crop out the pitcher to remove unnecessary noise.
- We introduce a novel mechanism for automated labeling of the pitch type in each clip by aligning the pitch speed information included in the broadcast game and that in the PitchF/X (Statcast) data.
- Our dataset incorporates skeleton information of the pitcher generated from OpenPose pose estimation [9]. This would save considerable amount of time for training.

2   Related Work

There are four main datasets used as benchmarks in activity recognition currently according to a recent survey paper [2]. Both the HMDB-51 [10] (http://serre-lab.clps.brown.edu/resource/hmdb-a-large-human-motion-database/) and the UCF-101 [11] (https://www.crcv.ucf.edu/data/UCF101.php) are relatively old datasets with 51 and 101 classes respectively. However, when combined, the datasets have hardly more than 15,000 videos, and not all of the classes overlap. Additionally, the Sports-1M (https://cs.stanford.edu/people/karpathy/deepvideo/) has been around since 2014, and contains 487 classes and links to 1 million Youtube clips of sports activities. Together, these three datasets have traditionally formed the standard, though there are a number of other datasets that are used, and the standard is hardly as organized as something like Imagenet [12].

In 2017, the Kinetics dataset was published [13]. It contains 400 classes and a minimum of 400 clips per class, and a total of almost 307,000 clips. This dataset is high quality and collected by hand, and has been expanded to 600 classes recently, with over 495,000 clips [14]. This dataset is becoming the standard for video activity recognition tasks, though currently it is not yet the definitive standard.

It is important to bring up the Imagenet dataset [12]. This is a dataset of still images that started with 1000 classes. It now has over 14 million images in it. Though not directly useful for video activity recognition, many networks first train on the Imagenet dataset to get a good spatial understanding, and then perform transfer learning to learn temporal features, or fine tune their spatial understanding to the specific task (and temporal understanding will be handled by another network).

Finally, for our specific task the mlb-youtube dataset (https://github.com/piergiaj/mlb-youtube) exists [15]. We have declined to use this dataset for the following reasons. This dataset includes a bit under 4000 videos of pitches taken from the 2017 MLB postseason. This dataset was collected with the primary goal of looking at the broadcast and deciding what is going on (i.e. in play, no activity, swing, etc). However, they also included pitch types to clips when applicable. The paper spends very little time on their data collection process and their use of the pitch classifications, so we were doubtful of the accuracy of the dataset. Furthermore, we felt that only 4000 clips would not be sufficient even for transfer learning, and the clips were all 7 s long, and often included activity other than just the pitch (like the batter swinging right after). Additionally, their dataset did not crop out the pitcher, which we chose to do. Finally, to incorporate the skeleton information into the clip, we would still have to run OpenPose [9] on their dataset, as the authors did. Due to all of these factors, we felt that in the long run it would be most effective to generate our own dataset with a largely automated process.

3   Dataset Creation Process

The goal of the data collection and processing process towards a high quality dataset is to collect accurate clips of pitches, along with a label of what pitch type that pitch was. Our number one goal was label accuracy, so whenever there was a tradeoff between accuracy and amount of data collected, we preferred accuracy. The entire process is largely automated driven a python script. As mentioned previously, there are only two steps that require manual intervention, both for the purpose of ensuring high quality of the dataset. At the beginning of the data collection process, we inspect each full game that we have downloaded to make sure that it contains infographic for pitching speed at the right place and at the right frames via a tool that we developed, which we call infographic labeler. This is because we would extract the pitch type for automatic labeling of the clips from the infographic information using optical character recognition (OCR). Towards the end of the process, we again inspect the clips to identify and manually remove erroneous clips.

Overall, the dataset creation process contains seven steps after we have downloaded the games, which will be elaborated in details below. Each step along the way saves out checkpoint files so we can resume if something goes wrong in the run. A baseball game is usually around 3 h, and it takes the script about 1–2 h to process a 3 h game (depending on a few factors). Games tend to have around 200–300 pitches, and we tend to extract around 90–130 pitches from each game. There are a few scripts that supplement this main script and make the manual portions of the data collection process much easier.

Over about two weekends and a week of this data collection pipeline running continuously, we collected close to 10,000 clips. We decided to stop data collection for the time being at that point. We spent about 1 h per day tending to the scripts, so this data collection process is quite sustainable if more data needs to be collected in the future. The dataset currently contains 6 different pitch types, namely, fastball, slider, sinker, curveball, changeup, and kuncle-curve. Their distribution is shown in Fig. 1.

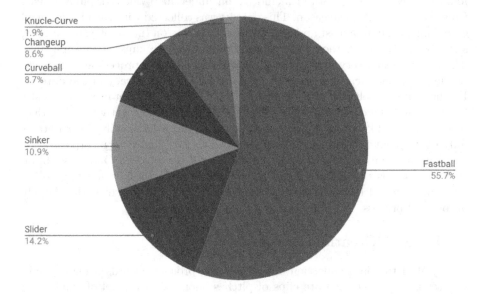

Fig. 1. The class (pitch type) distribution of our dataset.

3.1 Download the Full Videos

Most MLB games are available on Youtube. We use youtube-dl (https://ytdl-org.github.io/youtube-dl/index.html), a command line program for downloading Youtube videos, to download these videos. For our process, we manually selected videos. We primarily selected games from the 2011–2015 seasons, at Baltimore Stadium (Orioles Park). We chose these parameters because the Baltimore broadcast includes the speed of each pitch in the same location on the stream in (almost) every broadcast.

3.2 Clip the Videos

Next, we generated cut points based on jump cuts in the broadcast. This is done using a fairly simple process with OpenCV. We calculated the average pixel difference between consecutive frames, and if this difference exceeds a certain threshold, we mark that frame as a cut point. This method works very reliably in determining the jump cuts. These cut frames are saved out to a file.

3.3 Detect Pitch Speeds

This step serves two purposes. First, it serves to determine which clips are of actual pitches, and which clips are not (these clips are useless to us). It relies on an unexpected element of the broadcast: the pitch speed in the infographic shown on screen. We chose to use the Tesseract OCR engine [16] for optical character recognition of the speed data appear in the relevant frames. In particular, we used the Python port pytesseract (https://pypi.org/project/pytesseract/) and identified the speed information at a specific location in the infographic. If the text of the form ## MPH is detected and a few other criteria are met to help reduce false positives, the clip is marked as a clip of a pitch and the pitch speed is recorded to be used later, as shown in Fig. 2. A few Tesseract settings were changed to make the engine better tuned for digital recognition. Also, we do not run OCR every frame as that would be excessively resource-intense. Instead, we just run it every 5th frame during the later 55% of every clip.

Fig. 2. An example frame showing where and what the OCR is looking for.

3.4 Download PitchF/X Data

What makes this whole data collection possible is the PitchF/X data collection system (this system was replaced by Statcast in 2015, and it is slightly different but similar enough that the details still all apply). This system uses sensors throughout the stadium to detect the exact movement of every pitch. Up to 81 data points about every pitch are automatically recorded and available online. We used pybaseball (https://github.com/jldbc/pybaseball), which draws its data from BaseballSavant (https://baseballsavant.mlb.com/). The central task of data collection was then boiled down to associating every clip of a pitch we picked out in the previous step with its respective PitchF/X entry.

We considered a number of ways of doing this. We considered using the timestamp embedded in PitchF/X data. However, we found first that this timestamp would drift from the broadcast timing significantly (even though the broadcasts we have included commercial breaks). Additionally, we found that the timestamp formatting depended on the stadium. In some stadiums it was relative from the start of the game, others UTC, others based in the time zone of the stadium. We considered a few other ways to associate the clips, but we finally settled on using the pitch speed, as it was easily identifiable both on the Youtube video and in the PitchF/X data. We found that different sources of the PitchF/X data had slightly different pitch speeds, but the release_speed column in BaseballSavant data always matched the speed that was shown on the broadcast. Figure 3 shows an example of some of the data downloaded from PitchF/X. Though the PitchF/x system is very powerful, it is not very well documented. Some of the common terms in both PitchF/X and Statcast are explained at https://fastballs. wordpress.com/2007/08/02/glossary-of-the-gameday-pitch-fields/.

pitch_typ	game_da	release_:	release_;	release_;	player_na	batter	pitcher	events	descripti	spin_dir	sp
SI	9/2/2014	96.6	null	null	Zack Britt	453943	502154	grounded	hit_into_play		
SI	9/2/2014	94.8	null	null	Zack Britt	453943	502154	null	called_strike		
SI	9/2/2014	96.2	null	null	Zack Britt	571740	502154	single	hit_into_play_no_out		
SI	9/2/2014	95.4	null	null	Zack Britt	421124	502154	double	hit_into_play_no_out		
SI	9/2/2014	95.2	null	null	Zack Britt	421124	502154	null	called_strike		
SI	9/2/2014	95	null	null	Zack Britt	421124	502154	null	ball		
SI	9/2/2014	95.8	null	null	Zack Britt	421124	502154	null	ball		
SI	9/2/2014	96.2	null	null	Zack Britt	502117	502154	field_out	hit_into_play		

Fig. 3. This is an example of some of the data downloaded from PitchF/X. There are many more columns to this data, but some of the columns are null as they may not apply.

We should mention that we actually found that a number of data points in the PitchF/X data could change slightly depending on the stadium. Most notably, some blogs report that the pitch classification algorithm, which is the same code everywhere, can produce slightly different results depending on the stadium. Thus, this could introduce a small bit of bias in our dataset as we were pulling mostly from games at Baltimore, though some of the more recent games we looked at were at other stadiums. Again due to lack of formal documentation of PitchF/X, we do not have too much hard data to back that up. In theory we could do some basic statistical analysis to verify this inconsistency to some extent, though likely we would not be able to characterize it very well.

3.5   Align OCR Speeds with PitchF/X Speeds

We were left with two lists, one list of pitch speeds as extracted from the broadcast using the OCR, and one list of pitch speeds from the PitchF/X data. We wish to align these two lists, which means associating every OCR speed with its respective PitchF/X speed (there will be some PitchF/X speeds left unassociated). The PitchF/X speeds are the ground truth, and are known to be accurate. On the other hand, the OCR pitch speeds have a number of omissions (the OCR missed the pitch or one of the checks failed and so the pitch was not added to the list) and a small number of additions, or erroneous extra detections, mostly due to broadcast inconsistencies. These additions made the task of alignment quite hard, even when we used a method that threw out upwards of 75% of pitches, we could not avoid erroneous additions.

We initially experimented with methods of alignment that relied on having no erroneous additions. However, when we determined that it would be impossible to have no erroneous additions whatsoever, we developed a statistical method of aligning the two series based on the Needleman-Wunsch (NW) algorithm [17]. We intentionally chose not to use dynamic time warping (DTW) [18] to align the two sequences because the way to calculate the distance (also called score or cost) between the two sequences as required by DTW is not relevant in our case. Specifically, this task is slightly different from aligning two time series data. For example, sensor time series data, s1, s2, s3... si generally has some type of characteristic where si depends on s(i-1), or all of the previous data points. Also, these series are generally assumed to be approximating some true differentiable function s(t). On the other hand, pitch speeds do not really depend on the previous pitch speeds and it is not approximating any true function s(t), let alone a differentiable one.

The NW algorithm is a dynamic programming algorithm that was originally created in order to solve a similar problem, *i.e.*, the alignment of gene sequences. The algorithm takes in two series, along with a gap penalty and a mismatch penalty. In the final alignment, the algorithm can generate both gaps and mismatches. In the case of genes, the two input sequences might look like: (1) GCATGCU, and (2) GATTACA. And the generated alignment might look like: (1) GCATG_CU, and (2) G_ATTACA. The algorithm tries to maximize the score of the alignment. Matches increase the score by 1, mismatches decrease it by mismatch_penalty, and gaps decrease it by gap_penalty. In this alignment, the algorithm generated 2 gaps and 2 mismatches. In our case, we want a high mismatch penalty compared to the gap penalty (we used 3 and 1 respectively in our code because the combination gives us the best accuracy). We did this as we throw out mismatches anyways, as we are trying to reduce error as much as possible during the data collection process.

Nonetheless, the NW algorithm is the step which introduces the most error to the data collection and processing process. Most of this error comes when there are strings of same pitch speeds in either the OCR or the PitchF/X lists (i.e. 94 94 94 94). While we experimented with various forms of throwing out these strings, or the last number in these strings, all of these methods both threw out an incredible amount of data (sometimes upwards of 75%) and did not increase accuracy very much still. Our analysis shows that this process contributed around 5–7% error to the final dataset (95–98% accuracy). Some of this error is pruned by the manual deletion step. This error rate is still just an estimate based on fairly crude assumptions about the OCR detected and PitchF/x pitch speeds and could be looked at further. The detailed analysis is given in Sect. 4.

Once this alignment process is done, we then just match up all of the other PitchF/X data (as we saved that with the PitchF/X speed list). Now, the hard part is done. The PitchF/X data includes a pitch type entry, so we now have every clip of a pitch labeled with its pitch type (or at least every clip we can confidently match up). As noted before, the PitchF/X data includes much more than just pitch type and speed.

3.6   Run OpenPose on the Clips

Next we run OpenPose on each of the clips. OpenPose is a computer vision library developed by the CMU Perceptual Computing Lab, and it estimates the pose of people in a picture frame. It both saves this skeleton to an array, with each joint position and the confidence in that position as a row, as well as drawing the skeleton onto the frame. We investigated using the raw joint positions, along with the hand detection feature that OpenPose also has. We found that this method simply does not provide enough data to train an accurate neural network, therefore, although we save the raw OpenPose data, we choose not to use the data at this stage of our project. Instead, we opt to use the broadcast frames with the OpenPose skeleton drawn on top. The work in [15] chose to use the frame annotated with OpenPose heatmap as input, which is rather similar to our approach.

3.7   Crop the Pitcher and Transform the OpenPose Data

With the final goal of presenting the machine learning system to a sports team for use to analyze potential talent, we believed that including the whole broadcast including the infographics would be unrealistic. The full broadcast would introduce some noise (pitcher in different locations on screen, different backgrounds, etc), but it would also show the movement of the ball as well as any infographics shown on screen, which the algorithm might pick up on.

In order to focus just on the pitcher, we developed a fairly simple algorithm based off the fact that the camera angle of pitches is almost always from the back, around 3rd base position. We used the OpenPose data to find the bottom most person on the screen, which we assumed to be the pitcher, found the average bounding box for this pitcher, and then cropped the frame based off that. We wanted to preserve a consistent aspect ratio across all crops, so we included some fixed offsets so the pitcher would be well centered in the crop even if he is taller or shorter than average. This step is also why we do the manual delete checking at the end. It is at this point that we actually save the clips to the disk.

Additionally, we transform the OpenPose data so that it reflects the new cropped clips. Thus, if we want to use this OpenPose data in the future it is relative to the cropped frame, not the whole broadcast frame. An example cropped frame is shown in Fig. 4.

Fig. 4. This is an example frame from a cropped clip. The pitcher is the main focus of the video and the OpenPose skeleton is drawn on top.

3.8 Manual Delete Check

It is at this point that we run the deleter tool, as shown in Fig. 5, which makes it easy to find any clips that are not from the right angle or are not of a pitch. We manually select these clips and the deleter tool removes them from all of the final records (the cropped clip, the aligned PitchF/X entry, and the cropped OpenPose data).

4 Analysis

The most critical step in our dataset creation process is the alignment of the pitch speed information included in the video frames and that of the PitchF/X data using the NW algorithm. Hence, it is necessary to analyze the NW algorithm's impact to the labeling accuracy. We wanted to characterize and estimate the error introduced by this process. This analysis is done in four steps.

4.1 Generate Ground Truth

We generate the ground truth pitch speed sequence (this simulates the PitchF/x pitch speed sequence). We simply pick a length, 175–300, then generate that many samples using a normal distribution with a mean of 88 and a standard deviation of 6.2, which matches the mean and standard deviation of the true PitchF/X speeds.

4.2 Generate Detected Sequence

Next, we generated a simulated OCR detected sequence. We did this using two parameters: drop rate and erroneous addition rate. First we applied erroneous additions. This is to simulate when the algorithm detects a clip as a pitch, even

Fig. 5. The screenshot of the deleter tool in play. Four clips that indicated by the red rectangle are erroneous and are deleted. (Color figure online)

though it might be the clip after and the infographic may still be on the screen. Thus, for every item in the original pitch speed list (we start off with the ground truth), there is a chance equal to the erroneous addition rate that an erroneous addition is added. This operation is done first so that there is more independence between the two rate parameters. Second we apply dropping. The OCR script will often either miss pitches altogether or the erroneous addition code that we added to try to reduce the number of erroneous additions is triggered and so a pitch is not added to the final list. To simulate this, each speed has a chance equal to the drop rate of being deleted from the list. Additionally, for each entry in the detected list that was in the ground truth list, we keep track of it.

4.3 Alignment with NW Algorithm

After generating the two lists, we simply run the NW algorithm and get the alignment sequences.

4.4 Calculate Accuracy

We determine the alignment accuracy in terms of the fraction of errors we make. A match is wrong if the alignment returned from the NW algorithm has two aligned values as equal, but in reality they are not the samevalue (as we kept track of this when we generated the detected list), but just happened to be the same value or one was an erroneous addition.

4.5   Repeat for Different Conditions

We could not characterize the drop rates or erroneous addition rates, as this would take significant manual work by looking at the outputs of the data collection script and matching it up with the actual pitch clips. Thus, we experimented with a range of drop rates and erroneous addition rates. We chose a range of 0.2 to 0.8 to look at for the drop rate and 0.1 to 0.2 as a range for the erroneous additions rate, as these seemed like reasonable ranges considering our aggressive erroneous addition detection code. We ran every sample 10 times and took the average accuracy. We also tested 5 mismatch and gap penalty combinations.

4.6   Results

The accuracy result is visualized using a 3D scatter graph. Two of the axes represent the NW parameters we used, mismatch penalty and gap penalty. Figure 6 shows the result for the parameters that we have used in creating our dataset, *i.e.*, mismatch penalty 3 and gap penalty 1. We can see that, as expected, increasing drop rate or addition rate reduces accuracy, with addition rate contributing more. However, what is important to note is that even at the likely overestimated bounds of 0.8 drop rate (which is absurdly high) and 0.2 erroneous addition rate (also likely quite high), we still get an accuracy of about 93%.

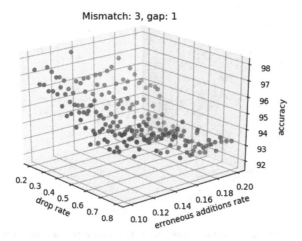

Fig. 6. 3D scatter graph for the accuracy with mismatch penalty of 3 and gap penalty of 1.

It is interesting to look at the graphs for other parameters as shown in Figs. 7, 8, 9 and 10. It seems that the parameters of mismatch penalty at 3, gap penalty at 1 performed the best, but was looking to drop off quite a bit. Each set of parameters produces its own distinct graph with unique features and concavity.

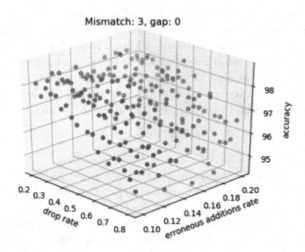

Fig. 7. 3D scatter graph for the accuracy with mismatch penalty of 3 and gap penalty of 0.

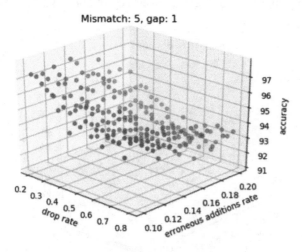

Fig. 8. 3D scatter graph for the accuracy with mismatch penalty of 5 and gap penalty of 1.

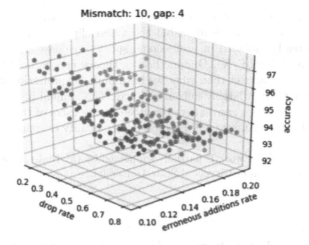

Fig. 9. 3D scatter graph for the accuracy with mismatch penalty of 10 and gap penalty of 4.

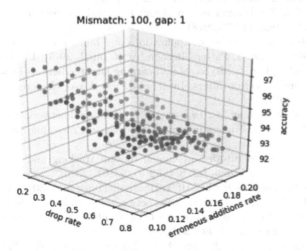

Fig. 10. 3D scatter graph for the accuracy with mismatch penalty of 100 and gap penalty of 1.

5   Conclusion

In this paper, we presented our work on building a new dataset for baseball pitch type recognition in a largely automated manner. Manual intervention is required only at the beginning and at the end of a fully automated process, and both manual steps are meant to improve the label accuracy of the dataset. As will be presented in a separate paper submitted to the same conference, we managed to achieve a significantly better pitch type accuracy [19] using our dataset than that reported in [15]. We attribute the improvement at least partially to the larger and higher quality dataset. Perhaps more importantly, we pointed a novel procedure that can automatically label the video clip data by exploiting the information already included in professional sport broadcast and other related resources (such as PitchF/X). Our method in theory can be used to construct much more massive dataset than what we have built for professional baseball as well as other sports in a largely automated way. The manual inspection needed to remove erroneous clips can be crowdsourced to the users of the datasets, and hence, gradually, the datasets will become higher quality overtime.

Acknowledgment. This work is partially supported by the Undergraduate Summer Research Award program at Cleveland State University.

References

1. Chen, M., Li, Y., Luo, X., Wang, W., Wang, L., Zhao, W.: A novel human activity recognition scheme for smart health using multilayer extreme learning machine. IEEE Internet of Things J. **6**(2), 1410–1418 (2018)
2. Kong, Y., Fu, Y.: Human action recognition and prediction: a survey. arXiv preprint arXiv:1806.11230 (2018)
3. Lun, R., Zhao, W.: A survey of applications and human motion recognition with Microsoft Kinect. Int. J. Pattern Recogn. Artif. Intell. **29**(5), 1555008 (2015)
4. Piergiovanni, A., Fan, C., Ryoo, M.S.: Learning latent subevents in activity videos using temporal attention filters. In: Thirty-First AAAI Conference on Artificial Intelligence (2017)
5. Simonyan, K., Zisserman, A.: Two-stream convolutional networks for action recognition in videos. In: Advances in Neural Information Processing Systems, pp. 568–576 (2014)
6. Zhao, W.: A concise tutorial on human motion tracking and recognition with Microsoft Kinect. Sci. China Inf. Sci. **59**(9), 93101 (2016)
7. Fast, M.: What the heck is PITCHf/x. Hardball Times Ann. **2010**, 153–158 (2010)
8. Carreira, J., Zisserman, A.: Quo vadis, action recognition? A new model and the kinetics dataset. In: Proceedings of the IEEE Conference on Computer Vision and Pattern Recognition, pp. 6299–6308 (2017)
9. Cao, Z., Simon, T., Wei, S.E., Sheikh, Y.: Realtime multi-person 2D pose estimation using part affinity fields. In: Proceedings of the IEEE Conference on Computer Vision and Pattern Recognition, pp. 7291–7299 (2017)
10. Kuehne, H., Jhuang, H., Garrote, E., Poggio, T., Serre, T.: HMDB: a large video database for human motion recognition. In: 2011 International Conference on Computer Vision, pp. 2556–2563. IEEE (2011)
11. Soomro, K., Zamir, A.R., Shah, M.: UCF101: a dataset of 101 human actions classes from videos in the wild. arXiv preprint arXiv:1212.0402 (2012)
12. Deng, J., et al.: ImageNet: a large-scale hierarchical image database. In: 2009 IEEE Conference on Computer Vision and Pattern Recognition, pp. 248–255. IEEE (2009)
13. Kay, W., et al.: The kinetics human action video dataset. arXiv preprint arXiv:1705.06950 (2017)
14. Carreira, J., Noland, E., Banki-Horvath, A., Hillier, C., Zisserman, A.: A short note about kinetics-600. arXiv preprint arXiv:1808.01340 (2018)

15. Piergiovanni, A., Ryoo, M.S.: Fine-grained activity recognition in baseball videos. In: Proceedings of the IEEE Conference on Computer Vision and Pattern Recognition Workshops, pp. 1740–1748 (2018)

16. Smith, R.: An overview of the Tesseract OCR engine. In: Ninth International Conference on Document Analysis and Recognition (ICDAR 2007), vol. 2, pp. 629–633. IEEE (2007)

17. Gotoh, O.: An improved algorithm for matching biological sequences. J. Mol. Biol. **162**(3), 705–708 (1982)

18. Berndt, D.J., Clifford, J.: Using dynamic time warping to find patterns in time series. In: KDD Workshop, Seattle, WA, vol. 10, pp. 359–370 (1994)

19. Chen, R., Siegler, D., Fasko Jr., M., Yang, S., Luo, X., Zhao, W.: Baseball pitch type recognition based on broadcast videos. In: Ning, H. (ed.) CyberDI 2019/CyberLife 2019. CCIS, vol. 1138, pp. 328–344. Springer, Singapore (2019)

Ford Vehicle Classification Based on Extreme Learning Machine Optimized by Bat Algorithm

Yile Zhao[✉] and Zhihai Lu

School of Education Science, Nanjing Normal University, Nanjing 210023, Jiangsu, China
zhaoyile1144@163.com, 1030609013@qq.com

Abstract. The application of automobile identification in life is more and more extensive, so research on related technologies is receiving widespread attention. This article focuses on research on Ford vehicle identification, the theoretical method of identification is proposed and its effectiveness is verified in experiments. We first obtain the side-view image of the Ford car. Secondly, we use gray level co-occurrence to extract the feature of Ford car. Third, we use extreme learning machine as the classifier. Finally, we use bat algorithm to optimize the algorithm, and employ 10-fold cross-validation to ensure the validity of the data. The results of the research indicate that in the same kind of research, the method we employ has the highest accuracy ($84.92 \pm 0.64\%$).

Keywords: Gray-level co-occurrence matrix · Extreme learning machine · Bat algorithm · Identification · Cross-validation

1 Introduction

Identification has been an important application in modern social life. For example, flower recognition [1], currency recognition, etc. In the automotive field, it is no exception. License plate recognition and vehicle identification have a very important role in life. License plates and models are the most important signs of a car. This plays a great role in police crime detection, vehicle access management, and traffic police command traffic.

de Souza et al. [2] employed the method of Singular Value Decomposition (SVD) to extract image feature. Jia [3] used convolutional neural network (CNN) to improve the classification performance for car identification. Tao and Lin [4] used logistic regression (LR) to identify car brands. Suchkov [5] utilized naïve Bayesian classifier (NBC) to detect Ford vehicle.

Compared with traditional neural networks [6–10] and support vector machines [11–15] extreme learning machine has a fast learning speed and good generalization performance. Therefore, it has been widely used in pattern recognition and other fields in recent years, but random assignment mechanism of extreme learning machine has a great influence on its accuracy. Bat algorithm is a new type of optimization algorithm with good global search capability, which plays an important role in the optimization

© Springer Nature Singapore Pte Ltd. 2019
H. Ning (Ed.): CyberDI 2019/CyberLife 2019, CCIS 1138, pp. 360–370, 2019.
https://doi.org/10.1007/978-981-15-1925-3_26

of extreme learning machine input weight and threshold. Although convolutional neural network [16–20] may get better results, it requires a large amount of data.

We first obtain the side-view image of the Ford car, simply processed the original image. Secondly, we use gray level co-occurrence to extract the feature. Third, we use extreme learning machine as the classifier. Finally, we use bat algorithm to optimize the algorithm, and employ 10-fold cross-validation to ensure the validity of the data.

2 Dataset

We shot 260 pictures of car, including 130 Ford pictures, 130 Non-Ford pictures. We perform a series of processing on the original image, such as the background color is set to black, and the image size is uniform. Figure 1 shows the samples of dataset. Figure 1(a) shows a picture of Ford vehicle, and Fig. 1(b–d) display non-Ford vehicles.

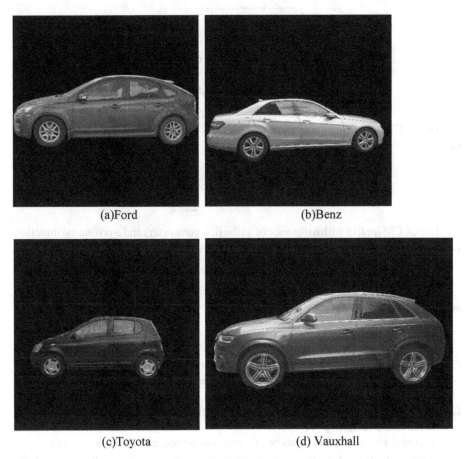

(a)Ford   (b)Benz

(c)Toyota   (d) Vauxhall

Fig. 1. Samples of dataset

3 Methodology

3.1 GLCM

GLCM, the gray level co-occurrence matrix, GLCM is an L*L square matrix, and L is the gray level of the source image. It described a joint distribution of two pixels with a certain spatial positional relationship, which can be regarded as a joint histogram of two pixels grayscale pairs, which is a second-order statistics. There are four directions of movement: 0°; 45°; 90°; 135°. Suppose the texture matrix P of the image is shown in Fig. 2.

0	1	2	0	2	1
1	2	1	1	1	1
2	1	0	1	2	0
2	2	0	0	1	2
1	2	1	2	2	1
0	1	0	1	0	2

Fig. 2. Texture matrix P of the image

The GLCM matrix with a distance of 1 (the first parameter) and a positional direction of 0° (the second parameter) is as follows:

$$P(d = 1, \theta = 0°) = \begin{matrix} 1 & 5 & 2 \\ 3 & 3 & 6 \\ 2 & 5 & 2 \end{matrix} \tag{1}$$

The GLCM matrix with a distance of 1 (the first parameter) and a positional direction of 45° (the second parameter) is as follows:

$$P(d = 1, \theta = 45°) = \begin{matrix} 0 & 6 & 4 \\ 6 & 10 & 6 \\ 4 & 6 & 6 \end{matrix} \tag{2}$$

Commonly used GLCM features:

(1) Energy: The magnitude of the energy is related to the magnitude of the gray level co-occurrence matrix element values. It reflects the distribution of grayscale of a certain image and the thickness of the texture.
(2) Contrast: Contrast is positively correlated with the sharpness of the image and the depth of the texture.
(3) Entropy: Entropy is the concept of thermodynamics, and in an image, entropy is a representation of all information about an image, which indicates whether the image texture is complex and uniform.

3.2 Extreme Learning Machine

Classifiers play an important role in machine learning, and the classifier used in this paper is Extreme Learning Machine [21–25]. ELM is developed from a single hidden layer neural network. But the difference from the traditional BP algorithm is that it does not require reverse iteration to adjust the weight of the hidden layer. Specifically, in the feedforward neural network of a single hidden layer, for the input feature vector, the weight of the input layer to the hidden layer is randomly assigned, but the weight of the hidden layer to the output layer needs to be obtained according to the least square method [26]. ELM learns very fast and avoids the thorny issues of proper learning, over-fitting, and so on. The architecture of ELM is shown in the Fig. 3.

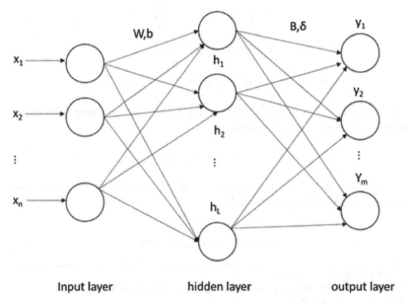

Fig. 3. Structure of extreme learning machine

In ELM, the output of the feedforward neural network of g(x) can be expressed as

$$\sum_{i=1}^{L} \beta_i g\left(w_i \cdot x_j + b_i\right) = h_j, \, j = 1, \cdots, N \tag{3}$$

where wi is the input weight of the input layer to the ith hidden layer, bi is the deviation of the ith hidden layer node. The biggest innovations of ELM are:

(1) In ELM, input weights and hidden layer deviations are given randomly, and no adjustment is needed after setting. This is not the same as BP neural network, BP needs to constantly reverse the adjustment of weights and thresholds. So here you can reduce the amount of computation by half.
(2) The connection weight β between the hidden layer and the output layer does not need to be iteratively adjusted, but is determined once by solving the equations.

3.3 Bat Algorithm

There are many global optimization algorithms [27–31], among which the bat algorithm is an algorithm proposed by Yang in 2010 to search for the global optimal solution, which has high accuracy and effectiveness. General speaking, the bat algorithm is a mechanism that simulates bat echo emission and detection [32]. All bats use echolocation to sense the distance, and can automatically adjust the frequency or wavelength of the transmitted pulse, and the frequency f is between [f min, f max] and the wavelength λ is between [λmin, λmax]. According to an article published by Yang in 2010, there are some formulas we need to simulate the position and speed update of the bat in each time step t:

$$f_i = f_{min} + (f_{max} - f_{min})\beta \tag{4}$$

$$v_i^{t+1} = v_i^t + (x_i^t - x_*) f_i \tag{5}$$

$$x_i^{t+1} = x_i^t + x_i^{t+1} \tag{6}$$

where $\beta \in [0, 1]$, x_* is global optimal solution. After determining a solution, use a random walk to generate a new solution:

$$x_{new} = x_{old} + \varepsilon * A^{(t)} \tag{7}$$

where $\varepsilon \in [-1, 1]$, A(t) is the average loudness of all bats in the current time step. The pseudo-code of BA is shown in Table 1.

Table 1. Pseudo-code of BA

Initialize populations x_i and v_i
Initialization frequency f_i, pulse emissivity r_i and loudness A_i
While(t<MAX_ITER)
Generate new solutions by adjusting the frequency
Update speed and position according to the above formula
if (rand>r_i)
Select one from the best solution and generate a local solution
end if
Random flight produces random solution
if(rand<A_i&&f(x_i)<f(x_*))
Accept the new solution and update the loudness and pulse emissivity according to the formula
end if
Find the current optimal solution x_*
end while

3.4   10-Fold Cross-validation

The classifier is an important part of machine learning, so the performance of the classifier will affect the entire algorithm structure. Therefore, we need to know the performance of the classifier. This paper uses 10-fold cross-validation to verify the performance of the classifier [33–37]. The steps of 10-fold Cross-Validation are as follows [38]:

(1)   Randomly divide the data set into 10.
(2)   Take 1 of them as a test set and the other 9 as a training set.
(3)   Repeat process (2) 9 times.

Figure 4 shows an illustration of cross validation. The black block represents the training data, and other colors represent different test folds. This 10-fold cross validation will perform 10 runs, each run of which the dataset split is randomly initialized.

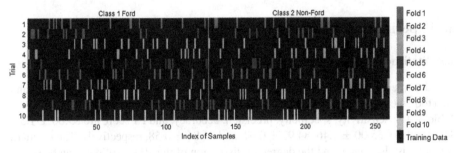

Fig. 4. Illustration of cross validation

The following are some evaluation indicators of the classification algorithm. We need to understand their meaning and calculation formula (Table 2).

Table 2. Confusion matrix

Predicted	Actual	
	Positive	Negative
Positive	True Positive (TP)	False Positive (FP)
Negative	False Positive (FP)	True Negative (TN)

Accuracy is the number of correctly classified samples divided by the total number of samples. Accuracy is our most common evaluation indicator.

$$\text{Accuracy} = (TP + TN)/(P + N) \qquad (8)$$

Sensitive represents the proportion of all positive cases that are correctly divided, and measures the ability of the classifier to identify positive cases.

$$\text{Sensitive} = TN/P \qquad (9)$$

Specificity represents the proportion of all negative cases that are correctly divided, and measures the ability of the classifier to identify negative cases.

$$Specificity = TN/N \tag{10}$$

Precision represents the proportion of the case that is actually positive, in the cases that is divided into positive cases.

$$Precision = TP/(TP + FP) \tag{11}$$

F1 score is a measure of the classification problem, which is the harmonic mean of the accuracy rate and the recall rate.

$$F1score = 2 * precision*recall/(precision + recall) \tag{12}$$

4   Experiment Results and Discussions

4.1   Statistical Results

In order to ensure a more accurate estimate, we made 10 10-fold cross-validation and then took the mean. The 10 runs of 10-fold cross validation results were shown below in Table 3. The sensitivity, specificity, precision, accuracy, and F1 score are 85.00 ± 1.51, 84.85 ± 1.76, 85.00 ± 1.46, 84.92 ± 0.64, and 84.84 ± 0.58, respectively. The standard deviation is also given, and the degree of dispersion of the 10 sets of data can be seen.

Table 3.   Statistical results of proposed GLCM-ELM-BA algorithm

Run	Sensitivity	Specificity	Precision	Accuracy	F1
1	82.31	87.69	87.06	85.00	84.51
2	83.08	85.38	85.56	84.23	84.15
3	86.15	83.85	83.99	85.00	84.77
4	86.15	82.31	82.98	84.23	84.38
5	84.62	86.92	86.74	85.77	85.51
6	84.62	84.62	84.62	84.62	84.62
7	86.15	84.62	84.97	85.38	85.22
8	84.62	83.08	83.41	83.85	83.99
9	84.62	86.92	87.01	85.77	85.70
10	87.69	83.08	83.66	85.38	85.51
Mean ± SD	85.00 ± 1.51	84.85 ± 1.76	85.00 ± 1.46	84.92 ± 0.64	84.84 ± 0.58

4.2 Training Algorithm Comparison

We compared BA algorithm with other swarm intelligence training methods. Diker, Avci [39] used genetic algorithm (GA) to train extreme learning machine. According to the theory of evolution, that is, the survival of the fittest and the elimination of the weak, GA gradually finds the optimal solution. Abdelaal and El Tobely [40] used particle swarm optimization (PSO) train ELM. In short, PSO uses a particle to constantly change its speed and position to search for the optimal solution. The results of GA [39] and PSO [40] are shown in Tables 4 and 5. The comparison of three training algorithms was shown

Table 4. Statistical results of using GA

GA [39]	Sensitivity	Specificity	Precision	Accuracy	F1
1	84.62	80.00	81.01	82.31	82.73
2	86.15	82.31	82.95	84.23	84.40
3	80.00	83.08	82.56	81.54	81.23
4	80.77	83.85	83.37	82.31	81.99
5	82.31	83.85	83.63	83.08	82.92
6	83.85	83.08	83.36	83.46	83.55
7	81.54	81.54	81.61	81.54	81.53
8	84.62	83.08	83.41	83.85	83.99
9	79.23	86.15	86.03	82.69	82.20
10	82.31	82.31	82.42	82.31	82.29
Mean ± SD	82.54 ± 2.12	82.92 ± 1.53	83.04 ± 1.28	82.73 ± 0.87	82.68 ± 0.99

Table 5. Statistical results of using PSO

PSO [40]	Sensitivity	Specificity	Precision	Accuracy	F1
1	83.08	86.15	85.87	84.62	84.39
2	83.08	83.85	84.11	83.46	83.45
3	79.23	86.92	86.08	83.08	82.33
4	83.85	84.62	84.49	84.23	84.15
5	86.15	79.23	80.75	82.69	82.98
6	82.31	82.31	82.42	82.31	82.29
7	83.08	82.31	82.55	82.69	82.75
8	81.54	80.77	81.01	81.15	81.21
9	83.08	85.38	85.16	84.23	84.06
10	88.46	81.54	82.88	85.00	85.48
Mean ± SD	83.38 ± 2.36	83.31 ± 2.36	83.53 ± 1.81	83.35 ± 1.13	83.31 ± 1.18

in Table 6. We can find our proposed BA can achieve better performance than GA [39] and PSO [40].

Table 6. Training algorithm comparison

Training algorithm	Sensitivity	Specificity	Precision	Accuracy	F1
GA [39]	82.54 ± 2.12	82.92 ± 1.53	83.04 ± 1.28	82.73 ± 0.87	82.68 ± 0.99
PSO [40]	83.38 ± 2.36	83.31 ± 2.36	83.53 ± 1.81	83.35 ± 1.13	83.31 ± 1.18
BA (Ours)	85.00 ± 1.51	84.85 ± 1.76	85.00 ± 1.46	84.92 ± 0.64	84.84 ± 0.58

4.3 Comparison to State-of-the-Art Approaches

We compare this proposed GLCM-ELM-BA algorithm with state-of-the-art approaches: LR [4], NBC [5]. The comparison results of various algorithms are shown in Table 7. Our research algorithm steps include feature extraction, classifier design and search optimal solution. It shows that the algorithm we proposed is much more accurate than state-of-the-art approaches. In addition, our research can also be applied to other fields, such as flower recognition, text recognition and car license plate recognition.

Table 7. Ford-vehicle identification algorithm comparison

Approach	Accuracy
LR [4]	72.20 ± 0.79
NBC [5]	75.25 + 0.67
GLCM-ELM-BA (Ours)	84.92 ± 0.64

5 Conclusions

In the experiment, the Ford car photos were processed according to the above algorithm flow, and the accuracy was 84.92 ± 0.64%, which indicates that the method we used is not only effective but also accurate. However, the sample size of the photos we processed is not large enough and this method requires a high degree of processing power of the computer. Therefore, our next task is to continuously improve the algorithm. Undoubtedly, this research will bring great progress to the society, especially to the traffic police. Although our research can accurately identify Ford, what we need to do in the future is to apply this algorithm to more car brands. For example, designing a set of algorithms can identify the brand of any car.

References

1. Cibuk, M., et al.: Efficient deep features selections and classification for flower species recognition. Measurement **137**, 7–13 (2019)
2. de Souza, J.C.S., et al.: Data compression in smart distribution systems via singular value decomposition. IEEE Trans. Smart Grid **8**(1), 275–284 (2017)
3. Wang, S.-H., Jia, W.-J., Zhang, Y.-D.: Ford motorcar identification from single-camera side-view image based on convolutional neural network. In: Yin, H., et al. (eds.) IDEAL 2017. LNCS, vol. 10585, pp. 173–180. Springer, Cham (2017). https://doi.org/10.1007/978-3-319-68935-7_20
4. Tao, Y., et al.: Vehicle identification method based on wavelet energy and logistic regression. Adv. Comput. Sci. Appl. **5**(1), 579–582 (2018)
5. Suchkov, M.: Motor side-view recognition system based on wavelet entropy and Naïve Bayesian classifier. Adv. Eng. Res. **127**, 78–82 (2018)
6. Zhang, Y.: Stock market prediction of S&P 500 via combination of improved BCO approach and BP neural network. Expert Syst. Appl. **36**(5), 8849–8854 (2009)
7. Wei, G.: A new classifier for polarimetric SAR images. Progress Electromagnet. Res. **94**, 83–104 (2009)
8. Wei, G.: Color image enhancement based on HVS and PCNN. Sci. China Inf. Sci. **53**(10), 1963–1976 (2010)
9. Wu, L.: A hybrid method for MRI brain image classification. Expert Syst. Appl. **38**(8), 10049–10053 (2011)
10. Wu, L.: Optimal multi-level thresholding based on maximum tsallis entropy via an artificial bee colony approach. Entropy **13**(4), 841–859 (2011)
11. Wu, L.: An MR brain images classifier via principal component analysis and kernel support vector machine. Progress Electromagnet. Res. **130**, 369–388 (2012)
12. Wu, L.: Classification of fruits using computer vision and a multiclass support vector machine. Sensors **12**(9), 12489–12505 (2012)
13. Dong, Z.: Classification of Alzheimer disease based on structural magnetic resonance imaging by kernel support vector machine decision tree. Progress Electromagnet. Res. **144**, 171–184 (2014)
14. Sun, P.: Pathological brain detection based on wavelet entropy and Hu moment invariants. Bio-Med. Mater. Eng. **26**(s1), 1283–1290 (2015)
15. Parhizkar, E., et al.: Partial least squares - least squares - support vector machine modeling of ATR-IR as a spectrophotometric method for detection and determination of iron in pharmaceutical formulations. Iran. J. Pharm. Res. **18**(1), 72–79 (2019)
16. Lu, S.: Pathological brain detection based on AlexNet and transfer learning. J. Comput. Sci. **30**, 41–47 (2019)
17. Muhammad, K.: Image based fruit category classification by 13-layer deep convolutional neural network and data augmentation. Multimed. Tools Appl. **78**(3), 3613–3632 (2019)
18. Xie, S.: Alcoholism identification based on an AlexNet transfer learning model. Frontiers in Psychiatry **10** (2019). Article ID 205
19. Zhao, G.: Polarimetric synthetic aperture radar image segmentation by convolutional neural network using graphical processing units. J. Real-Time Image Proc. **15**(3), 631–642 (2018)
20. Ameri, A., et al.: Regression convolutional neural network for improved simultaneous EMG control. J. Neural Eng. **16**(3), 11 (2019). Article ID 036015
21. Lu, S.: A pathological brain detection system based on extreme learning machine optimized by bat algorithm. CNS & Neurol. Disord. - Drug Targets **16**(1), 23–29 (2017)
22. Wu, Y.: Extreme learning machine used for focal liver lesion identification. J. Gastroenterol. Hepatol. **32**(S3), 168 (2017)

23. Muhammad, K.: Ductal carcinoma in situ detection in breast thermography by extreme learning machine and combination of statistical measure and fractal dimension. J. Ambient Intell. Humaniz. Comput. (2017). https://doi.org/10.1007/s12652-017-0639-5

24. Lu, S.: Pathological brain detection in magnetic resonance imaging using combined features and improved extreme learning machines. J. Med. Imaging Health Inf. **8**, 1486–1490 (2018)

25. Zhao, G.: Smart pathological brain detection by synthetic minority oversampling technique, extreme learning machine, and jaya algorithm. Multimed. Tools Appl. **77**(17), 22629–22648 (2018)

26. Karkkainen, T.: Extreme minimal learning machine: ridge regression with distance-based basis. Neurocomputing **342**, 33–48 (2019)

27. Gorriz, J.M.: Multivariate approach for Alzheimer's disease detection using stationary wavelet entropy and predator-prey particle swarm optimization. J. Alzheimers Dis. **65**(3), 855–869 (2018)

28. Li, Y.-J.: Single slice based detection for Alzheimer's disease via wavelet entropy and multilayer perceptron trained by biogeography-based optimization. Multimed. Tools Appl. **77**(9), 10393–10417 (2018)

29. Qian, P.: Cat swarm optimization applied to alcohol use disorder identification. Multimed. Tools Appl. **77**(17), 22875–22896 (2018)

30. Hou, X.-X.: Alcoholism detection by medical robots based on Hu moment invariants and predator-prey adaptive-inertia chaotic particle swarm optimization. Comput. Electr. Eng. **63**, 126–138 (2017)

31. Li, P., et al.: Pathological brain detection via wavelet packet tsallis entropy and real-coded biogeography-based optimization. Fundamenta Informaticae **151**(1–4), 275–291 (2017)

32. Bangyal, W.H., et al.: Optimization of neural network using improved bat algorithm for data classification. J. Med. Imaging Health Inf. **9**(4), 670–681 (2019)

33. Lu, H.M.: Facial emotion recognition based on biorthogonal wavelet entropy, fuzzy support vector machine, and stratified cross validation. IEEE Access **4**, 8375–8385 (2016)

34. Yang, J.: Pathological brain detection in MRI scanning via Hu moment invariants and machine learning. J. Exp. Theor. Artif. Intell. **29**(2), 299–312 (2017)

35. Nayak, D.R.: Detection of unilateral hearing loss by stationary wavelet entropy. CNS & Neurol. Disord. - Drug Targets **16**(2), 15–24 (2017)

36. Hou, X.-X.: Voxelwise detection of cerebral microbleed in CADASIL patients by leaky rectified linear unit and early stopping. Multimed. Tools Appl. **77**(17), 21825–21845 (2018)

37. Li, Y.: Detection of dendritic spines using wavelet packet entropy and fuzzy support vector machine. CNS & Neurol. Disord. - Drug Targets **16**(2), 116–121 (2017)

38. Oh, Y.J., et al.: Understanding location-based service application connectedness: model development and cross-validation. Comput. Hum. Behav. **94**, 82–91 (2019)

39. Diker, A., et al.: A new technique for ECG signal classification genetic algorithm wavelet kernel extreme learning machine. Optik **180**, 46–55 (2019)

40. Abdelaal, M., et al.: Bone cancer detection using particle swarm extreme learning machine neural networks. J. Med. Imaging Health Inf. **9**(3), 508–513 (2019)

Design and Implementation of a Wearable System for Information Monitoring

Qi Zhao[✉] and Tongtong Zhai

Beihang University, Beijing, China
zhaoqi@buaa.edu.cn

Abstract. In view of the limited measurement accuracy and short data transmission distance of existing smart wearable devices, this paper designs a wearable multi-information monitoring system. It achieves multi-directional observation of human body via energy consumption, vital signs and location-tracking information. The overall system consists of three parts: data terminal, LoRa communication network and remote monitoring center. Considering the constraint of power and memory cost, the system adopts a set of appropriate chips for hardware design. As the core of the system, data terminal contains a main control module, a signal acquisition module, and a display and transmission module. The microcontroller controls each sensor to collect different original signals which are performed noise reduction, and then these signals are processed by the algorithm embedded on chips to create the final information for monitoring human body. Test results show that the wearable multi-information monitoring system meets the demand for the measurement accuracy of human body information.

Keywords: Wearable equipment · Embedded system · Energy consumption · Physiological signs · Position tracking

1 Introduction

The advent of the era of intelligent wearable devices means not only the extension of hardware intelligence, but also the realization of information perception through software support, data interaction and cloud interaction [1, 2]. Currently, the popular wearable products focus on measuring steps, monitoring sleep and other aspects, which are of great value for the mass market, but the progress is still limited in terms of the application scenarios such as medical health, smart city, and future military war. Take future military demand as an example. Timely acquiring the soldier's energy expenditure is helpful for scientific assessment of their physical condition and combat capability [3]. The monitoring of soldiers' physiological data can diagnose and treat the wounded promptly [4]. Tracking soldiers' position information real-timely can have a better understanding of the overall battlefield situation. In addition, with the development of mobile Internet technology and the maturity of core hardware technology of wearable devices, low-

© Springer Nature Singapore Pte Ltd. 2019
H. Ning (Ed.): CyberDI 2019/CyberLife 2019, CCIS 1138, pp. 371–388, 2019.
https://doi.org/10.1007/978-981-15-1925-3_27

power chips and flexible circuit boards are becoming available. And the problems of limit measurement accuracy and short transmission distance existing in wearable equipment are expected to be solved. Thus, it is urgent for researchers to promote the innovation and development in this area.

The existing steps detection methods can be divided into two categories: dynamic threshold method and peak detection method. Among them, the dynamic threshold algorithm requires a large waveform amplitude, otherwise it will miss the step. In addition, the peak detection algorithm demands the sinusoidal waveform to be very smooth, or else it may cause the peak misjudgment. Heart rate and blood oxygen saturation are easily interfered by baseline drift in the process of measurement, and there is no good method to solve this problem. The interpolation fitting method is difficult to extract the interpolation points when the pulse wave signal is unstable. In addition, if the wavelet transform method is used, the baseline drift of the low-frequency layer can be removed by dividing it into ten layers or more.

To handle with the above application limitations and technical problems, we design a wearable multi-information monitoring system. By optimizing the hardware structure and software algorithms, we realize multi-scenario intelligent application, which is able to process information more efficiently and achieve seamless communication. Step counting is realized by detecting the variation of acceleration during motion. To obtain the heart rate and blood oxygen saturation value by detecting the period of the Photo Plethysmo Graphy pulse wave and the AC-DC signals. Longitude and latitude information is obtained by extracting and transforming coordinate data of navigation message signal, and the current altitude value is calculated by atmospheric pressure. All functions mentioned above enable supervisors to timely acquire the status information of the user. In addition, to achieve acceptable communication distance and quality with limited power consumption, the system will exchange data via LoRa long-distance transmission technology.

2 Hardware

The overall system structure contains three parts: data terminal, LoRa communication network and remote monitoring center.

2.1 Data Terminal

The data terminal worn on the user performs information acquisition and transmission. It adopts a microcontroller to combine various sensors and LoRa communication architecture, as shown in Fig. 1. Specifically, the terminal uses STM32F429 processor to dispatch the sensors for obtaining the original data and to transmit the human body data via the LoRa remote transmission module.

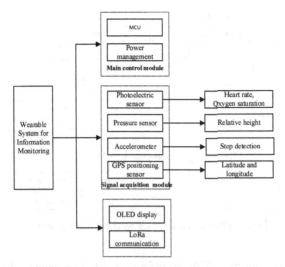

Fig. 1. Data terminal architecture diagram

Main Control Module

The main control module takes microcontroller as the core, and its peripheral circuit consists of 5 parts, which are serial debugging interface, low level reset, start mode setting, key and power management.

The power management sub-module provides sufficient energy for the entire terminal. Since the voltage for various types of chips is commonly selected as 5 V and 3.3 V, a lithium battery module is designed as a power input unit to make sure that the voltage can be regulated.

Signal Acquisition Module

The signal acquisition module is the key part of terminal equipment, which is responsible for getting the original signals from each sensor. It is divided into four parts: inertial sensor, photoelectric sensor, pressure sensor and GPS positioning sensor.

The inertial sensor MPU9250 is in charge of the raw acceleration data gathering. It integrates a three-axis accelerometer which adopts digital I2C serial interface to communicate with the main control module. Besides, we set the full range of the sensor as ± 2 g.

For photoelectric sensor, we adopt MAX30105 as its main chip. Its peripheral circuit integrates three LEDs lying close to the skin on arms. LEDs first emit light through the skin and get reflected light which is converted into an electrical signal. Then AD unit is applied to transform the electrical signal into quantized data. Moreover, the photoelectric sensor adopts the SPI serial communication interface to send quantized data to the main control module for signal processing and operation. The block diagram of its module is shown in Fig. 2.

Pressure sensor LPS22HB is responsible for the acquisition of original atmospheric pressure data. It uses digital I2C serial interface to exchange data with microcontroller. The range of pressure measures by the sensor is 260–1260 hpa.

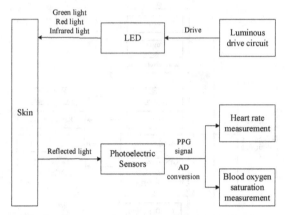

Fig. 2. Photoelectric sensor working block diagram

The GPS positioning chip UC6226 performs real-time geographic location information collection. We adopt the GNSS low-noise amplifier to amplify the navigation signal
which is obtained by the ceramic antenna, and then use the surface acoustic wave filter
to denoise the position information.

2.2 LoRa Communication Module

The LoRa part works in the unlicensed free frequency bands including 433 MHz,
868 MHz and 915 MHz. In this paper, 433 MHz is selected as the communication
frequency band. In addition, the peripheral circuit of the communication module is
composed of SX1278 chip, antenna, crystal oscillator circuit and transceiver switch
circuit.

The LoRa module exists in data terminal and monitoring center. In data terminal, it
mainly operates in the sending mode to transmit information to the monitoring center,
while in monitoring center, it mainly works in the receiving mode to receive data terminal
information. Since the SX1278 is a half-duplex transceiver, the transmitter and receiver
required different configurations.

2.3 Remote Monitoring Center

The main function of the monitoring center is to visualize the received data, and the
center staff can perform scheduling and give advice to the person who wears the device.
All the operations in monitoring center are real-time.

The monitoring center is similar to the data terminal in structure, which aims to
manage and collect data, so it does not include a signal acquisition module.

3 Software

The software design of the system implements data acquisition and processing, liquid
crystal display and LoRa transmission.

3.1 Step Detection

This section firstly constructs a human motion model to reveal the relation between acceleration and step number, and then uses the inertial sensor to monitor the changing information of acceleration during the human motion. Finally, we reduce the noise of the data by preprocessing to provide effective input for the step detection algorithm.

Human Motion Model

Human body movement process is analyzed by three components: forward, lateral and vertical [5, 6]. In the process of movement, the acceleration of the arm will alternately oscillate at the highest and lowest points, which can be decomposed into tangential and normal analysis to obtain component acceleration and combined acceleration [7, 8].

$$a_t = g \sin \theta \tag{1}$$

$$a_c = 2g(\cos \varphi - \cos \theta) \tag{2}$$

$$a = \sqrt{a_c^2 + a_t^2} = g\sqrt{3\cos^2 \theta - 8\cos \varphi \cos \theta + 4\cos^2 \varphi + 1} \tag{3}$$

where θ is the angle between the arm and the vertical direction in the process of motion, and φ is the angle when the arm swings to the highest point.

According to the formula (3), the variation trend of the combined acceleration waveform is similar to the sine wave, so the motion steps is consistent with the number of sine wave periods.

Original Data Acquisition and Preprocessing

Considering the power consumption and memory occupancy, the sampling frequency is set as 50 Hz to obtain more reasonable original acceleration data which can reflect the motion relation.

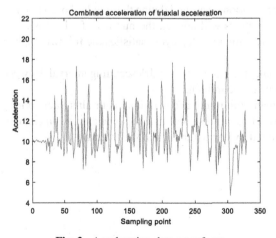

Fig. 3. Acceleration data waveform

Since the direction of motion is uncertain, only considering the data of one axis will affect the measurement results. In this paper, the signal vector amplitude of the triaxial acceleration is calculated as the final data. The result is shown in Fig. 3.

As the noise will affect the measurement accuracy, we first preprocess the signal. In general, the human step frequency will not exceed 5step/s, so the signal above that level will be considered as noise. We use a low pass filter to remove the high frequency noise in the combined acceleration signal. However, there are burrs in the filtered signal, and three-point sliding average filtering is adopted in the subsequent processing.

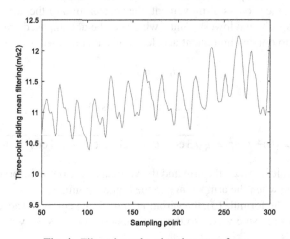

Fig. 4. Filtered acceleration data waveform

As can be seen from Fig. 4, the high-frequency noise can be reduced after low-pass filtering, and the waveform is suitable.

Step Counting Law Setting

We analyze the acceleration curve and define the standard of a step. In a certain interval, a pair of peaks and valleys will appear, the number of which is consistent with steps. The change of acceleration in a step cycle satisfies the following laws:

1. A step cycle contains an ascending and descending interval. However, it is not strictly monotonous due to the influence of noise.
2. The maximum and minimum acceleration occur alternately during motion. Nevertheless, the existence of noise will lead to some extremum can not be found, which needs to be judged by characteristics of the interval.

Algorithm 1 Step detection

Inputs: Original triaxial acceleration,ax, ay, az
Outputs: $StepNumber$

1: Calculate the combined acceleration of the triaxial acceleration, $ax, ay, az \rightarrow a$
2: Perform low pass and sliding mean filtering on a, $a \rightarrow Afil$
3: Set slide window N
4: **if** $Afil$ is the largrest value in window **then**
5:　　**if** $Afil$ is in the rising range **then**
6:　　　　$PeakValue \leftarrow Afil$.
7:　　　　$peak[] \leftarrow PeakValue$.
8:　　**end if**
9: **end if**
10: **if** $Afil$ is the smallest value in window **then**
11:　　**if** $Afil$ is in the descending range **then**
12:　　　　$ValleyValue \leftarrow Afil$.
13:　　　　$valley[] \leftarrow ValleyValue$.
14:　　**end if**
15: **end if**
16: Set the threshold for PeakValley amplitude, $ThreAlti$
17: Set the threshold for PeakPeak interval, $ThreInteri$
18: **if** $PeakValue - ValleyValue > ThreAlti$ **then**
19:　　**if** interval of adjacent peaks $> ThreInteri$ **then**
20:　　　　$StepNumber \leftarrow StepNumber + 1$.
21:　　**end if**
22: **end if**

The conditions of the effective pace we determine are as follows. Find the local peak value, and then further verify the existence of maximum via rising and falling characteristics of the interval. Set the threshold for peak-valley amplitude difference, and the step is certified when a pair of extremes are detected whose threshold condition is satisfied. In this paper, the threshold is set as 0.2 g.

3.2 Measurement of Heart Rate and Blood Oxygen Saturation

Blood flow produces regular pulse waves, from which we can get a lot of physiological information [9, 10]. Photo Plethysmo Graphy (PPG) tracing method uses single-wavelength light as the incident light source, which illuminates the subcutaneous tissue from the human body surface and reaches the photoelectric sensor through reflection. The sensor converts the intensity of the reflected light into an electrical signal to obtain pulse information.

Establishment of Mathematical Model
The pulse wave is periodic and consistent with the heart beat in frequency [11]. We indirectly measure the real-time heart rate (HD) in formula (4) by detecting the pulse wave cycle after obtaining the PPG signal.

$$HR = 60/T \tag{4}$$

Where T represents a complete pulse wave period in units of s. And the unit of HR is sub/min.

Oxygen saturation is the percentage of oxyhemoglobin in blood to total hemoglobin capacity [12, 13]:

$$SpO_2 = \frac{C_{HbO_2}}{C_{HbO_2} + C_{Hb}} \tag{5}$$

where SpO_2 is oxygen saturation, HbO_2 represents oxygenated hemoglobin and Hb denotes reduced hemoglobin.

When the light enters the subcutaneous tissue, the attenuation effect of the non-pulsating component on the incident light is relatively constant, and the reflected light constitutes the DC component of the PPG signal. The absorption and reflection of the pulsating component of light vary with the heartbeat period, and the reflected light constitutes the AC component of the PPG signal.

According to lambert-beer law, the variety rate of light intensity can reflect the change in blood volume [14, 15]. We select two incident light sources with wavelengths of λ_1 and λ_2, and the ratio parameters D_{λ_1} and D_{λ_2} of AC and DC components reflected from the two channels are:

$$D_{\lambda_1} = \frac{I_{AC}^{\lambda_1}}{I_{DC}^{\lambda_1}}, \quad D_{\lambda_2} = \frac{I_{AC}^{\lambda_2}}{I_{DC}^{\lambda_2}} \tag{6}$$

Through theoretical derivation, we can get:

$$\frac{D_{\lambda_1}}{D_{\lambda_2}} = \frac{I_{AC}^{\lambda_1}/I_{DC}^{\lambda_1}}{I_{AC}^{\lambda_2}/I_{DC}^{\lambda_2}} = \frac{\varepsilon_{HbO_2}^{\lambda_1} C_{HbO_2} + \varepsilon_{Hb}^{\lambda_1} C_{Hb}}{\varepsilon_{HbO_2}^{\lambda_2} C_{HbO_2} + \varepsilon_{Hb}^{\lambda_2} C_{Hb}} \tag{7}$$

$$SpO_2 = \frac{C_{HbO_2}}{C_{HbO_2} + C_{Hb}} = \frac{\varepsilon_{Hb}^{\lambda_2}(D_{\lambda_1} - D_{\lambda_2}) - \varepsilon_{Hb}^{\lambda_1}}{(\varepsilon_{HbO_2}^{\lambda_1} - \varepsilon_{Hb}^{\lambda_1}) - (\varepsilon_{HbO_2}^{\lambda_2} - \varepsilon_{Hb}^{\lambda_2})(D_{\lambda_1} - D_{\lambda_2})} \tag{8}$$

Where ε represents the absorption coefficient and c denotes the concentration. We select λ_2 as the point of equal absorption coefficient point, then

$$\varepsilon_{HbO_2}^{\lambda_2} = \varepsilon_{Hb}^{\lambda_2} \tag{9}$$

$$SpO_2 = \frac{\varepsilon_{Hb}^{\lambda_1}}{\varepsilon_{Hb}^{\lambda_1} - \varepsilon_{HbO_2}^{\lambda_1}} - \frac{\varepsilon_{Hb}^{\lambda_2}}{\varepsilon_{Hb}^{\lambda_1} - \varepsilon_{HbO_2}^{\lambda_1}} \times \frac{D_{\lambda_1}}{D_{\lambda_2}} \tag{10}$$

where $\varepsilon_{Hb}^{\lambda_1}, \varepsilon_{Hb}^{\lambda_2}, \varepsilon_{HbO_2}^{\lambda_1}$ are all constants. Set

$$A = \frac{\varepsilon_{Hb}^{\lambda_1}}{\varepsilon_{Hb}^{\lambda_1} - \varepsilon_{HbO_2}^{\lambda_1}}, \quad B = \frac{\varepsilon_{Hb}^{\lambda_2}}{\varepsilon_{Hb}^{\lambda_1} - \varepsilon_{HbO_2}^{\lambda_1}}, \quad r = \frac{D_{\lambda_1}}{D_{\lambda_2}} \tag{11}$$

Then the formula for calculating the oxygen saturation can be obtained as follows:

$$SpO_2 = A + Br \tag{12}$$

Where r is the eigenvalue. In actual measurement, considering the difference of human body and measurement accuracy, we redefine the formula (12) by quadratic regression fitting [16, 17]:

$$SpO_2 = A + Br + Cr^2 \tag{13}$$

where A, B, C are undetermined coefficients, which need to be calibrated by standard oximeter. We selected ten subjects for calibration experiments, and imported the $SpO_2 - r$ data results into the curve fitting toolbox cftool in MATLAB. The result is as follows:

$$SpO_2 = 93.684 + 30.353r - 45.07r^2 \tag{14}$$

PPG Original Signal Acquisition and Preprocessing
The external noise mainly appears in two parts when acquiring the original signa: the high-frequency noise caused by external electromagnetic interference and the low-frequency noise results from slight changes in the human body surface, leading to burrs and baseline drift. Taking infrared light as an example, the waveform and spectrum of its original PPG signal are shown in Figs. 5 and 6.

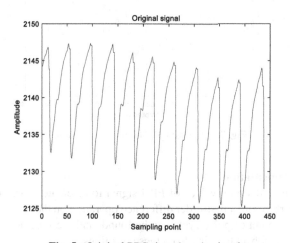

Fig. 5. Original PPG time domain signal

We utilize the first-order lag filtering method to remove the small-amplitude high-frequency electronic noise and alleviate the burr phenomenon. Based on the low frequency features of baseline drift noise, 50 FIR high-pass filter and median filter are utilized to restrain noise. The waveform and spectrum of processed PPG signal are shown in Figs. 7 and 8.

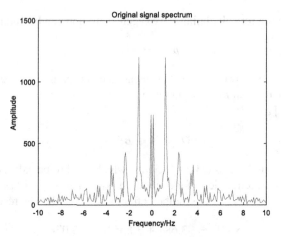

Fig. 6. Original PPG frequency domain signal

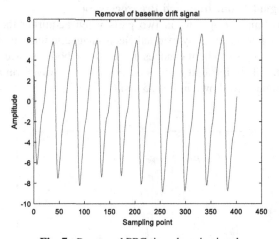

Fig. 7. Processed PPG time domain signal

Algorithm Implementation

We seek the peak and valley values of PPG signal to calculate heart rate and oxygen saturation. By performing a first-order difference operation on the pre-processed signal, the fastest change point of pulse cycle can be found, and the specific position of wave crest can be found by backtracking from these positions.

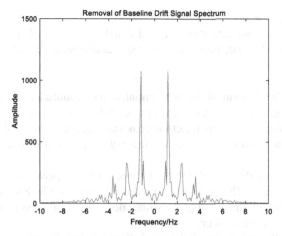

Fig. 8. Processed PPG frequency domain signal

Algorithm 2 Peak search algorithm, take green light PPG signal as an example

Inputs: Green light PPG signal,$GreenData$; Data length,N;

Outputs: Peak amplitude array,$alti[]$; Peak position array,$locs[]$;

1: $GreenMean \leftarrow \sum_{i=1}^{n} GreenData/N$

2: $AnGreen \leftarrow GreenData - GreenMean$

3: Filter angreen to remove baseline drift and get $AnGreenfil$

4: $Dgreen[n] \leftarrow AnGreenfil[n] - AnGreenfil[n-1]$

5: $thre \leftarrow \sum_{i=1}^{n} |Dgreen(i)|/len(Dgreen)$

6: **while** $Dgreen[i] < thre$ **do**

7:    remove $Dgreen[i]$

8: **end while**

9: **if** $Dgreen[i]$ is larger than the first and last three points **then**

10:    $locs[] \leftarrow i$

11:    $anti[] \leftarrow Dgreen$

12: **end if**

13: Set the interval between adjacent peak values ,$threpeakpeak$

14: Descend order $alti[]$

15: Detect the distance between other peaks and the current peak,$pdist$, the current

16: peak is starting from the maximum peak

17: **if** $pdist < threpeakpeak$ **then**

18:    remove the contrast peak

19: **end if**

20: Update $locs[]$ and $alti[]$

Heart Rate Measurement

In this paper, the sampling rate of the original signal is 50 Hz. The heart rate is measured based on the time difference between several successive peaks. The algorithm steps are as follows:

1. After taking the 1-norm of the differential signal amplitude, the average value is taken as the detection threshold screening signal.
2. The sliding window is set to detect the peak value of the region, and the whole input sequence is traversed to get the amplitude array and position array composed of peak points.
3. Set the minimum distance threshold between the two peaks to screen and judge the obtained peaks. The peaks are sorted by the magnitude, and starting with the maximum peak, eliminate all peaks whose distance from it less than the threshold. Traverse all the peaks and update the array.
4. According to the result of peak detection, the corresponding position of PPG signal after preprocessing is backtracked to find M real peak points and position array $locs[M]$.
5. Through the above steps, HR is calculated via formula (15):

$$HR = 3000 \times (M - 1)/(locs[M - 1] - locs[0]) \tag{15}$$

Blood Oxygen Saturation Measurement

The peak point of each pulse cycle of PPG signal can be obtained from above section. The signal is segmented at intervals of peak points. And the DC and AC components of each section are calculated to obtain blood oxygen saturation. The process is as follows:

1. Find the minimum amplitude point between the two peaks as the valley value, traverse all the peaks according to the position, and preliminarily obtain the valley value amplitude array and position array.
2. Set the peak-valley value range difference threshold, remove the peak-valley value points that do not meet the conditions, and update the array.
3. After the signal is re-segmented, the amplitude difference between the adjacent peaks and valleys is taken as the AC quantity, and the average value of the absolute value of the whole signal is taken as the DC quantity, and the signal component ratio D_i of the segment can be obtained; All component values of PPG signals in two channels are obtained respectively, and the average is taken as the final component ratio.
4. The eigenvalue is obtained and the blood oxygen value is calculated according to the curve expression of the quadratic regression fitting.

3.3 Positioning Monitoring

The data protocol of positioning chip conforms to the specification of Unicore Protocol. which is a special communication protocol for location and is used for data analysis and extraction.

Various positioning message formats are defined in the protocol. After initial configuration, the navigation signal of the message is transmitted to the main control chip through the UART mode. By selecting, calculating and format conversion, the latitude and longitude information can be obtained. The statement used in this paper is in the following format:

$GNGLL, <latitude ddmm.mmmmm>, <N (north latitude) or S (south latitude)>, <longitude dddmm.mmmmm>, <E (east longitude) or W (west longitude)>, <UTC time>, <state, A = positioning, V = not positioned>, <mode identification>, <checksum>.

And the parsed data is expressed as:

$GNGLL, 3998.571837, N,11634.731368, E,072034.000, A, A*4C.

After receiving the navigation signal, the main control module will judge the validity of the data. If the data is valid, the relevant information will be extracted. Otherwise, the waiting state will be returned. Selection and format conversion are performed on the valid data.

3.4 Height Measurement

Numerically, the atmospheric pressure is equal to the mass of the entire column of air up to the upper limit of the atmosphere. Obviously, with the rise of altitude, the atmospheric pressure gradually decreases. Based on the related research before, there is a corresponding function relationship between them within a certain range of altitude [18]:

$$H = 44330 \times \left[1 - \left(\frac{P}{P_0} \right)^{1/5.255} \right] \tag{16}$$

where P_0 denotes the standard atmospheric pressure value. Then the altitude can be obtained by calculating the current air pressure value.

Original Data Acquisition and Preprocessing

After on-chip reset and system initialization, the original temperature and pressure data are acquired by the pressure sensor [19]. By reading the temperature correction parameters in the EEPROM storage, we can calculate the actual pressure and temperature values.

In order to make the system more stable and avoid noise interference as possibly, we use the median and low-pass filters to process the air pressure data. Among them, the low-pass filtering adopts the first-order lag method, and the equation is as follows:

$$Y_n = (1 - k)X_n + kY_{n-1} \tag{17}$$

where Y_n is the current filtering output value, X_n denotes the current read-out data, Y_{n-1} represents the filtering output value last time, and k is 0.75.

Altitude Calculation

Due to the non-linear function relationship between pressure and altitude, there are problems such as slow operation speed and high hardware storage resource occupation in the method of direct calculation. Therefore, in this paper, we adopt linear interpolation method to reduce the complexity, and consider the independent variable and dependent variable as linear relationship for a period of time. Discretize the formula, create a pressure-altitude table and put it into the microcontroller memory. When the pressure is known, the corresponding altitude can be obtained by looking up the table.

4   Results

The software and hardware of the system described above are tested jointly to verify the accuracy of each information parameter.

4.1   Step Counting

We recorded the steps of the subjects under walking and running conditions respectively, and compared with the actual steps to calculate the measurement error. Set the actual steps as 500 and 1000, and the results are shown in Table 1.

Table 1. Step count measurement result

Subject	Testing steps (walk/run)	Error (walk/run)	Accuracy (walk/run)
A	517/522	17/22	96.70%/95.70%
	1030/1040	30/40	97%/96%
B	515/518	15/18	97%/96.30%
	1032/1036	32/36	96.80%/96.40%
C	520/525	20/25	96%/95%
	1035/1044	35/44	96.50%/95.60%
D	516/517	16/17	96.80%/96.70%
	1029/1032	29/32	97.10%/96.80%
E	518/522	18/22	96.40%/95.70%
	1034/1038	34/38	96.60%/96.20%

It can be seen from the table that the detection accuracy is above 95% in both walking and running states.

4.2   Monitoring of Heart Rate and Blood Oxygen Saturation

Heart rate and blood oxygen saturation were monitored at the fingers and wrists respectively, and compared with the synchronous data of the medical pulse oximeter. The measurement results and errors are shown in Fig. 9.

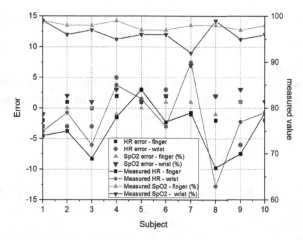

Fig. 9. Heart rate and blood oxygen saturation measurement results

From Fig. 9, we can observe that the results of heart rate and oxygen saturation measurement at the finger are ideal, which verifies the effectiveness of the algorithm. However, the skin difference, such as skin color, thickness and other factors, between people leads to the unsatisfactory effect at the wrist. For a few people, the quality of the original PPG signals at the wrist is poor, resulting in high measurement errors. In the future, we will continue to optimize the photoelectric signal acquisition circuit and optimize the measurement results.

4.3 Positioning Results

Latitude and Longitude Measurement
Go to the open space for testing, and compare the collected data with the high-precision positioning receiver board T326. Multiple groups of data are continuously measured at the same fixed point, and the test results are given in Table 2.

Table 2. Latitude and longitude measurement results

	Test terminal		T326	
Time	Longitude	Latitude	Longitude	Latitude
1	116.34731	39.98572	116.34720	39.98580
2	116.34762	39.98567	116.34720	39.98580
3	116.34756	39.98534	116.34720	39.98580

From Table 2, it is clear that error between the experimental value and the standard value of longitude and latitude is varying from 10^{-3} to 10^{-5}.

Altitude Monitoring

For the actual measurement in the new main building of Beihang university, the experimental results each floor are taken as the average of five measuring points The relative heights of each layer are calculated, as shown in Table 3.

Table 3. Altitude measurement results

Floor	Average atmospheric pressure/hPa	Average altitude/metre	Relative height/meter
1	1008.18	42.2	0
2	1007.62	46.9	4.7
3	1007.07	51.5	4.7
4	1006.53	56.5	5.0
5	1006.01	61.3	4.8
6	1005.48	66.5	5.0
7	1004.96	71.0	4.6

According to the table above, we can see that with the decrease of atmospheric pressure, the altitude will gradually increase. Besides, the relative height between each floor varies from 4.6 to 5.0 m, and the error with the real height is within the acceptable range.

5   Conclusion

This paper completed the hardware and software design to extract the useful information which can meet the stringent requirement of accuracy and coverage of data. Experimental results show that the proposed system can achieve better monitoring performance. Firstly, the step detection can be realized by calculating the number of acceleration cycles during motion. The experimental results show that the detection accuracy reaches 95% under walking and running conditions. Secondly, the photoelectric volume pulse wave signal can be obtained and the single-wavelength green light is measured for a certain period of time to obtain a real-time heart rate. The real-time oxygen saturation can be obtained by DC-AC calculation of two wavelength reflected light. The experimental results show that the error of heart rate at finger is within ±3 and the error of oxygen saturation is within ±1. Thirdly, the positioning signal can be obtained by ceramic antenna. After amplification and filtering, the latitude and longitude values in the analytical information can be extracted. According to the experimental test, the error between the experimental value and the standard value of longitude and latitude is varying from 10^{-5} to 10^{-3}. Finally, the value of altitude can be calculated through the functional relationship between altitude and atmospheric pressure within a certain height range. The complexity can be reduced by the linear interpolation method, and then the relative height value can be obtained.

Due to the low operation speed and precision of the microcontroller, the accuracy and real-time performance of the whole system are limited. Hardware and software structure will be further optimized to improve the system performance.

References

1. Wu, J., Li, H., Lin, Z., et al.: Competition in wearable device market: the effect of network externality and product compatibility. Electron. Commer. Res. **17**, 335–359 (2017)
2. Zornoza, J., Mujica, G., Portilla, J., et al.: Merging smart wearable devices and wireless mesh networks for collaborative sensing. In: Conference on Design of Circuits & Integrated Systems (2017)
3. Egbogah, E.E., Fapojuwo, A.O.: Achieving energy efficient transmission in wireless body area networks for the physiological monitoring of military soldiers. In: Military Communications Conference (2013)
4. Devara, K., Ramadhanty, S., Abuzairi, T.: Design of wearable health monitoring device. In: Biomedical Engineerings Recent Progress in Biomaterials, Drugs Development, & Medical Devices: International Symposium of Biomedical Engineering (2018)
5. Bui, D.T., Nguyen, N., Jeong, G.M.: A robust step detection algorithm and walking distance estimation based on daily wrist activity recognition using a smart band. Sensors **18**(7), 2034 (2018)
6. Chien, J.C., Hirakawa, K., Shieh, J.S., et al.: An effective algorithm for dynamic pedometer calculation. In: International Conference on Intelligent Informatics & Biomedical Sciences (2015)
7. Zhao, F., Li, X., Yin, J., Luo, H.: An adaptive step detection algorithm based on the state machine. In: Sun, L., Ma, H., Fang, D., Niu, J., Wang, W. (eds.) CWSN 2014. CCIS, vol. 501, pp. 663–672. Springer, Heidelberg (2015). https://doi.org/10.1007/978-3-662-46981-1_62
8. Tang, M., Gao, T., Zhou, S., et al.: The design of network pedometer based on Bluetooth 4.0. In: International Conference on Intelligent Transportation (2018)
9. Hertzman, A.B.: The blood supply of various skin areas as estimated by the photoelectric plethysmograph. Am. J. Physiol.-Legacy Content **124**(2), 328–340 (1938)
10. Hertzman, A.B.: Observations on the finger volume pulse recorded photoelectrically. Am. J. Physiol. **119**, 334–335 (1937)
11. Shao, D., Liu, C., Tsow, F., et al.: Noncontact monitoring of blood oxygen saturation using camera and dual-wavelength imaging system. IEEE Trans. Biomed. Eng. **63**(6), 1091–1098 (2016)
12. Bhat, S., Adam, M., Hagiwara, Y., et al.: The biophysical parameter measurements from PPG signal. J. Mech. Med. Biol. **17**(7), 1740005 (2017)
13. Phillips, C., Liaqat, D., Gabel, M., et al.: WristO2 – reliable peripheral oxygen saturation readings from wrist-worn pulse oximeters (2019)
14. Jarosz, A., Kolacinski, C., Wasowski, J., et al.: Design and measurements of the specialized VLSI circuit for blood oxygen saturation monitoring. In: Experience of Designing & Application of CAD Systems in Microelectronics (2017)
15. Preejith, S.P., Ravindran, A.S., Hajare, R., et al.: A wrist worn SpO2 monitor with custom finger probe for motion artifact removal. In: 2016 38th Annual International Conference of the IEEE Engineering in Medicine and Biology Society (EMBC), pp. 5777–5780. IEEE (2016)

16. Yan, L., Hu, S., Alharbi, S., et al.: A multiplexed electronic architecture for opto-electronic patch sensor to effectively monitor heart rate and oxygen saturation. In: Optical Diagnostics & Sensing Xviii: Toward Point-of-Care Diagnostics (2018)

17. Mohan, P.M., Nisha, A.A., Nagarajan, V., et al.: Measurement of arterial oxygen saturation (SpO2) using PPG optical sensor. In: 2016 International Conference on Communication and Signal Processing (ICCSP), pp. 1136–1140. IEEE (2016)

18. Ingvalson, R.: Systems and methods for differential altitude estimation utilizing spatial interpolation of pressure sensor data (2012)

19. Maniraman, P., Chitra, L.: Comparative analysis of capacitive type MEMS pressure sensor for altitude sensing. In: Emerging Trends in New & Renewable Energy Sources & Energy Management (2014)

A Review of Internet of Things Major Education in China

Yuke Chai[1], Wei Huangfu[1], Huansheng Ning[1], and Dongmei Zhao[2(✉)]

[1] School of Computer and Communication Engineering,
University of Science and Technology Beijing, Beijing 100083, China
Chaiyke@163.com
[2] College of Economic and Management, China Agriculture University, Beijing 100083, China
zhaodongm@vip.163.com

Abstract. With the continuous expansion of the Internet of things industry, more and more problems are exposed, among which professionals shortage is one of the most important problems. In order to solve the shortage of IoT professionals, the Chinese government encourages colleges and universities to open IoT major. Due to the short opening time, there are still some problems in professionals training. This paper reviews the situation of Internet of things major education in China. First, we introduce the background of development of Internet of things major, and then through the analysis of the layer models, we put forward a universally used IoT Major's knowledge structure and related courses list. Finally, this paper classifies the universities offering Internet of things major in China and analyzes their educational status from the perspective of education level, geographical location and curriculum system.

Keywords: Cyberspace · Internet of Things · Education

1 Introduction

In recent years, cyberspace is developing rapidly. As a connection between cyberspace and physical space, the scale of Internet of Things (IoT) is also growing. According to Internet Data Center (IDC) [1], the number of IoT connections reached 11.5 billion in 2018, and it is predicted that nearly 30 billion will be connected by 2020. With the accelerated construction of 5G infrastructure, on June 6, 2019, the Ministry of Industry and Information Technology officially issued 5G commercial licenses, marking China's formal entry into the 5G era. Billions of new devices and components are put into the network and connected. In the future, about 20% of 5G facilities will be used for communication between people, and 80% will be used for communication between things and people [2]. The combination of IoT and artificial intelligence (AI), cloud computing, big data and other technologies will deepen the application of IoT and promote the orderly and healthy development of IoT [3]. As the market becomes more and more dependent on the IoT, more and more problems are exposed. The shortage of professionals is one of the most serious problems. With the continuous expansion of IoT application industry, there is a growing demand for IoT professionals [4]. However,

© Springer Nature Singapore Pte Ltd. 2019
H. Ning (Ed.): CyberDI 2019/CyberLife 2019, CCIS 1138, pp. 389–404, 2019.
https://doi.org/10.1007/978-981-15-1925-3_28

according to a report by the national bureau of statistics (NBSPRC), only 20% of people working in relevant areas of IoT have professional skills, which is far from meeting the current market demand. Therefore, to solve this problem, it is necessary to actively cultivate IoT professionals.

Since 2010, many universities in China have offered the IoT major. However, due to the short opening time, there are still some problems in the current education of the IoT major. In this paper, we mainly analyze the IoT major education situation of in China from the school level and geographical location. This paper is organized as follows: Sect. 2 introduces the background of the development of the specialty of IoT in China; Sect. 3 through the analysis of layer models of IoT, we propose a general knowledge structure of IoT major; Sect. 4 from the perspective of education level, geographical location and curriculum system, we classify the universities that set up IoT major in China, and analyze the current education situation of them; Sect. 5 conclusion.

2 Background of IoT Education

In 2009, Premier Wen Jiabao wrote the IoT development into the "Government Work Report" and established the Perception Center to vigorously develop IoT. In 2010, at the third session of the Eleventh National People's Congress, the Premier proposed to accelerate the development of IoT again, which was officially taken as the next generation of information development industry in September of the same year.

In order to strengthen the cultivation of IoT professionals and create a good environment for the development of IoT, the government has provided many national-level projects, such as the National Natural Science Foundation of China (NSFC), the National High-tech Research and Development Plan of China (863 Plan), the main national basic research and development plan of China (973 Plan) and so on. In addition, government departments provide educational policies to promote the development of IoT professionals. Encouraged by the national policy, many universities in China have set up IoT major [5].

However, due to the short opening time, IoT education has many problems: There is no standard curriculum system, so the IoT courses in many colleges and universities are miscellaneous and unclear, and the curriculum system is unreasonable. IoT is an interdisciplinary subject, whose knowledge covers computer technology, communication technology, automatic control technology, sensing information processing technology and so on. However, in many colleges and universities, the direction of students' training is ambiguous, and there is insufficient connection between courses. Practical courses are out of line with the needs of enterprises, which make students have poor practical ability. More seriously, it is difficult for the graduates to find a job. The above problems need to be solved and further improved.

3 IoT Knowledge Structure

Knowledge structure [7] is very important for universities to formulate training plans. Knowledge structure is represented by curriculum system, so knowledge system is the core of curriculum system. At present, there is a general situation in colleges and universities, where different courses have the same name, or different courses have the same content [8]. It is very unreasonable. So, it is necessary to organize the knowledge system of IoT major. In this section, we will introduce and analysis some layered models of IoT, and chose the most appropriate model as the basis to establish the IoT knowledge structure.

3.1 IoT

IoT [9, 10] is a concept based on the Internet, it is considered a giant complex intelligent system. According to various sensors, IoT can connect Human to Human, Human to Things, Things to Things, for exchange information and communication. It was considered an information industrial revolution after computers and the Internet [11].

3.2 Some Layered Models of IoT

IoT structure is the cornerstone of the IoT [12]. It has great significance for studying the intrinsic characteristics of IoT. To formulate the IoT knowledge system, it is necessary to analyze the structure of IoT in depth. As a complex system, the layered model has been widely used in building IoT [13, 16]. So far. there are various kinds layered models of IoT proposed by researchers [14, 15], such as the three-layer model, four-layer model, IBM eight-layer model [17], and six-layer model which is proposed by Ning etc. [18].

Three-Layer Model. According to the definition, IoT is divided into three layers, perception Layer, network Layer and application Layer, as shown in Fig. 1. The three-layer model is the most widely recognized at present [18]. First, the perception layer can automatically sense external physical information through RFID, sensors, smart appliances, etc. And It is often used with the automatic induction device. Network layer (also called transmission layer), uses wireless and wired networks to encode, authenticate and transmit the collected data, to ensure data transmission high-speed, safe and high-quality. The application layer is the fundamental goal of the IoT development. It combines the actual requirements with the IoT technologies to achieve specific applications [20].

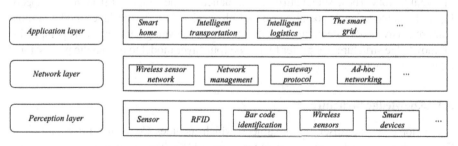

Fig. 1. Three-layer mode of IoT.

Four-Layer Model. Similar to the three-Layer model, according to the service function of IoT, IoT is subdivided into perception layer, network layer, management layer and application layer. The difference is that the four-layer model divides the application layer into supporting layer and application layer. The supporting layer mainly uses cloud computing, cloud storage, big data and other technologies to process the data from the network layer. The four-layer model is a subdivision of the traditional three-layer model of the IoT, which is essentially consistent (Fig. 2).

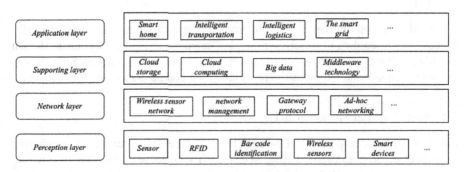

Fig. 2. Four-layer mode of IoT.

IBM Eight-Layer Model. It includes Sensor actuator Layer, Sensor network Layer, Sensor gateway Layer, Wide area network layer, Application Gateway Layer, Service Platform Layer, Application Layer, Analysis and Optimization Layer.

U2IoT Six-Layer Model. It includes the sensor-actuator layer, network layer, application layer, service integration layer, national management layer, and international coordinator layer.

Analyzing four kinds of layered models, these models are similar, four-layer and eight-layer models are subdivisions of the traditional three-layer model, and six-layer model is on the basis of the three-layer model and add the influence and supervision of social factors. These social attributes influence IoT development from many aspects. From the perspective of social world and physical world, the six-layer model is more comprehensive and more in line with the current IoT development. From the perspective of knowledge structure, we prefer to use the four-layer model to design the professional knowledge structure of IoT.

3.3　Knowledge Structure

IoT major is an interdisciplinary subject, which includes Computer, Information and Communication, Automation and other majors [18]. The main courses are as follows: Computer Basis, Programming, Circuit and Electronics, Discrete Mathematics , Introduction to IoT, Signal and System, Data Structure, Computer Composition Principle, Communication Principle, Operating System, Computer Network, Database,

Network Security, Wireless Sensor Network, RFID, Cloud Computing, Multimedia Technology, etc.

As an applied course, the IoT professional education should follow "general knowledge, specialty knowledge and application knowledge". The IoT curriculum system focuses on professionals training objectives and follows the principle of coordinated development of theoretical knowledge and applied skills, constructing a curriculum system consisting of three knowledge domains: general education, professional education and practical application. And each knowledge field has its own emphasis. So according to the model structure of "Field - Module -Unit - Point" [21], we subdivide the IoT major knowledge content.

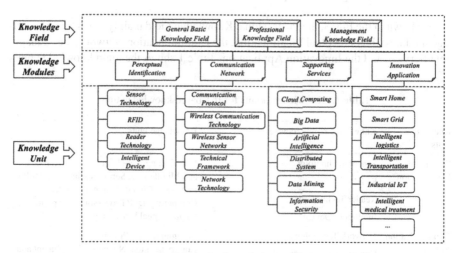

Fig. 3. IoT major knowledge structure.

- *Knowledge Field:* We regard Basic knowledge field, Management knowledge field and Professional knowledge field as the three major knowledge fields of IoT major [21]. Among them, the basic general knowledge mainly enables students to master the most basic common knowledge; the management knowledge field, including vocational training, psychological quality and other comprehensive courses; the professional knowledge field related to the discipline foundation, technical skills, and the related fields of IoT. Because the distinction between general and practical courses of each subject are not too obvious, this paper mainly discusses the core knowledge of IoT, the professional knowledge field.
- *Knowledge Module:* According to the four-layered model of IoT, which mentioned in the previous section, we divide the field of professional knowledge of IoT into four knowledge modules: perception and identification layer, communication network layer, management and service layer, and innovation application layer. Each knowledge module interacts with each other to form the professional knowledge field of IoT.

- *Knowledge Unit:* According to the characteristics of each knowledge module, it is divided into several knowledge units, which represent a specific category of technology or knowledge.
- *Knowledge Points:* Each knowledge unit or several knowledge units covers a lot of knowledge points. These knowledge points can be regarded as a professional course of IoT, or a separate chapter in a course, which is the smallest unit of knowledge units. For example, the perception identification module includes four knowledge units: sensor technology, radio frequency identification technology (RFID), reader technology and intelligent terminal equipment. These four knowledge units also contain several knowledge points, such as EPC coding, label technology, antenna and middleware, sensor technology, integrated circuit design, embedded computing, DSP processor and application, digital signal processing, etc.

As shown in the Fig. 3, the Professional Knowledge Field mainly covers four knowledge modules: the Perceptual Identification, the Communication Network, the Management Service, and the Innovation Application. Table 1 shows the main knowledge units

Table 1. Knowledge points for IoT based knowledge units.

Knowledge units	Knowledge points (courses)	Knowledge units
Perceptual Identification	Sensor Technology	EPC Coding, Label Technology, Antenna and Middleware, Sensor Technology, RFID, Integrated Circuit Design, Embedded Computing, DSP Processor and Application, Digital Signal Processing, etc
	RFID	
	Reader Technology	
	Intelligent Device	
Communication Network	Network Technology	Communication Principle, Mobile Communication, Network Management and Security, Multimedia Communication Technology, Computer Network, Communication Technology, Signal and System, Gateway Protocol, etc
	Communication Protocol	
	Technical Framework	
	Wireless Communication Technology	
	Wireless Sensor Network	
Supporting Service	Cloud Computing	Data Structure, Operating System, Database, Software Engineering, Advanced Programming Technology, Cloud Computing, Fog Computing, Dew Computing, Database, AI, Big Data, Data Mining, Data Processing, Information Security, Authentication, Message Integrity
	Artificial Intelligence	
	Big Data	
	Distributed System	
	Data Mining	
	Information Security	
Innovation Application	Smart Home	Intelligent Transportation Systems, Smart Vehicle-to-Vehicle Communications, Social Intelligence, Deep Learning, Smart home-to-home interactions, Smart Cities and Collaborations, IoT based Healthcare, AI, Middleware Technology, Innovative Practice, Other Applied Courses
	Smart Grid	
	Intelligent Logistics	
	Intelligent Transportation	
	Industrial IoT	
	Intelligent Medical	

of IoT and their knowledge points. For example, the perception identification module includes four knowledge units: sensor technology, RFID, reader technology and intelligent terminal equipment. These four knowledge units also contain several knowledge points, such as EPC coding, label technology, antenna and middleware, sensor technology, integrated circuit design, embedded computing, DSP processor and application, digital signal processing, etc. Figure 3 shows IoT Major Knowledge structure.

IoT major is an interdisciplinary subject, its content is complex. Therefore, it is very important to grasp the main technical context if we want to make a reasonable knowledge structure. As the core of IoT education, knowledge structure, based on undergraduate general education, divides and describes the basic and professional knowledge of IoT horizontally and vertically, in terms of subject categories, knowledge levels, classroom teaching and practical teaching. Each university IoT major can make different teaching plans according to its own discipline characteristics and professional direction. When making teaching plans according to its own actual situation, each university should also design teaching contents in close combination with the knowledge system of IoT major.

The knowledge system structure proposed in this paper is established on the basis of undergraduate general education and comprehensive education. According to the classification of professional disciplines and the knowledge content, four levels of knowledge field, knowledge module, knowledge unit and knowledge point are formed. Moreover, knowledge unit is the basic unit of teaching.

4 IoT Major Situation in China

In China, the Ministry of Education approved 40 universities to set up IoT major in 2010 and 27 universities in the second batch in 2011. Since then, many universities have set up IoT major and started to recruit the undergraduate students. So far, more than 700 universities and colleges in China have opened IoT major. However, because of the characteristic of IoT, the education planning, curriculum settings and teaching methods of IoT major are still in the exploratory stage. The education of IoT professionals should meet the needs of the development of national strategic emerging industries.

In this section, we will classify the colleges and universities that set up IoT major in China from the perspective of education level, geographical location and curriculum system analysis their characteristics.

4.1 From the Perspective of Education Level

In China, according to the educational level and training nature of institutions of higher learning, we divide the institutions of higher learning offering IoT major into Academic Undergraduate Universities and Vocational Colleges. Among them, academic undergraduate universities generally refer to key or comprehensive universities with higher educational level; Vocational college aims to cultivate technical talents, and its educational level is not as good as that of academic undergraduate university. We make a comparison between Academic Undergraduate Universities and Vocational Colleges in four aspects, training objectives, school system, training methods and course features, as shown in the Table 2.

Table 2. Comparison between Academic Undergraduate Universities and Vocational Colleges.

	Academic Undergraduate Universities	Vocational Colleges
Training objectives	Cultivating high-level applied professionals	Cultivate technical personnel with strong practical ability to meet the needs of enterprises
School system	4 years	3–4 years
Training methods	2 + 2 culture mode. The first two years pay attention to theoretical teaching, after two years pay attention to practical ability	Employment-oriented, vocational ability training as the core, project guidance, step by step
Course features	(1) The curriculum is Comprehensive; (2) Different schools focus on different content	(1) Focusing on the cultivation of students' skills; (2) Practical and applied courses account for a large proportion, while theoretical and general courses account for a small proportion

Academic Undergraduate Universities

Talent Training Objectives. The mode of cultivating IoT professionals in academic under-graduate universities is based on professional knowledge, with professional ability as the core and teaching practice as the basic way to cultivate high-level applied profession-als. At the same time, students are required to have a strong sense of dedication, social responsibility and innovation. Its purpose is to meet the needs of regional and local IoT industry development.

Among many academic undergraduate universities, the national key universities account for the majority. For example, Harbin Polytechnic University has five series of IoT courses: theory series, system series, tool series, Engineering Series and management series. Through these five series, students are trained step by step, devoted to cultivating and mastering the basic theory, method and technology of IoT, and can flexibly transform the knowledge they have learned into application. They want to cultivate high-tech professionals who can meet the needs of IoT industry and information industry, as well as those engaged in the research, teaching, design and development of IoT.

Training Methods. School system: 4 years. Most of the academic undergraduate uni-versities adopt the 2 + 2 training mode. In the first two years, they pay attention to the theoretical teaching, and in the second two years, they pay attention to promote and train students' practical ability. On the basis of basic theoretical research, pay more attention to cultivate students' application ability and practical ability. Through practical courses, school-enterprise cooperation and innovation competitions, students' practical ability can be improved.

Course Features. Due to the cross-disciplinary characteristics of IoT, in terms of cur-riculum design, the courses system of IoT in universities is mainly divided into general education courses, public basic courses, professional basic courses, professional compulsory courses, professional elective courses and practical innovation courses. As shown in the Fig. 4.

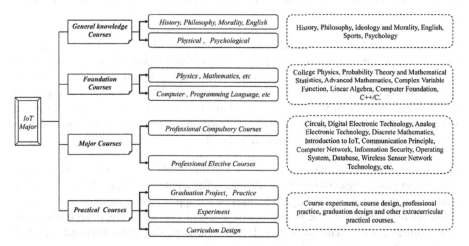

Fig. 4. Course system of IoT major in universities.

- General knowledge courses are integral part of higher education. It is "nonprofessional and non-professional education". Through studying this part of knowledge, students can master the most basic common knowledge. Including history, philosophy, ideology and morality, English, sports, psychological quality and so on.
- Foundation courses are compulsory courses for all majors. This part of knowledge enables students to have basic theoretical knowledge. Including: University physics, probability theory and mathematical statistics, advanced mathematics, complex function, linear algebra, computer foundation, C++/C.
- Major Course: Major Course includes professional compulsory courses and professional elective courses. Professional compulsory courses are basic courses for professionals. Including: circuit, digital electronic technology, analog electronic technology, discrete mathematics, data structure, computer composition principle, IoT introduction, communication principle, computer network, information security, operating system, database, wireless sensor network technology, etc. Professional elective courses refer to a kind of elective courses related to professional courses. According to students' knowledge level, ability and interest, it is not necessary for all students. By learning this part of knowledge, students can master more professional and specific knowledge. Including, embedded operating system, microcontroller, Java, Linux, software engineering, algorithm design, network programming, Web application development technology, network security, TCP/IP protocol, AI, digital image processing, multimedia technology, cloud computing, RFID technology.
- Practical courses mainly cultivate students' practical ability, including experiment, curriculum design, professional practice, graduation design and other extracurricular practical courses.

Vocational Colleges

Talent Training Objectives. Vocational colleges serve for the enterprises, industries and provide qualified technical and skilled professionals for them. In the process of training, these schools focus on the needs of enterprises, and strive to cultivate technical professionals with strong practical ability.

Training Methods. School system: 3–4 years. Compared with the academic undergraduate universities, vocational colleges pay more attention to students' practical ability in the process of training. Gradually formed a "employment-oriented, vocational ability training as the core, project guidance, progressive stage" professionals training mode [22]. For example, Xi'an Vocational and Technical College in the training of Internet of Things talents, enterprise demand-oriented, based on existing teaching resources, with Internet of Things technology as the core to build.

Course Features. Vocational colleges are oriented to application services. The curriculum system mainly emphasizes the cultivation of students' practical ability. Under the guidance of theoretical knowledge, it focuses on the application of skills courses to meet the needs of enterprises. The proportion of practical courses is much larger than that of theoretical courses. The curriculum lays emphasis on the application and students' practical abilities at the operational level.

Most of the colleges' curriculum system can be summarized as the curriculum system shown in Fig. 5. The curriculum system includes three levels: Fundamentals of computer, Application direction, Job purpose. According to the different employment directions of IoT, courses in different application directions are offered for students, such as: IoT software development, embedded hardware development, network technology, information security, sensor technology, RFID and other courses. At the same time, in order to cultivate students' solid basic knowledge, basic courses such as computer program foundation, hardware foundation and computer application are offered. Each basic course contains theory and practice, while the lower course is a prerequisite course for the upper course. It prepares knowledge and skills for the later teaching content. General knowledge pays attention to the cultivation of students' innovative ability in each course.

4.2 From the Perspective of Geographical Location

At present, China has initially formed the spatial pattern of four industrial clusters of IoT. The four industrial clusters are the Bohai Rim, the Yangtze river delta, the pearl river delta and the Midwest regions with Beijing, Shanghai, Shenzhen and Chongqing as the core. At the same time, through the geographic location analysis (incomplete statistics) of 533 universities offering IoT major in China, we find that there are more universities in the Bohai Sea, Yangtze River Delta, Pearl River Delta and the central and western regions, which are highly consistent with the four industrial clusters of IoT, there may be a close relationship between IoT professionals training and local regional economy. We will review the industrial characteristics of these four regional clusters and the characteristics of IoT professionals training.

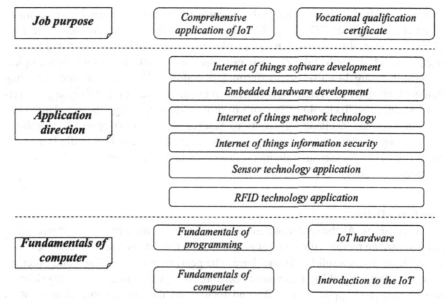

Fig. 5. Course system of IoT major in vocational colleges.

Bohai Rim Region

Industry Analysis. The Bohai Rim region as the most important urban cluster in northern China, it integrates politics, economy and culture. Bohai Rim is an important research, design and equipment manufacturing base for IoT industry in China. The region has strong research and development strength, wide penetration of the perception industry, fast network transmission mode and diversified integrated platform. It has basically formed a perfect system framework for the development of IoT.

University Analysis. Relying on Tianjin, Beijing and Hebei, there are 84 undergraduate universities offering IoT majors in this area. In addition, there are also Tsinghua University, Peking University, Chinese Academy of Sciences and other key national universities which are not offering IoT majors but also engaged in the projects of IoT. They have a large number of universities and research institutes. The region has excellent scientific research strength and industrial innovation technology capability. Professionals culture in this area focuses on training high-end skilled professionals with professional ability and quality, such as the application and maintenance of IoT system, the development and management of IoT application system.

Yangtze River Delta

Industry Analysis. The Yangtze River Delta is the birthplace of IoT, and has the industrial foundation in the field of electronic information. The orientation of IoT industry in this area is mainly concentrated in the high-end industrial chain. The Yangtze River Delta has a leading edge in the perceptual layer technology.

University Analysis. Universities in the Yangtze River Delta are mainly concentrated in Shanghai, Jiangsu and Zhejiang. Zhejiang Province is one of the provinces where the IoT industry started earlier, which has a solid industrial foundation and strong technological research strength. Take Zhejiang province for example, according to incomplete statistics, there are 14 universities offering IoT major in Zhejiang Province, including Zhejiang University of Technology, Hangzhou Dianzi University and Zhejiang University of Science and Technology. In terms of professionals training, Zhejiang University of Technology and Zhejiang University of Science and Technology focus on training advanced engineering and technical professionals in the field of IoT, while Hangzhou Dianzi University focuses on training professionals in the development, integration and management of IoT.

Pearl River Delta

Industry Analysis. As the center of China's manufacturing industry, the pearl river delta region is also the region with the most mature marketization and the most complete system in China. It has a solid industrial base. The pearl river delta region mainly focuses on the manufacturing of IoT equipment, software and system integration, network operation services and application fields, and focuses on the innovation and application of IoT, the construction of IoT infrastructure equipment, and the improvement of the informatization level of urban management.

University Analysis. In the Pearl River Delta region, Guangdong and Fujian are the main areas, with 34 universities offering IoT major. Taking Guangdong Province for example, these universities are distributed in Guangzhou, Foshan, Zhuhai, Zhongshan and other developed cities in the Pearl River Delta of Guangdong Province [23]. From the point of view of specialty setting and development, the IoT major in the above-mentioned schools is set up under the information engineering department or the computer department. It can be seen that the schools attach importance to the construction and development of IoT major based on network information technology. According to the characteristics of the IoT industry in Guangzhou, it is required that universities take professional technology as the orientation, strengthen the cultivation of students' knowledge, skills and attitudes, and cultivate development-oriented, compound and innovative technical skills professionals [24].

Midwestern Regions

Industry Analysis. Although the economic and electronic information industries in the midwestern regions lag behind the other three IoT industrial clusters as a whole, due to the driving effect of the IoT industry, the midwestern regions are closely combined with their own technological advantages, and the IoT industry has developed rapidly. For example, Wuhan has advantages in optoelectronics industry, Xi'an has a large number of universities and research institutes, which build a complete industrial chain and industrial system of IoT, and set up more production and office alliances or organizations. Such as Chongqing "China Mobile National M2 M Operating Center", "National M2 M Industrial Base", Wuhan "Radio Frequency Identification Innovation Technology Alliance" and so on.

University Analysis. There are many colleges and universities in the midwestern regions, which have strong scientific research ability. At the same time, each region has its own technological advantages. The professionals training plan of IoT major in each region is based on the local talent needs. Taking Hunan Province as an example, as a major province of education, Hunan has abundant higher education resources. As of 2016, 17 undergraduate colleges and universities of IoT have opened in Hunan Province, accounting for 33% of the universities in Hunan Province [25]. 8 higher vocational colleges of IoT application technology have opened, accounting for 11% of Higher Vocational Colleges in Hunan Province. In the universities that have set up IoT major, they have obvious specialty characteristics and advantages in related fields. For example, Central South University and Nanhua University. Therefore, the major universities in Hunan Province embedding their own professional characteristics and advantages in various IoT application fields, reforming the personnel training mode, forming a unique professionals training mode of IoT.

Through the analysis of the industrial characteristics of regional clusters and the characteristics of professionals cultivation of IoT, we find: (1) In the areas where the IoT industry concentrated, there are more IoT universities, which shows that IoT professionals cultivation has a close relationship with regional economy. (2) Although the emphasis of IoT professionals cultivation in this area is related to the discipline advantages of the school itself, it is more influenced by the economic development direction and the characteristics of IoT industry in this area.

Whether Academic Undergraduate Universities and Vocational Colleges the orientation of professionals cultivation should be regional, focusing on application, to promote regional and industrial development and innovation. Professionals training in Colleges and universities should be connected with regional economy to achieve open teaching and scientific research. By this way, a large number of applied research results and advanced applied professionals can be better provided for regional, industrial and local economic and social development.

4.3 From the Perspective of Curriculum System

IoT is an interdisciplinary subject, which based on computer science, computer and communication, electronics, automation and other disciplines. Its application involves various industries and fields [26]. Because each school's professional background and subject characteristics are different, there will be differences in the curriculum system and students' knowledge structure. Through the analysis of 27 universities, which offering IoT major earlier in China, 22% of them focused on perceptual control, 15% of them focused on communication network, 41% focused on software application, and the rest preferred comprehensive.

Emphasis on the Direction of Perception and Control. For example, Shandong University of Science and Technology, China's key universities in Shandong Province. It sets the IoT major in the College of Electrical Engineering, and takes the courses of sensor and networking technology, embedded system, IoT control technology and application as the main courses.

It emphasis on students' practical abilities in IoT applications, intelligent information processing and embedded systems. Graduates are required to have the basic ability of design and development. This kind of university includes Beijing Technology and Business University, Wuhan University of Technology and so on, which set IoT major in automation department or electrical engineering department,.

Emphasis on Transmission and Network Direction. For example, the University of Electronic Science and Technology, the key university of "985 Project", "211 Project", "World-class University and First-class discipline" in China. Its communication major and network engineering major are the national characteristic major. The advantages of communication major are obvious, so its IoT major is set up in the College of Communication Engineering, focusing on IoT transmission and network. They pay attention to cultivating students' solid basic knowledge of communication, network and information perception and processing. Students need to master IoT structure, communication transmission networking protocols and other knowledges.

Emphasis on Internet of Things Software and Service Direction. Most universities set up IoT major in computer colleges, such as University of Science and Technology Beijing, Beijing Institute of Technology, Harbin Institute of Technology, Jilin university and so on. This type of school's computer major has long history, for computer software applications and services have their own discipline characteristic and the deep background, so even opened IoT major, the course of IoT is not much different from the computer course. It just adds some IoT professional courses to the core courses of computer major, such as introduction to IoT, sensor principle and application, RFID principle and application, wireless sensor network principle and so on. So these universities are more focused on IoT software and services.

In a word, the current professional course system of IoT lacks its own discipline characteristics.

5 Conclusion

Based on the current cultivation situation of undergraduate IoT major, This paper analyzes the current status of IoT major education in China from the perspectives of school level, geographical location and curriculum system. The contributions of this paper are as follows:

(1) After analyzing and comparing the current IoT layer models, we propose a widely used IoT Major's knowledge structure and course list.
(2) Analyzing the differences between Academic Undergraduate Universities and Vocational Colleges in terms of training purposes, educational system, training methods and curriculum design. We also find that although most universities combine their own disciplinary advantages with the IoT curriculum, the curriculum system is still unreasonable, miscellaneous and not deep enough.

(3) Through the analysis of the geographical location of IoT colleges and universities and the industrial characteristics of the local region, we find that the cultivation of IoT professionals has a close relationship with regional economy. It is also influenced by the direction of economic development and the characteristics of IoT industry in terms of training programs and curricula.

(4) By analyzing the curriculum system of some colleges and universities, we find that most colleges and universities are divided into three types of curriculum Settings, which tend to be perceptual control. Network transmission; Software services. Lack of Internet of things professional characteristics.

Through the above conclusions, we suggest that universities can refer to the IoT major knowledge structure (Fig. 5) and the list of related courses (Table 1.) proposed in this paper and combine the discipline advantages of universities with the integration of regional economy when setting up IoT, so as to provide a large number of applied research results and excellent professionals.

At present, IoT major in Chinese universities is in the exploratory stage. In addition to the curriculum, there are still many problems in IoT education, such as the lack of professional teachers and experimental equipment. Therefore, there are still many problems worthy of improvement in the IoT major.

In a word, with the advent of 5G and intelligent age, the workforce training for IoT professionals will become more and more important. Only through continuous reform and innovation, clear training objectives, combined with industrial characteristics and the advantages of colleges and universities to set up a reasonable curriculum system, can we cultivate more outstanding IoT professionals.

References

1. Analysis of the development status and trend of the Internet of things. http://www.iotcn.org.cn/2018/06/14/4125/
2. Li, S., Li, D.X., Zhao, S.: 5G internet of things: a survey. J. Ind. Inf. Integr. **10**, 1–9 (2018)
3. Xu, L.D., He, W., Li, S.: Internet of things in industries: a survey. IEEE Trans. Ind. Inf. **10**(4), 2233–2243 (2014)
4. Ning, H.S., Xu, Q.Y.: Research on global internet of things' developments and it's construction in China. Dianzi Xuebao (Acta Electronica Sinica) **38**(11), 2590–2599 (2010)
5. Kshetri, N.: The evolution of the internet of things industry and market in China: an interplay of institutions, demands and supply. Telecommun. Policy **41**(1), 49–67 (2017)
6. Ning, H.: Unit and Ubiquitous Internet of Things. CRC Press, Boca Raton (2014)
7. Silva, B.N., Khan, M., Han, K.: Internet of things: a comprehensive review of enabling technologies, architecture, and challenges. IETE Tech. Rev. **35**, 205–220 (2017)
8. Ning, H., Hu, S.: Technology classification, industry, and education for future internet of things. Int. J. Commun. Syst. **25**(9), 1230–1241 (2012)
9. Xia, F., Yang, L.T., Wang, L., et al.: Internet of things. Int. J. Commun. Syst. **25**(9), 1101–1102 (2012)
10. Gubbi, J., Buyya, R., Marusic, S., Palaniswami, M.: Internet of things (IoT): a vision, architectural elements, and future directions. Future Gener. Comput. Syst. **29**(7), 1645–1660 (2013)

11. Kopetz, H.: Internet of things. In: Kopetz, H. (ed.) Real-Time Systems. Real-Time Systems Series. Springer, Boston (2011). https://doi.org/10.1007/978-1-4419-8237-7_13

12. Yun, M., Yuxin, B.: Research on the architecture and key technology of Internet of Things (IoT) applied on smart grid. In: International Conference on Advances in Energy Engineering. IEEE (2010)

13. Borgia, Eleonora: The internet of things vision: key features, applications and open issues. Comput. Commun. **54**, 1–31 (2014)

14. Ning, H.S., Liu, H.: Cyber-physical-social-thinking space based science and technology framework for the Internet of Things. Sci. China (Inf. Sci.) **58**(3), 1–19 (2015)

15. Ning, H., Wang, Z.: Future internet of things architecture: like mankind neural system or social organization framework? IEEE Commun. Lett. **15**(4), 461–463 (2011)

16. Wu, M., Lu, T.J., Ling, F.Y., Sun, J., Du, H.Y.: Research on the architecture of internet of things. In: International Conference on Advanced Computer Theory & Engineering (2010)

17. Ma, P.: IBM-The eight-layer architecture of the Internet of things (2010). http://www.51CTO.com

18. Ning, H., Sha, H.: Internet of things: an emerging industrial or a new major? In: Internet of Things (2011)

19. Shi, Y.R., Hou, T.: Internet of things key technologies and architectures research in information processing. Appl. Mech. Mater. **347–350**, 2511–2515 (2013)

20. Yang, Z., Yue, Y., Yu, Y., et al.: Study and application on the architecture and key technologies for IOT. In: International Conference on Multimedia Technology (2011)

21. J.Y. Wang, Z.: Internet of things engineering professional knowledge system and curriculum planning in colleges and universities (2011). (in Chinese)

22. Sun, Q.B., Liu, J., Shan, L.I., Fan, C.X., Sun, J.J.: Internet of things: summarize on concepts, architecture and key technology problem. J. Beijing Univ. Posts Telecommun. **33**(3), 1–9 (2010)

23. Chun-Xiang, W.U., Luo, J.F., Zhang, Z.Y.: Reflections on IoT major construction in colleges and universities in Guangdong. J. Dongguan Univ. Technol. (2014)

24. Jeong, S., et al.: Enabling transparent communication with global ID for the internet of things. In: Sixth International Conference on Innovative Mobile and Internet Services in Ubiquitous Computing (2012)

25. Jia, Q., Ericson, D.P.: Equity and access to higher education in China: lessons from Hunan province for university admissions policy. Int. J. Educ. Dev. **52**, 97–110 (2017)

26. Zanella, A., Bui, N., Castellani, A., et al.: Internet of things for smart cities. IEEE Internet Things J. **1**(1), 22–32 (2014)

CyberLife 2019: Cyber Health and Smart Healthcare

Research on the Influence of Internet Use on Adolescents' Self

Huimei Cao and Jiansheng Li[✉]

School of Education Science, Nanjing Normal University, Nanjing 210097,
Jiangsu, People's Republic of China
170602125@stu.njnu.edu.cn, 42056@njnu.edu.cn

Abstract. The paper explored the influence of Internet use on adolescents' self-construal and body esteem. Experiment 1 investigated the relationship between Internet use behavior and self-constructional types of 886 adolescents by questionnaire and Self-constructional Scale. The results show that the interdependent-self have a stronger motivation to use the Internet to communicate in order to keep in touch with others and integrate into the group. And they pay more attention to their privacy. However, the independent-self tend to update their status on the Internet, who don't care too much about the self-image management on the Internet or whether the published status leaking personal information. But none of the result reached significant levels. Experiment 2 investigated the influence of Internet use on body esteem of 240 middle school students by questionnaire and Body Esteem Scale. The results show that there is a significant difference in body esteem on genders. The correlation between boys' Internet use and body esteem is significant and positive, especially in the dimension of physical attractiveness and upper body strength. There is no significant correlation between Internet use and body esteem in girls.

Keywords: Internet use · Self-construal · Body esteem · Adolescents

1 Introduction

Self, also known as self-awareness or self-concept, mainly refers to the cognition of one's own state of existence; it is the result of self-evaluation on one's social role. Specifically, self-awareness is the understanding of oneself, the comprehension of the world relations around him, and the realization of his presence. Self-awareness includes the following three aspects: First, the personal understanding and evaluation of his physiological state, including the weight, height, figure, appearance and other body image and gender, as well as the feeling of pain, hunger, tiredness and so on. Second, the personal understanding and evaluation of his psychological state, mainly including the ability, knowledge, emotion, temperament, personality, ideals, beliefs, interests, hobbies, and other aspects. The third is of the relationship with the surrounding, including the status and role in certain social relations, as well as the comprehension and assessment of the relationship with others.

As an emerging information media, the Internet has a deep effect on politics, economics, culture and other fields at an unprecedented speed, causing great changes in

© Springer Nature Singapore Pte Ltd. 2019
H. Ning (Ed.): CyberDI 2019/CyberLife 2019, CCIS 1138, pp. 407–416, 2019.
https://doi.org/10.1007/978-981-15-1925-3_29

people's lifestyles, production methods and people's ideology. It has also brought a large quantity of influences and impacts to the contemporary teenagers' ideology, value and behavior. The development of adolescents' self-awareness will inevitably be influenced by the network culture and its use behavior [1].

When a person pays attention to his status and role in social relations, as well as the understanding and evaluation of relationship with others, self-construal is formed. Meanwhile, body esteem is formed when an individual focus on his weight, height, figure, appearance, etc. This study mainly investigated the influence of Internet use on self-construal and body esteem of adolescents.

2   Self-construal

2.1   Concept of Self-construal

Self-construal, proposed by Markus and Kitayama in 1991, is a cognitive structure for understanding the self from the perspective of the relationship with others [22]. While individuals know themselves in a reference system, they either regard themselves as an independent entity separated from others, or place themselves in a social network [7].

Markus and Kitayama divided the construal of the self into two types: the independent construal and the interdependent construal. Individuals of independent self-construal are egocentric, independent, autonomous, and unaffected by others, the behavior of whom mainly based on their own inner thoughts or feelings but without considering the relationship with the outside. On the contrary, the interdependent, seeing themselves as the indispensable part of the collective, focus more on the collective, consider the whole connection, regard themselves as a part of the collective, and establish contact actively with others to reduce differences [4].

In 1996, Brewer and Gardner further developed the theory of Markus et al., extending the interdependent construal of the self to two parts: self-definition from their own intimate relationship with others and self-definition from the group they belonging. Self-construal consists of three components: individual self, relative self and collective self, which were named "the tripartite model of self-construal" by Sedikides et al. [5–7].

The independent self and the interdependent self coexisted in each individual. Brewer et al. pointed out clearly that the activation of dominant self-construal tendency (independent or interdependent) in the current situation was influenced by situational factors. Therefore, from the perspective of stability, the construal of the self can be divided into chronic self-construal and situational self-construal [7].

2.2   The Effect of Internet Use on Self-construal

With the development of network technology, situational factors of self-construal for adolescents have also changed. For example, the impression of an individual on the Internet is based on the recognition and identification of Internet avatars (avatars, icons and labels representing individuals on the network) and Internet behaviors (status, comments, thumb-up). The selection of the online avatar image is a process for individuals to reconstruct their real personality [8]. Moreover, the behavior of adolescents on network communication also has the characteristics of self-construal type (independent or interdependent construal).

Self-disclosure and self-presentation are important aspects of communication. With the development of interpersonal relationship, there will be deeper and more sincere communication, making individuals more likely to show themselves to others. In the social network environment, individuals tend to disclose more private information about themselves, presenting more content [9]. Also, the anonymity of the Internet is a factor that affects individuals' communication behavior, making users forget the communication code of conduct and resulting in extreme remarks or slanderous words [10]. Therefore, the behavior of teenagers on the Internet is a reflection, supplement or reconstruction of their real personality.

3 Body Esteem

3.1 Concept of Body Esteem

"Body esteem", also called "Physical self-esteem", refers to "the individual evaluation in different aspects of his own body", or whether they are satisfaction or dissatisfaction [11]. "Body esteem" is a complex phenomenon. It is a multi-dimensional in nature which include perception, affect, cognition, behavioral and so on [12–14].

There are two structures with physical self-esteem: implicit and explicit. They are not only independent of each other but also have a low correlation. Implicit body esteem is objective. However, as the measurement is easily affected by many situational factors, it is unstable to measure the implicit body esteem in a certain index [15].

Body esteem involves a multifaceted psychological experience with a cognitive, behavioral and affective expression of one's own body, which is often used to describe the level of individual satisfaction or dissatisfaction with their body or appearance. Mendelson believes that physical assessment includes at least three aspects: the overall perception of one's own body, the perception and assessment of body weight, and the evaluation of the appearance from others. To a large extent, it is based on a feedback from the social environment to assess a person's body [3].

3.2 The Effect of Internet Use on Body Esteem

In today's online society, it is becoming increasingly significant that there's a huge effect of Internet use on body esteem. Researchers found that the body images of the female in the field of fashion models, cartoon characters, film and television actors, and the winner of miss America pageant were thinner than in the past few decades [2]. Bing thin is consistently emphasized as a symbol of beauty for female in movies, magazines and television shows. That is to say, the female figure presented in current Internet use are thinner than in the past, and even thinner than most of the actual female. The influence of the developing Internet use in gender is also changing. For instance, female's ideal body shape is being thinner, while male's is likely to want to be heavier, to become more muscular and to achieve the male's ideal of a V-shape figure, as to be thinner [16]. The formation of the ideal figure is closely related to the aesthetic standards of appearance in social culture, which represents individual expectation of their own physical appearance [17].

According to communications theories, exposing to the content used on the web repeatedly invites viewers to accept its description as a reflection of reality. In this case, the Internet use of a consistent description of thinness makes it attractive for female to view this belief as a norm and expectation. However, this description is very distorted. If female take the female images presented by the Internet as their real ideal, they will lower their satisfaction with their own body and take actions, such as dieting, anorexia, binge eating or taking weight-loss drugs, etc. [2]. In this way, adolescents' physical self-esteem is reduced when compared with the ideal shape from the Internet use [18].

Currently, teenagers have been becoming the main users of the Internet. The use of the Internet, with its popularity among teenagers, has been a hot issue for parents and society [19]. Based on this background, we investigated the influence of Internet use on adolescents' self-construal and body esteem.

4   Experiment 1: Research on the Effect of Internet Use on Self-construal

4.1   Participants

A total of 886 adolescents, aged 9–22, were surveyed by online questionnaire, including 428 boys (48.3%) and 458 girls (51.7%).

4.2   Research Instruments

Questionnaire of the Internet Use. A self-designed questionnaire was used to investigate the tendency of adolescents' online behaviors, including general demographic variables such as gender, age and Internet use behavior. Therein social behavior of Internet includes the use of different social platforms and status update of them.

Self-constructional Scale. According to the theory of self-construal proposed by Markus and Kitayama (1991), Singelis compiled the Self-constructional Scales (SCS) for students. A total of 24 items were included in the scale, 12 of which were independent self-constructional measurement and the other 12 measured interdependent self-construal. Each item was scored at 7 points (1 = strongly disagree, 7 = strongly agree). In addition, the internal consistency coefficients of the independent self-subscales and the interdependent self-subscales were respectively 0.69 and 0.73 [20]. In 2008, Wang et al., testing the Chinese version of the self-constructional scale with the students at age of 17–25, found that the internal consistency coefficients of the subscales were respectively 0.76 and 0.81, and 0.88 that of total scale [21]. Hardin et al. [22] adapted the SCS and added 6 items, totaling 30 items, which had good validity. It can be divided into 2 parts: 15 items for the independent construal of the self, while the rest for interdependent self-construal. It can also be divided into 6 dimensions: autonomy/assertiveness, individualism, behavioral consistency, primacy of self, esteem for group and relational interdependence. Each item was rated at 7 points (1 = strongly disagree, 7 = strongly agree). In this study, we evaluated the self-construal of students on the basis of a revised scale by Hardin et al.

4.3 Data Analysis and Results

Internet Use and Self-construal Types of Adolescents. Of the 886 adolescents surveyed, 27.7% were the independent construal and 72.3% were interdependent construal. 87.9% of teenagers expressed that they were used to communicate with others through the Internet while the rest was in contrast. Among the teenagers who used Internet to communicate, only 72.9% of them updated their status and mood on social networking sites, while 27.1% didn't or even didn't upload photos.

Relationship Between Internet Use and Self-construal. In order to investigate the relationship between different types of self-construal and the use and update of Internet, we conducted cross-table analysis (see Table 1).

Table 1. Relationship between self-construal and Internet use (unit: %)

Self-construal		Independent self	Interdependent self	Total
The Internet	Use	85.7[a]	88.8[a]	87.9
	Update	68.2[a]	67.4[a]	67.6

Note: in the table, letter a means there is no significant difference between the data.

It can be seen from Table 1 that among the independent construal of the self, the proportion of teenagers who use Internet to communicate is 85.7%, slightly less than the interdependent self of 88.8%, with no significant difference. In terms of updating status on the Internet, the proportion of independent self is 68.2%, slightly higher than the interdependent self of 67.4%, also, with no significant.

5 Experiment 2: Research on the Effect of Internet Use on Body Esteem

5.1 Participants

The participants in the present study were from two middle schools in Anqing, Anhui Province, including the first and second grade of junior high school, and first grade of high school. There were 80 students in each grade ($N = 240$), with the ages of 13 to 18 years and average age of 14.9($\pm$1.416). After excluding omissions, multiple-choice and the questionnaires with obvious regularity, 232 valid questionnaires remained: 140 boys and 92 girls; 77 first-year, 76 second-year students in junior high school and 79 first-year students in high school. There were no overweight participants, and the socioeconomic status of them was not significantly different.

5.2 Research Instruments

Body Esteem Scale. Compiled and tested by Franzoi and Shields [23, 24], the Body Esteem Scale (BES) was revised into Chinese version consisted of 32 items in 2002 [25], which was divided into three dimensions according to the three-dimensional structure model of body esteem (see Fig. 1). For male, it was divided into Physical Attractiveness ($\alpha = 0.81$), Upper Body Strength ($\alpha = 0.85$) and Physical Condition ($\alpha = 0.86$) while it was divided into Sexual Attractiveness ($\alpha = 0.78$), Weight Concern ($\alpha = 0.87$) and Physical Condition ($\alpha = 0.82$) for female.

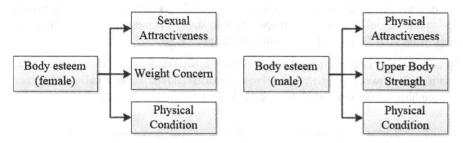

Fig. 1. Three-dimensional structure model of body esteem

Effect of Internet Use Questionnaire. We used a self-designed questionnaire to investigate the extent to which adolescents are affected by Internet use in life. It includes how long you have been using the Internet and the degree of shape change affected by Internet use. The items such as "You are influenced by using Internet such as TV or computer, and often participate in physical exercise to change your body figure." "You agree that the body shape and appearance of idols on the Internet are ideal" etc.

5.3 Data Analysis and Results

Gender Differences in Physical Self-esteem. Table 2 shows the average and standard deviation of body esteem for all participants and boys and girls. It can be concluded from Table 2 that the score of male's body esteem ($M = 102.83$, $SD = 17.79$) is higher than that of female's ($M = 91.97$, $SD = 15.211$). According to the sampled data, one-way ANOVA was used to test the difference of body esteem of adolescents of different genders, and the results are shown in Table 3 ($F = 23.148$, $p = 0.000$). It can be seen that there is a significant gender difference in physical self-esteem. Independent sample t-test also showed the same results ($t = 4.969$, $p = 0.000$).

Influence of Internet Use and Body Esteem. Table 4 shows the correlation between male and female Internet use and body esteem. There is no significant correlation between the influence of Internet use and body esteem for female ($r = -0.127$, $p = 0.228$), while there is a significant correlation for male ($r = 0.315$, $p = 0.000$). Moreover, the effect of Internet use is significantly correlated with Physical Attraction ($r = 0.291$, $p = 0.000$) and Upper Body Strength ($r = 0.357$, $p = 0.000$) for male.

Table 2. Means, standard deviation of physical self-esteem

Gender	Male ($N =$ 140)		Female ($N =$ 92)		Total ($N =$ 232)	
	M	SD	M	SD	M	SD
Body esteem	102.83	17.795	91.97	15.211	98.52	17.608

Table 3. One-way ANOVA of gender and physical self-esteem

		Sum of squares	df	Mean square	F	p
Body esteem	Between groups	6549.104	1	6549.104	23.148	0.000
	Within groups	65072.788	230	282.925		
	Total	71621.892	231			

Table 4. Correlation analysis of Internet use and body esteem

Gender		Male			
		Physical attractiveness	Upper body strength	Physical condition	Body esteem
Effect of Internet use	r	0.291**	0.357**	0.211*	0.315**
	p	0.000	0.000	0.012	0.000
Gender		Female			
		Sexual attractiveness	Weight concern	Physical condition	Body esteem
Effect of Internet use	r	−0.109	−0.131	−0.048	−0.127
	p	0.299	0.212	0.649	0.228

Note: * means the difference is significant at .05 level; ** means the difference is significant at .01 level.

6 Conclusion

6.1 Self-construal and Internet Use

Studies on self-construal and Internet use found that the adolescents of the interdependent construal of the self are more likely to use social media to communicate and maintain relationships with others but less likely to update their status on social platforms.

Compared with the independent self-construal, the interdependent have stronger motivation to use the Internet to contact with others and integrate into groups, through which they can obtain greater satisfaction [26]. The independent seldom care about the self-image management on the Internet or whether the published status exposed the information, while the interdependent construal pay more attention to self-privacy and seldom disclose their information on the Internet relatively [27].

Chen and Marcus also found that participants mainly used the Internet to communicate with others and maintain existing relationships, with no difference in the amount and authenticity of online and offline communication. There was no significant difference between the on and off-line communication behaviors of the independent self-construal. However, in order to maintain the relation with others and reduce self-disclosure and self-presentation, the interdependent construal tend to publish more information related to others or show more protective behavior [28].

6.2 Body Esteem and Gender

We found that there is a gender difference in body esteem, with men having significantly higher scores of body esteem than women. This conclusion is consistent with previous research, in which Duncan et al. [29] also found there is a gender difference in body esteem, and the level of body esteem of boys is significantly higher than that of girls.

6.3 Body Esteem and Internet Use

There is a significant and positive correlation between male Internet use and body esteem, which is consistent with the research of Mcgee et al. [30]. But the result also shows that female's Internet use had little effect on their body esteem.

In a summary, the network environment has an effect on adolescents' self-construal and body esteem. Therefore, we must guide the youth to use the Internet correctly, making it a useful tool for the formation of adolescents' self-awareness.

Acknowledgements. This work was supported by the National Social Science Foundation of China (13BRK026).

References

1. Wang, M.: Network action and self-awareness of college students – a survey and analysis of 428 college students in Huazhong University of Science and Technology. M. A. thesis. Huazhong University of Science and Technology, Wuhan (2004)
2. Grabe, S., Ward, L.M., Hyde, J.S.: The role of the media in body image concerns among women: a meta-analysis of experimental and correlational studies. Psychol. Bull. **134**(3), 460–476 (2008)
3. Lipowska, M., Lipowski, M., Olszewski, H., Dykalska-Bieck, D.: Gender differences in body-esteem among seniors: Beauty and health considerations. Arch. Gerontol. Geriatr. **67**, 160–170 (2016)
4. Markus, H.R., Kitayama, S.: Culture and the self: Implications for cognition, emotion, and motivation. Psychol. Rev. **98**(2), 224–253 (1991)
5. Brewer, M.B., Gardner, W.: Who is this we? Levels of collective identity and self representations. J. Pers. Soc. Psychol. **1**, 83–93 (1996)

6. Sedikides, C., Brewer, M.B.: Individual, relational and collective self: partners, opponents, or strangers? In: Individual Self, Relational Self, Collective Self. Psychology Press. London (2002)
7. Liu, Y.: Self-construal: review and prospect. Adv. Psychol. Sci. **19**(3), 427–439 (2011)
8. Luo, T., Zhou, Z.J.: The influence of online avatar on teenagers' identity construction. China Youth Study **1**, 84–87 (2013)
9. Hew, K.F.: Students' and teachers' use of Facebook. Comput. Hum. Behav. **27**(2), 662–676 (2011)
10. Mckenna, K.Y.A.: Plan 9 from cyberspace: the implications of the internet for personality and social psychology. Pers. Soc. Psychol. Rev. **4**(1), 57–75 (2000)
11. Hu, Z.H., Xie, G.D.: Experimental analysis of explicit & implicit self-esteems between obese university students and common ones. J. Mianyang Norm. Univ. **30**(5), 107–109 (2011)
12. Thompson, J.K.: Body image disturbance: assessment and treatment. In: Body Image Disturbance: Assessment and Treatment. Pergamon Press. London (1990)
13. Banfield, S.S., Mccabe, M.P.: An evaluation of the construct of body image. Adolescence **37**(146), 373–393 (2002)
14. Parks, P.S., Read, M.H.: Adolescent male athletes: body image, diet, and exercise. Adolescence **32**(127), 593–602 (1997)
15. He, B., Tang, S.J.: Measurement research on relation between explicit and implicit self-esteem of college students. Bull. Sport. Sci. Technol. **17**(12), 123–125 (2009)
16. Furnham, A., Badmin, N., Sneade, I.: Body image dissatisfaction: Gender differences in eating attitudes, self-esteem, and reasons for exercise. J. Psychol. Interdiscip. Appl. **136**(6), 581–596 (2002)
17. Chen, X., Jiang, Y.J., Ye, H.S.: The relationships between media influence body mass index and body esteem. Chin. J. Appl. Psychol. **13**(2), 119–124 (2007)
18. Wilcox, K., Laird, J.D.: The impact of media images of super-slender women on women's self-esteem: identification, Social comparison, and self-perception. J. Res. Pers. **34**(2), 278–286 (2000)
19. A survey report on Chinese teenagers' online behavior in 2012. http://www.cnnic.net.cn/hlwfzyj/hlwxzbg/qsnbg/201312/P020131225339891898596.pdf. Accessed 25 Dec 2013
20. Singelis, T.M.: the measurement of independent and interdependent self-construals. Pers. Soc. Psychol. Bull. **20**(5), 580–591 (1994)
21. Wang, Y.H., Yuan, Q.H., Xu, Q.M.: A preliminary study on self-constructional scales (SCS) of Chinese-version. Chin. J. Clin. Psychol. **16**(6), 602–604 (2008)
22. Hardin, E.E., Leong, F.T.L., Bhagwat, A.A.: Factor structure of the self-construal scale revisited: Implications for the multidimensionality of self-construal. J. Cross-Cult. Psychol. **35**(35), 327–345 (2004)
23. Franzoi, S.L., Shields, S.A.: The body esteem scale: multidimensional structure and sex differences in a college population. J. Pers. Assess. **48**, 173–178 (1984)
24. Franzoi, S.L.: Further evidence of the reliability and validity of the body esteem scale. J. Clin. Psychol. **50**, 237–239 (1994)
25. He, L., Zhang, L.W.: Relationship between evaluation modes of abstract body self-esteem, concrete body self-esteem and life satisfaction. J. Beijing Univ. Phys. Educ. **25**(3), 320–323+330 (2002)
26. Kim, J.H., Kim, M.S., Nam, Y.: An analysis of self-construals, motivations, Facebook use, and user satisfaction. Int. J. Hum. Comput. Interact. **26**(11), 1077–1099 (2010)
27. Long, K., Zhang, X.: The role of self-construal in predicting self-presentational motives for online social network use in the UK and Japan. Cyberpsychol. Behav. Soc. Netw. **17**(7), 454–459 (2014)
28. Chen, B., Marcus, J.: Students' self-presentation on Facebook: an examination of personality and self-construal factors. Comput. Hum. Behav. **28**(6), 2091–2099 (2012)

29. Duncan, M.J., Al-Nakeeb, Y., Nevill, A.M.: Body esteem and body fat in British school children from different ethnic groups. Body Image 1(3), 311–315 (2004)
30. Mcgee, B.J., Hewitt, P.L., Sherry, S.B., Parkin, M., Flett, G.L.: Perfectionistic self-presentation, body image, and eating disorder symptoms. Body Image 2(1), 29–40 (2005)

Breast Cancer Risk Assessment Model Based on sl-SDAE

Xueni Li and Zhiguo Shi[✉]

University of Science and Technology Beijing, Beijing 100083, China
szg@ustb.edu.cn

Abstract. In recent years, the incidence of breast cancer among women in China has increased year by year and it has become the most common malignant tumor in women in China. There are already breast cancer risk assessment models for women in Europe and the United States. However, there is no effective breast cancer risk assessment model suitable for women in China. The paper established an effective breast cancer risk assessment model. It selected the survey data of breast cancer population in China as a data set. The paper combines SDAE and LSTM to build a model based on deep learning methods. It uses the roc curve as an indicator of the experimental results. Experiments show that the model has better performance than traditional machine learning algorithms.

Keywords: Breast cancer · SDAE · LSTM · ROC curve · Machine learning

1 Introduction

Breast cancer has become the most common malignant tumor in women worldwide and also the most common malignant tumor in women in China [1]. In 2018, China's cancer registration area data showed that the incidence of breast cancer among Chinese women increased from 31.90/100,000 in 2000 to 63.40/100,000 in 2014 [2]. Breast cancer rose to the sixth leading cause of cancer death in women from 2004 to 2005 [3], accounting for 5.9% of the total number of female cancer deaths. The situation of breast cancer prevention and control is grim [4].

At present, the most popular and comprehensive of breast cancer screening is developed countries such as Europe and the United States. Since the 1980s, various breast cancer risk screening models have been developed for white female populations. For example, Gail proposed the Gail model in 1989 [5]. It incorporates five risk factors for breast cancer into risk assessment models such as women's age at menarche, number of breast biopsies, primiparity, number of breast cancers in first-degree families, and age of patients [6]. At present, the Gail model has been widely used in Europe and the United States through the continuous development of other scholars such as Gail.

The Claus model first proposes a model of single-gene advantage inheritance in the population. The BRCAPRO model is the most widely used genetic model for calculating the probability that family history is associated with BRCA gene mutations. It takes into account the age of first- and second-generation relatives of breast cancer patients,

© Springer Nature Singapore Pte Ltd. 2019
H. Ning (Ed.): CyberDI 2019/CyberLife 2019, CCIS 1138, pp. 417–430, 2019.
https://doi.org/10.1007/978-981-15-1925-3_30

unilateral or bilateral breast cancer or breast cancer [7]. The BOADICEA model is also used for the screening of BRCA gene mutation carriers and for the detection of cancer risk [8]. The BRAC1/2 analysis adds a multi-gene complex analysis that affects gene junction effects. The Tyrer-Cuzick model is mainly for British, Australian and New Zealand people. It is a breast cancer risk prediction model that integrates family history of breast cancer, hormonal status and benign breast disease. It has been confirmed by many confirmatory studies. The model included BRCA status, height, weight, hormone replacement therapy, age of first live birth, age of first-episode cancer in relatives, presence or absence of ovarian cancer, maternal and paternal second- or third-generation family history. This model is mainly used to detect the probability of breast cancer within 10 years, but it is insufficient to predict breast cancer in people with dysplasia after 10 years. The GWAS model found a number of single-nucleotide polymorphisms (SNPs) sites associated with breast cancer pathogenesis, using the genetic information contained in these sites to model. The breast cancer risk prediction model based on GWAS can help to carry out genetic screening for high-risk populations, develop interventions, guide disease treatment and predict disease prognosis. However, at present, this kind of research is still in its infancy, and there are still some limitations. For example, the site effect found at present is weak, and the interaction between genes and the environment is not considered. Therefore, the application of large-scale screening for female population still requires further extensive and in-depth research. At the same time, it is necessary to discover and verify more positive genetic predisposition sites through breast cancer GWAS in the future, which will be beneficial to promote Chinese women's breast cancer. Individualized risk prediction and prevention.

The advantages and disadvantages of each model are shown in Table 1:

Table 1. Comparison of advantages and disadvantages of different breast cancer prediction models

Different models	Advantages	Disadvantages
Gail module	Suitable for white female groups, fast, simple and effective	Not suitable for the Chinese population, the risk factors adopted are not comprehensive enough
Claus module	Forecast for family history	No history information related to the onset of breast cancer, while ignoring non-family history information
BRCAPRO module	Comprehensive consideration of information about relatives with and without breast cancer	Ignore non-family history and do not apply to people without a family history of breast cancer
BOADICEA module	More accurate than the BRCAPRO model	Also did not consider the risk factors of non-family history
Typer-Cuzick module	The predictive factors involved are comprehensive and the predictions are good	Insufficient prediction of breast cancer with dysplasia after 10 years
GWAS module	Helps to carry out targeted genetic testing of high-risk populations	The site effect found so far is weak, and the interaction between genes and environment is not considered

Most of the above models are applicable to white female groups. The prediction accuracy of Chinese population is low. However, China currently lacks large-scale domestic breast cancer data, and there is no effective and extensive breast cancer prediction and evaluation model suitable for Chinese population. The breast cancer predictive evaluation model proposed in this paper mainly studies the models applicable to the Chinese population, including comprehensive predictive factors involving family history and non-family history information, using deep learning to model, avoiding machine learning to consider selection predictors The problem of missing important information.

The organizational structure in the later part of this paper: Sect. 2 introduces some of the basic concepts involved in the model and the main organizational structure of the paper; Sect. 3 is the specific model construction process; Sect. 4 is the experimental results and the results of this experiment and other models; Sect. 5 is the conclusion, summarizing the work of this paper.

2 Basic Concepts and Framework

Auto encoders (AE), an automatic encoder consists of an encoder and a decoder. The encoding process refers to the mapping process from the input layer to the hidden layer. The decoding process refers to the hidden layer to the output. An automatic encoder network structure with a single hidden layer is (Fig. 1):

Its encoder and decoder formula is:

$$h = S_f(W_{xh}x + b_{xh}) \tag{2-1}$$

$$z = S_g(W_{hx}x + b_{hx}) \tag{2-2}$$

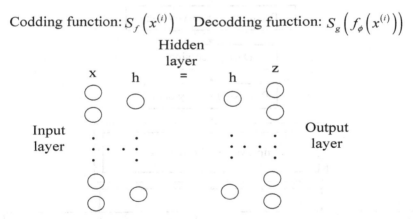

Codding function: $S_f\left(x^{(i)}\right)$ Decodding function: $S_g\left(f_\phi\left(x^{(i)}\right)\right)$

Fig. 1. Single hidden layer self-encoder network structure

The encoder represented by Eq. (2-1) represents a non-linear mapping between the input vector x and the hidden layer feature representation h. The decoder represented by Eq. (2-2) represents a non-linear mapping from the hidden layer feature representation h to the original input vector x. Where W, b represent the weight and offset of the neural network. S represents a nonlinear transfer function. S_f is the activation function of the encoder. S_g is the activation function of the decoder. He activation function is usually the sigmoid function.

The denoising auto-encoder (DAE) is implemented by superimposing noise on the original feature vector [8]. In fact, noise robustness constraints are added to the autoencoder. The input vector after superimposing noise is represented by $\tilde{x}$. For the noise reduction autoencoder, the difference between the reconstructed feature representation of $\tilde{x}$ as the input vector and the original feature vector is called the reconstruction error. The denoising auto-encoder can be used to process data sets with more internal disturbances. By setting a certain Gaussian noise or a reasonable damage probability [9], a new more stable and robust feature representation of the original input vector can be obtained, and the obtained features are more conducive to classifier classification.

SDAE is a variant of a deep self-encoding network that uses a DAE instead of an Auto-Encoder (SAE), stacked by a multi-layer DAE stack. Implementing the extraction of high-dimensional features by establishing a SDAE [10]. Let $h_j(x)$ denote the degree of activation of the jth neuron on the hidden layer when x is entered. At the same time, $h_j(x)$ is the jth component of the vector h. The role of sparse auto-encoders is mainly two-fold: on the one hand, sparse auto-encoders can limit the refactoring features obtained as sparse as possible. Because the general sparse expression in deep learning is more effective than other forms of expression. On the other hand, if there are too many hidden layer neuron nodes, it still requires the network to output a compressed representation of the feature. We need to add sparse punishment to hidden layer neurons. Make sure that the auto-encoder network can still find interesting structures in the original data even in the presence of a large number of hidden layer neurons [11].

The overall framework of this study is as follows (Fig. 2):

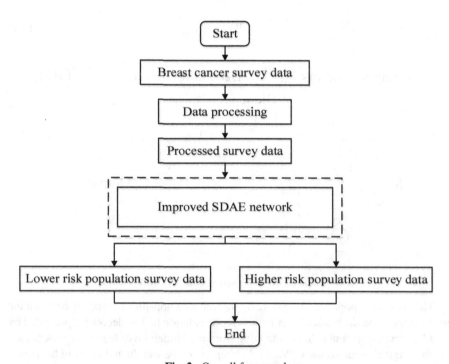

Fig. 2. Overall framework

3 Multilayer Neural Network Based on SDAE Improvement

The neural network designed in this paper is a stacked noise reduction automatic encoder with 3 hidden layers and the network has been improved to make it have better performance on the data set used.

3.1 Building a Predictive Model

s-DAE is a deep network formed by stacking DAEs. The training process uses a layer-by-layer greedy unsupervised method [12]. Each layer is used as an independent DAE, training one layer at a time [13], and output coding feedback of the lower layer DAE as the input of the current layer by layer. As the input of the current layer. Once the front k layer is trained, the k+1th [14] layer can be trained. In this paper, a stacking noise reduction automatic encoder with 3 hidden layers is selected as the network structure of the model. The specific schematic diagram is shown in Fig. 3.

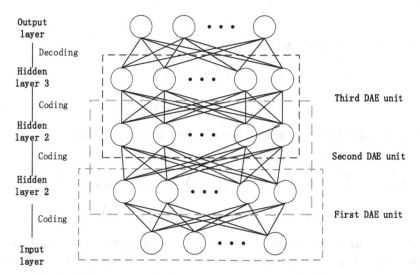

Fig. 3. Stacked noise reduction autoencoder structure with 3 hidden layers

However, it is worth noting that the input data x needs to be damaged. It can randomly set $\underline{X}$ to 0 according to a certain probability distribution or randomly add Gaussian whitened noise in $\underline{X}$. Considering that the data may be incompletely entered [15], information is wrong, etc. These can be regarded as noise generated on the original data. In this paper, noise reduction is performed by randomly adding Gaussian noise to the input data. That is $\tilde{x} = x + \lambda \times \varepsilon, \varepsilon \sim N(0, \sigma^2 I)$, the input vector $\tilde{x}$ after the interference is obtained. Where λ represents the noise figure.

The basic core of layer-by-layer training is:

(1) Training the first layer of DAE_1, and input data $\tilde{x}$ trains parameter (W_1, b_1) through repeated iterations. So that the error function $J_{AE}^1(\varphi)$ uses the squared error loss

as the loss function of the encoder. That is, the sum of squared errors of the reconstructed signal and the original signal is the smallest. As far as possible, ensure that $g(f(\tilde{x}))$ is similar to $\tilde{x}$. This is the perfect refactoring (Fig. 4).

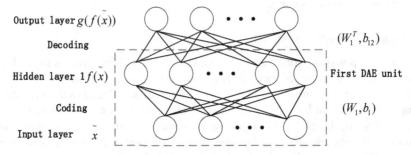

Fig. 4. The first hidden layer structure of the stack noise reduction automatic encoder

In Fig. 6, the output of each layer is as follows:

$$f(\tilde{x}) = \sigma\left(W_1\tilde{X} + b_1\right) \tag{3-1}$$

$$g(f(\tilde{x})) = W_1^T f(\tilde{x}) + b_{12} \tag{3-2}$$

The weight and offset are updated using gradient descent. The original cross entropy loss function is as follows [16]:

$$J_{AE}^1(\varphi) = \sum_{x \in S} L\left(x_i, g_\theta\left(f_\theta\left(\tilde{x}^i\right)\right)\right) = \sum_{x \in S} \left\| x^i - g_\theta\left(f_\phi\left(\tilde{x}^i\right)\right)\right\|_2^2 \tag{3-3}$$

$$(W_1, b_1, b_2) \leftarrow argmin\left(J_{AE}^1(W_1, b_1, b_2)\right) \tag{3-4}$$

However, such a model may have a risk of overfitting, the sparseness limit and regularization are added to reduce the risk of overfitting. The improved loss function is:

$$J = -\frac{1}{m}\sum_{i=1}^m \sum_{j=1}^n \left\{x_j^i lnh_\theta\left(g_\theta(\tilde{x}_j^{(i)})\right) + (1 - x^i)ln\left[1 - h_\theta\left(g_\theta(\tilde{x}_j^{(i)})\right)\right]\right\} + \frac{\lambda}{2}\|W\|^2 + \beta\sum_{j=1}^{S_L} KL(\rho\|\hat{\rho}_j^i) \tag{3-5}$$

Where β is the weighting factor that controls the sparsity penalty term, $\frac{\lambda}{2}\|W\|^2$ is as follows:

$$\frac{\lambda}{2}\|W\|^2 = \frac{\lambda}{2}\sum_{l=1}^{L-1}\sum_{i=1}^{S_t-1}\sum_{l=1}^{S_t}\left(W_{jt}^{(l)}\right)^2 \tag{3-6}$$

(2) Fix the first layer that has been trained, discard the decoding process of the DAE_1, and retain the weight and offset of the DAE_1 input layer and the hidden layer. The hidden layer of the DAE_1 is used as the input layer of DAE_2. DAE_2 takes $f(\tilde{x})$ as the input data of DAE_2, and passes the $f(\tilde{x})$ input data through iterative training parameter (W_2, b_2) repeatedly, so that the error function $J_{AE}^2(\varphi)$ is minimized. Make sure [17] $g(f(\tilde{x}))$ is similar to $f(\tilde{x})$ as much as possible (Fig. 5).

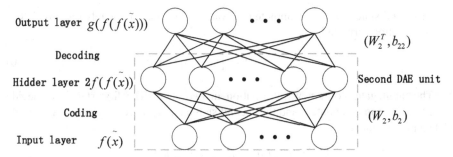

Fig. 5. Stacked noise reduction auto encoder second hidden layer structure

(3) Repeat the process in (2), fix the layer 2 that has been trained, discard the decoding process of DAE_2, retain the weight and offset of the DAE_2 input layer and the hidden layer, and use the hidden layer of DAE_2 as the input layer of DAE_3. Taking $f(f(\tilde{x}))$ as the input data of DAE_3, passes the $f(f(\tilde{x}))$ input data through iterative training parameter (W_3, b_3) repeatedly, so that the error function $J^3_{AE}(\varphi)$ is minimized. It is guaranteed that $g(f(f(f(\tilde{x}))))$ is approximated to $f(f(\tilde{x}))$ as much as possible. That is, the perfect reconstruction of $f(f(\tilde{x}))$ (Fig. 6).

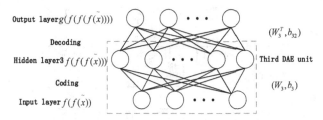

Fig. 6. Stacked noise reduction auto encoder third hidden layer structure

Output is as follows:

$$f(f(f(\tilde{x}))) = \sigma(W_3 f(f(\tilde{x})) + b_b) \tag{3-7}$$

$$g(f(f(f(\tilde{x})))) = W_3^T f(f(f(\tilde{x}))) + b_{32} \tag{3-8}$$

$$(W_3, b_3, b_{32}) \leftarrow argmin\left(J^3_{AE}(\varphi)(W_3, b_3, b_{32})\right) \tag{3-9}$$

(4) Until the third layer is trained, the last layer of output value z can be considered as a deep unsupervised feature representation of the original breast cancer etiology data by SDAE.

(5) The output of SDAE is used as the input of LSTM after data conversion. First decide what information to discard from the cell. This decision is done through a forgotten gate layer that extracts the sum and outputs a value between 0 and 1 to the value in

each cell state. 1 means complete reservation, 0 means completely discarded. The calculation formula is:

$$f_t = \sigma\left(W_f \cdot [h_{t-1}, x_t]\right) + b_f \tag{3-10}$$

The input gate stores new information in two steps. The first step is to decide which values can be updated, and the second step tanh layer creates candidate vectors. Calculation formula:

$$i_t = \sigma\left(W_i \cdot [h_{t-1}, x_t]' + b_i\right) \tag{3-11}$$

$$C_t = \tanh\left(W_C \cdot [h_{t-1}, x_t] + b_C\right) \tag{3-12}$$

Where i_t is the output of the sigmoid function indicating whether an input is generated, the value range is 0–1, and C_t is the newly generated candidate vector. Forgetting gate $f_t \cdot C_{t-1}$ is to forget the early information, and the result plus $i_t \cdot C_t$ is how much the candidate vector is updated by E after scaling the state value. The output gate first determines the scaling of the output, then the cell state passes the tanh function, and the result is multiplied by ot (Fig. 7).

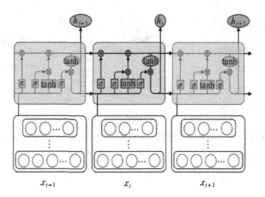

Fig. 7. Network structure

4   Experimental and Performance Analysis

4.1   Feature Selection

The data set used in this paper is derived from the national key research and development plan 2016YFC09013. It is led by the China Center for Disease Control and Prevention, the Center for Chronic Diseases, the First Hospital of Peking University, and the Second Hospital of Shandong University. A standardized 80,000 community cohort and a 10,000 clinical cohort for Chinese women's breast cancer were investigated. The survey contains a large amount of information about the respondents, mainly in the following aspects:

basic information (including the location of the respondent, height and weight, living conditions, etc.), menstrual history, birth history, family history, personal illness and examination and treatment History, vitamins and minerals, smoking status, diet, alcohol consumption, tea drinking, physical activity, attitudes and cognition about breast cancer, sleep scale, psychological scale, height and weight waist circumference hip measurement table, blood pressure measurement Table, blood sample table, breast clinical checklist, ultrasound report template. Each survey object contains 350 features. Some of these features have missing values. Some features have been recommended by medical experts to be independent of breast cancer. Some features have strong correlations, so not all features are used. For prediction. First, the data is cleaned and processed to remove the features that are not worthy of breast cancer, and then the correlation is judged by the corr function. The strongly related feature selection is left and the remaining features are removed.

The data is simply processed and converted into a DataFrame of pandas. Correlation score calculations were performed using the corr function that comes with the pandas library. As shown in 4-1:

$$DataFrame.corr\left(method ='\ person',\ \text{min\_periods} = 1\right) \qquad (4\text{-}1)$$

Then use the seaborn function to draw the heat map [18] as shown in Fig. 8. The darker the color, the stronger the correlation between the two random variables, and the lighter the color, the weaker the linear correlation between the two random variables.

According to the above characteristics, there are 250 features left in this paper.

4.2 Experimental Configuration

The calculation framework for this article uses TensorFlow. TensorFlow is highly flexible and highly portable. This flexible architecture helps you deploy computing tasks on the CPU and GPU of desktops, servers, and mobile devices without having to rewrite the code to fit each platform. As a deep learning framework for Google's open source, TensorFlow has the most Fork and Star numbers on Github. It is also rich in applications such as graphics classification, audio processing, recommendation systems, and natural language processing. Supports both Python and C++ programming languages, running smoothly on both CPU and GPU [19]. This article not only uses the TensorFlow calculation framework as a calculation tool, but also uses the Jupyter notebook to retain the visual experiment results obtained in this paper. Jupyer notebook is a powerful interactive tool that supports more than 40 programming languages, supports Markdown syntax, and supports mathematical formulas. It can be used to record experimental results, combined with Python-related visualization libraries such as matplotlib, seaborn, etc. to present data rules in various forms of charts, finally to write beautiful interactive documents, or to make teaching materials and so on. The main uses of Jupyter notebook include: data cleaning and exploration, statistical analysis, visual analysis, machine learning analysis and more.

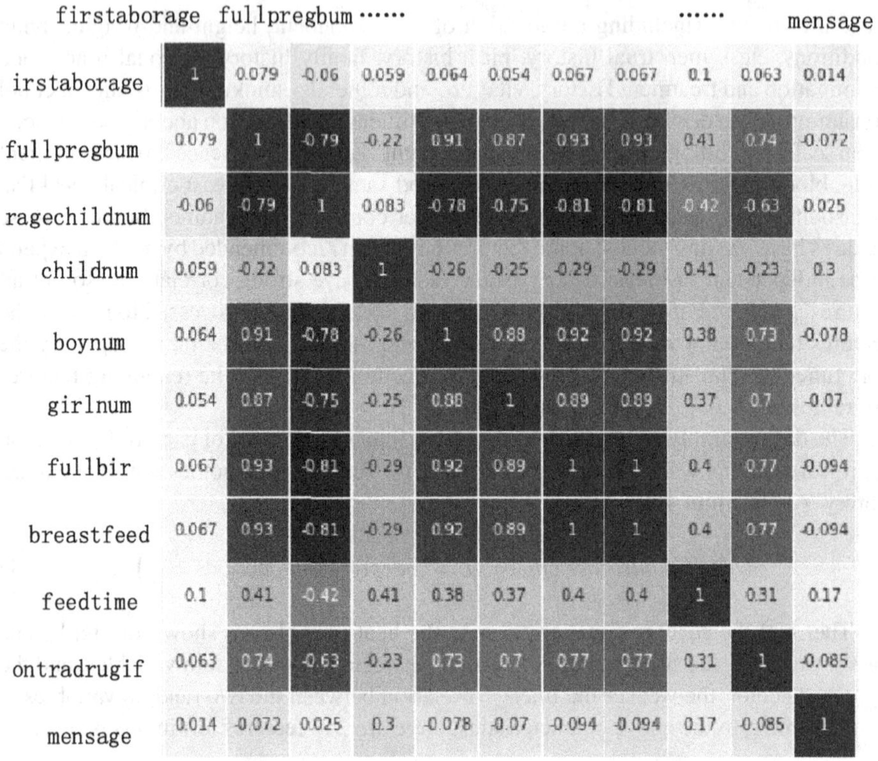

firstaborage	fullpregbum	......								......	mensage

irstaborage	1	0.079	-0.06	0.059	0.064	0.054	0.067	0.067	0.1	0.063	0.014
fullpregbum	0.079	1	-0.79	-0.22	0.91	0.87	0.93	0.93	0.41	0.74	-0.072
ragechildnum	-0.06	-0.79	1	0.083	-0.78	-0.75	-0.81	-0.81	-0.42	-0.63	0.025
childnum	0.059	-0.22	0.083	1	-0.26	-0.25	-0.29	-0.29	0.41	-0.23	0.3
boynum	0.064	0.91	-0.78	-0.26	1	0.88	0.92	0.92	0.38	0.73	-0.078
girlnum	0.054	0.87	-0.75	-0.25	0.88	1	0.89	0.89	0.37	0.7	-0.07
fullbir	0.067	0.93	-0.81	-0.29	0.92	0.89	1	1	0.4	0.77	-0.094
breastfeed	0.067	0.93	-0.81	-0.29	0.92	0.89	1	1	0.4	0.77	-0.094
feedtime	0.1	0.41	-0.42	0.41	0.38	0.37	0.4	0.4	1	0.31	0.17
ontradrugif	0.063	0.74	-0.63	-0.23	0.73	0.7	0.77	0.77	0.31	1	-0.085
mensage	0.014	-0.072	0.025	0.3	-0.078	-0.07	-0.094	-0.094	0.17	-0.085	1

Fig. 8. Screenshot of the feature vector heat map

4.3   Experimental Result

The confusion matrix of the experiment is shown in Table 2: where the positive class represents cancer data and the negative class represents non-cancer data.

Table 2. Confusion Matrix

Category	Forecast is positive	Forecast is negative
Actually positive	260	6
Actually negative	2	30562

The evaluation indicators of the experiment are shown in Table 3: The correct rate refers to the percentage of the correct classification of the breast cancer classification model in which all samples are established; the missed alarm rate is misclassified into the low-risk category in the actual cancer sample; the sensitivity is that the actual cancer sample is correctly judged as the percentage of cancer patients according to the criteria of the classification model; specificity is the proportion of samples that are actually not cancerous and are correctly classified as not having cancer.

Table 3. Evaluation indicators

Model	Correct rate	Missed alarm rate	Sensitivity	Specificity
Breast cancer risk assessment model	0.99	0.02	0.97	0.99

4.4 Comparative Analysis of Experiments

In order to verify the validity of the model, naive Bayes, automatic encoder and noise reduction automatic encoder were selected as the comparison algorithm.

The four models were compared by four indicators: correct rate, miss alarm rate, sensitivity and specificity. The comparison results are shown in the following table (Table 4):

Table 4. Evaluation of model effect

Model	Correct rate	Missed alarm rate	Sensitivity	Specificity
Naive Bayes	0.82	0.06	0.94	0.81
Automatic encoder	0.92	0.04	0.96	0.92
Noise reduction auto-encoder	0.96	0.03	0.97	0.96
Research model	0.99	0.02	0.98	0.99

The ROC curve can remain unchanged as the distribution of positive and negative samples in the test set changes. Therefore, the roc curve is generally used to judge the classification effect of unbalanced data. The cancer samples are far less than the non-cancer samples in data sets, so the roc curve can be used to more clearly determine the classification effect of the four models. The pyplot module in the matplotlib library is used to plot the ROC curves of the four different algorithms in the same coordinate. The combined calculation of the AUC values is used to verify the performance of the model. Among them, the true positive rate (sensitivity) is taken as the ordinate, and the false positive rate (1-specificity) FPR is taken as the abscissa, and the value range is 0–1. The blue curve represents the model of this paper, the red curve represents the naive Bayes algorithm, the green curve represents the automatic encoder, and the purple curve represents the noise reduction auto-encoder (Fig. 9).

The abscissa of the above figure is the pseudo-positive rate is the proportion of the sample that is actually not cancerous is misclassified as cancer, and the true ratio of the ordinate is the proportion of the actual cancer sample correctly classified as the cancer sample.

The AUC value calculated by the Naive Bayes algorithm is 0.84. At the same time, the confusion matrix shows that the model has a missed alarm rate of 0.06 and a specificity of 0.81. The classification effect of the model is not good.

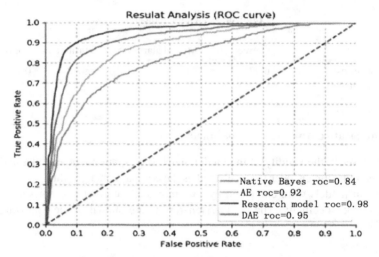

Fig. 9. ROC comparison (Color figure online)

The automatic encoder calculates an AUC value of 0.92. At the same time, the confusion matrix shows that the model has a missed alarm rate of 0.04, a sensitivity of 0.96 and a specificity of 0.92. The overall classification of the model is not effective. The confusion matrix distribution of the model is shown in the following figure: where the label value in the confusion matrix is 1 representing cancer data, representing non-cancer data if the label value is 0 (Fig. 10).

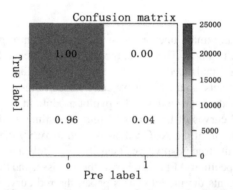

Fig. 10. Automatic encoder confusion matrix distribution

The noise reduction automatic encoder has an accuracy of 0.95, a false alarm rate of 0.03, a sensitivity of 0.97, and a specificity of 0.96. The model is better than the classification of the above two models, but it is still worse than the model of this study.

According to the above analysis, the model of this paper is the most accurate compared with the naive Bayes, automatic encoder and noise reduction auto-encoder. It shows that the model has a good classification effect in breast cancer prediction. Because of the complexity of the model, the amount of training set data and data noise, the models obtained by the automatic encoder and the noise reduction auto-encoder often have the risk of over-fitting, so that the classification effect is not as good as the model.

5 Conclusions

This paper establishes an effective breast cancer risk assessment model. The model consists of a stacked noise reduction automatic encoder network with three hidden layers and LSTM network. The dataset adopts the survey data of female breast cancer in China provided by the data collection and analysis platform of breast cancer disease cohort in China, and then selects the characteristics of many features, and inputs the filtered data into the risk assessment model for breast cancer prediction.

The experimental results show that the model constructed by this paper has a missed rate of 0.02, and the overall correct rate is up to 0.99, which has a good classification effect on breast cancer in China. Compared with traditional machine learning algorithms such as, naive Bayes algorithm, automatic encoder and noise reduction auto-encoder, it has better performance.

Acknowledgement. This study was supported by State's Key Project of Research and Development Plan (No. 2018YFC0810601, No. 2016YFC0901303). The work was conducted at University of Science and Technology Beijing.

References

1. Allemani, C., Weir, H.K., Carreira, H., et al.: Global surveillance of the cancer survival 1995-2009: analysis of the individual data for 25,676,887 patients from 279 population-based registries in 67 countries (CONCORD-2). Lancet **385**(9972), 977–1010 (2015)
2. Fan, L., Strasser-Weippl, K., Li, J.J., et al.: Breast cancer in China. Lancet Oncol. **15**(7), e279–e289 (2014)
3. Chen, W., Zheng, R., Zeng, H., Zou, X., Zhang, S., He, J.: Analysis of the incidence and mortality of malignant tumors in China in 2011. Chinese Tumor (01), 1–10 (2015)
4. Sun, K., Zheng, R., Gu, X., et al.: Analysis of the incidence and age of breast cancer in women with cancer registration in China from 2000 to 2014. Chin. J. Prev. Med. (6) (2018)
5. Gail, M.H., Brinton, L.A., Byar, D.P., et al.: Projecting individualized probabilities of developing breast cancer for white females who are being examined annually. J. Natl. Cancer Inst. **81**(24), 1879–1886 (1989)
6. Li, J., Wang, W., Li, S.: Preliminary study on the application of Gail breast cancer risk assessment model. Contemp. Chin. Med. **16**(14), 40–41 (2009)
7. Li, X., Yang, X., Li, M.: Research progress and clinical application of breast cancer risk assessment model. Cancer Prev. Res. **38**(5), 604–606 (2011)
8. Zheng, W., Wen, W., Gao, Y.T., et al.: Genetic and clinical predictors for breast cancer risk assessment and stratification among Chinese women. J. Natl Cancer Inst. **102**(13), 972–981 (2010)
9. Rumelhart, D.E., Hinton, G.E., et al.: Learning representations by back-propagating errors. Cogn. Model. **323**(6088), 399–421 (1986)
10. Zhang, Q., Zhang, F.: Fault diagnosis method of press bearing based on SDAE-SVM. Digit. Manuf. Sci., 203–208 (2018)

11. Vincent, P., Larochelle, H., Bengio, Y., et al.: Extracting and composing robust features with denoising autoencoders. In: International Conference on Machine Learning, pp. 1096–1103. ACM (2008)

12. Krogh, A., Hertz, J.A.: A simple weight decay can improve generalization. In: International Conference on Neural Information Processing Systems, pp. 950–957. Morgan Kaufmann Publishers Inc. (1991)

13. Luo, Y., Wan, Y.: A novel efficient method for training sparse auto-encoders. In: International Congress on Image and Signal Processing, pp. 1019–1023. IEEE (2014)

14. Bengio, Y., Lamblin, P., Popovici, D., et al.: Greedy layer-wise training of deep networks. Adv. Neural. Inf. Process. Syst. **19**, 153–160 (2007)

15. Vincent, P., et al.: Stacked denoising autoencoders: learning useful representations in a deep network with a local denoising criterion. J. Mach. Learn. Res. **11**(6), 3371–3408 (2010)

16. Xie, G., Zhang, T., Liu, M.: Collaboraive topic regression recommendation model based on SDAE feature representation. Comput. Eng. Sci., 924–932 (2019)

17. Hinton, G.E., Salakhutdinov, R.R.: Reducing the dimensionality of data with neural networks. Science **313**, 504–507 (2006)

18. Petracei, E., Deearli, A., Schairer, C., et al.: Risk factor modification and projections of absolute breast cancer risk. J. Natl Cancer Inst. **103**(13), 1037–1048 (2011)

19. Bengio, Y., Courville, A., Vincent, P.: Representation learning: a review and new perspectives. IEEE Trans. Pattern Anal. Mach. Intell. **35**(8), 1798–1828 (2013)

Prediction Model of Scoliosis Progression Bases on Deep Learning

Xiaoyong Guo[1], Suxia Xu[2], Yizhong Wang[2(✉)], Jason Pui Yin Cheung[3], and Yong Hu[3]

[1] School of Science, Tianjin University of Science and Technology, Tianjin 300457, China
[2] College of Electronic Information and Automation, Tianjin University of Science and Technology, Tianjin 300457, China
yzwang@tust.edu.cn
[3] Department of Orthopaedics and Traumatology, University of Hong Kong, Pok Fu Lam, Hong Kong

Abstract. By deep learning technique, we present a new approach to model idiopathic single curve scoliosis. We leverage the advanced version of the recurrent neural network, that is, the long short-term memory network, to achieve the goal. We frame scoliosis as a classification problem and a regression problem. A network for classification is designed first. We perform the training and testing with real clinic records that are imputed by various tricks. Using this model, one can classify the current level of scoliosis into three predefined groups via a few publicly measurable indictors, such as body height or arm span. We also design a regression network that can predict the future progression of spine curvature. This model can infer the development in spine curvature at a certain time span according to the changes of other indictors. Both of these models are evaluated by various metrics. The experiment shows that the quantitative picture of the scoliosis can be captured by our models giving a significant performance boost. Hence, the resulting decision-support system can help to decide the necessity of a further intervene both for physicians and patients.

Keywords: Scoliosis · Deep learning · Recurrent neural network · Long short-term memory

1 Introduction

In recent years, many aspects in artificial intelligence community have been enlightened by the emergence of the deep learning (DL) techniques [1]. DL algorithms are based on the deep neural networks (DNNs) and optimized by training on the dataset blessed with an enormous amount of labelled data. As a representation learning method, a DL algorithm is able to learn intricate patterns via multiple levels of distribution of representations and does not require hand-engineered features. In an image classification network, for example, features

© Springer Nature Singapore Pte Ltd. 2019
H. Ning (Ed.): CyberDI 2019/CyberLife 2019, CCIS 1138, pp. 431–440, 2019.
https://doi.org/10.1007/978-981-15-1925-3_31

are grouped together layer by layer. The convolution kernels at bottom layer learn such as edges and motifs, which are often called texture features. The features learned at higher layers are more complex, and the features learned in the vicinity of the very top layers can almost describe an object, which are often referred to as semantic features. In such a level of abstraction, a classifier can do classification interactively and engagingly. A well trained DNN is often attributed as DL model that can be deployed for very complex tasks with splendid efficiency and accuracy. State of the art DL models not only beat records in field of computer vision [2,3] and speech recognition [4,5], but also outperform conventional machine learn techniques applied in scientific community, such as analyzing particle accelerator data [6] and reconstructing brain circuits [7].

In this paper, we employ a DNN architecture, i.e. recurrent neural network (RNN), to model scoliosis, which is an idiopathic spine distortion annoying millions of people in the world [8]. The severity of scoliosis is usually measured by the so called cobb angle that is the angle between the two most tilted vertebrae in the spinal curves [9]. Most scoliosis patients with minor curves may have the symptoms including poor personal image, poor truncal balance, and susceptibility to back pain [10]. Those with moderate or worse spinal deformity may cause cardiopulmonary compromise resulting in significant health complications or even death [11,12]. For any circumstance, periodical examinations are necessary to determine which scoliosis condition may progress and which needs further anticipation such as bracing or surgery. More often than not, physicians make their decision depending on experience. Storing such valuable experience in an expert system is somehow a formidable task. Fortunately, clinic data are constantly accumulated during the years when the computer based databases become available. This makes the data-driven method to be a promising candidate to predict the scoliosis progression so as to help doctors determine which treatment method is most appropriate.

There is increasing number of publications devoted to achieve this goal. In these studies the conventional statistical and machine learning techniques, such as linear regression [10], clustering [9,12], and support vector machine [13], are adopted. In this work, the decision support system for scoliosis prognosis assumes a similar form as in our previous work [10] (For details please see Figs. 1 and 2 in that reference and we do not intend to reproduce these figures here). However, the prediction module is now powered by DL technique. Our network architecture is simple and straightforward. It is shown that our proposed methods outperform the existing schemes by a wide margin, and make the decision support system more accurate and reliable.

The remainder of this paper is organized as follows. Section 2 elaborates the entire model design including data acquisition and network architecture. After experiments shown in Sect. 3, we conclude this work in Sect. 4.

2   Model

In this section, we give the main indications which are necessity to understand our investigation. Details about the formulism can be found in the references.

2.1 Recurrent Neural Network and Long Short-Term Memory Network

RNNs are very powerful dynamic systems. Once unfold in time, RNNs can be seen as very deep feed-forward networks in which all the layers share the same weights. They process an input data sequence one element at a time, maintaining information about the history of all the past elements of the sequence in their hidden units. With this virtue, RNNs have the potential advantages in the field of natural language processing, data sequence classification, and predicting the next element of a time series [1]. The so called long short-term memory (LSTM) network is a modified architecture of RNN [14]. Owing to LSTM and ways of training it, the problem of gradient explode or vanish, which may arise during the training of the conventional RNNs, is solved, and the information over a long period of time is now memorable. The main ingredient of the LSTM network is the LSTM cell that is demonstrated in Fig. 1. The inputs of LSTM network are the same as RNN, that is, a series of data $x = \{x_0, x_1, \cdots, x_t, \cdots, x_T\}$ where x_T is the last element of the sequence with T being the length of the sequence.

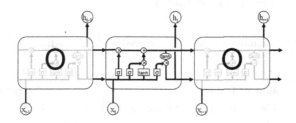

Fig. 1. Sketch of LSTM cell. Here, σ denotes the sigmoid function, $\times$ and $+$ represent the Hadamard product and addition between two matrixes with same size. Each hidden unit grouped under node O can map an input sequence with element x_t, into an output sequence with element h_t, with each h_t depending on all the previous $x_{t'}$ (for $t' \leq t$).

LSTM network is introduced in this study not only to estimate the current status of scoliosis but also to predict its future progression. To train the network, prospective anthropometry records of scoliosis patients at the Duchess of Kent Children's Hospital at Hong Kong, between years 1975 and 2014 are screened. The data is constituted by a series of anthropometry records that are measured during clinic. Using these data, we build the dataset on which the network can be trained and tested. Learning representations from raw input data is an advantage of DL algorithms. To make use of this property, we do not intent to rank the significance of those indicators intuitively or statistically. The network will perform feature extraction and contextual reasoning concurrently. Since the records are very different among patients and a single patient with many records is very rare, we use various tricks to handle the data in the sensible fashion and examine the model performance.

2.2   Dataset and Network Design

To reduce the complexity of the scoliosis data, so as to be modeled easily for prognosticating, the following inclusion criteria and exclusion criteria are instituted. The patient diagnosed as thoracic major curved scoliosis and in the mean time without any spine surgery history is included. The exclusion criteria includes the records which have the missing values in the indicators that are acknowledged related to the progression of scoliosis, and the records that have had brace treatment. Each involved record usually contains multiple spine curvatures. We select the largest one as the main curvature and omit other insignificant curvatures. There are totally 10 indicators: Gender, Age, FamilyHistory, BodyHeight, SittingHeight, ArmSpan, RiserSign, Radius, Ulna, and CobbAngle. The first nine are features associated with particular record, and the last one is used for designing the label in the classification task or taken as the ground truth in the regression task. With the label, one can classify the existing samples into certain groups characterized by different conditions. We define three levels of scoliosis basing on the standard: (1) Mild, cobb angle is less than 20°; (2) Moderate, cobb angle lies between 20 and 40°; (3) Severe, cobb angle is greater than 40°. For a full description of indicators see Table 1.

Table 1. Indicators in each involved scoliosis medical record.

Indicator	Description	Values
Gender	Male or female	Category {F, M}
Age	The actual age at the appointment date	Intger
FamHistory	Whether the family members have scoliosis	Category {+, −}
BodyHeight	Body height (cm)	Float
SittingHeight	Sitting height (cm)	Float
ArmSpan	Arm span (cm)	Float
RisserSign	Relevant to the maturity of skeleton	Float {0~11}
Radius	Relevant to the maturity distal radius	Float {0~11}
Ulna	Relevant to the maturity distal ulna	Float {1~9}
CobbAngle	The standing cobb angle	Float
Grade	The grade of the scoliosis progress	Category {0, 1, 2}

To feed these indictors into the recurrent net, male is denoted by 0 while female is by 1. No family history of scoliosis is denoted by 0 while yes is by 1. The Age, BodyHeight, SittingHeight, ArmSpan, RisserSign, Radius, and Ulna are retained as their original form. Finally, for the labels, class mild is attributed as label 0, moderate is 1, and severe is 2. We also note that some entries have partially missing indicators, such as BodyHeight, SittingHeight, ArmSpan, RisserSign, and etc. In order to ensure the authenticity of the dataset, supplementing these missing values is not considered at first. Records without missing value are selected for training, and then regular tricks for missing value imputation are conducted according to the existing records.

Table 2. Network architectures.

Network	Layers	Configuration	Trainable parameters
Classification	LSTM	64 LSTM cells	18944
	Dropout	dropout rate 0.2	None
	Fully connected	3 output neurons	195
	Activation function	softmax	None
Regression	LSTM	64 LSTM cells	19200
	Dropout	dropout rate 0.2	None
	Fully connected	1 output neuron	65
	Activation function	None	None

Due to the advantage of recurrent structure in processing sequence information, LSTM cells are used as the basic component to build our deep learning algorithm. The designed networks are illustrated in Table 2. Here, two architectures are considered. The first one is built for classification. The input of this network is a one-dimensional feature sequence composed by indicators where the cobb angle is not included. The output is three probabilities indicating that the current scoliosis level of the patient should be attributed into which predefined grades. The second one is built for regression. It takes the changes of the indictors together with the corresponding time span as input, and output only one quantity representing the progressing of the cobb angle during that span. Both of these two networks contain a single LSTM layer with 64 LSTM cells. Since number of samples for training is not large, a dropout layer with rate 0.2 is introduced after the LSTM layer to alleviate the risk of over-fitting. Following the dropout layer, we also add a fully connected layer which links the final hidden state h_T to the output space. The activation function of the fully connected layer is different to fulfill different tasks.

3   Experiments and Discussions

In this section, we are in a position to perform the training, visualize the results, and evaluate the performance of the resulting model. Training and testing are performed entirely on CPU, because the data throughput of our network is not large. On a computer with ubuntu16.04 and Intel Core i5-7400 CPU, we use the deep learning framework TensorFlow (release 1.3.0). The loss functions are categorical cross entropy for classification and mean square error for regression. The weights are randomly initialized. Throughout training we use the Adam optimizer with constant learning rate 0.01 and a batch size of 20. The network is trained for about 200 epochs.

3.1   Evaluating Current Status

To classify the existing samples so as to evaluating current condition for a particular patient, we first train the classification network. When the training is complete, a mapping between the indictors and the grade of cobb angle is learned. Using this model, we can infer the current condition of patients via certain conveniently accessible indictors, such as body height or arm span. It is applicable for circumstance where the pivotal equipment for cobb angle measurement is unavailable. This model can also cluster the patients into three predefined groups, and the patients in same group may assume similar condition that is curable with similar therapy.

A straightforward strategy to remedy missing indicators is to remove the samples that contain any missing indicators. By this way, only samples that are fully recorded are included in the training dataset. We find totally 250 samples fulfilling this requirement. Within these samples, 82 samples are labeled as mild, 153 samples are labeled as moderate, and 15 samples are labeled as severe. Since there are many samples with only one or two missing indicators, we also try to fix up manually. For example, for a certain record, the SittingHeight in the first appointment is missing. In other two consecutive appointments, its value is 84.3 for the second appointment while 86.5 for the third appointment. According to the context data, the SittingHeight for the first appointment shall be hypothesized as 82.1. By this strategy, we can collect 460 samples in which 143 samples are labeled as mild, 203 samples are labeled as moderate, and 114 samples are labeled as severe. Our third strategy is to use all the records. There are 3216 available samples which contain 845 mild samples, 2103 moderate samples, and 268 severe samples. We complement the missing indicators by simply attributing *null*. Our final strategy is to incorporate the local linear interpolation method (LLI) as well as global statistic approximation (GSA) that have been used in our prior work [10].

Table 3. Accuracy of our classification model. Here, training is performed on dataset with various methods of missing data imputation.

Missing data imputation	Number of samples	Accuracy
Complete records	250	67%
Manually imputation	460	87%
Attributing *null*	3216	35%
LLI	3216	59%
GSA	3216	71%

The accuracy is an important metrics to evaluate a classifier. Table 3 shows comparative performance between various imputation methods. The strategy we used to constitute the training dataset is highly influential on accuracy. The highest accuracy is obtained by using the dataset with hand-written missing indictors. This method can remedy the records that are closest to the real data distribution. In addition, the improvement of accuracy depends on the distribution of three degrees in the dataset. Even in the manually supplemented dataset, there are only 114 samples labeled as severe. This leads to insufficient training in such class and affects the improvement of the overall accuracy.

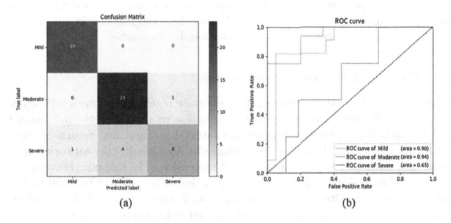

Fig. 2. (Color online) Confusion matrix and ROC curves corresponding to model trained on manually imputation dataset. Here, 18 samples belong to class mild, 25 samples belong to class moderate, and 13 samples belong to class severe. These samples are chosen randomly from training dataset.

When the number of samples in each category is not equal, it is difficult to gain insight into the performance of a classifier via accuracy. To further evaluation, we calculate the confusion matrix and the receiver operating characteristic (ROC) curves for the model trained with manually imputation dataset. Since using this dataset we can achieve a highest accuracy, evaluating the corresponding model should be more useful. The results are plotted in Fig. 2. Figure 2(a) is the confusion matrix which can provide the model's classification precision for each category and also gain insight into certain characteristics of the dataset itself. In the confusion matrix, the numbers in horizontal lines indicate the predicted label, and these displayed in vertical lines are the true label. Thus, the larger the numbers on the diagonal are, the better precision in this category is achieved. Each integer that are not on the diagonal are the false predictions. It is shown that the model struggles on the third category. In this category, 2 samples are classified as mild and 3 are classified as moderate. Figure 2(b) shows the corresponding ROC curve. Here, the x-axis represents the false positive rate (FPR), and the y-axis represents the true positive rate (TPR). The ROC curve describes the ratio between TPR and FPR. The worst classifier has a ROC curve lying on the diagonal line from bottom left to up right, while the best one's ROC curve is a broken line linking the points $(0, 0)$, $(0, 1)$, and $(1, 1)$. The closer to the point $(0, 1)$ of the ROC curve is, the better classifier we get. The area under the

ROC curve is the so called area under curve (AUC). The AUC is a quantitative metric for a classifier, and a greater AUC means a better performance. It can also be seen from Fig. 2(b) that the performance of the model on category severe is worst. The intuition we got from Fig. 2(b) is accordance with Fig. 2(a). The pool performance of our classifier on category severe can be attributed to the lack of data. In fact, most patients would have brace treatment or surgery before their condition becomes severe. The classification on category moderate is the best of all. Since there are 203 samples in this class, the network is well trained and most effective features of this category can be abstracted.

3.2   Predicting Future Progression

To predict the future progressing of the scoliosis, a regression network is designed which takes the time interval between two successive appointments and changes of other indicators corresponding to that interval as input and returns a single value that means the change in the cobb angle. As a result, this network has an input dimension of 10 and a single output dimension. The mapping between the changing of indicators and the progressing of the cobb angle can be learn by this network, whose architecture is shown in Table 2 where the last fully connected layer has no activation function and only one output neuron. To train this model, we exclude the samples that contain only one appointment. Similar trick in handling remaining records as in the last subsection is used. We transform the categorical features to numeric features. Time interval between two successive appointments is calculated by making subtraction between the corresponding two ApptDates. Other indicators, such as BodyHeight, SittingHeight, ArmSpan, RisserSign, Radius, Ulna, and CobbAngle, are all represented by their difference between two successive appointments. After these procedures, there are 1655 samples among which 17 samples are perfectly complete and 68 samples are nearly complete. Due to lack of complete sample, LLI method is used for data imputation. Using these data, we assemble a training dataset with 1324 samples and a testing dataset with 331 samples.

To evaluate our regression model, we calculate the root mean square error (RMSE) and mean absolute percentage error (MAPE). These metrics are defined as

$$RMSE = \sqrt{\frac{1}{N}\sum_{i=1}^{N}(y_i - \hat{y}_i)^2}, \tag{1}$$

$$MAPE = \frac{1}{N}\sum_{i=1}^{N}|\frac{y_i - \hat{y}_i}{y_i}|. \tag{2}$$

Here, N is the number of samples, y_i is the actual progressing in cobb angle between two successive appointments, and $\hat{y}_i$ is the predicted value of y_i. To calculate RMSE and MAPE, we randomly select 32 samples from testing dataset. It is shown that our model can achieve a performance with $MAPE = 0.216$ and $RMSE = 1.229$. This performance is better than conventional machine learning method as we previously provided in Ref. [10]. In Fig. 3, we plot the predicted progression of cobb angle versus its true value in a histogram. Here, the predicted values are plotted as red bars, while blue bars indicate the real values. Notice that not only the future tendencies of the progression of cobb angle are correctly predicted, but the predictions are also very close to their real value. This means that our model is proper trained and acquires the capability of quantitative prognosis.

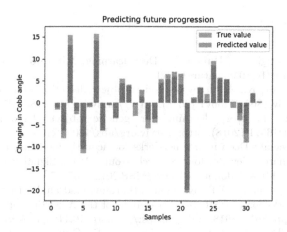

Fig. 3. (Color online) Predicting future cobb angle progression. These 32 samples are chosen randomly from testing dataset.

4   Conclusion

In this paper, DL technique is wielded to model single curve scoliosis. Basing on RNN architecture with LSTM cells, two networks are designed and trained with different strategies. The initial LSTM layer of the network extracts features from the input data while the fully connected layer predict the output probabilities. A classification model is introduced first. It feeds with indictors of a particular record and outputs the severity of curvature in spine. The experiments demonstrate the effectiveness of this classifier and also prove that the network is suitable to learn the complex structure such as the scoliosis. This model can be used for constantly tracking the curve progression in untreated idiopathic scoliosis during growth, when the equipment for cobb angle measurement is not available. Patient can even use this model at home and estimate their current condition by measuring just body height or arm span. The second one is designed for regression. It feeds with time span of the period between two adjacent appointments and the changes of indictors during that span. The model can predict the future progression of cobb angle. It is shown that the regression model indeed captures the quantitative picture of scoliosis evolution. Therefore, both patient and doctor can get benefit from the study.

Through the investigation, it can be seen that the accuracy of the model is greatly limited by the dataset. As a result, it is necessary to appropriately increase the data and apply more reasonable methods to impute the existing data, so as to obtain a better performance. Due to the refreshing simplicity of our model, it is memory saving and extremely fast making it ideal for the applications that can deploy on an embedded system or even on a smart phone.

References

1. LeCun, Y., Bengio, Y., Hinton, G.: Deep learning. Nature **521**, 436–444 (2015). https://doi.org/10.1038/nature14539
2. Chahal, K., Dey, K.: A survey of modern object detection literature using deep learning. arXiv:1808.07256v1 (2018). http://arxiv.org/abs/1808.07256
3. Liu, L., et al.: Deep learning for generic object detection: a survey. arXiv:1809.02165v1 (2018). http://arxiv.org/abs/1809.02165
4. Hinton, G., et al.: Deep neural networks for acoustic modeling in speech recognition: the shared views of four research groups. IEEE Signal Process. Mag. **29**, 82–97 (2012). https://doi.org/10.1109/MSP.2012.2205597
5. Sainath, T., Mohamed, A.R., Kingsbury, B., Ramabhadran, B.: Deep convolutional neural networks for LVCSR. In: Proceedings of the Acoustics, Speech and Signal Processing, pp. 8614–8618 (2013). https://doi.org/10.1109/ICASSP.2013.6639347
6. Ciodaro, T., Deva, D., de Seixas, J., Damazio, D.: Online particle detection with neural networks based on topological calorimetry information. J. Phys. Conf. Ser. **368**, 012030 (2012). https://doi.org/10.1088/1742-6596/368/1/012030
7. Helmstaedter, M., et al.: Connectomic reconstruction of the inner plexiform layer in the mouse retina. Nature **500**, 168–174 (2013). https://doi.org/10.1038/nature12346
8. Scoliosis in Depth Report. http://www.nytimes.com/health/guides/disease/scoliosis/print.html. Accessed 8 Oct 2015
9. Chalmers, E., et al.: Predicting success or failure of brace treatment for adolescents with idiopathic scoliosis. Med. Biol. Eng. Comput. **53**, 1001–1009 (2015). https://doi.org/10.1007/s11517-015-1306-7
10. Deng, L.M., Hu, Y., Cheung, J.P.Y., Luk, K.D.K.: A data-driven decision support system for scoliosis prognosis. IEEE Access **5**, 7874–7884 (2017). https://doi.org/10.1109/ACCESS.2017.2696704
11. Kuroki, H.: Brace treatment for adolescent idiopathic scoliosis. J. Clin. Med. **7**, 136 (2018). https://doi.org/10.3390/jcm7060136
12. Wu, H., et al.: Prediction of scoliosis progression in time series using a hybrid learning technique. In: Proceedings of the 27th Annual International Conference on Engineering Medicine and Biology Society, pp. 6452–6455 (2006). https://doi.org/10.1109/IEMBS.2005.1615976
13. Ajemba, P.O., Ramirez, L., Durdle, N.G., Hill, D.L., Raso, V.J.: A support vectors classifier approach to predicting the risk of progression of adolescent idiopathic scoliosis. IEEE Trans. Inf. Technol. Biomed. **9**, 276–282 (2005). https://doi.org/10.1109/titb.2005.847169
14. Hochreiter, S., Schmidhuber, J.: Long short-term memory. Neural Comput. **9**, 1735–1780 (1997). https://doi.org/10.1162/neco.1997.9.8.1735

A Hybrid Intelligent Framework for Thyroid Diagnosis

Zhuang Li[1,3], Jingyan Qin[2], Xiaotong Zhang[1,3(✉)], and Yadong Wan[3,4]

[1] Beijing Advanced Innovation Center for Materials Genome Engineering,
University of Science and Technology Beijing, Beijing 100083, China
zxt@ies.ustb.edu.cn
[2] School of Mechanical Engineering,
University of Science and Technology Beijing, Beijing 100083, China
[3] School of Computer and Communication Engineering, University of Science and Technology
Beijing, Beijing 100083, China
[4] Beijing Key Laboratory of Knowledge Engineering for Materials Science,
University of Science and Technology Beijing, Beijing 100083, China

Abstract. Thyroid disease exists across the whole world and many people are
suffering from this disease. The diagnosis of thyroid disease is of great importance to human life. Although there are already some researches that introduces
various methods for thyroid diagnosis and achieves good results, the performance
of diagnosis still needs to be improved. Therefore, a hybrid intelligent framework,
in which an optimal support vector machine (SVM) based on a hybrid optimization algorithm and a recursive feature elimination (RFE) method are incorporated,
is proposed to predict thyroid disease in this paper. The hybrid optimization algorithm combines the teaching-learning based algorithm (TLBO) and differential
evolution (DE), contributing to the parameter optimization of SVM. And the RFE
method is introduced to obtain the optimal feature subsets for thyroid diagnosis.
A thyroid dataset collected from UCI repository is utilized to evaluate the performance of the proposed framework. The experimental results demonstrate that
the proposed framework achieves better and more stable performance than other
compared methods.

Keywords: Thyroid diagnosis · Support vector machine · Teaching-learning
based optimization algorithm · Differential evolution · Feature selection

1 Introduction

Thyroid disease is an interminable and complex infection which happens because the
thyroid-simulating hormone (TSH) is at an inappropriate level [1]. TSH consists of two
iodine containing hormones, Triiodothyronines (T3) and Thyroxine (T4). When low level
of the hormones produced by the thyroid gland, hypothyroidism occurs. On the contrary,
hyperthyroidism happens due to the over secretion of the hormones. Thyroid disease is
widespread throughout the world, especially in India and many people are suffering from

© Springer Nature Singapore Pte Ltd. 2019
H. Ning (Ed.): CyberDI 2019/CyberLife 2019, CCIS 1138, pp. 441–451, 2019.
https://doi.org/10.1007/978-981-15-1925-3_32

the side effects of this disease, such as melancholy, tiredness, apprehension, tremors and so on [2]. Therefore, the diagnosis of thyroid disease is very significant to human life.

There are already some great researches on thyroid diagnosis and the machine learning based methods plays an essential role among them. In 2016, Razia and Rao [1] reviewed different neural network modeling methods for thyroid diagnosis and discussed some parameter estimation methods and execution methods of the distinctive neural network models. Chandel et al. [3] used different classifier, K-nearest neighbor (KNN) and Naive Bayes (NB) to detect thyroid disease and the results showed that the accuracy of KNN is better than NB. Geetha and Baboo [4] proposed a hybrid DE kernel based NN algorithm for the dimension reduction of thyroid data and the reduced data was provided to SVM with radial basis function kernel (RBE) for classification. The results demonstrated that the proposed method is superior to the Kernel based NB classifier. Shankar et al. [2] also realized the importance of feature selection for thyroid diagnosis and used an improved gray wolf optimization algorithm to select the optimal features. Then a multi-kernel based SVM was utilized to detect the thyroid disease.

In this paper, a hybrid framework is proposed to improve the accuracy of thyroid disease diagnosis, in which an optimal support vector machine (SVM) based on a hybrid optimization algorithm (TLDE) and a recursive feature elimination (RFE) method are incorporated. The TLDE algorithm combines TLBO algorithm and DE algorithm, contributing to the parameter optimization of SVM. And the RFE method based on least absolute shrinkage and selection operator [5] (Lasso) estimator is introduced to obtain the optimal feature subsets for thyroid diagnosis. A thyroid dataset collected from UCI repository [6] is utilized to evaluate the performance of the proposed framework. The experimental results demonstrate that the proposed framework achieves better and more stable performance than other compared methods.

The rest of this paper is organized as follows. Background materials are introduced in Sect. 2. In Sect. 3, the detailed information of the proposed hybrid intelligent framework is described. The experimental design and discussion are demonstrated in Sect. 4. Finally, conclusions are drawn in Sect. 5.

2 Background Materials

2.1 Support Vector Machine

SVM, proposed by Vladimir and Vapnik in 1995 [7], has been wildly applied to solve problems in various fields, such as human actions recognition [8], disease diagnosis [9, 10], and so on. The most important advantage of SVM is that the model parameters obtained are definite global optimal solution because of the convex optimization theory. The detail of SVM was described as follows.

Given a training dataset $\{(\mathbf{x}_1, y_1), \ldots, (\mathbf{x}_l, y_l)\}$, where l is the size of the training data, $\mathbf{x}_i \in \mathbf{R}^n$ is an input data point and $y_i \in \{-1, 1\}$ is the corresponding target label of $\mathbf{x}_i$. The data points are separated with a hyper plane $\mathbf{w}^T \phi(\mathbf{x}) + b = 0$, where $\mathbf{w}$ is the coefficient vector that is normal to the hyper plane, b is the offset from the origin, and $\phi(\mathbf{x})$ denotes a transformation of feature-space, which transform the original feature space $\mathbf{R}^n$ to a high dimensional space $\mathbf{R}^m$ ($m > n$). The objective of SVM is to find an optimal separating margin by minimizing Eq. (1), where $\xi_i, i = 1, 2, \ldots, l$ is slack variables.

$$L(\mathbf{w}, b, \xi_i) = \frac{1}{2}||\mathbf{w}||^2 + C \sum_{i=1}^{l} \xi_i$$
$$\text{s.t.,} \quad y_i(\mathbf{w}^T \phi(\mathbf{x}) + b) \geq 1 - \xi_i \qquad (1)$$
$$\xi_i \geq 0, i = 1, 2, \ldots, l$$

By introducing Lagrange multipliers $\alpha_i \geq 0$ and $u_i \geq 0$, the problem can be reformulated to the Lagrangian problem as shown in Eq. (2), from which $\mathbf{w}$ can be solved in the form of $\sum_{i=1}^{l} \alpha_i y_i \phi(\mathbf{x}_i)$. And the Lagrange dual problem can be obtained as shown in Eq. (3), a quadratic optimization problem with linear constraints.

$$L(\mathbf{w}, b, \xi_i) = \frac{1}{2}||\mathbf{w}||^2 + C \sum_{i=1}^{l} \xi_i - \sum_{i=1}^{l} \alpha_i \{y_i(\mathbf{w}^T \phi(\mathbf{x}) + b) - 1 + \xi_i\} - \sum_{i=1}^{l} u_i \xi_i \quad (2)$$

$$\max_{\alpha} \sum_{i=1}^{l} \alpha_i - \frac{1}{2} \sum_{i=1}^{l} \sum_{j=1}^{l} \alpha_i \alpha_j y_i y_j \phi(\mathbf{x}_i)^T \phi(\mathbf{x}_j)$$
$$\text{s.t.,} \quad \sum_{i=1}^{l} \alpha_i y_i = 0 \qquad (3)$$
$$\alpha_i \geq 0, i = 1, 2, \ldots, l$$

From the corresponding Karush-Kuhn-Tucker (KKT) conditions, shown in Eq. (4), $\mathbf{w}$ can be reduced to $\sum_{i \in K} \alpha_i y_i \phi(\mathbf{x}_i)$, where K denotes the set of indices of data points having $0 \leq \alpha_i \leq C$, known as support vectors. Then, b can be calculated from $y_i(\mathbf{w}\phi(\mathbf{x}_i) + b) - 1 = 0$, where $\mathbf{x}_i$ are support vectors. After that, the discriminant function can be obtained by Eq. (5).

$$\begin{cases} \alpha_i \geq 0, u_i \geq 0, \\ y_i(\mathbf{w}\phi(\mathbf{x}_i) + b) - 1 + \xi_i \geq 0, \\ \alpha_i(y_i(\mathbf{w}\phi(\mathbf{x}_i) + b) - 1 + \xi_i) = 0, \\ \xi_i \geq 0, u_i \xi_i = 0. \end{cases} \qquad (4)$$

$$f(\mathbf{x}) = \text{sgn}(\sum_{i \in K} \alpha_i y_i \phi(\mathbf{x}_i)^T \phi(\mathbf{x}) + b) \qquad (5)$$

2.2 Teaching-Learning Based Optimization

Teaching-learning-based optimization algorithm is a population-based algorithm for optimization problems introduced by Rao et al. [11]. Similar to other population-based algorithm, TLBO algorithm needs the common controlling parameters, such as population size and number of generations, but it does not require any specific controlling parameters. Because of the simple and effective characteristics, TLBO algorithm has been successfully applied to solve problems in various fields, such as mechanical design [12], pattern recognition [13], chemistry engineering [14] and so on.

TLBO algorithm uses a metaphor of learning process consisting of two phases, teacher phase and learner phase. In TLBO algorithm, the population is a group of learners in a class and the fitness value is the learner's outcome based on different subjects offered to the learner which is considered as various design variables. During the teacher phase, the teacher that has the best fitness value diffuses its knowledge among learners, and then learners exchange the knowledge during the learner phase.

Teacher Phase. This phase is the first phase of TLBO, in which learners increase their knowledge with the help of the teacher. And the teacher tries to improve the mean of learners to a certain level depending on the capability of the class. At any iteration, let $s_{i,j}$ ($i = 1, 2 \ldots, NP$, $j = 1, 2, \ldots, D$) represents the result of i_{th} learner in j_{th} subject, where NP and D are the number of learners and subjects respectively. Besides, let $s_{gbest,j}$ be the result of the best learner in j_{th} subjects and **Mean**$_j$ be the mean result of the learners in j_{th} subject. Then solution is modified to be a candidate solution of the next iteration by Eq. (6), where r is a random number in the range of [0.1] and T_F is the teaching factor that determined the value of mean to be changed. The value of T_F is either 1 or 2 and randomly given by Eq. (7).

$$s_{i,j}^{cand} = s_{i,j} + r \cdot (s_{gbest,j} - T_F \cdot \textbf{Mean}_j) \tag{6}$$

$$T_F = \text{round}(1 + \text{rand}(0, 1)) \tag{7}$$

After that, greedy selection strategy that always prefers the solution with better function value, is adopted to select the solution from $s_{i,j}^{cand}$ and $s_{i,j}$ for the next iteration.

Learner Phase. Learners enhance their knowledge not only by learning from the teacher but also by learning from other learners. In this phase, a learner learns new knowledge from the random interaction with another leaner. Randomly select a learner $s_{q,j}$ for the learner $s_{i,j}$, where $q \neq i$, and modify the learner $s_{i,j}$ by Eq. (8). Then the same greedy selection strategy as in teacher phase is used to update the solution for the next iteration.

$$s_{i,j}^{cand} = \begin{cases} s_{i,j} + r \cdot (s_{i,j} - s_{q,j}), & f(s_{i,j}) < f(s_{q,j}) \\ s_{q,j} + r \cdot (s_{q,j} - s_{i,j}), & f(s_{i,j}) > f(s_{q,j}) \end{cases} \tag{8}$$

2.3   Differential Evolution

Differential evolution, proposed by Storn and Price [15], is a simple and efficient optimization algorithm for continuous optimization problems. It includes mutation, crossover and selection, and the crucial part of DE is the mutation scheme, which perturbs a vector based on the difference of the other two randomly selected population vectors.

DE begins with a randomly initialized population of NP D-dimensional real value vectors and each vector is a candidate solution. Let $s_i = [s_{i,1}, s_{i,2}, \ldots, s_{i,D}]$ represents the i_{th} vector of the population at any iteration. For each vector, a new corresponding mutant vector v_i is generated based mutation operator which is shown in Eq. (9), where,

$r1, r2$ and $r3$ are three randomly chosen integers which are different from i and F is a constant factor which controls the amplification of the difference $(\mathbf{s}_{r2} - \mathbf{s}_{r3})$.

$$\mathbf{v}_i = \mathbf{s}_{r1} + F \cdot (\mathbf{s}_{r2} - \mathbf{s}_{r3}) \tag{9}$$

After the mutation, the crossover is conducted to select a vector from the original vector and the mutant vector as shown in Eq. (10), where rand$(0, 1)$ is a random number from $[0,1]$, CR is the crossover probability, a predefined constant, and randint$(1, D)$ is a random integer from $[1, D]$.

$$\mathbf{u}_{i,j} = \begin{cases} \mathbf{u}_{i,j}, & \text{if rand}(0, 1) \le CR \text{ or } j = \text{randint}(1, D) \\ \mathbf{s}_{i,j}, & \text{otherwise} \end{cases} \tag{10}$$

Then, the same greedy selection strategy is adopted to select the solution from $\mathbf{u}_i$ and $\mathbf{s}_i$ for the next iteration.

2.4 Recursive Feature Elimination

Recursive feature elimination is a widely used feature selection method, which utilizes an external estimator to evaluate the subsets of features and select features from smaller and smaller sets of features recursively. Therefore, the performance of RFE relies heavily on the estimator. Guyon et al. [16] proposed a new gene selection method using linear support vector machine based on recursive feature elimination in 2002 and the experiment shown that the genes selected by this method obtained better performance for cancer classification. In 2010, Meinshausen [17] applied stability selection to the Lasso estimator shown in Eq. (11) that includes an $l1$ penalty bounding the absolute sum of all coefficients. One advantage of the Lasso estimator is that it can shrink the coefficients of certain features exactly to zero. In this paper, the lasso estimator is adopted as the external estimator of RFE method.

$$\hat{\boldsymbol{\beta}} = \underset{\beta \in \mathbb{R}^p}{\arg\min}(||\mathbf{Y} - \mathbf{X}\boldsymbol{\beta}||_2^2 + \lambda \sum_{k=1}^{p} |\boldsymbol{\beta}_k|) \tag{11}$$

where, $\mathbf{X}$ is the input data matrix, and $\mathbf{Y}$ is the corresponding target label matrix. p is the number of features, $\boldsymbol{\beta}_k$ is the weight of k_{th} feature and λ is a regularization parameter.

3 The Proposed Intelligent Framework

3.1 The TLDE Algorithm

Although TLBO algorithm has been successfully applied in various fields, it has an inherent origin bias within teacher phase [18]. Zou et al. [19] pointed that it is one of the topics that are worth studying in the next years. In order to solve the origin bias and enhancing the ability for global optimization, a hybrid optimization is proposed by incorporating the DE algorithm in the TLBO algorithm.

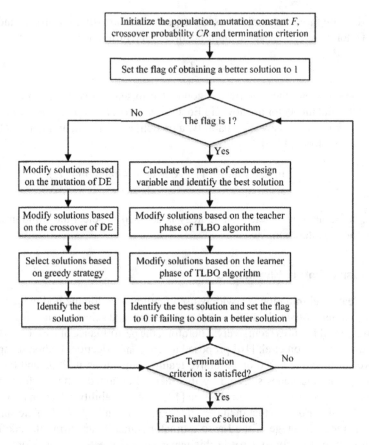

Fig. 1. Flowchart of the TLDE algorithm

As the origin bias occurs at convergence where $s_i \approx s_{gbest} \approx$ **Mean** and DE algorithm is a simple and efficient optimization algorithm for global optimization, it is reasonable to propose the TLDE algorithm that uses TLBO algorithm for searching firstly and replaces TLBO algorithm with DE algorithm when TLBO algorithm fails to obtain a better solution. When TLBO algorithm cannot get a better solution, it is almost time for the occurrence of origin bias. Therefore, the TLDE algorithm can avoid the occurrence of origin bias and maintain the global optimization ability. The flowchart of the TLDE algorithm is shown in Fig. 1.

3.2 The Hybrid Intelligent Framework

In order to improve the performance of thyroid diagnosis, a hybrid framework named RFE-TLDE-SVM is proposed to by combining the SVM optimized by TLDE algorithm with RFE method. Firstly, RFE method based on Lasso estimator, a popular feature selection method, is utilized to obtain an optimal feature subset because the performance of machine learning methods is greatly influenced by the features. Besides, as the performance of SVM is significantly affected by its parameters, TLDE algorithm, a simple and effective optimization algorithm combining the TLBO algorithm and the DE algorithm, is used to optimize the key parameters of SVM automatically.

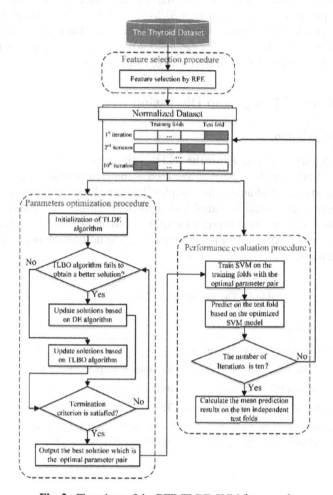

Fig. 2. Flowchart of the RFE-TLDE-SVM framework

There are many kernels can be used for SVM, such as linear kernel, polynomial kernel, sigmoid kernel, radial based kernel and so on. Among these kernels, radial based kernel shown in Eq. (12) is wildly used in various fields. Therefore, this paper utilizes TLDE algorithm to optimize the two key parameters of SVM with radial based kernel, regularization parameter C and kernel width γ. The flowchart of the proposed framework is demonstrated in Fig. 2.

$$k(\boldsymbol{x}, \boldsymbol{x}_i) = \exp(-\gamma ||\boldsymbol{x} - \boldsymbol{x}_i||^2) \tag{12}$$

In the proposed framework, there are three primary procedures: feature selection procedure, parameters optimization procedure and performance evaluation procedure. During the feature selection procedure, RFE method based on Lasso estimator, a classic feature selection algorithm, is used to select the optimal feature subsets. In the parameters optimization procedure, the parameters of SVM are optimized by TLDE algorithm, and the average classification accuracy on the 5-fold cross validation is considered as the fitness function. And in the performance evaluation procedure, the objective is to evaluate the overall performance of SVM model with optimal parameters.

4   Experimental Design and Discussion

4.1   Data Description

The thyroid dataset is collected from the UCI repository, consisting of 215 samples and three classes. Each sample is labeled based on a complete medical record, including anamnesis, scans, etc. And among the samples, 150 samples are euthyroidism, 30 samples are hypothyroidism and 35 samples are hyperthyroidism. Five features which are all continuous values, are used to predict whether a patient's thyroid can be classified as euthyroidism, hypothyroidism, or hyperthyroidism. The description of the five features is given in Table 1.

Table 1. The five features of the thyroid dataset.

Features	Description
F1	T3-resin uptake test (a percentage)
F2	Total serum thyroxin as measure by the isotopic displacement method
F3	Total serum triiodothyronine as measured by radioimmunoassay
F4	Basal thyroid-stimulating hormone (TSH) as measured by radioimmunoassay
F5	Maximal absolute difference of TSH value after injection of 200 mg of thyrotropin-releasing hormone as compared to the basal value

4.2   Experimental Setup

The involved methods were implemented on MATLAB platform and the LIBSVM package [20] was utilized for the SVM algorithm. Before the modeling procedure, the data was normalized to interval $[-1, 1]$. And the experiment was conducted on a computer with Intel (R) Core (TM) i7-6700 CPU (3.4 GHz) and 16 GB of RAM.

For fair comparison, the same number of function evaluations and population size were used for GA, PSO, DE, TLBO and TLDE algorithms. The number of function evaluations and population size are set to 2000 and 8 respectively. The searching ranges of the parameters $C \in [2^{-5}, 2^{-3}, \ldots, 2^{15}]$ and $\gamma \in [2^{-15}, 2^{-13}, \ldots, 2]$ were adopted for Grid search method. And the same searching ranges of parameters $C \in [2^{-5}, 2^{15}]$

and $\gamma \in [2^{-15}, 2]$ were utilized for GA, PSO, DE, TLBO and TLDE algorithms. Besides, the crossover probability and mutation probability were set to 0.8 and 0.05 respectively for GA. For PSO, the inertia weight was set to 1 and the acceleration coefficients were set to 2.05. The mutation constant and crossover probability were set to 0.5 and 0.9 respectively for DE and TLDE algorithms.

4.3 Measure for Performance Evaluation

The performance of RFE-TLDE-SVM was evaluated by classification accuracy (ACC) which is shown in Eq. (13), where TP and TN denotes the number of thyroid patients and healthy people that are correctly classified respectively, FP represents the number of healthy people misclassified as thyroid patients, and FN denotes the number of thyroid patients that are misclassified as healthy people.

$$ACC = \frac{TP + TN}{TP + FP + FN + TN} \times 100\% \tag{13}$$

4.4 Results and Discussion

Table 2 shows the detailed statistical results of ACC scores on 10-fold cross validation of different methods on the thyroid dataset with the optimal feature subset and Fig. 3 shows the trend of different methods on the 10 folds. The results show that RFE-TLDE-SVM outperforms other methods on most folds. And RFE-TLDE-SVM performs better compared to other methods with the largest average result of 97.17% and the smallest standard deviation of 0.038. It indicates that the proposed framework can achieve better and more stable results than other compared methods. Besides, the average ACC scores on 10-fold cross validation in the studies of Shankar et al. [2] and Shen et al. [9]

Table 2. The detailed ACC of different methods with the optimal feature subset.

Folds	RFE-Grid-SVM (%)	RFE-GA-SVM (%)	RFE-PSO-SVM (%)	RFE-DE-SVM (%)	RFE-TLBO-SVM (%)	RFE-TLDE-SVM (%)
1	100.00	95.24	100.00	100.00	100.00	100.00
2	100.00	100.00	100.00	100.00	100.00	100.00
3	100.00	100.00	100.00	100.00	1.0000	100.00
4	95.45	100.00	95.45	95.45	95.45	1.0000
5	100.00	100.00	100.00	100.00	100.00	100.00
6	95.45	95.45	95.45	95.45	95.45	95.45
7	95.24	90.48	95.24	90.48	90.48	90.48
8	100.00	100.00	100.00	100.00	100.00	100.00
9	85.71	85.71	85.71	95.24	90.48	95.24
10	85.71	90.48	90.48	85.71	85.71	90.48
Mean	95.76	95.74	96.23	96.23	95.76	97.17
Std.	5.41	4.97	4.66	4.66	4.97	3.80

Bold values indicates the best results.

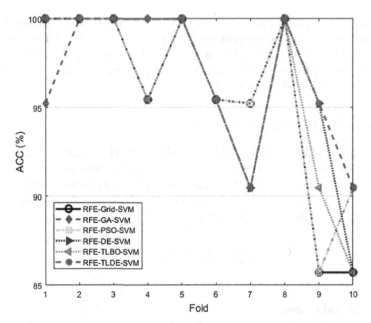

Fig. 3. Line chart of *ACC* scores of different methods with the optimal feature subset.

are 96.22% and 96.38% respectively, so the performance of RFE-TLDE-SVM is also superior to the existing two methods.

From the above analysis, it can be seen that the proposed RFE-TLDE-SVM method shows an excellent performance and it is the most powerful method for thyroid diagnosis compared to other methods.

5   Conclusions

In this paper, a hybrid framework was developed to improve the accuracy of thyroid disease diagnosis, in which an optimal support vector machine (SVM) based on a hybrid optimization algorithm (TLDE) and a recursive feature elimination (RFE) method are incorporated. The TLDE algorithm unites the TLBO algorithm and the DE algorithm, contributing to the parameter optimization of SVM. And the RFE method based on Lasso estimator is introduced to obtain the optimal feature subsets for thyroid diagnosis. The performance of the proposed framework is evaluated by a thyroid dataset collected from UCI repository and the experimental results demonstrate that the proposed framework is superior, achieving better and more stable performance than other compared methods.

The proposed framework can also be applied to solve other disease diagnosis problems in the future. And another future work that is worth investigating is to conduct the feature selection and parameter optimization simultaneously by TLDE and other optimization algorithms.

Acknowledgement. This paper is sponsored by National key research and development program of China (No. 2017YFB0702300).

References

1. Razia, S., Rao, M.N.: Machine learning techniques for thyroid disease diagnosis - a review. Indian J. Sci. Technol. **9**(28), 1–9 (2016)
2. Shankar, K., Lakshmanaprabu, S.K., Gupta, D., Maseleno, A., de Albuquerque, V.H.C.: Optimal feature-based multi-kernel SVM approach for thyroid disease classification. J. Supercomput. 1–16 (2018)
3. Chandel, K., Kunwar, V., Sabitha, S., Choudhury, T., Mukherjee, S.: A comparative study on thyroid disease detection using K-nearest neighbor and Naive Bayes classification techniques. CSI Trans. ICT **4**(2–4), 313–319 (2016)
4. Geetha, K., Baboo, C.S.S.: Efficient thyroid disease classification using differential evolution with SVM. J. Theor. Appl. Inf. Technol. **88**(3), 410–422 (2016)
5. Tibshirani, R.: Regression shrinkage and selection via the Lasso. J. Roy. Stat. Soc.: Ser. B (Methodol.) **58**(1), 267–288 (1996)
6. Dua, D., Graff, C.: UCI Machine learning repository. University of California, School of Information and Computer Science, Irvine (2019). http://archive.ics.uci.edu/ml
7. Vladimir, V.N., Vapnik, V.: The Nature of Statistical Learning Theory. Springer, New York (1995). https://doi.org/10.1007/978-1-4757-3264-1
8. Laptev, I., Caputo, B.: Recognizing human actions: a local SVM approach. In: Proceedings of the 17th International Conference on Pattern Recognition, pp. 32–36. IEEE (2004)
9. Shen, L., Chen, H., Yu, Z., et al.: Evolving support vector machines using fruit fly optimization for medical data classification. Knowl. Based Syst. **96**, 61–75 (2016)
10. Cai, Z., Gu, J., Chen, H.L.: A new hybrid intelligent framework for predicting Parkinson's disease. IEEE Access **5**, 17188–17200 (2017)
11. Rao, R.V., Savsani, V.J., Vakharia, D.P.: Teaching-learning-based optimization: a novel method for constrained mechanical design optimization problems. Comput. Aided Des. **43**(3), 303–315 (2011)
12. Zou, F., Wang, L., Hei, X., et al.: Multi-objective optimization using teaching-learning-based optimization algorithm. Eng. Appl. Artif. Intell. **26**(4), 1291–1300 (2013)
13. Murty, M.R., Naik, A., Murthy, J.V.R., et al.: Automatic clustering using teaching learning based optimization. Appl. Math. **5**(8), 1202 (2014)
14. Chen, X., Mei, C., Xu, B., et al.: Quadratic interpolation based teaching-learning-based optimization for chemical dynamic system optimization. Knowl. Based Syst. **145**, 250–263 (2018)
15. Storn, R., Price, K.: Differential evolution – a simple and efficient heuristic for global optimization over continuous spaces. J. Global Optim. **11**(4), 341–359 (1997)
16. Guyon, I., Weston, J., Barnhill, S., et al.: Gene selection for cancer classification using support vector machines. Mach. Learn. **46**(1–3), 389–422 (2002)
17. Meinshausen, N., Bühlmann, P.: Stability selection. J. Roy. Stat. Soc. Ser. B: Stat. Methodol. **72**(4), 417–473 (2010)
18. Pickard, J.K., Carretero, J.A., Bhavsar, V.C.: On the convergence and origin bias of the teaching-learning based optimization algorithm. Appl. Soft Comput. **46**, 115–127 (2016)
19. Zou, F., Chen, D., Xu, Q.: A survey of teaching-learning based optimization. Neurocompting **335**, 366–383 (2019)
20. Chang, C.C., Lin, C.J.: LIBSVM: a library for support vector machines. ACM Trans. Intell. Syst. Technol. (TIST) **2**(3), 27 (2011)

Geriatric Disease Reasoning Based on Knowledge Graph

Shaobin Feng[1], Huansheng Ning[1], Shunkun Yang[2], and Dongmei Zhao[3(✉)]

[1] School of Computer and Communication Engineering,
University of Science and Technology Beijing, Beijing 100083, China
[2] School of Reliability and Systems Engineering,
Beihang University, Beijing 100191, China
[3] College of Economic and Management, China Agriculture University,
Beijing 100083, China
zhaodongm@vip.163.com

Abstract. The lack of health care for ageing has become one of China's most serious challgengs. The main work of this paper is building a database of a geriatric knowledge graph and proposing three inference rules based on Bayesian algorithm, which can effectively help the elderly to understand their health better and find out the abnormal condition as soon as possible. At the same time, it can assist doctors make auxiliary medical decisions and improve the cure rate. This article introduced a complete process of building a knowledge graph, from schema structure design to data acquisition, and processing the data until it fits the standard. Before applying to disease reasoning, we imported knowledge data into the Neo4j graph database to make full use of the inference flexibility and accuracy of the knowledge graph.

Keywords: Knowledge graph · Reasoning algorithm · Geriatric disease · Neo4j

1 Introduction

Aging is a worldwide problem. As the world's most populous country, China is under great pressure on the issue of aging, which seriously affects the development of China's society and economy [3]. According to the National Bureau of Statistics of China [1], by the year 2018, there was 166.58 million population over 65 years in China, accounting for 11.9% of it whole, which is far exceeding the 7% defined by international standards, as shown in Fig. 1. It is inevitable for the older to suffer from illness for the body's aging and resistance decline. The disease characteristics of the older are concealed, sudden, easy to recurrence, and cause complications [5]. Therefore, the key to curing geriatric diseases is to detect the condition in time in the early stage of illness, and pay attention to the occurrence of complications during the treatment period. In China, the direct challenge brought by the dramatic increase in the elderly population is the serious shortage of medical resources. On the one hand, there is a gap between the

© Springer Nature Singapore Pte Ltd. 2019
H. Ning (Ed.): CyberDI 2019/CyberLife 2019, CCIS 1138, pp. 452–465, 2019.
https://doi.org/10.1007/978-981-15-1925-3_33

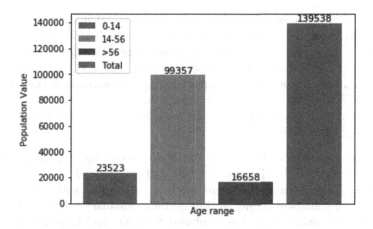

Fig. 1. The statistics of Chinese age level in 2018 [1]

public institutions and the private in terms of medical security, public hospitals cost low fee but difficult to make an appointment, while private hospitals and clinics have higher medical levels and good services, but the fees are unacceptable to ordinary people. On the other hand, medical disparities between different levels of development still exist. The medical facilities in major urban centers on the East Coast are close to the level of developed countries, while the medical level in the central and western regions needs to be improved [16].

In the context of the rapid development of Internet technology, smart healthcare is a way to combine modern technology and medical services with big data and artificial intelligence technology to achieve self-help, scientific and data-based medical care, which provides an effective way to solve shortage of elderly medical resources. Smart healthcare combines every aspects of medical services, achieving timely detection of the disease, improving the rates of treatment and the recovery. Maranesi et al. [7] proposed to use gait detection to identify elderly people with Parkinson's disease. Urŏsević et al. [13] built a developed interactive environment for geriatric risk assessment to assist geriatricians and physicians in treating. Mohammed et al. [8] developed a Mobile device based smart medication reminder program to keep elderly people to take the correct dose of the drug on time. In view of the lack of medical resources in China, this paper proposes a model of geriatric disease reasoning based on knowledge graph (KG), which can play role through all the processes of medical treatment. First, elder people can import their historical medical case of illness into the KG model to have a clear understanding of their health, then, keep their health consciously. Secondly, in the course of treatment, the KG model can be combined with the patient's symptoms and historical cases to reason out the possible conditions of the disease and complications, to assist the doctor in making the disease decision.

Some related work based on knowledge graph about medical has been proposed in recent years. Jiang et al. [4] proposed a data-driven methodology for creating a medical knowledge network (MKN), which combined Markov logic network (MLN) for medical diagnosis. Ruan et al. [11] designed a novel tool (QAnalysisto) to extract the structure of electronic medical record (EMR) data which can be used to built KG. It help doctors to get statistical results directly. In the work of Xiong et al. [15], the process of building the KG of Chinese patent medicine has been introduced in detail, including data extraction, corresponding Atlas mode, knowledge alignment, knowledge fusion.

Compared with the above work, the main contribution of our paper lies in the following three aspects:

1. Completed process of Chinese geriatric KG bases construction, including schema structure, knowledge acquisition and integration.
2. Personal medical record evaluation model based on KG, which not only help doctors get statistical results directly but also auxiliary medical decision.
3. Geriatric reasoning algorithm combined with KG and Bayesian algorithm, which is a data-knowledge driven method.

The rest of the paper is organized as follows: Sect. 2 discusses related research. In Sect. 3, we describe geriatric KGs building processes, and assessment models and algorithms. Section 4 conducts model evaluation and discussion. Finally, Sect. 5 consists of the conclusion and future work.

2   Related Research

2.1   Introduction of Knowledge Graph

KG was first proposed by Google [2], to mainly optimize existing search engine performance, and promote keyword-based search to relationship-based related search. KG is a graph model composed of nodes and edges, where nodes represent entities in the real world, and edges represent relationships between entities. The essence of KG is the knowledge-based semantic web, it pays more attention to discovering the connections between entities and hiding knowledge compared with the semantic web. To ensure the rigor of knowledge, ontology technology is used to construct the data schemas of KG with the standard of entities, concepts, attributes, and relationships [9].

In this paper, we follow the RDF (Resource Deion Framework) standard and construct geriatric diseases knowledge in a binary relationship, as the (subject, predicate, object) (SPO) triple, where the subject and object are entities, and the predicate is the relationship between them. Based on the above criteria, we can extract the fact, "Zhang San suffers from heart disease, he often feels angina, and the doctor Li Si recommends him to take aspirin when he is treated in the cardio-cerebral vascular department", as shown in Table 1.

2.2   Reasoning Algorithm Based on Knowledge Graph

In this paper, the knowledge graph-based reasoning algorithm we discuss is actually based on the triple prediction completion or knowledge completion [6] under

Table 1. Extracted SPO triple from the fact

Subject	Predicat	Object
Zhang San	Feeling	Angina
Heart Disease	Symptoms	Angina
Zhang San	Suffering from	Heart Disease
Heart Disease	Belong to	Cardio-cerebral Vascular Department
Li Si	Curing	Zhang San
Aspirin	Treatment	Heart Disease

the open world assumption (OWA), meaning a non-existing RDF triple is interpreted as unknown instead of false relationship [9]. As shown in the Fig. 2, the missing edge does not mean that this patient is absolutely not suffering from this disease, for diseases has have local interactions.

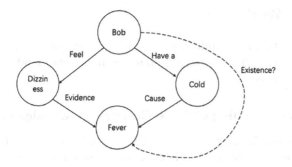

Fig. 2. The missing relationship model

This example of disease KG model is called homogeneity, a trend in which an entity is associated with other entities with similar characteristics. In addition, graph paths can also exhibit local dependencies, they can span ternary chains and involve different types of dependencies [17]. Based on the characteristics of geriatric diseases and the homogeneity rules of knowledge KG (such as transitivity and type constraints), and inspired by the naïve Bayesian algorithm, we have proposed three rules for geriatric reasoning:

1. The disease that the patient has ever suffered may have recurred.
2. The patient's concentrated diseases in a certain part of the body will lead to other complications.
3. The more the patient's symptoms correspond to a disease, the more likely it is to develop the disease.

The reasoning algorithm proposed in this paper is to maximize the possibility of disease between patients and diseases under known conditions in the KG. The detailed part will be introduced in Sect. 3.

2.3 Neo4j Gpaph Database

Neo4j is currently the most popular graphics database. It is a special type of the NoSQL (not only SQL) database, which is used to store large relational data, supporting complete database transactions, RDF standards and various programming languages such as Java, Python and C++, etc [14]. The graph constructed by Neo4j is a directed graph with nodes and edges instead of row and column structure from traditional database. The nodes and edges can be set to one or multiple properties. In this paper, the geriatric disease data is stored in Neo4j, each patient, disease, department, symptom, treatment method, etc. are stored as single nodes, and the node is expanded into a vast KG network based on various relationships. The adjacency list structure of Neo4j ensures that database performance won't decrease as the increased amount of data, and its unique Cypher language makes it easy and fast to traverse the entire KG and retrieve the target nodes [12]. All of the above features make Neo4j as an excellent tool for prototyping.

3 Specific Work

In this section, we will introduce our work in detail, including the process of building Chinese geriatric KG bases, patient medical record models, and geriatric reasoning algorithms.

3.1 Process of Building Chinese Geriatric Knowledge Graph Bases

Building Model. As shown in the Fig. 3, the building model of KG is mainly divided into the following parts. Schema is an important guarantee for the scientific and completeness of KG. Obtaining perfect and accurate data is the premise of comprehensiveness of KG. Reasonable processing of data is the requirements of high accuracy for KG. We use a top-down method to build a KG model based on a pre-set schema structure, and then import data that includes high-quality knowledge information from encyclopedic website.

Schema Structure. In scheme, there are six types of entity concepts and seven types of entity relationship concepts, as shown in Table 2 and Fig. 4.

- "Same" refers to the alias relationship between disease and disease, such as the relationship between Alzheimer's disease and Parkinson's disease.
- "Cause" refers to the concomitant relationship between disease and disease, such as high blood pressure may cause a stroke.
- "Describe" represents some symptoms used to describe the disease, such as difficulty breathing may be a sign of cerebral obstruction.
- "Occur in" represents symptoms are reflected of parts of the body, such as chest congestion may occur in the chest.
- "Suffer from" indicates which part of the human body the disease is in, such as eyes may have cataracts.

- "Treat" indicates which method the disease can be used to treat, such as surgical treatment can treat acute appendicitis.
- "Belong to" means the disease belongs to a certain department, such as cirrhosis needs to be treated in hepatobiliary department.

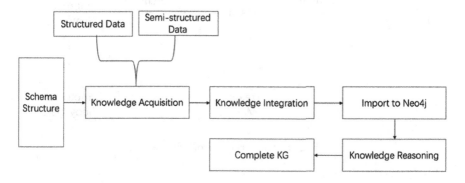

Fig. 3. The building model of KG

Table 2. The concepts of entity and their relationships in schema

Entity concepts	Entity relationship concepts
Disease	Same
Performance symptom	Cause
Treatment department	Describe
Treatment method	Occur in
Symptomatic body-part	Suffer from
Diseased Body-part	Treat
	Belong to

Although "Symptomatic Body-part" and "Diseased Body-part" in the entity concepts have some repetitiveness, such as eyes, chest, stomach and other parts, if only one entity is used to describe the location and symptom of the disease at the same time, the accuracy of KG cannot be reflected.

Knowledge Acquisition and Integration. The geriatric knowledge is obtained by two ways in this paper, supplementary data from structured data (Chinese medical KG from OpenKG.CN [10]) and main data from semi-structured data (data in html format from Chinese medical website). The Java-based Jena package is used to parse the structured data, and the Python-based RE module and the Beautifulsoup4 module to parse the semi-structured data

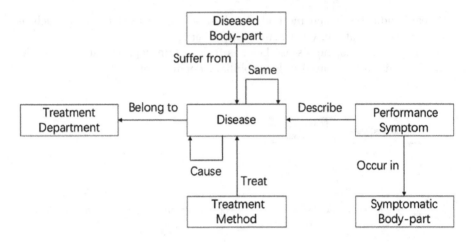

Fig. 4. The structure of schema

in the html format. By now, the task of knowledge acquisition has been completed. The knowledge about the entities and their relationships mainly comes from semi-structured data. However, the semi-structured data coms from different institutions and contains a lot of redundancy and error information, and the relationship between entities may be vague, lacking hierarchy and logic. Therefore, it is necessary to integrate knowledge from different sources and formats, including cleaning and alignment.

In order to clean the data, the error data caused by network anomalies and format errors is removed or fixed by program and manual review, and the remaining 15 types of entities are displayed in the Table 3.

Table 3. Remaining entity types

Disease	Alias	Infectiousness
Medical insurance	Diseased body-part	Treatment method
Treatment department	Infectiousness	Cure rate
Treatment cycle	Infected object	Treatment cost
Performance symptom	Clinical examination	complication

According to the schema model introduced above, we extract "Disease", "Performance Symptom", "Treatment Department", "Treatment Method", "Diseased Body-part" as our target entities, and transform the two entities "Alias" and "Complication", into the entity "Disease" and two relationships "Same" and "Cause". The next task is to unify the description of the symptoms, meaning entity alignment, to ensure the high KG efficiency. There are a large number of repeated expressions in "Performance Symptom" due to different medical institutions describe the symptoms differently. For example, symptoms of cataract includes "hearing loss", "inaudible", and "hearing disorder", which is a same express. Based on the Chinese word segmentation tool, we divide the entities with the similarity of characters into a group, and manually covert the original data to a standard entity attribute.

The disease has an abstract mapping for "Diseased Body-part" (such as osteoporosis in the elderly is a bone and systemic disease), but the performance symptom doesn't. The description part of the symptoms is necessary to describe the symptoms of the disease more easily for the user. And the creation of the entity "Symptomatic Body-part" is a difficult point in the construction of the geriatric disease KG. Based on relevant medical information and "Diseased Body-part", 20 types of "Diseased Body-part" are set up to map the body organs and mental, including the brain, eyes, nose, ears, mouth, face, neck, heart, chest, abdomen, lower body, lower back, limbs, blood vessels, skin, muscles, bones, emotions, state, physiological parts, involving symptoms such as nausea, fatigue, night sweats, etc.

Knowledge Graph Import and Display. The geriatric KG model has been built, and the next step is using Cypher to import KG bases to the Neo4j database through CSV file for reasoning and visualization, part of the KG bases are shown in the Fig. 5.

Fig. 5. The partial KG with different entities and relationships

3.2   Patient Electronic Medical Record Based on Neo4j

EMR is an electronic patient record based on a specific system that provides users with access to complete and accurate data, alerts, alerts, and clinical decision support systems. In this paper, we propose an EMR based on Neo4j, which makes full use of Neo4j's data storage features. By creating a new entity concept "Patient" and a new relationship "Once Suffered" between "Patient" and "Disease" to achieve the function of recording patient-related information, as shown in Table 4. Patients can access their EMRs by their username and password, and easily view their historical cases and related information through a visual interface from Neo4j. Doctors can use the super account to find patients in the EMR database based on the combination of various records, or to gather patients with similar parameters to analyze their historical cases and treatments. In conclusion, Neo4j-based EMR can help patients better understand their disease history, help doctors reduce workload and making decisions.

Table 4. Overview of EMR based on Neo4j

Record	Description	Storage location
Username (Unique)	Log in to the account and identify the patient	Properties of Node "Patient"
Password	Protect account	
Name	Patient information	
Sex		
Age		
Telephone		
Address		
Character		
Hobby		
Allergies		
Date	Historical disease information	Properties of Edge "Once Suffered"
Symptom		
Treatment hospital		
Therapist		
Treatment method		
Treatment cycle		
Cost		

3.3   Reasoning Algorithm

In the reasoning algorithm, the three rules proposed in Sect. 2.2 are combined with the Naive Bayesian algorithm to perform geriatric reasoning in the KG. First, based on the knowledge-driven model, the disease parameters from the patients will be sent into the KG, and the graph traversal algorithm will be performed to obtain the relevant connected entities according to the three inference rules summarized, as shown in Algorithm 1. The three algorithmic inference results will be returned as three disease sets $\{X, Y, Z\}$. According to the thinking of Naive Bayesian, the three inference results are independently of each other, and set the weight of X to 2, the weight of Y to 1.5, and the weight of Z to 1 according to medical knowledge. Then, the calculation formula of the geriatric reasoning algorithm can be given:

Algorithm 1. Reasoning Algorithm Based on KG

In: $S = \{S_1, S_2...S_s\}$ ← Patient described symptoms in "Performance Symptom"
   $D = \{D_1, D_2...D_d\}$ ← Patient's historical diseases in "Disease"
Out: $X = \{X_1, X_2...X_x\}$ ← "Diseases" reasoned based on Rule 1
   $C = \{C_1, C_2...C_c\}$ ← "Diseased Body-part" that needs care based on Rule 2
   $Y = \{Y_1, Y_2...Y_y\}$ ← "Diseases" reasoned based on Rule 2
   $Z = \{Z_1, Z_2...Z_z\}$ ← "Diseases" reasoned based on Rule 3

function RULE 1(S, D)
   for $i = 0 \rightarrow s - 1$ **do**
      for $j = 0 \rightarrow d - 1$ **do**
         if S_i connected with D_j **then**
            X appends Dj
   return X

function RULE 2_1(D)
   for $i = 0 \rightarrow d - 2$ **do**
      for $j = i + 1 \rightarrow d - 1$ **do**
         if D_i and D_j connected to the same "Diseased Body-part" D_B **then**
            C appends D_B
   return C

function RULE 2_2(S, C)
   for $i = 0 \rightarrow s - 1$ **do**
      for $j = 0 \rightarrow c - 1$ **do**
         if S_i and C_j connected to the same "Diseased" B **then**
            Y appends B
   return Y

function RULE 3(S)
   for $i = 0 \rightarrow s - 2$ **do**
      for $j = i + 1 \rightarrow s - 1$ **do**
         if S_i and S_j connected to the same "Diseased" B **then**
            Z appends Dj
   return Z

$$P(W|S, D) = \frac{Weight_W}{\sum_{K \in \{X,Y,Z\}} Number_K * Weight_K} \quad \text{for } W \text{ in } \{X, Y, Z\}. \quad (1)$$

4   Model Evaluation and Simulation Experiment

To ensure that it's easy and convenient for patients and doctors to use our reasoning models, the model provides two major functions, including disease search and conditional reasoning. Disease search supports clue search, fuzzy search and comprehensive search, which will return all information about the disease, including "name", "department", "diseased part", etc, and disease reasoning mainly includes two aspects, EMR management and disease reasoning.

Clue Search. Return the corresponding disease research results by selecting the given research option such as in the "Treatment Department", "Diseased Body-part", "Performance Symptom", etc, Table 5 shows the experiment of clue search, which searches about diseases that belong to both the "heart" and "cardiothoracic surgery" department and the exhibit symptoms is "right heart failure".

Table 5. The clue search experiment

Parameter	Result
Disease name	Cardiac infarction
Disease alias	Acute myocardial infarction, Myocardial infarction, Myocardial infarction
Department	Department of Cardiology, Cardiothoracic Surgery
Diseased part	Muscle, Heart
Related symptoms	Chest pain, Nausea, Vomiting, Palpitations, Retrosternal pain, Left heart failure, Right heart failure, Difficulty breathing
Treatment	Physical therapy, Medication
Complications	Pneumonia, Pleurisy, Shock, Sudden death, Pericarditis, Arrhythmia

Fuzzy Search. Match the disease symptom or disease name with the entered keyword and return the disease.

Comprehensive Search. Combine the above two query modes, and the search accuracy is increased while the search range is relatively reduced, Table 6 summarizes the experiment of three search methods.

EMR Management. The first use of reasoning function need to register and fill in relevant information, including username (unique), password, name, sex, age, Historical diseases, address, etc.

Table 6. The introduction of search function

Search Function	Input	Output
Clue	Heart from "Disease"	Myocardial infarction
	Cardiothoracic Surgery from "Treatment Department"	
	Right heart failure from "Performance Symptom"	
Fuzzy	Thyroid	Hypothyroidism
		Hyperthyroidism
		Hyperthyroidism crisis
Comprehensive	Thyroid	Hyperthyroidism
	Eye protrusion "Performance Symptom"	

```
The conditional reasoning list as follows
Probability              Disease
---Recurrence related diseases and probability---
0.032                    Sepsis
0.143                    Hypertension
---Location related diseases and probability---
0.111                    Arteriosclerosis
0.048                    Aortic dissection
0.048                    Hypotension
0.048                    Diabetes heart disease
0.048                    Stroke
0.048                    High blood fat
0.143                    Hypertension
0.0048                   Atrial fibrillation
0.048                    Hyperthyroid heart disease
0.111                    Atrial flutter
0.048                    Cardiac conduction obstruction
0.111                    Myocardial infarction
0.048                    Pericarditis
--- Symptom associated diseases and probabilities ---
0.111                    Arteriosclerosis
0.063                    Vertigo
0.143                    Hypertension
0.111                    Atrial flutter
0.111                    Myocardial infarction
Health Tips: Please protect the health of your heart, blood and blood vessels!
```

Fig. 6. The results of geriatric reasoning based on KG

Disease Reasoning. After establishing EMR, the patients can enter their own symptoms to the system and find which diseases they are suffering from. The KG system will calculate the disease probability according to the user's past medical history and the symptoms combined with the reasoning algorithm, and achieves the result of the patient-specific reasoning, Fig. 6 shows a complete reasoning simulation experiment, the patient used to suffer from hypertension, stroke, coronary heart disease, and shows three symptoms of dizziness, vomiting and palpitations.

5   Conclusion

With the advent of the era of big data, how to mine the association between data is particularly important. The rise of the KG has enabled the computer field to found a new way or idea to process the acquired knowledge and standardize it to serve humans in combination with big data. Based on the perspective of smart older care and smart medical care, this paper constructs geriatric disease knowledge graphs and proposes three rules of reasoning. By the reasoning model and EMR model based on KG, It can be effective to provide self-service health check for the elderly, and can detect the condition in time to improve the cure rate. At the same time, help doctors to provide medical decision-making and medical record management, which can effectively improve treatment efficiency, and alleviate the problem of imbalance of medical resources.

The next work is to intelligently and automatically extract entities and relationships from irregular stylistic data based on machine learning and natural language processing techniques. In addition, using deep learning and knowledge embedding technology can greatly improve the accuracy of knowledge reasoning and the compatibility with big data.

References

1. National Bureau of Statistics of China: The 2018 Population Age Structure of China. http://data.stats.gov.cn/easyquery.htm?cn=C01&zb=A0301&sj=2018
2. Google: Introducing the Knowledge Graph: Things, Not Strings. https://googleblog.blogspot.com/2012/05/introducing-knowledge-graph-things-not.html
3. Hong, L.: The influence of aging population on China's economy in the information society. In: 2010 2nd IEEE International Conference on Information Management and Engineering, pp. 264–267 (2010). https://doi.org/10.1109/ICIME.2010.5478053
4. Jiang, J., Li, X., Zhao, C., Guan, Y., Yu, Q.: Learning and inference in knowledge-based probabilistic model for medical diagnosis. Knowl.-Based Syst. **138**, 58–68 (2017)
5. Kabboord, A.D., Van Eijk, M., Buijck, B.I., Koopmans, R.T., van Balen, R., Achterberg, W.P.: Comorbidity and intercurrent diseases in geriatric stroke rehabilitation: a multicentre observational study in skilled nursing facilities. Eur. Geriatr. Med. **9**(3), 347–353 (2018)
6. Lin, Y., Liu, Z., Sun, M., Liu, Y., Zhu, X.: Learning entity and relation embeddings for knowledge graph completion. In: Twenty-Ninth AAAI Conference on Artificial Intelligence (2015)
7. Maranesi, E., et al.: A stereophotogrammetric-based method to assess spatio-temporal gait parameters on healthy and Parkinsonian subjects. In: 2015 37th Annual International Conference of the IEEE Engineering in Medicine and Biology Society (EMBC), pp. 5501–5504 (2015). https://doi.org/10.1109/EMBC.2015.7319637
8. Mohammed, H.B.M., Ibrahim, D., Cavus, N.: Mobile device based smart medicationreminder for older people with disabilities. Qual. Quant. **52**(2), 1329–1342 (2018). https://doi.org/10.1007/s11135-018-0707-8

9. Nickel, M., Murphy, K., Tresp, V., Gabrilovich, E.: A review of relational machine learning for knowledge graphs. Proc. IEEE **104**(1), 11–33 (2016). https://doi.org/10.1109/JPROC.2015.2483592

10. OpenKG.CN: Chinese symptom library. http://openkg.cn/dataset/symptom-in-chinese

11. Ruan, T., Huang, Y., Liu, X., Xia, Y., Gao, J.: QAnalysis: a question-answer driven analytic tool on knowledge graphs for leveraging electronic medical records for clinical research. BMC Med. Inform. Decis. Making **19**(1), 82 (2019)

12. Stark, B., Knahl, C., Aydin, M., Samarah, M., Elish, K.O.: Betterchoice: a migraine drug recommendation system based on neo4j. In: 2017 2nd IEEE International Conference on Computational Intelligence and Applications (ICCIA), pp. 382–386. IEEE (2017)

13. Urošević, V., Paolini, P., Tatsiopoulos, C.: Configurable interactive environment for hybrid knowledge- and data-driven geriatric risk assessment. In: 2017 25th International Conference on Software, Telecommunications and Computer Networks (SoftCOM), pp. 1–7 (2017). https://doi.org/10.23919/SOFTCOM.2017.8115520

14. Vukotic, A., Watt, N., Abedrabbo, T., Fox, D., Partner, J.: Neo4j in Action. Manning Publications Co. (2014)

15. Xiong, W., Zeng, Z., Xie, Y., Nie, B., Zhou, X.: Study on taboo knowledge map of Chinese patent medicine compatibility. In: AIP Conference Proceedings, p. 020052. AIP Publishing (2019)

16. Yu, P., Liu, X., Wang, J.: Geriatric medicine in China: the past, present, and future. Aging Med. **1**(1), 46–49 (2018)

17. Zhou, J., Cui, G., Zhang, Z., Yang, C., Liu, Z., Sun, M.: Graph neural networks: a review of methods and applications. arXiv preprint arXiv:1812.08434 (2018)

Image Analysis Based System for Assessing Malaria

Kyle Manning[1], Xiaojun Zhai[1] ⓘ, and Wangyang Yu[2](✉)

[1] School of Computer Science and Electronic Engineering,
University of Essex, Colchester CO4 3SQ, UK
[2] Key Laboratory of Modern Teaching Technology, Ministry of Education,
School of Computer Science, Shaanxi Normal University, Xi'an, China
ywy191@snnu.edu.cn

Abstract. Malaria, a not only widespread but also a potentially fatal disease that can be found mainly in tropical regions of the world, with the World Health Organization reporting an estimated 219 million cases worldwide as of 2017 of which 435,000 were mortal. Diagnosis currently involves taking a blood sample from a patient who is presumed to be infected, which is examined under a microscope by trained experts, although reliable the process is tedious. This disease is an ever-increasing problem thereby creating a need for an automated solution to the diagnosis of malaria. The primary objective of this project is to design a tool that can diagnose malaria from an image of a blood smear that has been stained with the commonly used Giemsa stain (which highlights the parasites in a red blood cell by turning them dark purple). In this paper, we have developed a graphical user interface to assist with the separation of red blood cells and extraction of the cells infected with the malaria parasite as well as an ANN (Artificial Neural Network) for cell classification. The graphical user interface allows the user to analyse the blood sample by running a series of image processing techniques followed by the extraction of infected cells, the results have shown that these techniques could be potentially used to detect malaria. Currently, the achieved results shown that the proposed system has 92% accuracy of a database contains a large number of ground-truth images.

Keywords: Malaria · Image analysis · Artificial intelligence

1 Introduction

Malaria is a global disease with its existence potentially dating back to sixth century BC [1] after being studied by many civilisations including the likes of Ancient Egypt and Ancient Greece. Although the disease has been eliminated in certain climates and progress has been made to eradicate malaria it still kills millions of people every year. Advancements in the healthcare industry means that it is both preventable and curable [2] provided that a person suspected to have the disease is diagnosed and treated as soon as possible. The current gold standard is microscopic analysis which is a manual process and Neural Network (NN) have been incorporated to aid this step within diagnosis.

© Springer Nature Singapore Pte Ltd. 2019
H. Ning (Ed.): CyberDI 2019/CyberLife 2019, CCIS 1138, pp. 466–486, 2019.
https://doi.org/10.1007/978-981-15-1925-3_34

However, this counts towards one of the underlying problems and it costs time for all parties involved as well as performance for the NN when using a complex architecture.

As mentioned earlier the problem with the manual process is that it takes long to complete and requires highly trained professionals and good equipment to complete the task accurately. Aside from the time, the accuracy of this process will also be determined by the persons current state, including tiredness, nausea, headaches and any other conditions that could impact the practitioner's ability to correctly diagnose a patient.

The second challenge is usability, it's not currently possibly for all practitioners to do this task because it requires additional training, using a Graphical User Interface (GUI) would improve the ease of use as it wouldn't require someone to know how to code or be trained how to undertake microscopic analysis. In addition, a NN would decrease the time taken as well as the number of patients that can be diagnosed within a period of time. However, although performance is improved it must be noted that the complexity of the NN architecture will determine the networks runtime, this challenge will be overcome through using a basic Feed-Forward Neural Network (FFNN) called *patternnet*.

In this paper, we study two common ways of automated malaria diagnosis, the first is through Machine Learning (ML) using an FFNN a subset of Artificial Neural Networks (ANN) and the second is a separate process to segment and extract features from images of a larger scale. These are a couple of the many approaches that have been taken by malariologists; others include Rapid Diagnostic Tests (RDT), Serology and Simian Malaria Species Confirmation Service (SMSCS) for testing the additional non-classic species. The purpose is to improve the usability of additional tools such as NN to allow more medical practitioners to diagnose patients faster. Malaria detection in any form is an open-ended process because there are multiple ways to go about it by using various tools. In terms of automation, there isn't a set algorithm that works well enough to be suited for every situation due to different factors such as the stages of malaria, stains used in blood smears, the viscosity of blood smears, image quality and colour to name a few.

This paper aims to demonstrate that there are ways to improve malaria diagnosis by using automation while taking as many challenges as possible into consideration as the issues impact the quality of the diagnosis and the level of trust patients have in medical practitioners presenting an accurate result. Segmenting Red Blood Cells (RBC) in images of microscopic slides known as blood smears by developing algorithms and then classifying the cells as either parasitized or uninfected. Automating the process of diagnosis would drastically shorten the amount of time it would take to obtain a test result. Thereby allowing medical practitioners to conduct more tests and find out how severe a patient's condition is by viewing how many infected cells they have. The major contributions of this work are highlighted as follows:

- The improved segmentation and NN classification algorithm would significantly improve the performance of diagnosis of the malaria and would overcome the potential human inaccuracies which the program would not face.
- A GUI has been created that aids in the detection of malaria parasites. As trained professionals typically undertake a manual examination, a user interface is to be developed to cater to those such as medical practitioners that would otherwise send off the blood samples for testing.

The rest of the paper is organized as follows. The next section provides a brief discussion on important related works conducted. This is followed by a discussion on the system architecture used in this work in Sect. 3. In Sect. 4, we present analysis of experimental result in detail. The article finally concludes in Sect. 5.

2 Literature Review

2.1 Malaria

Malaria caused by the protozoan pathogen known as Plasmodium, is not able to survive outside of their host. In total, in humans five main species of malaria can be found, these are *P. vivax, P. falciparum, P. ovale, P. malariae* with the addition of *P. knowlesi* which commonly infects animals but has been seen in humans in recent years although rare [3, 4]. According to the NHS [4], *P. falciparum* predominately found in Africa is the deadliest species of malaria, being accountable for most deaths worldwide. *P. vivax* found in Asia and South America shows less severe symptoms but is also harder to detect and can remain dormant in the liver for up to three years [5], this may then lead to relapses. *P. ovale* mostly found in West Africa is similar to *P. vivax* as it can also exist in the liver for several years without any signs. Both *P. vivax* and *P. ovale* can be classed as asymptomatic [6] because the patient infected by the parasite doesn't display any symptoms; however, it can relapse. Last but not least is, *P. malariae* representing only a small portion of infections in Kenya as of 2014 to 2015 [7], because *P. malariae* is harder to detect.

A genus (which is described by Maximum Yield [8] as a biological classification of living organisms) of mosquitoes known as Anopheles spreads the parasite in females because the females feed on blood to produce eggs. Only Anopheles mosquitoes also called "night-biting" mosquitoes because they bite between dusk and dawn [4], can transmit malaria as this is the most common method. It is possible however to become infected through other means. The process starts when a mosquito bites a malaria-infected person, upon biting the person the mosquito ingests a small amount of blood that contains the parasite [2] when it next feeds the parasites in the mosquito mix with its saliva and are then injected directly into the person bitten. Sharing needles and the likes which contain blood, blood transfusions and organ transplants can also spread the disease, but this is rare.

2.2 Image Processing

The images need to be analysed in order to extract information such as features and characteristics or be enhanced to be able to be passed onto other actions, this is performed by algorithms. It is standard to extract and remove information from an image however it is never acceptable to add to an image, because removing information can enhance the vital information and make those features more accessible [9].

In digital image processing, the idea is to change an image so that it meets the desired output. Nonetheless, a problem with this is the final output is dependent on the steps taken to obtain the result. It is highly likely that different computers, software and hardware alter images in ways not wanted by the user which is an issue that is neither discussed in great detail or at all. Medical professionals will rely on this to produce the same output regardless of computer specification.

If the resulting images were sent through an email client or added to a word report from copy and pasting, the image resolution and potentially the image colour can change. As a result, image quality is reduced [9]. Depending on where the images are to be used the file format must be considered such as the difference between JPEG or PNG in terms of both input and output image. JPEGs use lossy compression, whereas PNGs use lossless; this means that PNG images retain more information at the expense of file size.

The process of segmentation is to divide an image into smaller subsections (set of pixels) [10], the level of depth in which the segmentation continues is dependent on the application. The full image is the blood smear, and the segmentation is to stop once it reaches the region of interest, thereby making the smaller subsections the individual cells. In turn, the classification can work on a focused area of the full image and should produce better results. Attempting to classify cells on a full image could result in misclassifications because every blood smear and cell is different. Solely using a full image would either require more pre-processing or a different approach entirely such as fine-tuned Convolutional Neural Network (CNN), this is to ensure that what is inside the cell determines the classification.

The most common techniques used are ML, Mathematical Morphology (MM), Watershed and Thresholding, however there are many others which makes choosing the best one the hardest stage in image processing. Thresholding works on a pixel by pixel basis in which every pixel's intensity in an image is compared to the threshold value, it is assigned one of two classes: background or foreground based on whether the intensity is higher or lower than the threshold. Watershed is a form of region-based segmentation that separates the image into catchment basins, a location in a lower region which collects water as a single water body from a higher region as it falls as mentioned by [11] and these are the objects to be identified by the program. MM contains a set of operations which are used in conjunction with other methods to enhance an image based on shapes. The purpose is to remove irrelevant artifacts in an image using a set of transformations; the most common methods are dilation and erosion which are normally used together.

ML is field within Artificial Intelligence (AI) which aims to learn patterns in data to classify them into different categories [12]. There are two main classes of ML, Unsupervised Learning (UL) and Supervised Learning (SL). UL states that a collection of data is analysed by the algorithm without any prior knowledge of the data and therefore lets the algorithm decide how to sort the data. SL is the process of taking input data and the desired target as output and using an algorithm to learn the best function to map the input to the output.

2.3 Discussion

There is a plethora of different techniques that can be used to segment images; however, most of them have benefits, and drawbacks so it comes down to choosing the right one to be adapted. Table 1 discusses the main problems with today's state of the art algorithms currently used in computer vision for image segmentation.

Table 1. Comparison of segmentation techniques

Ref.	Algorithm	Pros	Cons
[13–16]	Watershed	• Fast • Simple • Ease of use • Better handling of gaps • Better boundary placement with high contrast	• Suffers from over-segmentation • Noise sensitivity • Poor boundary placement of important regions with low contrast • Poor detection of thin objects
[17–22]	K-Means	• Efficient • Scalable for larger datasets • Unsupervised learning • Low computational costs	• "K" is difficult to estimate • Accuracy is dependent on the initial "K" value • Doesn't guarantee continuous areas
[10, 17, 21, 22]	Thresholding	• Improved region detection with images of high contrast • Simple calculation	• Sensitive to noise because of greyscale overlapping • Intensity homogeneity
[10, 19, 20, 22]	Fuzzy C-Means	• Generates good results • Efficient • Always converges • Unsupervised learning	• Takes long to compute • Sensitive to initial cluster centre • Sensitive to noise
[19, 22]	Neural networks	• Automatically learn new patterns and features • Scalable • Can be efficient based on implementation • Achieves the state-of-the-art results • Work well on large datasets • Robust	• Complex implementations take longer to run and require large training data • Can suffer from overtraining/overfitting • Underfitting where it can't learn the training data and therefore can't adapt to new data

Based on the findings in Table 1 it is reasonable to determine which segmentation technique to use for each application, if the intention is to use the algorithm on a large dataset using a NN might prove to be more accurate as it can work on more complex patterns. Additionally, if the intention is to improve an existing method to use it for more application, it can be seen that modifying the threshold technique could prove beneficial if also used in conjunction with another method such as the Watershed or KM. Further modifications can be made to accommodate both NN and other segmentation techniques; this is useful when segmenting images first and then using a NN to classify the image.

3 System Architecture

In order to extract the infected cells, the blood smears first have to be separated through the use of segmentation; the base algorithm used in this paper is the Watershed technique. However, this technique alone is not sufficient enough to produce accurate results which is why it has been adapted to accommodate its shortcomings discussed previously. Similarly, for classification, a combination of an FFNN and a separate function for the blood smears should aid the identification of benign and infected cells. The basic FFNN does not work well when the input is the entire image; therefore, features detected should be extracted first and then applied to the network. The pseudo code for the initial segmentation is discussed in later sections.

3.1 Initial Segmentation

The original watershed technique (see Algorithm 1) usually results in over-segmentation if the image is not preprocessed first. Lines 1–3 involve reading the image into the program, converting it to greyscale and finally has Watershed called on it. The image displayed over-segmentation, regardless of conversion, including binary, inversion and greyscale.

Algorithm 1 Original Watershed Segmentation

Input: *im* = image selected by user
Output: segmented image

---

read *im* into program
convert image from RGB to grayscale
call: watershed

The adapted method (see Algorithm 2) improves this; however, it fails to identify lighter cells and highlights cells on the edge of the image, but these are cut-off so the full cell cannot be seen. Lines 1–3 involve reading, converting and inverting the image, this was so that the regions of interest are in white and the background in black. Lines 4–8 run several operations on the image to make the objects more distinct, the gradient magnitude shows how the grey levels change in image by default the function uses Sobel (an operator used for edge detection). A set of MM calls namely *imopen, imerode* and *imreconstruct* which removes objects smaller than a set number of pixels, removes small holes and identify the areas of the image with high intensity respectively. Lines 6 and 7 convert the image to binary using a threshold then calculates the distance between each "0" and non-zero pixel.

Algorithm 2 Marker based Watershed Segmentation
Input: **im** = image selected user Output: **stats** = the bounding area of the extracted cells
read *im* into program convert image from RGB to grayscale invert image compute gradient magnitude **call:** mathematical morphology operations convert image to binary compute distance transform **call:** watershed convert matrix of labels to RGB image *stats* = calculate coordinates and bounding regions for objects (cells) in image show RGB image with labels layered over **for** every bounding region in *stats, kk* from *1* to *height(stats)* draw bounding region as rectangle add cell count next to rectangle **end for** **return** *stats*

3.2 Initial Classification

The MATLAB Machine Learning Toolbox is used to create a basic FFNN using *patternnet*, which requires an array of the image elements, but to retrieve better results from the algorithm the input should be features of the image. The tool creates a basic and an advanced script files named *NN_Basic_Script.m* and *NN_Advanced_Script.m* respectively. Both files have been altered to change the ratio of training, testing and validation images and the size of the hidden layer.

Initially, the entire image was passed to the NN, which required the images to be reshaped from 2D to 1D, as the photos were also larger, they were resized to 64×64 which is why the total size of the image is 4096. The preprocessing stage was updated and replaced the 1D image matrix with features of the image as this generated a better result. The updated process can be seen in Sect. 3.4.

3.3 Image Pre-processing

For the algorithms to run successfully on the images they first need to be preprocessed to format that is uniform across different kinds of images, this means that after preprocessing all images should look the same regardless of whether the initial image was too bright or too dark. In all cases the images should be converted to greyscale, from there the images can then be adjusted if grey intensity is too high or low to warrant a good output.

3.4 Improved Classification

An extraction method along with a NN is used to classify images, the NN should work on both blood smears and images of single cells, this way the NN can be used for multiple purposes. To use the NN the images needed to be preprocessed first, this includes creating a location for where the images are to be saved once they have been preprocessed. The FFNN used consists of three layers, an input layer with 1 input containing 5 elements, a hidden layer using 10 neurons and an output layer with 2 outputs, which are the classes (see Fig. 1 for a view of the network).

Algorithm 3 Improved NN Classification

Input: malariaInputs = matrix of n by 5 (5 is the features of the images and n is the total number of images),
malariaOutputs = classification targets represented as n by 2 matrix (2 is the amount of classifications possible and n is the number of images presented)
Output: confusion matrix of performance evaluation

create destination location
create datastore for healthy and unhealthy cells
shuffle datastores to randomise data
if user input == null or != digit
set image count to 1000
else
set image count to input
end if
initialMalariaInputs = matrix of all zeros (image count x total features)
for i = 1 to image count **do**
generate output file name
if i <= image count divided by 2
generate input file name from unhealthy datastore
else
generate input file name from healthy datastore
end if
copy input file to destination with output file name
read output image file into program
resize and save output file to 64 x 64
convert output file from RGB to grayscale
adaptively threshold image with high sensitivity
convert image to binary using threshold
call: mathematical morphology operations
convert image from uint8 to double
features = generate array of image features
append *features* to *initialMalariaInputs*
end for
randPerm = generate random permutation of image count
fHalf = generate malaria targets in the form 0 1
sHalf = generate malaria targets in the form 1 0
initialMalariaTargets = vertically concatenate the two matrices *fHalf* and *sHalf*
shuffle *initialMalariaInputs* and *initialMalariaTargets* using *randPerm*
save *initialMalariaInputs* into file
save *initialMalariaTargets* into file

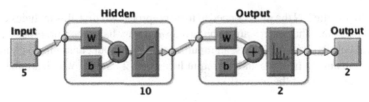

Fig. 1. FFNN architecture.

3.5 Implementation

The overall program flowchart is shown in Fig. 2, and the overall process can be described in the following steps:

The first step is to upload an image of a blood smear, *upload* calls *uploadbutton_Callback* which displays a file browser
Select technique then allows the user to select how they wish to segment the image, it calls *popupmenu1_Callback* which sets the variable *handles.first_menu* to the value from the dropdown
Run calls *techrunbutton_Callback* and that executes the technique, stores the segmented image and table of regions then enables the run button for extensions
Select extension contains the classification techniques, it calls *popupmenu2_Callback* which sets the variable *handles.second_menu* to the value from the dropdown
Run calls *extrarunbutton_Callback* and runs the extraction unless NN is chosen which then validates the user input and generates the required files using the first dataset
Reset calls *resetbutton_Callback* which then calls *Reset_Overlay(handles)*

Segmentation_Watershed_Threshold
This block provided the functions such as *imopen*, *imerode* and all other MM operations, which is used to enhance the processing of images for NNs. Additionally, the statistical information is collected to analyse the data.

NN Block
The FFNN uses MATLAB's Deep Learning Toolbox as it contains all the types of NNs, the one implemented uses the default parameters, the modifiable parameters are listed as follows:

- Training function – "*trainscg*", how the network is trained
- Processing function – "*removeconstantrows*" which removes rows, which have the same, value throughout as it does not provide any extra information to the NN. "*mapminmax*" which maps the matrix values in the range −1 to 1
- Dividing function – "*divider and*", how the input data is split in the network
- Performance function – "*cross entropy*", how the performance of the network is calculated

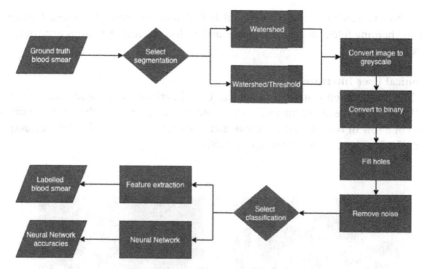

Fig. 2. Program flowchart.

- Goal – "0", how well the NN needs to perform
- Gradient – "$1e^{-6}$", minimum performance gradient which is how well the performance needs to stop improving by before it's stopped
- Epochs – "1000", the number of times the data is shown to the network to be learned
- Max Fail – "6", how many times the validation needs to fail before the NN stops training.

Additional Functions

To aid code generation and general usability additional functions were created, the following functions provide the ability to reset parts of the GUI and debugging for the segmentation techniques.

Reset_Overlay

Focusing on the second image panel using the handles object, this function locates all objects such as the rectangles and text including the image itself and removes them from the panel. The objects are removed so that the segmentation or extraction technique can either be rerun or have a different technique run without having the objects overlap the objects currently displayed. The method resets the image zoom level to default, disables the run button for Select extension because the required image no longer exists, and calls *Reset_Command_Window()* clears the command line.

Reset_Command_Window

In case any function generates command line output this function is called at the start of each function to clear the command line from any output produced.

Find_Rectangles

This function is used for debugging the segmentation techniques, this allowed for the comparison of the labelled images with the image that has been extracted to see how close the bounding boxes of segmented cells are to the bounding boxes in the tagged image. To do this the function uses the unlabelled image to locate the labelled image in the neighbouring folder using the filename and performs a few operations to retrieve the bounding boxes.

Graphical User Interface

The GUI was created using the Graphical User Interface Development Environment (GUIDE). To run the algorithms a GUI is created to aid its users, with two image panels on top of a row of buttons and selections aids the user in selecting the right techniques and methods to be used for the images selected.

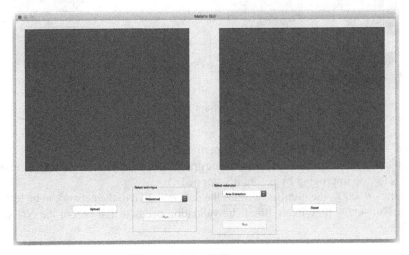

Fig. 3. Program GUI on start-up.

When the program is initially started Fig. 3 illustrates the first state of what the program looks before any user input has been made. The GUI and current algorithms only work on blood smears as the segmentation is used to separate the images into individual images. Where the single cell images are used in a different area of the program, (see Fig. 3).

Fig. 4. Program GUI command panel.

The south-central portion of the GUI contains the command panel (see Fig. 4), which includes all the functionality for the program. This panel is the only part of the program the user will interact with unless editing the code to add additional methods for segmentation and extensions. The program naturally flows from left to right of the command panel, but the user can run sections multiple times such as the segmentation or extraction algorithms, but these are run on the original image again and not on the previous state of the picture. It is possible to add more methods without much hassle by editing the GUI figure, creating a new function for the process and finally adding a few lines of code to the GUI file.

To prevent errors the ability to run specific functions is restricted through simple logic checking which checks if previous steps have been completed first to enable the run buttons. The ability to run any of the segmentation techniques requires an image to be uploaded to the program first, to run the area extraction the image needs to be segmented first, so the run button is only enabled once the segmentation has completed as extraction requires the output.

4 Experimental Results

4.1 Database

To assess the program created, a combination of different images from two datasets built up the database. Having a reliable database would ensure that when testing the program, it would account for the many kinds of images potentially presented. The main two types of images used were (A) individual cells and (B) blood smears, which show multiple cells together, which include both infected and non-infected cells. The two types of images being used are stored in different folders to make it easier to identify which type of image is selected when the program is running. The database is set up with two folders each with their subfolders as follows:

- Cells

  - Parasitized
  - Uninfected

- Blood Smears

  - Labelled
  - Unlabelled

Individual cell images obtained in the first dataset were provided by the National Institutes of Health (NIH) [23] and the blood smears in the second dataset were presented by University College London [24]. Both sets display similar characteristics; some of these can also be seen in similar photos found online. Not all the cells have a 100% circular shape, most of the cells have a rounded appearance but this is not always the case as seen in Figs. 5 and 6. Some common features are:

- Although RBCs they don't all appear as the same colour which is a factor due to the staining and how the images were uploaded to be made digital. The cell colour varies from a light purple to a vibrant pink

- Cell texture varies between each cell, and some appear to be one colour whereas others have distinct shading (Fig. 5)
- Blood smears also aren't always the same luminosity; some images appear very bright while others appear dark
- The entire database is ground-truthed; this is to focus on the regions of interest and to keep the images consistent.

4.2 Cells

The first dataset provided consists of 27,560 images that are only individual cells with a resolution of 72 ppi, a width of 121 pixels and height of 133 pixels resulting in a total of 121 × 133 = 16,093 pixels or 16.1 kilopixels. The ground-truthing of these single cells involved separating the background and foreground to make a clear distinction as to what is the region of interest. In this case, the background is black, and the foreground is the cell, however, in real scenarios, this will not always be the case and the images would need additional processing.

Examples of both infected and uninfected cells are shown below in Figs. 5 and 6 respectively. Organising the dataset into two separate folders made the creation of the NN more efficient as the program is coded in a supervised manner. Of the two different folders:

- 13,780 are parasitized cells
- 13,780 are uninfected cells.

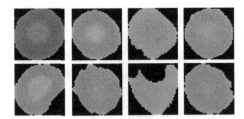

Fig. 5. Uninfected cells. (Color figure online)

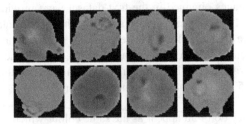

Fig. 6. Infected cells. (Color figure online)

4.3 Blood Smears

The second dataset contains approximately 60 blood smears; these images are common for a blood smear where all blood cells appear, yet to be segmented into separate images ready to be examined. As before, the pictures of the blood smears have also been ground-truthed, so every image has a resolution of 150 ppi, a width of 1300 pixels and height of 1030 pixels with a total of $1300 \times 1030 = 1,339,000$ pixels or 1.3 megapixels.

In order to have enough data to work with each blood smear shows enough cells that are infected and uninfected by the parasite to allow the program to handle both cases without having an abundance of cells. The dataset is organised as in two folders as follows:

- 30 are unlabelled (Fig. 7)

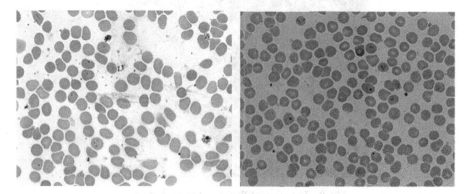

Fig. 7. Unlabelled blood smears.

- 30 are labelled (Fig. 8)

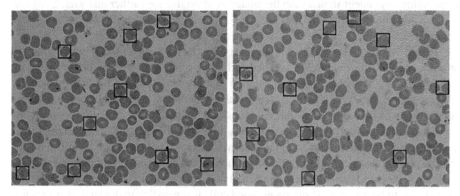

Fig. 8. Labelled blood smears.

4.4 Performance Evaluation

Segmentation

Figure 9 shows how the original Watershed performs without any additional operations, as you can see it does not perform very well. The plain use of the Watershed technique results in over-segmentation, which are all the lines and tiny regions. In this case running the area extraction on this image would produce terrible results because it is capturing all the areas of the image, which provide no information because all the cells have been split.

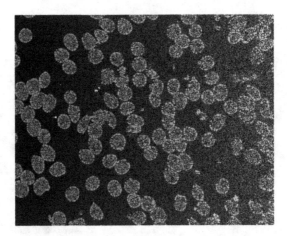

Fig. 9. The result of Watershed segmentation.

The problem with this is that too many regions in the image are being found, which the biggest problem with the original technique is, although correct segmentation is application dependent in this case the image is of no use for further analysis. Cells can be seen in white although this does not provide any correlation to what has actually been segmented.

Figure 10 shows the results of Algorithm 2 but the problem with this is that although it works better with than the Watershed technique alone it is still unable to detect all cells separately. It is clear that segmentation did not work too well for cells, which are close together, they were either merged together with neighbouring cells or not detected at all. This was solved by the Modified Classification at the expense of extra noise.

The problem with this implementation is that it has the potential to group important cells together through incorrect segmentation, for example infected and no infected cells can be classed as one as seen in the big blue bounding boxes. It is important that the cells are separate so that the extraction can run on a cell-by-cell basis in order for the cell boundaries to be accurate, if run on grouped cells it is possible to detect darker regions, which are not inside cells leading to inaccurate results.

Once testing the previous two version of the Watershed technique, the modified version that shows a more accurate result. The solution was to use a combination of the

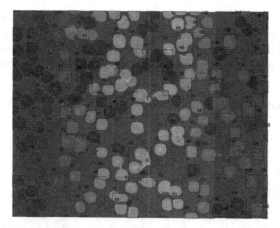

Fig. 10. Marker-based Watershed segmentation example (Color figure online)

Watershed technique with thresholding. The images used have all been stained using Giemsa, so it is possible that other staining techniques display different results.

Comparing Figs. 10 and 11, the latter captures more cells. This modified version combats the over-segmentation issue in Fig. 9 and the issue joining multiple together in Fig. 10. This then means that the cells are ready to be analysed individually without the risk of any incorrect merging.

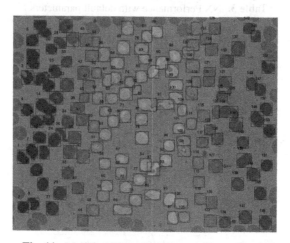

Fig. 11. Modified Watershed segmentation example

4.5 Neural Network

Using the correct inputs and methods in the NN was vital to the project producing results. To test the performance of the NN it was run 5 times with different parameters to see which inputs produced the best results. The first set of configs (configurations) (see Table 2) were using the default parameters created by the NN, the second set of config (see Table 4) used the best config from Table 2 and modified default parameters. Table 4 was used to see whether changing the defaults would improve the network.

Table 2. Configurations with default additional parameters

Configuration	Input layer	Hidden layer(s)	Output layer
1	4096	10	2
2	4096	50 20	2
3	4096	5 20 10	2
4	4	10	2
5	4	50 20	2
6	4	5 20 10	2
7	5	10	2
8	5	50 20	2
9	5	5 20 10	2

Table 3. NN Performance with default parameters

Configuration	Overall avg accuracy (training, validation, testing) – 5 runs
1	69.1%
2	69.9%
3	67.9%
4	91.9%
5	92.0%
6	91.9%
7	91.4%
8	91.3%
9	91.4%

In total there are 9 configs, 1–3 uses raw images 64 × 64, 4–6 uses 4 features of the images and the last 3 use 5 features of the image. The features calculated are:

- White count – In a binary image the white area is the region of interest, therefore using this made sense because non-infected cells shouldn't have any white count

Table 4. NN Config with modified parameters

Config.	Input layer	Hidden layer	Output layer	Additional parameters
4(a)	4	10	2	Max Fail = 100; Training = trainbr; Performance = mse;
4(b)	4	10	2	Max Fail = 50; Training = trainlm; Performance = sse;
4(c)	4	10	2	Max Fail = 30; Training = traingd; Performance = mse;

- Skewness – Used to determine the surface of a cell, because it can detect darker areas in images, which in this case would be the parasites in the grey. It looks at the histogram of the image and compares how balanced the grey levels are across the image which shows how symmetrical it is
- Entropy – Determines how much information is in the image so an image filled with many black areas will have a higher entropy than an image with one colour
- Area – Calculates the area of the objects in the image, an infected cell will have an area because the parasite is of a different colour. In non-infected cells, the value is next to none because no objects will appear
- Variance – Shows how each pixel differs from one another, in a cell the parasite is of a darker purple therefore will show an increased variance around the cell because neighbouring pixels will be lighter

The initial runs show that configs 4–6 on average perform the best with a slightly higher accuracy than configs 7–9, however the first 3 performed quite badly in comparison only achieving at most a high of 69% accuracy. Changing to features improved the NN accuracy drastically, looking at Fig. 12 moving from a 1D image to features allowed the NN to identify patterns easier. As it is a basic FFNN and not of a complex CNN which does take the entire image, it is wiser to use pre-calculated features as input. In a CNN the convolutional layers are responsible for detecting the image features using the RGB channels of an image.

Configs 1–3 did not perform well because the 1D image is the flattened form of the 3D image, therefore loses spatial information making it harder to compute features. The convolutional layer in a CNN uses a filter/kernel which prevents this from happening by capturing multiple features, this is the reason why this layer is normally used multiple times at different stages such as in layer 1 and then again in layer 4 after other operations. Basic features such as edges would be detected in the first convolutional run, and in further runs, it's able to distinguish more complex features.

The following parameters were changed to further verify the system:

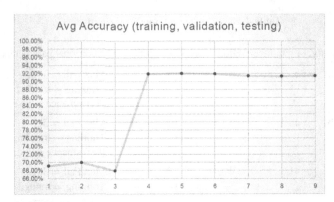

Fig. 12. Visualisation of default configuration accuracies.

- Max Fail – To see whether changing this value either incrementing would allow the NN to perform better. It would give the NN more a chance to improve its performance before failing completely
- Training function – whether the program being trained differently would affect the performance determining the best way to train the NN based on the features
- Performance function – As the training function changes, the required performance function also needed to change with it

Comparing the results from Tables 3 and 5 changing the parameters made a negligible difference. Allowing the NN to fail later and using a different training function made a difference of only 0.1% to 0.3% which is not enough to warrant the increase in time it takes to run the NN. It seems that it would be beneficial to test the NN again using double the amount of input elements to test whether the results would also change. Although the results didn't improve when using 1 additional input it is possible for it to be different when multiple features are added altogether.

Table 5. NN Performance with modified parameters

Configuration	Overall avg accuracy (training, validation, testing) – 5 runs
4(a)	92.1%
4(b)	92.2%
4(c)	87.5%

Viewing Config 4(c), it shows that the training function didn't perform as well as the others including tests using the default configuration. The default parameter *trainscg* is an improvement over *traingd* which are both gradient descents but *traingd* is the original and *trainscg* is able to converge quicker, this means that the algorithm is able to run faster and thereby reach the closest value to the goal within a shorter time.

5   Conclusion

Achieving a 92% accuracy by updating the parameters in the NN shows promise considering the use of a FFNN. The intention was to create a single process from start to finish that allowed a user to diagnose a patient from a blood smear. It is clear from the work undertake that Neural Networks and segmentation techniques have come a long way but there are still many improvements to be made which can be seen from the results.

The limitations of these techniques are that they don't work in all cases and require additional tuning for them to work as expected. In the grand scheme of things, it means that the implementations can't be used widely because of the equipment and staining methods used which differ between medical facilities. Although this is a problem, it has highlighted the potential for the future.

The segmentation doesn't work well on cells which aren't clearly defined such as faint cells which blend in with the background, additional thresholding and morphological operations could help to separate foreground from the background at the risk of including irrelevant artifacts. The NN is very basic using only the features, which have been manually extracted, to achieve results higher than 92% a CNN could be used instead because it can run a more complex architecture which would determine the best features from ground truth images. The CNN is likely to work better at the expense of time, in order for the FFNN to perform better more features would need to be extracted manually.

Overall, with the increase in popularity for NNs and techniques for image processing, we move that step closer to creating fully automated processes that will aid the eradication of malaria. The use will see the change from a supplementary aid to a state-of-the-art tool that will become the gold standard in malaria diagnosis.

Acknowledgement. This work is supported by the National Natural Science Foundation of China (Grant No.: 61602289), the Fundamental Research Funds for the Central Universities of China (Grant No.: GK201803081).

References

1. Cox, F.E., et al.: History of the discovery of the malaria parasites and their vectors (2010)
2. Mace, K.E., Arguin, P.M., Tan, K.R.: CDC - Malaria. https://www.cdc.gov/mmwr/volumes/67/ss/pdfs/ss6707a1-H.pdf. Accessed Apr 2019
3. World Health Organization: Disease information - Malaria. https://www.who.int/ith/diseases/malaria/en/. Accessed Apr 2019
4. NHS: Malaria - Causes. https://www.nhs.uk/conditions/malaria/causes/. Accessed Apr 2019
5. WWARN - Worldwide Antimalarial Resistance Network: Plasmodium vivax and drug resistance. https://www.wwarn.org/about-us/malaria-drug-resistance/plasmodium-vivax-and-drug-resistance. Accessed Apr 2019
6. MMV - Medicines for Malaria Venture: Definitions and symptoms of Malaria. https://www.mmv.org/malaria-medicines/definitions-and-symptoms. Accessed Apr 2019

7. Lo, E., et al.: Plasmodium malariae prevalence and csp gene diversity, Kenya, 2014 and 2015. Emerg. Infect. Dis. **23**, 601–610 (2017). https://doi.org/10.3201/eid2304.161245
8. MaximumYield: What is Genus? - Definition from MaximumYield
9. Russ, J.: The Image Processing Handbook, 6th edn. CRC Press, Boca Raton (2011)
10. Zaitoun, N.M., Aqel, M.J.: Survey on image segmentation techniques. Procedia Comput. Sci. **65**, 797–806 (2015)
11. What Is a Catchment Area of a River or Lake? - WorldAtlas.com
12. Castle, N.: An Introduction to Machine Learning Algorithms (2017)
13. Ng, H.P., Ong, S.H., Foong, K.W.C., Goh, P.S., Nowinski, W.L.: Medical image segmentation using K-Means clustering and improved watershed algorithm (2006)
14. Nguyen, H.T., Worring, M., Van Den Boomgaard, R.: Watersnakes: energy-driven watershed segmentation. IEEE Trans. Pattern Anal. Mach. Intell. **25**, 330–342 (2003)
15. Grau, V., Mewes, A.U.J., Alcañiz, M., Kikinis, R., Warfield, S.K.: Improved watershed transform for medical image segmentation using prior information. IEEE Trans. Med. Imaging **23**, 447 (2004). https://doi.org/10.1109/TMI.2004.824224
16. Areeckal, A.S., Sam, M., David, S.: Computerized radiogrammetry of third metacarpal using watershed and active appearance model (2018)
17. Yuheng, S., Hao, Y.: Image segmentation algorithms overview (2017)
18. Dhanachandra, N., Chanu, Y.J.: Image segmentation method using K-Means clustering algorithm for color image (2015)
19. Kumari, N., Saxena, S.: Review of brain tumor segmentation and classification (2018)
20. Salihah, A., Nasir, A., Jaafar, H., Azani, W., Mustafa, W., Mohamed, Z.: The cascaded enhanced K-Means and fuzzy C-Means clustering algorithms for automated segmentation of malaria parasites. In: Malaysia Technical Universities Conference on Engineering and Technology (MUCET 2017), vol. 150 (2018). https://doi.org/10.1051/matecconf/201815006037
21. Sharma, N., Mishra, M., Shrivastava, M.: Colour image segmentation techniques and issues: an approach. Int. J. Sci. Technol. Res. **1**, 9–12 (2012)
22. Dass, R., Priyanka, S.D.: Image segmentation techniques. IJECT **3** (2012)
23. Malaria | NIH: National Institute of Allergy and Infectious Diseases
24. UCL - London's Global University: Malaria Database. www.ucl.ac.uk

Research in Breast Cancer Imaging Diagnosis Based on Regularized LightGBM

Chun Yang and Zhiguo Shi(✉)

University of Science and Technology Beijing, Beijing 100083, China
szg@ustb.edu.cn

Abstract. Breast cancer is the main cause of cancer death in women, and it is increasingly threatening women's health. The research of breast cancer imaging radiology aims to replace the traditional artificial diagnosis and promote the accurate diagnosis and treatment of breast cancer. Using deep learning technology to extract the features of breast images and to construct a breast cancer imaging diagnostic system is essential for promoting the development of diagnostic efficiency in the field of imaging medicine. Firstly, the existing mammography datasets are rotated 50 times at random angles (0°–360°) with the data enhancement technology. Then the images are encoded, and then the image features are extracted by ResNet50. In the process of classification, a regularized LightGBM model is added, the combination of the two models constitutes a model for breast cancer diagnosis. The addition of LightGBM improves the classification accuracy and performance of the model. The experimental results show that, the accuracy of the model on the two datasets of INbreast and DDSM is 91.7% and 93.6% respectively when doing the three-classification (normal/benign/malignant), when doing the binary classification (benign/malignant), the AUC on the two datasets are 0.942 and 0.962 respectively.

Keywords: Deep learning · Data enhancement · Regularization · LightGBM

1 Introduction

Breast cancer is the most common cancer in women worldwide and the leading cause of cancer death in women [1]. If breast cancer can be detected early and treated promptly, its therapeutic effect is also the best in malignant tumors [2]. Mammography is the primary means of screening and diagnosing of breast cancer, it can help clinicians detect and treat breast cancer early and in time, thus significantly reducing its mortality [3, 4]. Breast X-ray imaging diagnosis usually scans each image through a radiologist, identifying common abnormalities such as masses, calcifications, structural distortions and asymmetric dense shadows. It also requires clinical information from the patient, which consumes a lot of doctors' energy.

In recent years, Radiomics, which has emerged in the field of imaging medicine, has brought new opportunities for accurate diagnosis and treatment of breast cancer.

© Springer Nature Singapore Pte Ltd. 2019
H. Ning (Ed.): CyberDI 2019/CyberLife 2019, CCIS 1138, pp. 487–503, 2019.
https://doi.org/10.1007/978-981-15-1925-3_35

Radiomics uses a large number of automated data characterization algorithms to transform the image data of the Region of Interest (ROI) into high-resolution, extensible feature space data. Data analysis is a quantitative and high-throughput analysis of a large number of image data, and obtains high-fidelity target information to comprehensively evaluate various phenotypes of tumors, including tissue morphology, cellular molecules, genetic inheritance and other levels. Since 2007, some scholars [5] have used imaging features for the first time to predict the diagnosis and prognosis of liver cancer. Since then, researches on imaging genomics and clinical prediction have been published in succession [6, 7], confirming its huge clinical application potential.

At present, some patients' medical imaging results are false positive, which is related to the quality of the specific imaging technology, and may also be misdiagnosed by doctors. Radiology and imaging doctors rely on their personal experience to interpret medical images. It is inevitable that human error will occur, and relying on the doctors' naked eye to observe medical images will bring them a great burden. It is easy to be misdiagnosed due to fatigue. Some uncommon features in the image may also be ignored by doctors, such as irregular shape of tumors, small tumors and so on. Computer-aided diagnosis combined with deep learning technology can overcome the errors caused by various human factors and improve the accuracy of diagnosis to some extent [8]. At the same time, in recent years, along with the improvement of computer configuration and the development of parallel computing power of graphics computing unit (GPU), the efficiency of computer-aided diagnosis is much higher than that of manual diagnosis.

Using convolutional neural network to analyze medical images has become the choice of more and more researchers [9]. The convolutional neural network gradually acquires the characteristics of medical images through a series of convolutional and pooling layers, and finally aggregates into higher-order features as a basis for judging whether a patient is sick or not. But training a usable complex convolutional neural network model requires a large amount of qualified medical image data. One difficulty in current research in this field is the lack of sufficient qualified data.

Because of the obvious characteristics of image histology data such as small amount of data and high dimension, it is difficult to analyze image histology data directly by traditional deep learning model which requires a large number of parameters. Therefore, this paper proposes a method for breast cancer imaging diagnosis based on regularized LightGBM combined with the knowledge of data enhancement, which will be discussed from three aspects: research status at home and abroad, model design, experiment and verification.

2   Related Work

2.1   Advances in Breast Medical Imaging Diagnosis Based on Deep Learning

With the development of deep learning, many researchers have adopted deep learning methods to build new CAD systems [10, 11]. Like machine learning methods, deep learning is also divided into supervised learning and unsupervised learning. For example, the convolutional neural network (CNN) is a deep learning model under supervised learning, and the deep belief nets (DBN) is a deep learning model under unsupervised learning. CNN-based models often require detailed annotation of ROI [12, 13], which takes a lot of time and expense and creates great difficulties in supervised learning, especially medical imaging. Therefore, some researchers have studied weak supervised learning methods [14, 15], such as multiple instance learning (MIL), which only requires researchers to provide labels for the entire image, thus greatly reducing training costs.

CNN is a representative structure of the deep learning model and is also a research hotspot of deep learning. It is a feedforward artificial neural network with a multi-layer network structure, usually including input layer, convolution layer, activation function, pooling layer, and full connected layer.

The invention of CNN is primarily inspired by the human visual cortex. In 1959, David Hubel and Torsten Wiesel discovered the hierarchical structure of the visual cortex in animal experiments. Some cells in the visual cortex can only perceive signals in a certain area, which is called the receptive field. These cells are very sensitive to signals that resemble edges in the field. There are also some cells that are insensitive to location and more susceptible to the field. In the follow-up study, people gradually have a deeper understanding of the information processing mechanism of the brain vision system, and have a clearer understanding of the connections and functions of different regions of the brain.

Inspired by this, in 1980, Kunihiko Fukushima proposed neocognitron [16], neocognitron consists of S layer and C layer, which are composed of S cells and C cells respectively. The model is trained using similar Herb learning rules. The S layer is responsible for feature detection and the C layer is responsible for feature combination. For example, the S layer can detect local features such as edges, and the C layer expresses global features by connecting with the S layer. It is generally believed that neocognitron is the predecessor of CNN.

CNN has powerful feature extraction capabilities and can extract higher level features. Jiao et al. [17] developed a CNN-based CAD system in 2016 to classify breast cancer masses. It mainly contains a CNN and a decision mechanism. In the training process, CNN is used to extract high-level and medium-level features. After combining, the model is trained. The intensity information and depth features automatically extracted by CNN are combined to better simulate the doctor's diagnosis process, which achieve good results.

On the other hand, some researchers have developed variant technology based on CNN. In 2017, Al-Masni et al. [18] proposed a CAD system based on regional deep learning technology, which is a ROI-based CNN called YOLO (You Only Look Once). YOLO is a representative end-to-end training algorithm. It trains in mammogram datasets with ROI information and directly optimizes detection performance. In addition, the generalization of YOLO learning objectives is highly versatile and can detect multiple targets simultaneously. YOLO can also learn ROI and background at the same time, so their proposed CAD system can perform feature extraction, detect and classify breast masses in a CNN, which is a fast and accurate target detector [19].

Some researchers have proposed CNN that reduce the dependence on ROI annotations. Li et al. [20] developed an end-to-end training algorithm for breast cancer diagnosis of intact mammograms using a full convolution design. The algorithm uses CNN completely, so you can enter images of any size. In addition, it only needs to annotate the lesion in the first stage of training. After the training model recognizes partial patches, the weight of the complete image classification network can be initialized, and then the model can be migrated to a full-image classifier. End-to-end training can be carried out without ROI annotations, which greatly reduces the dependence on lesion annotations. Compared with the previous methods, this design is simple and the performance is superior.

2.2 Transfer Learning and Data Enhancement

The best way for convolutional neural networks to improve the generalization ability of the deep learning model is to use more data for training. Of course, in practice, the amount of data you have is limited. The solution to this problem is to use transfer learning or data enhancement.

In the case of supervised learning, the purpose of the training process is to obtain a model with accuracy and reliability for use in testing. Accuracy and reliability are based on two basic assumptions: (1) the samples of the training set and the test set are independently and identically distributed; (2) there are enough samples in the training set to ensure that the training is sufficient. In practical applications, these two conditions are often not strictly met. For example, in some applications data is time-sensitive. For example, stock data, models trained with large amounts of data in the past do not adapt well to future data. In addition, tagged data acquisition is much more difficult than unlabeled data, so in supervised learning, it is often difficult to satisfy condition 2.

Transfer learning [21–23] refers to a machine learning method that solves the problem of the target domain by migrating the knowledge learned from the source domain to a different but related target domain. The inspiration for transfer learning comes from human intelligence. People have strong abstraction ability, and solving different problems can be used to bypass the analogy. For example, tennis players can quickly get to know when they are learning table tennis. But for computer algorithms, the understanding and storage of knowledge is far less intelligent than humans. Changing scenarios often means that algorithms and models need to be rebuilt. For example, AlphaGo has the strength to defeat the top players in humans through the game of 30 million. But this training experience is hard to be used to train a computer mahjong player.

The theory of transfer learning is still one of the basic research topics in machine learning today, and many basic questions are not answered. For example, under what circumstances transfer learning is effective. In practice, transfer is not always effective. In general, in order to make transfer learning work, there needs to be a certain correlation between learning tasks, which is still lack of theoretical analysis. The lack of a measure of relevance means that the feasibility can only be verified by experimental results, and the performance that may be achieved cannot be predicted at the time of experimental design.

In deep neural networks, the application of transfer learning is mainly to use the model parameters trained on large datasets. The most common big dataset is the ImageNet dataset of natural image classification. From the feature analysis, the features learned by the CNN model are hierarchical features. The underlying convolutional layer responds to the edge information, and the intermediate convolutional layer output is the combination of the underlying convolutional layer output. Therefore, from bottom to top, the more the upper layer, the more the output of the convolutional layer is related to the data. CNN generally requires more data for full training, and the initialization of the model is also a factor that affects performance. From the perspective of parameter initialization, the strategy of migration learning can also be understood as using existing model parameters to specify a better initialization for new model parameters. The widespread use of transfer learning in CNN is based on a large number of experiments. According to the degree of correlation between the source domain and the target domain, the strategy of inheriting the underlying weight and discarding the top-level weight is generally used. Therefore, it can be flexibly selected, and the model structure of the target domain can also be different from the model structure of the source domain. In addition, some experiments have shown that training the depth model from its own data, sometimes the performance of the model is not as good as the transfer model of the general depth model trained on large datasets such as ImageNet.

Transfer learning and fine-tune are not the same concept. Transfer learning is a complete algorithmic system that is a method of applying a model from one dataset to another. Fine-tune is a means of processing, not only for transfer learning, but also an important step in the training of neural network models. The purpose of transfer learning is to use the old knowledge to serve new tasks. The knowledge here includes two parts of data and model, and fine-tune is only a fine-tuning of model parameters.

Data enhancement is particularly effective for problems such as object recognition. Because images are high-dimensional and include a variety of huge variables, many of them can be easily simulated. Even if the model has used convolution and pooling techniques to keep some of the comments unchanged, the operation of panning a few pixels in each direction of the training image can often greatly improve generalization.

There are usually these data enhancement methods:

(1) Data enhancement of color: color saturation, contrast and brightness.
(2) PCA Jittering, firstly calculate the mean and standard deviation according to the three color channels of RGB, normalize the input data of the network, calculate the covariance matrix on the whole training set, perform feature decomposition, obtain the feature vector and eigenvalue to make PCA Jittering.
(3) When the image is cropped and scaled, a random image difference method can be used.
(4) Crop Sampling, how to zoom and crop from the original image to get the input of the network. There are two methods that are commonly used: one is to use Scale Jittering, the VGG and ResNet models are trained using this method; the other is scale and aspect ratio enhancement transformation, which was first proposed by Google to train their Inception network.

3   Breast Cancer Imaging Diagnosis Model Based on Regularized LightGBM

3.1   The Overall Framework of Diagnostic Model

At present, there is no uniform standard for breast imaging equipment. When conducting a classification study on breast diseases, using a private dataset of a hospital or research institution, the performance of the constructed model is often controversial. Therefore, this paper will use the open source datasets for imaging studies of breast lesions, INbreast and DDSM (Digital Database for Screening Mammography) to train and test data. The cases in these two types of data were classified and evaluated after the diagnosis by biopsy according to the criteria of the Breast Imaging-Reporting and Data System (BI-RADS) with the participation of radiologists. The data are mainly divided into three categories: normal, benign and malignant.

In recent years, some very deep convolutional neural networks such as VGG, Inception and ResNet have achieved good results in many computer vision tasks. However, if we want to train these networks from scratch, we need a lot of data. If we use the datasets used in this paper to train, it will inevitably lead to over-fitting. One solution is to use transfer learning, which is to migrate the general depth model trained on large data sets such as ImageNet to train our own datasets, and then fine-tune the parameters. However, after doing the experiment, we found that the performance is not very good. Therefore, this paper attempts to use the method of using deep convolution feature to classify, and proposes a breast cancer imaging diagnosis model based on CNN and LightGBM, which combines convolutional neural network and gradient enhancement tree (Light-GBM). The convolutional neural network uses ResNet50 to extract the features of breast image. The regularized LightGBM is used as a classifier to classify them.

The framework of the breast cancer diagnosis model designed in this paper is shown in Fig. 1. The workflow is divided into two phases as follows:

The first stage: First, datasets are preprocessed, that is data enhancement and encoding. Then the ResNet50 is used to extract breast image features from the dataset, which effectively reduces the risk of over-fitting in the next stage of supervised learning.

The second stage: Regularized LightGBM gradient enhancement tree is used to supervise the classification. Gradient enhancement models are widely used in machine learning due to their good speed, accuracy and robustness to overfitting. The purpose of regularization is to facilitate feature selection and thus control the complexity of the model.

3.2   Preprocessing and Feature Extraction

In order to facilitate classification with LightGBM, data sets need to be preprocessed first. In order to solve the problem of the small sample size of the data set, this paper adopts the data enhancement technology to rotate 50 times at random angles ($0°$–$360°$) for each image, so the data set is greatly expanded. Next the image needs to be coded. Twenty regions (400×400) are randomly cropped from each rotated image, and each cropped region is encoded into one descriptor, that is to say, one image is encoded into

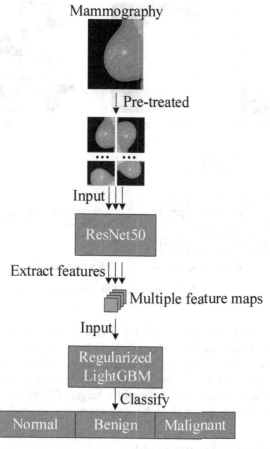

Fig. 1. Breast cancer imaging diagnosis model based on regularized LightGBM

20 descriptors. Finally, 20 descriptors are combined into a single descriptor through 3-norm pooling:

$$d_{pool} = \left(\frac{1}{N} \sum_{i=1}^{N} (d_i)^p \right)^{\frac{1}{p}} \tag{1}$$

Where d_{pool} represents the merge descriptor of the image, N is the number of cropped regions, d_i is the descriptor of the crop region, and the value of the hyperparameter p is 3. The p-norm of the vector gives the mean and maximum values of p as $p = 1$ and $p \to \infty$, respectively. Therefore, for each original image, 50 (the number of data enhancements) × 3 (CNN encoders) = 150 descriptors are finally obtained. The entire process of preprocessing is shown in Fig. 2.

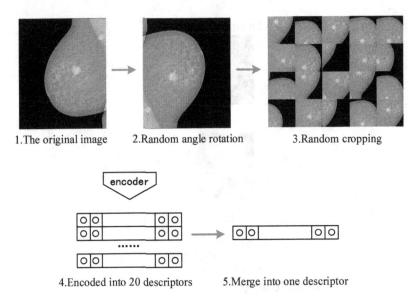

1.The original image   2.Random angle rotation   3.Random cropping

4.Encoded into 20 descriptors   5.Merge into one descriptor

Fig. 2. The process of preprocessing

After preprocessing, the features of the image are extracted. In this paper, the ResNet50 network inside Keras is adopted. In order to ensure that the network can use any size of images, the full connection layer is deleted. The Global Average Pooling (GAP) is used to convert the last convolutional layer containing 2048 channels in the model into a one-dimensional eigenvector with a length of 2048. The features extracted by the network are used as the input of the subsequent classification model.

3.3   The Regularized LightGBM Model

The Gradient Boosting Decision Tree (GBDT) is an integrated learning framework based on decision trees. It shows good classification accuracy in many fields, but it has some problems such as easy over-fitting and slow training speed. In response to these shortcomings, Guolin Ke et al. made corresponding improvements and proposed an efficient implementation of GBDT in 2017, namely LightGBM. Its main improvements include histogram algorithm and Leaf-wise decision tree growth strategy with depth limitation, which enhances the robustness to noise while ensuring good evaluation accuracy and training speed.

Gradient Boosting, the idea is: one-time iteration of variables, the sub-model is added one by one in the iterative process, and the loss function is guaranteed to decrease.

Assuming that $f_i(X)$ is a sub-model, the composite model is:

$$F_m(x) = \partial_0 f_0(x) + \partial_1 f_1(x) + \cdots + \partial_m f_m(x) \tag{2}$$

The loss function is $L[F_m(x), Y]$, and each time a new sub-model is added, the loss function is continuously reduced toward the gradient of the variable with the second highest information content.

$$L[F_m(x), Y] < L\left[F_{m-1}(x), Y\right] \tag{3}$$

LightGBM uses the histogram decision tree algorithm to discretize the successive floating-point eigenvalues from samples into k integers and construct a histogram of width k. In traversal, the discretized value is used as an index to accumulate statistics in the histogram, and then the optimal segmentation points are searched by traversal according to the discrete values of the histogram. This can effectively reduce memory consumption while reducing the time complexity.

LightGBM adopts a more efficient leaf growth strategy, namely Leaf-wise with depth limitation. The strategy traverses all the leaves before splitting, and then finds the leaves with the greatest splitting gain to split and cycle back and forth. Leaf-wise can get better accuracy under the same number of splits. At the same time, the maximum depth limit to prevent overfitting is added to Leaf-wise. Its growth strategy is shown in Fig. 3. In Fig. 3: white dots indicate the leaves with the highest split gain; black dots indicate leaves with the split gain not the largest.

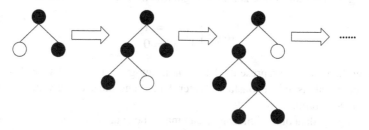

Fig. 3. Leaf-wise leaf growth strategy map

Another optimization of LightGBM is the histogram for differential acceleration. In general, a histogram of a leaf is constructed, and the widths of the histograms of the father node and the sibling node are both K. Therefore, only K times need to be calculated in the process of making difference, which greatly improves the speed of operation.

For the diagnosis of breast cancer imaging, because of the subtle differences between benign and normal data, classification with the standard LightGBM model may not be sufficient for data recognition of benign lesions. If the data of benign is misclassified into the data of normal, that will have a great impact on the classification results. In order to solve the problem more effectively, this paper improves the loss function of LightGBM.

Considering that regularization can optimize the model well, this paper decides to add regularization term after the loss function. There are two kinds of regularization term commonly used, namely L_1 regularization term and L_2 regularization term. There are differences between the two. The L_1 regular term can generate a sparse weight matrix, that is, a sparse model can be used for feature selection; the L_2 regular term can prevent model from overfitting. Since LightGBM can prevent overfitting by adjusting parameters, this paper introduces the L_1 regular term in the loss function, and then assigns higher weight to the data of benign lesions to deal with the misclassification problem. The specific improvement methods are as follows:

For the m-th tree, the original loss function is:

$$-\frac{1}{N}\sum_{i=1}^{N}L(y_i, F_{m-1}(x_i; A_{m-1})) \tag{4}$$

Where $F_{m-1}(x_i; A_{m-1})$ represents the predicted value of the input x_i of the model consisting of the former $m-1$ trees under the condition of the parameter A_{m-1}, where A_{m-1} includes the parameters $a_1, \cdots a_{m-1}$ of the former $m-1$ trees. $L(y_i, F_{m-1}(x_i; A_{m-1}))$ is a function describing the error between the true value y_i and the current model prediction value, which is a logarithmic loss function for LightGBM.

After the improvement, the loss function of the m-th tree is:

$$-\frac{1}{N}\left(\sum_{i=1}^{N}a_i L(y_i, F_{m-1}(x_i; A_{m-1})) + \lambda\sum_{i=1}^{N}|w_i|\right) \tag{5}$$

Where λ is the regularization coefficient, w_i represents the weight, and the purpose of obtaining the absolute value is for the regularization. The coefficient a_i is:

$$a_i = \begin{cases} c, & y_i = 1 \\ 1, & y_i = 0 \end{cases} \tag{6}$$

Among them, $y_i = 0$ represents data of non-benign lesions, and $y_i = 1$ represents data of benign lesions. c is a constant greater than 1, and the specific value is related to the data category ratio.

Weighting the data of benign lesions can make the data of benign lesions with larger weights get greater gradient and more attention in training, so as to balance the impact of confusion between benign and normal data, and improve the accuracy of the model in evaluating benign lesions. The L_1 regularization used in this paper is also called Lasso regression. It can generate a sparse weight matrix, that is, a sparse model for feature selection. At the same time, it also limits the degree of change of the model in the iterative training process, plays a role in controlling the complexity of the model, and makes the model after training have better smoothness.

The following is the classification process using the regularized LightGBM model:

(1) Normalize all the features extracted by ResNet50. Convert $\begin{bmatrix} v_{11} & \cdots & v_{1m} \\ \vdots & \ddots & \vdots \\ v_{n1} & \cdots & v_{nm} \end{bmatrix}$ to $\begin{bmatrix} b_{11} & \cdots & b_{1m} \\ \vdots & \ddots & \vdots \\ b_{n1} & \cdots & b_{nm} \end{bmatrix}$ to get p_i.

(2) Calculate the initial gradient value $g_m(x_i)$,

$$g_m(x_i) = \frac{2p_i \times learning\_rate}{1 + \exp[2p_i f_{m-1}(x) \times learning\_rate]} \tag{7}$$

In the formula, $f_{m-1}(x)$ is set to 0 or a random value, and then it is further converted

to: $\begin{bmatrix} b_{11} & \cdots & b_{1m} \\ \vdots & \ddots & \vdots \\ b_{n1} & \cdots & b_{nm} \end{bmatrix} \lambda \begin{bmatrix} \lambda_1 \\ \vdots \\ \lambda_n \end{bmatrix}.$

(3) Build a tree until the number of leaves is limited or all leaves cannot be split.

a. Calculate the histogram.

$$T = \begin{bmatrix} h_{11} & \cdots & h_{1m} \\ \vdots & \ddots & \vdots \\ h_{n1} & \cdots & h_{nm} \end{bmatrix} \tag{8}$$

In the formula, $h_{ij} = (c_{ij}, l_{ij})$, $c_{ij} = \sum_{k=1}^{n} 1|(b_{kj} = i - 1)$, $l_{ij} = \sum_{k=1}^{n} \lambda_k |(b_{kj} = i - 1)$.

b. Obtain the split gain from the histogram, and select the best splitting feature G and the splitting threshold I.

$$G_j = \max_{1 \le x < k} \left[\frac{\left(\sum_{i=1}^{x} l_{ij}\right)^2}{\sum_{i=1}^{x} c_{ij}} + \frac{\left(\sum_{i=x+1}^{k} l_{ij}\right)^2}{\sum_{i=x+1}^{k} c_{ij}} \right] \tag{9}$$

$$I_j = \arg \max \left[\frac{\left(\sum_{i=1}^{x} l_{ij}\right)^2}{\sum_{i=1}^{x} c_{ij}} + \frac{\left(\sum_{i=x+1}^{k} l_{ij}\right)^2}{\sum_{i=x+1}^{k} c_{ij}} \right] \quad 1 \le x < k \tag{10}$$

c. Establish a root node.

$$s = \arg \max(G_i) \quad 1 \le i \le m \tag{11}$$

$$Node = (s, G_s, I_s) \tag{12}$$

d. The data samples are segmented according to the best splitting feature G and the splitting threshold I of b.
e. Update the current output value for each sample.

(4) Update the gradient value $g_m(x_i)$ of the tree.
(5) Repeat steps (3) and (4) until all the trees are built.
(6) Adjust the parameters and perform classification experiments.

The algorithm flowchart of the regularized LightGBM is shown in Algorithm 1, where G, P represent two sample sets. In this algorithm, Step3 to Step8 need to be repeated continuously, in order to establish all the decision trees. In this process, the key is to select the best splitting feature and splitting threshold. Due to the addition of regularization of the algorithm, the feature selection can be optimized, and the algorithm itself has the characteristics of faster training rate, higher accuracy and low memory usage.

Steps of the algorithm:

Input: Features F^G , F^P ;
Output: Decision trees established;

Step1: The features F^G , F^P extracted from ResNet50 is normalized to get $p_i{}^g$ and $p_i{}^P$;

Step2: Calculate the initial gradient value $g_m(x_i)$ according to formula (7);

Step3: Calculate the histogram T according to formula (8);

Step4: The optimal splitting feature G and the splitting threshold I are selected according to formulas (9) and (10);

Step5: Establish the root node s according to formula (11);

Step6: Samples are segmented according to the optimal splitting feature G and the splitting threshold I ;

Step7: Update the output value of each sample;

Step8: Update the gradient value of the tree $g_m(x_i)$;

Step9: Repeat Step3~Step8 until all the trees are built.

Algorithm 1. The learning algorithm of regularized LightGBM

4   Experimental Verification

4.1   Data Collection and Environment Configuration

The data set in the experiment was selected from the INbreast and DDSM databases, which were open sourced by several medical institutions in Portugal and the United States. Information on the lesion area is marked by professional physicians, including the type of abnormality (structural disorder, calcification or mass), boundary contours, and abnormal benign and malignant. Among them, the INbreast data set included 115 cases, 90 of which had images of two individual breasts (MLO and CC) on each side, the remaining 25 cases are women who have breast resection and only two breast positions on one side. The dataset contains 410 images. By screening the data from DDSM, a total of 680 images of 172 patients are retained. The internal BI-RADS based rating distribution of the data is shown in Fig. 4.

The test platform hardware configuration is processor Intel Core i7-6700K@4.00 GHz * 8; graphics card GeForce GTX 1080; 4 GB of RAM and 8 GB graphics memory. The system uses the 64-bit version of Ubuntu 16.04 and uses Keras as the experimental platform. Keras is mainly used to build and retrain the depth model.

4.2   Experimental Results and Analysis

During the experiment, the original image set is first rotated by 50 random angles (0°–360°) for data enhancement, and then the rotated image is randomly cropped into

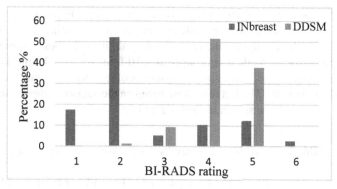

Fig. 4. Rating distribution based on BI-RADS in INbreast and DDSM

20 regions (400 × 400) and encoded into 20 descriptors. Then merge them into one descriptor. The ResNet50 in Keras is used for feature extraction. In order to remove the limitation of the input image size of the model, the fully connected layer is removed. In order to adapt LightGBM to the classification of the dataset in this paper, the standard LightGBM model is added to the L_1 regular term, and finally the regularized model is used for three classifications (normal/benign/malignant). In order to add classification labels to the INbreast dataset, according to the recommendations of professional doctors, the Bi-Rads category is 1 is defined as normal, and the Bi-Rads category is 2 or 3 is defined as benign. The Bi-Rads category is 4a and above is defined as malignant.

After continuously training the model and repeatedly adjusting the parameters, the ideal result is finally achieved. Table 1 shows the main optimal training parameter settings.

Table 1. The main optimal training parameters of regularized LightGBM

Parameter	Parameter setting
learnig_rate	0.001
objective	multiclass
metric	multi_logloss, multi_error
verbose	0
num_leaves	20
feature_fraction	0.46
bagging_fraction	0.69
bagging_freg	60
max_depth	7

Among the above parameters, num_leaves and max_depth is set to 20 and 7 respectively, which can control the model complexity and prevent overfitting. The learning_rate is set to 0.001, which can improve the generalization ability and robustness of the model.

In order to verify the method, this paper uses 10-fold cross-validation. The classification results are measured by accuracy. The results of the model training are compared with the ResNet50 model. The test results are based on the datasets INbreast and DDSM. The comparison results are shown in Tables 2 and 3.

Table 2. Comparison of the results of the two models of INbreast

Model	F1	F2	F3	F4	F5	F6	F7	F8	F9	F10	Mean
ResNet50	88.0	78.5	84.5	87.0	80.5	82.0	84.0	83.0	85.0	82.0	83.5
Regularized LightGBM	92.5	88.0	94.5	92.0	93.5	91.0	93.0	90.0	90.5	91.5	91.7

Table 3. Comparison of the results of the two models of DDSM

Model	F1	F2	F3	F4	F5	F6	F7	F8	F9	F10	Mean
ResNet50	91.5	81.0	86.0	86.5	92.0	85.0	83.0	82.0	87.0	84.0	85.8
Regularized LightGBM	95.0	91.5	94.5	96.0	92.5	92.0	93.0	95.0	90.5	95.5	93.6

Among them, F1–F10 represents the accuracy of 10 experimental results (unit: %), and mean represents the average of these 10 results. By comparison, it is not difficult to find that the average accuracy of INbreast and DDSM when training with ResNet50 model is 83.5% and 85.8% respectively, while the accuracy of training with regularized LightGBM model is 91.7% and 93.6% respectively.

If both normal and benign lesions are classified as benign lesions, the assessment can be converted to a binary classification problem of benign and malignant. The quality of the model can be measured by the area under the ROC curve (AUC). The performance of the model is compared with the existing classification models trained with INbreast and DDSM datasets, as shown in Fig. 5. In order to minimize the error of accuracy, the model is also verified by 10-fold cross-validation, and the diagnosis results are compared with the existing classification model, as shown in Table 4.

It can be seen from the comparison of the ROC curves in Fig. 5 that in the binary classification models of the INbreast data set, the model proposed in this paper is superior to the other two existing models in the AUC value. When the abscissa False Positive Rate is fixed in the [0, 0.4] interval, the ordinate TPR of the regularized LightGBM model is obviously higher than that of the two existing models. In the binary classification models of DDSM dataset, the model curve proposed in this paper is significantly higher than the other two models, which indicates that the performance of the model is significantly better. In order to facilitate more intuitive numerical comparison, the AUC values of these models and the results of previous studies are compiled into Table 4.

From the comparison of Table 4, we can see that in the classification models using the INbreast dataset, the diagnostic accuracy and AUC value of the existing ResNet50

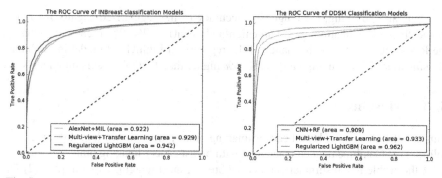

Fig. 5. Comparison of ROC curves for the two-classification models of INbreast and DDSM

Table 4. Comparison of several binary classification models of INbreast and DDSM

Model	Dataset	Accuracy(mean)	AUC
ResNet50	INbreast	86.4%	0.871
CNN + SVM	INbreast	89.1%	0.904
AlexNet + Multiple Instance Learning	INbreast	91.8%	0.922
Multi-view + Transfer Learning	INbreast	93.1%	0.929
Regularized LightGBM	INbreast	94.3%	0.942
ResNet50	DDSM	87.8%	0.879
CNN + SVM	DDSM	89.8%	0.895
CNN + RF	DDSM	91.4%	0.909
Multi-view + Transfer Learning	DDSM	93.8%	0.933
Regularized LightGBM	DDSM	96.9%	0.962

model are 86.4% and 0.871 respectively. The two values of the CNN combined with SVM classifier are 89.1% and 0.904 respectively. The two values of the AlexNet model combined with multi-instance learning are 91.8% and 0.922 respectively. The two values of the model which the multi-view input is combined with transfer learning are 93.1% and 0.929 respectively. However, the two values using the regularized LightGBM model can be increased to 94.3% and 0.942 respectively. In the classification models using the DDSM dataset, the diagnostic accuracy and AUC value of the existing ResNet50 model are 87.8% and 0.879 respectively. The two values of the CNN model combined with SVM classifier are 89.8% and 0.895 respectively. The two values of the CNN combined with RF are 91.4% and 0.909 respectively. The two values of the model which the multi-view input is combined with transfer learning are 93.8% and 0.933 respectively. However, the two values using the regularized LightGBM model can be increased to 96.9% and 0.962 respectively. The results show that the regularized LightGBM model can significantly improve the diagnostic results compared to existing models.

Generally speaking, the diagnostic accuracy of the regularized LightGBM model in the three-classification (normal/benign/malignant) is 7%–9% higher than that of the ResNet50, and the performance of the regularized LightGBM model in the binary classification (benign/malignant) is 2%–4% higher than that of the existing models.

5   Conclusions

Through the analysis of breast cancer imaging omics, the main work of this paper is as follows: (1) Combine data enhancement with breast cancer imaging omics diagnosis to solve the problem of insufficient number of breast cancer imaging case samples. (2) The ResNet50 in Keras is used to extract the features of the images, which is convenient for subsequent image classification. (3) The regularized LightGBM model is added in the process of classification, which improves the classification accuracy. The results show that the regularized LightGBM model has better accuracy than the ResNet50 when doing the three-classification (normal/benign/malignant); when doing the binary classification (benign/malignant), the regularized LightGBM model has better performance than the existing model. By comparing the AUC, the maximum values of existing models using INbreast and DDSM datasets are 0.929 and 0.933 respectively, while the two values of the model using the regularized LightGBM are respectively increased to 0.942 and 0.962.

Acknowledgement. This study was supported by State's Key Project of Research and Development Plan (No. 2018YFC0810601, No. 2016YFC0901303). The work was conducted at University of Science and Technology Beijing.

References

1. Torre, L.A., Islami, F., Siegel, R.L., et al.: Global cancer in women: burden and trends. Cancer Epidemiol. Biomark. Prev. **26**, 444–457 (2017)
2. Migowski, A.: Early detection of breast cancer and the interpretation of results of survival studies. Cien Saude Colet **20**, 1309 (2015)
3. Oeffinger, K.C., Fontham, E.T.H., Etzioni, R., et al.: Breast cancer screening for women at average risk. JAMA **314**, 1599–1614 (2015)
4. Weedon-Fekjaer, H., Romundstad, P.R., Vatten, L.J.: Modern mammography screening and breast cancer mortality: population study. BMJ **348**, g3701–g3708 (2014)
5. Segal, E., Sirlin, C.B., Ooi, C., et al.: Decoding global gene expression programs in liver cancer by noninvasive imaging. Nat. Biotechnol. **25**(6), 675–680 (2007)
6. Yamamoto, S., Korn, R.L., Oklu, R., et al.: ALK molecular phenotype in non-small cell lung cancer: CT radiogenomic characterization. Radiology **272**(2), 568 (2014)
7. Aerts, H.J., Velazquez, E.R., Leijenaar, R.T., et al.: Decoding tumour phenotype by noninvasive imaging using a quantitative radiomics approach. Nat. Commun. **5**, 4006 (2014)
8. Ying, Z., Lan, H., Huang, Y., et al.: CT-based radiomics signature: a potential biomarker for preoperative prediction of early recurrence in hepatocellular carcinoma. Abdom. Radiol. **42**, 1695–1704 (2017)
9. Ioffe, S., Szegedy, C.: Batch Normalization: Accelerating Deep Network Training by Reducing Internal Covariate Shift. Computer Science (2015)

10. Arevalo, J., Gonzalez, F.A., Ramos-Pollan, R., et al.: Convolutional neural networks for mammography mass lesion classification, Milan. IEEE (2015)
11. Chougrad, H., Zouaki, H., Alheyane, O.: Deep Convolutional Neural Networks for breast cancer screening. Comput. Methods Programs Biomed. **157**, 19–30 (2018)
12. Akselrod-Ballin, A., Karlinsky, L., Alpert, S., Hasoul, S., Ben-Ari, R., Barkan, E.: A region based convolutional network for tumor detection and classification in breast mammography. In: Carneiro, G., et al. (eds.) LABELS/DLMIA 2016. LNCS, vol. 10008, pp. 197–205. Springer, Cham (2016). https://doi.org/10.1007/978-3-319-46976-8_21
13. Becker, A.S., Marcon, M., Ghafoor, S., et al.: Deep learning in mammography: diagnostic accuracy of a multipurpose image analysis software in the detection of breast cancer. Invest. Radiol. **52**, 434–440 (2017)
14. Tang, P., Wang, X., Huang, Z., et al.: Deep patch learning for weakly supervised object classification and discovery. Pattern Recogn. **71**, 446–459 (2017)
15. Quellec, G., Lamard, M., Cozic, M., et al.: Multiple-instance learning for anomaly detection in digital mammography. IEEE Trans. Med. Imaging **35**, 1604–1614 (2016)
16. Fukushima, K., Miyake, S.: Neocognitron: a self-organizing neural network model for a mechanism of visual pattern recognition. In: Amari, S., Arbib, M.A. (eds.) Competition and Cooperation in Neural Nets, pp. 267–285. Springer, Heidelberg (1982). https://doi.org/10.1007/978-3-642-46466-9_18
17. Jiao, Z., Gao, X., Wang, Y., et al.: A deep feature based framework for breast masses classification. Neurocomputing **197**, 221–231 (2016)
18. Al-Masni, M.A., Al-Antari, M.A., Park, J.M., et al.: Detection and classification of the breast abnormalities in digital mammograms via regional Convolutional Neural Network, Seogwipo. IEEE (2017)
19. Redmon, J., Divvala, S., Girshick, R., et al.: You only look once: unified, real-time object detection, Las Vegas. IEEE (2016)
20. Li, S.: End-to-end training for whole image breast cancer diagnosis using an all convolutional design [J/OL]. arXiv, arXiv:1711.05775(2017). https://arxiv.org/ftp/arxiv/papers/1711/1711.05775.pdf. Accessed 15 Nov 2017
21. Pan, S.J., Yang, Q.: A survey on transfer learning. IEEE Trans. Knowl. Data Eng. **22**(10), 1345–1359 (2010)
22. Taylor, M.E., Stone, P.: Transfer learning for reinforcement learning domains: a survey. J. Mach. Learn. Res. **10**, 1633–1685 (2009)
23. Quionero-Candela, J., Sugiyama, M., Schwaighofer, A., et al.: Dataset Shift in Machine Learning. The MIT Press, Cambridge (2009)

Segmentation-Assisted Diagnosis of Pulmonary Nodule Recognition Based on Adaptive Particle Swarm Image Algorithm

Yixin Wang[1], Jinshun Ding[2(✉)], Weiqing Fang[1], and Jian Cao[2]

[1] Changshu No.2 People's Hospital, Changshu 215500, Jiangsu, China
[2] Changshu Meili Hospital, Changshu 215500, Jiangsu, China
769433552@qq.com

Abstract. The case characteristics of lung cancer are extremely complex, difficult to distinguish, and the rate of deterioration is rapid, and its early symptoms are not obvious. Early diagnosis and treatment of lung cancer is one of the main directions to reduce lung cancer mortality. The computer-aided detection and diagnosis system reduces the workload of the physician and improves the accuracy of image reading. In this paper, based on the analysis of the current lung nodule segmentation algorithm, in order to enhance the accuracy of lung nodule segmentation extraction, the adaptive particle swarm optimization algorithm is used to realize the simultaneous optimization of the number of mixed components and the model parameters, and finally realize the segmentation of lung nodules. The effectiveness and accuracy of segmentation of lung nodule recognition by adaptive particle swarm optimization algorithm is verified by adaptive particle swarm optimization and image model establishment. Provide new aids for the identification of pulmonary nodules.

Keywords: Adaptive particle swarm · Lung nodule · Image recognition · Auxiliary diagnosis

1 Introduction

The main causes of lung cancer are air pollution, chronic respiratory diseases, smoking, family inheritance, occupational exposure and diet. The case characteristics of lung cancer are extremely complex, difficult to distinguish, and the rate of deterioration is faster, and its early symptoms are not obvious. Missed diagnosis and misdiagnosis during early diagnosis are the direct cause of lower lung cancer cure rate. Clinical medical research shows that the cure rate of carcinoma in situ is close to 100%, the 5-year survival rate of patients with stage I lung cancer is as high as 60% to 90%, and the 5-year survival rate of patients with stage III and IV is only 5% to 20%. Due to the lack of an ideal early diagnosis, the early diagnosis rate of lung cancer is only about 14% [1]. Therefore, early diagnosis and treatment of lung cancer is one of the main directions to reduce lung cancer mortality.

© Springer Nature Singapore Pte Ltd. 2019
H. Ning (Ed.): CyberDI 2019/CyberLife 2019, CCIS 1138, pp. 504–512, 2019.
https://doi.org/10.1007/978-981-15-1925-3_36

The early manifestations of lung cancer are mainly pulmonary nodules. Pulmonary nodules refer to focal opaque lesions with a diameter of 3 to 30 mm in the lungs. The pulmonary nodules between 3 and 10 mm in diameter are called micro-nodules [2], and the pulmonary nodules between 10 and 20 mm in diameter are called Small nodules.

2   Research Status

The computer-aided detection and diagnosis system reduces the workload of physicians and improves the accuracy of image reading. It has a wide range of applications in medical imaging. The ability of CAD to use high-speed computer calculation and automatic processing has greatly helped early detection and diagnosis of lung cancer [3]. The system not only removes irrelevant information in the image, provides potential areas of the lesion, reduces the workload of the doctor, but also automatically analyzes the lesion area to obtain results and provide a relatively effective reference. The detection and segmentation of pulmonary nodules is the premise of feature extraction and classification identification of pulmonary nodules. Some scholars have done relevant research on automatic segmentation and semi-automatic segmentation of pulmonary nodules, but due to the complexity and diversity of lung nodules, the current segmentation methods are more or less insufficient [4] (Fig. 1).

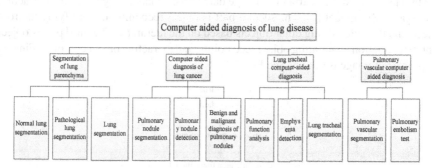

Fig. 1. Computer-aided detection and diagnosis

3   Introduction to Image Processing Methods

Based on the analysis of the current lung nodule segmentation algorithm, this paper improves the existing lung nodule segmentation algorithm, improves the accuracy of lung nodule segmentation, and improves the efficiency of lung nodule segmentation to some extent. In addition, the research of this subject is one of the important components of the lung CAD system research [5], which can effectively reduce the workload of doctors and provide a more accurate basis for judging the benign and malignant lung nodules, so as to improve the accuracy of lung cancer diagnosis.

3.1 Pretreatment of Chest CT Images

CT images may carry noise during imaging and transmission. The sources of noise are: First, during CT imaging, it is susceptible to various factors, such as the noise generated by the patient's own physiological motion, the quality of the scanning device itself, and the noise caused by the external environment. Second, in the CT image [6]. Noise may be generated during transmission. Therefore, the image needs to be smoothed and denoised, such as using a Gaussian filter to smoothly suppress artifacts in the chest CT image, or using a median filter to effectively suppress noise in the chest CT image.

3.2 Primary Segmentation of Lung Parenchyma

The initial segmentation of the lung parenchyma is an important step in achieving complete segmentation of the lung parenchyma. The initial segmentation of lung parenchyma mainly includes CT image OTSU-based image binarization, lung parenchymal extraction, filling of lung parenchyma and separation of left and right lung parenchyma, and finally the initial segmentation results of lung parenchyma [7].

3.3 Lung Nodule Image Extraction Process

In the process of lung nodule image extraction, it is mainly divided into three parts. The first part is model construction, image data transformation, organ outline extraction and organ fuzzy model [8]. The second part is object recognition, identifying the root object, and then identifying other objects based on the hierarchy. The third part is object contour extraction, generating object seed points, and extracting object contours. Finally, the processed image is output. See Fig. 2.

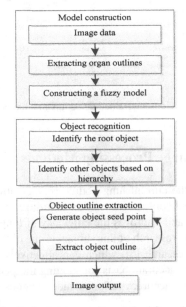

Fig. 2. Image processing method flow chart

4 Image Algorithm Model Research

In the lung nodule segmentation experiment, the number k of the mixed components is unknown, and often according to the empirical artificial setting, this process of excessively relying on artificial participation in image segmentation is unrealistic, so k often requires and blends proportions and components. The parameters are estimated together based on the sample. The pulmonary nodules vary according to the patient's condition and the condition, and the shape is different, which leads to the complexity of the gray distribution of the lung parenchyma. Therefore, the number of fixed distribution models can not fit all the cases well. In order to enhance the accuracy of segmentation and extraction of lung nodules, an adaptive particle swarm optimization algorithm was used to optimize the number of mixed components and model parameters, and finally the segmentation of lung nodules was achieved [9].

The inertia weight of the basic particle swarm optimization algorithm is generally fixed and cannot find the optimal solution well [10]. In this paper, the adaptive particle swarm optimization algorithm is used to adjust the inertia weight by nonlinear regression function, and the number of mixed components and parameters are optimized simultaneously by adjusting the vector length of the particle.

4.1 Adaptive Adjustment of Inertia Weight

In particle swarm optimization, diversity plays an important role in improving the effectiveness of particle optimization. The inertia weight is adjusted according to the condition of the particles to increase the diversity of the particles. Therefore, a diversity-based nonlinear adaptive strategy is introduced to adjust the inertia weight ω. In adaptive particle swarm optimization, diversity is defined as:

$$S(t) = f_{\min}(\mathbf{a}(t))/f_{\max}(\mathbf{a}(t)) \tag{4-1}$$

$$\begin{cases} f_{\min}(a(t)) = Min(f(a_i(t))) \\ f_{\max}(a(t)) = Max(f(a_i(t))) \end{cases} \tag{4-2}$$

Where $f(\mathbf{a}_i(t))$ is the fitness function value of the i-th particle, i = 1, 2...s. In the t-th iteration, $f_{\min}(\mathbf{a}(t)) \& f_{\max}(\mathbf{a}(t))$ is the minimum fitness value and the maximum fitness value in the current entire population [11].

Diversity $S(t)$ embodies the motion characteristics of the particles. In order to properly adjust the inertia weight, the local search ability and the global search ability are balanced, and the corresponding nonlinear regression function is introduced:

$$\gamma(t) = (L - S(t))^{-t} \tag{4-3}$$

Where L $\geq$ 2 is a predefined constant. In order to better update the speed, the ratio of change between the i-th particle and the global optimal particle is calculated:

$$A_i(t) = f(g(t))/f(\mathbf{a}_i(t)) \tag{4-4}$$

Where $f(g(t))$ is the fitness function value of the global optimal particle. The adaptive adjustment method of the final inertia weight ω is obtained from the above analysis, as shown in Equation E [12].

$$\omega_i(t) = \gamma(t)(A_i(t) + c) \qquad (4\text{-}5)$$

Where $\omega_i(t)$ i.s the inertia weight of the i-th particle in the t-th iteration, and $c \geq 0$ is a predefined constant, in order to improve the global recognition ability of the particle.

4.2 Adaptive Adjustment of the Number of Mixed Components and Parameters the Main Idea of the APSO Algorithm to Realize the Simultaneous Optimization of the Number of Model Components and Parameters Is as Follows

First, the upper bound k_{max} and the lower bound k_{min} of the mixed component number of the finite mixing model are respectively set. Then, in the process of particle swarm initialization, the parameters of each parameter and the number of model components in the distribution function are added to each particle to obtain particle groups with different vector lengths. In this simple way, the effectiveness of each particle can be determined. The number of mixed components.

Finally, in the particle optimization process, when the number of effective mixed components of the global optimal position and the individual optimal position is different from the number of components of other particles, the learning process is destroyed due to the difference in the length of the particle working vector, resulting in particle velocity update and The stop of the location update. To solve this problem, the maximum selection method is used to ensure that all particles in the update process have the same vector length. In the largest selection method, the current maximum effective mixing component number of all particles is selected and considered to be the effective number of all particle mixing components. At this point, each particle contains the same number of mixed components and the learning process continues. In order to demonstrate the adjustment of the parameters and model size in the mixed distribution model, Fig. 3-1 shows the iterative process of the particle swarm, assuming that the functions of the mixed distribution model are $k_1 = 2$, $k_2 = 5$ and $k_3 = 3$, and the first particle is the individual. Good particles, the third particle is the global best particle ($k_{best} = k_3$). In the adjustment process, when the number of components in the particle is greater than the optimal number of particles (for example, k_{best} is greater than $k3$), the number of components in the particles needs to be reduced; when the number of components in the particle is smaller than the optimal particle When the number is (for example, k_2 is less than k_2), the number of components in these particles needs to be increased. In the process of iterative process, the particles gradually approach the global optimal particle, so as to obtain the estimation result of the optimal model, and achieve accurate classification and acquisition of image pixels [13].

The optimal mixed component number k_{best} is judged by the formula (4-6), and the other particle size is updated by the formula (4-7).

$$k_{bsst} = MS(g), \ 1 \leq k \leq k_{max} \qquad (4\text{-}6)$$

$$K_i = \begin{cases} K_i - 1, & if\,(k_{best} < K_i) \\ K_i + 1, & if\,(k_{best} < K_i) \end{cases} \tag{4-7}$$

Among them, $MS(\bullet)$ is the obtained function of the optimal number of components under the current iteration of the particle swarm [14]. That is, the number of effective mixed components of the global optimal particle.

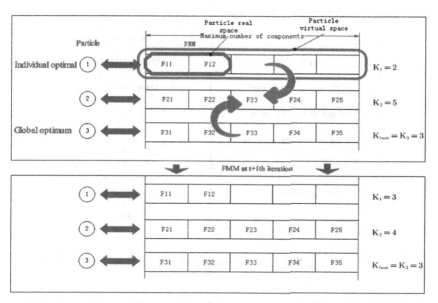

Fig. 3. APSO algorithm iterative process

5 Adaptive Particle Swarm Image Recognition Segmentation Algorithm Step Verification

Through the above method, the simultaneous estimation of the number of components and the model parameters in the process of segmentation of pulmonary nodules is achieved. The flowchart is shown in Fig. 4.

The algorithm steps are as follows:

Step 1: Initialize the parameters. The maximum number of iterations, the inertia weight, and the acceleration constants c1 and c2 are initialized, and the upper bound k_{max} and the lower bound k_{min} of the mixed component number are initialized.

Step 2: Initialize the population. Random (or regular) the parameters of the distribution function and the component number information of the model are added to each particle to obtain a particle group with different vector lengths. The initial individual optimal position of the population is the initial position of each particle, and the initial global optimal position is the global optimal particle in the initial particle group. And determining the current optimal mixed component number k_{best} according to the global optimal particle.

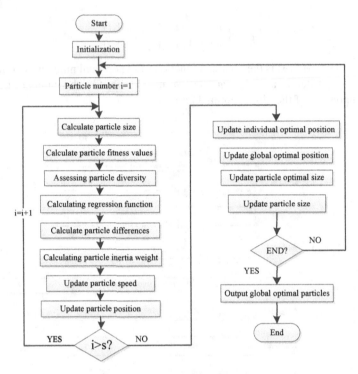

Fig. 4. APSO algorithm flow chart

Step 3: Update the effective mixing component number of the particles according to the formula (4-7).

Step 3: Update the effective mixing component number of the particles according to the formula (4-7).

Step 4: Calculate the current fitness value of each particle $f(a_i(t))$, evaluate the particle diversity $S(t)$, calculate the regression function $\gamma(t)$, calculate The particle difference $A_i(t)$, calculates the inertia weight $\omega_i(t)$.

Step 5: Determine the current maximum mixed component number according to the maximum selection method, and update the speed and position of each particle according to formulas (4-1) and (4-2).

Step 6: Update the individual optimal particles and the global optimal particles according to the formulas (4-3) and (4-4). And determine the optimal component number k_{best}.

Step 7: Determine whether the stop condition is satisfied. If it is the end iteration, execute step 7, otherwise return to step 3.

Step 8: Output the current global optimal position to produce the final data set. The estimation result of the optimal segmentation model model is obtained (Fig. 5).

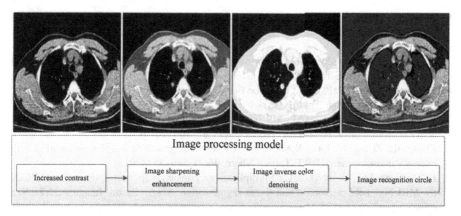

Fig. 5. Image processing model diagram

6 Summary

Image segmentation of lung nodules is an important part of lung cancer CAD and an important basis for judging the benign and malignant lung nodules. The main research content of this paper is based on the adaptive particle swarm image algorithm image lung nodule recognition segmentation algorithm, and the image adaptive particle model is used to realize the segmentation of lung nodules.

References

1. Faizal Khan, Z., Al Sayyari, A.S., Quadri, S.U.: Automated segmentation of lung images using textural echo state neural networks. In: International Conference on Informatics, Health and Technology, pp. 1–5 (2017)
2. Soltaninejad, S., Cheng, I., Basu, A.: Robust lung segmentation combining adaptive concave hulls with active contours. In: IEEE International Conference on Systems, Man, and Cybernetics (2016)
3. Zhao, Y., Bock, G.H.D., Vliegenthart, R., et al.: Performance of computer-aided detection of pulmonary nodules in low-dose CT: comparison with double reading by nodule volume. Eur. Radiol. **22**(10), 2076–2084 (2012)
4. Kumar, S.A., Ramesh, J., Vanathi, P.T., et al.: Robust and automated lung nodule diagnosis from ct images based on fuzzy systems. In: International Conference on Process Automation, Control and Computing, pp. 1–6. IEEE (2011)
5. Jacobs, C., van Rikxoort, E.M., Twelmann, T., et al.: Automatic detection of subsolid pulmonary nodules in thoracic computed tomography images. Med. Image Anal. **18**(2), 374–384 (2014)
6. Revel, M.P., Merlin, A., Peyrard, S., et al.: Software volumetric evaluation of doubling times for differentiating benign versus malignant pulmonary nodules. AJR Am. J. Roentgenol. **187**(1), 135–142 (2006)
7. Awai, K., Murao, K., Ozawa, A., et al.: Pulmonary nodules: estimation of mali gnancy at thin-section helical CT–effect of computer-aided diagnosis on performance of radiologists. Radiology **239**(1), 276–284 (2006)

8. Biernacki, C., Celeux, G., Govaert, G.: Assessing a mixture model for clustering with the integrated completed likelihood. IEEE Trans. Pattern Anal. Mach. Intell. **22**(7), 719–725 (2000)
9. Maulik, U., Bandyopadhyay, S.: Performance evaluation of some clustering algorithms and validity indices. IEEE Trans. Pattern Anal. Mach. Intell. **24**(12), 1650–1654 (2002)
10. Figueiredo, M.A.T., Jain, A.K.: Unsupervised learning of finite mixture models. IEEE Trans. Pattern Anal. Mach. Intell. **24**(3), 381–396 (2002)
11. Ji, D., Yao, Y., Yang, Q., et al.: MR image segmentation using graph cuts based geodesic active contours. Int. J. Hybrid Inf. Technol. **9**(1), 91–100 (2016)
12. Zhang, K., Zhang, L., Lam, K.M., et al.: A level set approach to image segmentation with intensity inhomogeneity. IEEE Trans. Cybern. **46**(2), 546–557 (2015)
13. Li, C., Gore, J.C., Davatzikos, C.: Multiplicative intrinsic component optimization (MICO) for MRI bias field estimation and tissue segmentation. Magn. Reson. Imaging **32**(7), 913 (2014)
14. Li, C., Xu, C., Gui, C., et al.: Distance regularized level set evolution and its application to image segmentation. IEEE Trans. Image Process. **19**(12), 3243–3254 (2010)

Auxiliary Recognition of Alzheimer's Disease Based on Gaussian Probability Brain Image Segmentation Model

Xinlei Chen, Dongming Zhao$^{(\boxtimes)}$, and Wei Zhong

Suzhou Guangji Hospital, Suzhou 215000, Jiangsu, China
13862339400@163.com

Abstract. Alzheimer's disease is an important disease that threatens the health of the elderly after cardiovascular disease, cerebrovascular disease and cancer. Early diagnosis and early intervention have an inestimable effect on disease control and treatment. Especially for China, which is facing the problem of population aging, early detection and early treatment are particularly important. According to the neuroimaging study of disease, by studying the degree of local brain loss in patients with Alzheimer's disease, the disease information of the disease manifested in the brain structure is revealed, such as the decrease of the volume of the hippocampus and the thickness of the medial frontal temporal cortex. Thin and so on. In this paper, the local Gaussian probability image segmentation model is used to segment and extract the brain nuclear magnetic image, and the image segmentation of the hippocampus structure is extracted. The local Gaussian probability algorithm of image segmentation extraction algorithm is designed and optimized. The maximal posterior probability principle and Bayes' rule are introduced to optimize the algorithm by grayscale processing of local image. Therefore, the Gaussian probability model is used to obtain the local mean and standard deviation as a function of spatial variation. Therefore, the probability model is more suitable for image segmentation with uneven gray scale than the probability model based on global hypothesis. Finally, experiments are carried out to verify the correctness of the theory and the robustness of Gaussian probability brain image segmentation.

Keywords: Alzheimer's disease · Nuclear magnetic resonance · Local Gaussian probability · Hippocampus · Image segmentation

1   Introduction

Symptoms of the disease: Alzheimer's disease (AD) is a chronic degenerative disease of the acquired nervous system. The early course of the disease is slow and only shows short-term memory impairment, but as the disease progresses, the subject will appear subjective. Abnormal behaviors such as memory loss, language barriers, time and spatial orientation disorders, poor judgment, emotional changes, and decreased self-care ability [1].

Status of disease development: Alzheimer's disease is not only a cardiovascular disease, cerebrovascular disease and tumor, but also an important disease that threatens

© Springer Nature Singapore Pte Ltd. 2019
H. Ning (Ed.): CyberDI 2019/CyberLife 2019, CCIS 1138, pp. 513–520, 2019.
https://doi.org/10.1007/978-981-15-1925-3_37

the health of the elderly. Since Dr. Alzheimer first reported a 51-year-old female case in the early 20th century (1906), Alzheimer's disease has been in the medical history for nearly 110 years. The discovery of this disease has opened up a new world for human understanding of dementia. There are about 7 million new cases each year, with an average of 1 case every 4 s (one stroke every 7 s), and the number of global dementia cases has doubled in about 20 years. According to epidemiological survey data, as of 2013, about 62% of dementia cases worldwide are concentrated in developing countries. By 2050, 71% of patients will be concentrated in developing countries, and the fastest progress in aging is China, India, and South Asia and the Western Pacific [2]. The current status of Alzheimer's disease is serious, but so far there is no effective way to cure or control the progression of the disease. Early diagnosis and early intervention have an inestimable effect on disease control and treatment. Especially for China, which is facing the problem of population aging, early detection and early treatment are particularly important.

2 Current Status of Disease Diagnosis

First, the biomarker diagnosis, and the emergence of the biomarker, to some extent, predicts the possibility of Alzheimer's disease. Currently, clinically used biomarkers based on cerebrospinal fluid extraction, and mainly include: protein (divided into T-tau, P-tau, P-tau231, P-tau181). Although the use of cerebrospinal fluid can achieve diagnostic purposes to a certain extent, it is generally not easily accepted by patients because it is often a traumatic procedure [3].

Second, genetic diagnosis. In the pathogenesis hypothesis of AD, genetic factors are considered as a risk factor for the onset (such as APOE4) or pathogenic factors (such as Presenilin-1, PS-1), and gradually become an active research area. Ten genetic loci related to AD have been identified, and more are being discovered [3].

Third, neuroimaging research [4]. Because neuroimaging techniques can assess the brain from a functional metabolic or structural perspective, the brains of AD patients can be studied early in the absence of morphological changes. Neuroimaging has gradually shown its unique advantages in the early prediction of AD and its intervention. There are mainly magnetic resonance imaging (MRI) studies, through the study of the extent of local brain damage in patients with Alzheimer's disease, revealing the disease information of the disease in the brain structure, such as the volume reduction of the hippocampus The thickness of the medial frontal temporal cortex is thinner. In the functional study of the brain, there is mainly functional magnetic resonance imaging (fMRI), which can display the activation site and extent of the brain region, so that it can further reflect the fusion of function and structure, which can be used as a primary diagnostic tool for clinical use; There are also studies of brain structure networks, such as MRI imaging techniques based on structure and DTI. By constructing a network of brain structures, the resulting structural network computing models are analyzed to study the network trends of the patient's brain. These neuroimaging studies have laid a solid foundation for the early prevention and detection of AD and the continuous improvement of clinical assistant diagnosis.

3   Disease Image Detection Principle

Plane measurement using MR1 plane measurement method to measure the brain structure of patients with Alzheimer's disease, compared with three-dimensional volume measurement, it is simpler, more convenient, faster, and more widely used clinically. DeLeon et al. [5]. performed MRI scans of cadaveric specimens from patients with Alzheimer's disease. The change in the width of the lateral ventricle can reflect the extent of hippocampal atrophy to some extent, and can simultaneously understand the extent of atrophy of the medial temporal lobe and the entire lobe. Tanabe et al. used MRI to measure the brain parenchyma, gray matter, cerebral sulcus, etc., and found that all indicators between the dementia group and the normal elderly control group have a large overlap, so the dementia can be suggested according to the sulcus measurement results, but Patients with Alzheimer's disease and normal elderly cannot be accurately identified. In view of this, it may be more appropriate to select the hippocampal structure and lateral ventricle width as an identification index. Therefore, it is more inclined to use MRI three-dimensional measurement method to study brain structural changes in patients with Alzheimer's disease. In some brain regions of patients with Alzheimer's disease, the intensity of functional magnetic resonance imaging scan signals increases and the range expands when they receive cognitive activation. Functional magnetic resonance imaging studies have found that patients with mild cognitive impairment developing Alzheimer's disease have a greater range of cerebral palsy signal strength in the right hippocampus during the memory phase of the memory test, which may be Al Pathological compensatory response in patients with Alzheimer's disease. In a cognitive activation test of patients with Alzheimer's disease, Johnson et al. found that the greater the degree of atrophy of the left frontal hippocampus, the greater the area activated and the stronger the signal. In this regard, he believes that in patients with memory problems, some of the remaining normal nerve tissue can replace the tissue that has already developed lesions, and thus the signal intensity of the activated brain region increases and the range expands.

4   Image Processing Methods Introduced

According to the requirements of the specific brain magnetic imaging technology analysis task, the Gaussian probability image segmentation model is used to segment the brain image. The purpose of segmentation is to divide or extract the original image into the hippocampus of the important disease analysis site of Alzheimer's disease. structure. By analyzing and comparing changes in hippocampal structure, it provides a basis for the diagnosis of Alzheimer's disease.

Brain image segmentation is mainly divided into the following steps:

(1) Image preprocessing, that is, image restoration. The image contains noise due to factors such as equipment. Removing noise without losing tissue information facilitates subsequent image processing;
(2) Tick off brain tissue. It is to remove non-brain tissue in the brain image, such as brain shell, fat and other tissues. Since the non-brain tissue and the background part

contain a large proportion in the brain image. Therefore, the removal of non-brain tissue and background parts can improve the processing accuracy of subsequent processes;

(3) Go to the offset field. Due to the influence of the imaging mechanism, the image will contain an offset field, resulting in uneven image gray scale, which makes the segmentation result inaccurate. Accurate offset field recovery model can greatly improve the accuracy of subsequent image processing;

(4) Split the effective area. The region of interest is segmented using an active contour model or the like to analyze the image. This paper attempts to integrate each problem into a unified image segmentation framework, which makes it highly intelligent.

The main work and research results and algorithm steps of this paper are as follows:

In order to make full use of the information of the local image, we first consider the grayscale distribution of the image in the neighborhood of each pixel.

For each point x in image domain Ω, there is a neighborhood of radius ρ : $\mathcal{O}_x \triangleq \{y : |x - y| \le \rho\}$.

Suppose the image consists of $\{\Omega_i\}_{i=1}^{N}$, satisfying the number of categories of $\Omega = \cup_{i=1}^{N} \Omega_i$, $\Omega_i \cap \Omega_j = \emptyset$ in which the N image is. For $\mathcal{O}_x$, $\{\Omega_i\}_{i=1}^{N}$ will divide the neighborhood $\mathcal{O}_x$ into $\{\Omega_i \cap \mathcal{O}_x\}_{i=1}^{N}$. For example, as shown, the figure (Fig. 1) has three parts: Ω_1, Ω_2, and Ω_3. These three parts divide x's neighborhood $\mathcal{O}_x$ into three sub-parts $\Omega_1 \cap \mathcal{O}_x$, $\Omega_2 \cap \mathcal{O}_x$ and $\Omega_3 \cap \mathcal{O}_x$

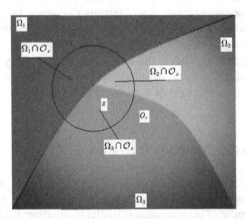

Fig. 1. Example of $\Omega_i \cap \mathcal{O}_x$. The green circle represents the neighborhood $\mathcal{O}_x$ of the current point x. The image consists of three parts: Ω_1, Ω_2, Ω_3; these three parts divide the neighborhood x of $\mathcal{O}_x$ into three sub-parts: $\Omega_1 \cap \mathcal{O}_x$, $\Omega_2 \cap \mathcal{O}_x$ and $\Omega_3 \cap \mathcal{O}_x$. (Color figure online)

According to the maximum a posteriori probability (MAP) [6], we will focus on how to optimally segment this neighborhood $\mathcal{O}_x$. We define $p(y \in \Omega_i \cap \mathcal{O}_x | I(y))$ as the posterior probability that the subpart y at a given $I(y)$ belongs to $\Omega_i \cap \mathcal{O}_x$.

According to the Bayes rule, we can get

$$p(y \in \Omega_i \cap \mathcal{O}_x | I(y)) = \frac{p(I(y) | y \in \Omega_i \cap \mathcal{O}_x) p(y \in \Omega_i \cap \mathcal{O}_x)}{p(I(y))} \tag{4.1}$$

Where $p(I(y)|y \in \Omega_i \cap \mathcal{O}_x)$ is the probability that the gray value $I(y)$ belongs to $\Omega_i \cap \mathcal{O}_x$, and is abbreviated as $p_{i,x}(I(y))$ for convenience; $p(y \in \Omega_i \cap \mathcal{O}_x)$ is the prior probability that y belongs to $\Omega_i \cap \mathcal{O}_x$, and $p(I(y))$ is the prior probability of gray value $I(y)$. Since $p(I(y))$ is independent of region $\Omega_i \cap \mathcal{O}_x$, $p(I(y))$ is treated as a fixed value.

Assuming that the probability of the prior probability $p(y \in \Omega_i \cap \mathcal{O}_x)$ is equal in each region, i.e.

$$p(y \in \Omega_i \cap \mathcal{O}_x) = \frac{1}{N}$$

then the prior probability $p(y \in \Omega_i \cap \mathcal{O}_x)$ can also be omitted. At the same time, we assume that the distribution of each pixel in each region is independent of each other, and by maximizing the posterior probability (the MAP) we can get:

$$\prod_{i=1}^{N} \prod_{y \in \Omega_i \cap \mathcal{O}_x} p_{i,x}(I(y)) \tag{4.2}$$

By doing an log transformation on Eq. (4.2), we can transform the maximized posterior probability to minimize the following energy ε_x^{LGDF}:

$$\varepsilon_x^{LGDF} = \sum_{i=1}^{N} \int_{\Omega_i \cap \mathcal{O}_x} -\log p_{i,x}(I(y)) dy \tag{4.3}$$

For the construction of probability $p_{i,x}(I(y))$ in Eq. (4.3), there is a large amount of literature to draw on, such as the Gaussian probability model based on variance with a fixed value proposed by Chan and Vese, or the proposed full Gaussian probability model, and the proposed Parzen-based window [7]. Nonparametric probability model. However, these probabilistic models are all built on the global space, and these probabilities do not change with the spatial position within each region. Therefore, these global-based probability models are not able to overcome the gray-scale inhomogeneities that often exist in images. In this section, we will use a local Gaussian probability model:

$$p_{i,x}(I(y)) = \frac{1}{\sqrt{2\pi}\sigma_i(x)} \exp\left(-\frac{(u_i(x) - I(y))^2}{2\sigma_i(x)^2}\right) \tag{4.4}$$

The probability model is defined at each point on the image and varies with the position of the space, where $u_i(x)$ and $\sigma_i(x)$ are local mean and standard deviation, respectively [8]. Since the local mean and standard deviation are functions that vary with spatial variation, the probability model is more suitable for dealing with gray-scale inhomogeneity than the probability model based on global hypothesis.

We further add the window function in Eq. (4.4) to get the following energy:

$$\varepsilon_x^{LGDF} = \sum_{i=1}^{N} \int_{\Omega_i \cap \mathcal{O}_x} -\omega(x - y) \log p_{i,x}(I(y)) dy \tag{4.5}$$

Where $\omega(x-y)$ is a non-negative window function that satisfies $\omega(x-y) = 0$, when $|x - y| > \rho; \int \omega(x-y)dy = 1.$

Mainly reflected in the following two aspects: (1) we introduce a local energy in the form of a Gaussian nucleus, (2) our energy is a double integral form, and ordinary energy is a one-fold integral.

Algorithm flow:

Brain MR image segmentation algorithm based on Gaussian probability segmentation model
Initialize the level set function p_1 and p_2
While the algorithm does not converge
according to (4.1) , (4.2) , Update Ω_i and O_x
according to (4.3) , Update level set function p_1
according to (4.4) , Update Ω_i and O_x
according to (4.5) , Update level set function p_2
end while

5   Experimental and Image Algorithm Experimental Results and Analysis

Image Preprocessing: Image preprocessing is a process before the brain magnetic image is sorted and submitted to the segmentation model for processing [9]. This process is called image preprocessing. In image analysis, the quality of the image directly affects the accuracy of the design and effect of the recognition algorithm. Therefore, before image analysis (feature extraction, segmentation, matching and recognition, etc.), pre-processing is required. The main purpose of image preprocessing is to eliminate irrelevant information in the image, restore useful real information, enhance the detectability of relevant information, and minimize data, thereby improving the reliability of feature extraction, image segmentation, matching and recognition.

The general pre-processing process is (Figs. 2 and 3):

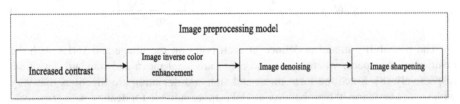

Fig. 2. Image preprocessing model.

Nuclear magnetic image offset field: Due to the existence of the offset field phenomenon in the original nuclear magnetic image, in addition, the boundary between the noise and the tissue in the image is also blurred [10]. The contrast between the gray

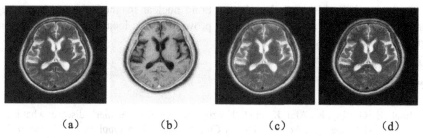

(a)   (b)   (c)   (d)

Fig. 3. Image preprocessing.

matter and the cerebrospinal fluid is lower, as in the vicinity of the ventricles in the image, the gray scale of the gray matter and the gray value of the cerebrospinal fluid are almost equal. Traditional image contrast based segmentation algorithms will find it difficult to get accurate results.

Figure 4(e) is a brain nuclear magnetic image of Alzheimer's disease. In order to show that the method can overcome gray unevenness and low contrast well, we add strong gray unevenness in Fig. 4(e). Sex, as shown in Fig. 4(f). Under the influence of gray scale inhomogeneity, the image contrast of the hippocampus region is very low. Figure 4(g) shows the initial curve, and the image segmentation is rough. The results of the Gaussian partition model are shown in Fig. 4(h). Since the Gaussian probability segmentation model is based on local hypotheses, the Gaussian probability segmentation model can correctly segment the image. On the other hand, the model results can also see that the Fig. 4 can Strongly overcome gray unevenness.

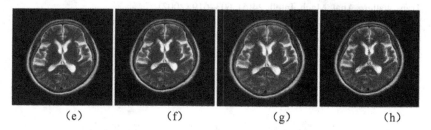

(e)   (f)   (g)   (h)

Fig. 4. Grayscale inhomogeneity and low contrast processing.

6   Summary

We propose a novel regional active contour model based on local Gaussian probability. The model effectively utilizes the local mean and local variance information of the image. Since the local mean and the local variance change with spatial changes, it can well overcome the gradation inequalities in the image, and the noise and Contrast is also better robust. It is worth mentioning that the local mean and local variance of the Gaussian probability model can be strictly derived from the variational principle, unlike other models that are explicitly defined.

The research in this paper has done a brain nuclear magnetic hippocampus segmentation experiment, and the experiment proved that the local Gaussian probability segmentation method is effective.

References

1. Braak, E., Griffing, K., Arai, K., et al.: Neuropathology of Alzheimer's disease: what is new since A. Alzheimer Eur. Arch. Psychiatry Clin. Neurosci. **249**(Suppl 3), 14–22 (1999)
2. Jagust, W.: Positron emission tomography and magnetic resonance imaging in the diagnosis and prediction of dementia. Alzheimer's Dement. **2**(1), 36–42 (2006)
3. deToledo-Morrell, L., Stoub, T.R., Bulgakova, M., et al.: MRI-derived entorhinal volume is a good predictor of conversion from MCI to AD. Neurobiol. Aging **25**(9), 1197–1203 (2004)
4. Blennow, K., de Leon, M.J., Zetterberg, H.: Alzheimer's disease. Lancet **368**(9533), 387–403 (2006)
5. Vese, L., Chan, T.: A multiphase level set framework for image segmentation using the Mumford and Shah model. Int. J. Comput. Vis. **50**(3), 271–293 (2002)
6. Rousson, M., Deriche, R.: A variational framework for active and adaptative segmentation of vector valued images. In: IEEE Workshop on Motion and Video Computing, pp. 52–62 (2002)
7. Zhu, S., Yuille, A.: Region competition: unifying snakes, region growing, and Bayes/MDL for multiband Image segmentation. IEEE Trans. Pattern Anal. Mach. Intell. **18**(9), 884–900 (1996)
8. Kim, J., Fisher, J., Yezzi, A., Cetin, M., Willsky, A.: Nonparametric methods for image segmentation using information theory and curve evolution. In: IEEE International Conference on Image Processing, vol. 3, pp. 797–800 (2002)
9. Comaniciu, D., Meer, P.: Mean shift: a robust approach toward feature space analysis. IEEE Trans. Pattern Anal. Mach. Intell. **24**(5), 603–619 (2002)
10. Meer, P., Georgescu, B.: Edge detection with embedded confidence. IEEE Trans. Pattern Anal. Machine Intell. **23**, 1351–1365 (2001)

Sleep Stage Classification Based on Heart Rate Variability and Cardiopulmonary Coupling

Wangqilin Zhao[1,3], Xinghao Wu[2,3], and Wendong Xiao[1](✉)

[1] University of Science and Technology, Beijing, China
wdxiao@ustb.edu.cn
[2] CAS Institute of Healthcare Technologies, Hefei, China
[3] National University of Singapore, Singapore, Singapore

Abstract. Sleep is a physiological process controlled by the autonomic nervous system. Autonomic activity differs in different stages of sleep. Heart Rate Variability (HRV) is a widely recognized indicator of autonomic activity and has been commonly used in sleep stage classification. However, HRV suffers from low repeatability and is very volatile. Cardiopulmonary Coupling (CPC) reflects autonomic activity from a different perspective and has been used to measure sleep quality. This paper explores the effect of using combination of HRV and CPC features in sleep stage classification. The experimental results using a decision-tree-based support vector machine (DTB-SVM) classifier on MIT-BIH polysomnographic database have shown that by adding three CPC features, the overall sleep stage classification accuracy has been raised from 95.74% (Kappa = 0.9257) to 96.89% (Kappa = 0.9449). The CPC features have shown to be superior in distinguishing deep sleep stage (with 3.69% increase). The classification accuracy of wake and light sleep has also improved with 1.67% and 0.89%, respectively.

Keywords: HRV · CPC · DTB-SVM · Sleep stage

1 Introduction

Sleeping takes up an average of a third of our lives, and it heavily impacts our physical and mental wellbeing. Studies have shown that poor sleep quality leads to an increased risk of depression, chronic illness and injury [1]. Sleep is a complex physiological process and can be divided into wake period, rapid-eye-movement (REM) period and non-rapid-eye-movement (NREM) period. The NREM period is further classified into Light Sleep (N1–N2) and Deep Sleep (N3–N4). Studying the distribution of sleep stage is an important basis for determining the quality of sleep. However, the classification of sleep stages remains a research issue.

Polysomnography (PSG) is commonly recognized as the benchmark for identifying sleep stages. PSG is a sleep test that measures a variety of physiological signals such as electrocardiogram (ECG), respiration, electro-oculogram (EOG), electroencephalogram (EEG) etc., and marks the sleep stage every 30-s epoch. PSG requires the subject to wear multiple sensors and be monitored in the observation room all night. PSG test is

© Springer Nature Singapore Pte Ltd. 2019
H. Ning (Ed.): CyberDI 2019/CyberLife 2019, CCIS 1138, pp. 521–527, 2019.
https://doi.org/10.1007/978-981-15-1925-3_38

inconvenient, and heavily disturbs the subject's sleep. Thus, there is a continuous effort to develop portable at-home sleep test techniques.

The autonomic nervous system (ANS) regulates cardiopulmonary system, as well as the sleep process [2, 3]. So, changes in sleep status are reflected in respiratory and ECG signals. Acquisition of respiratory signals and electrocardiograms can be achieved with comfortable wearable devices. Heart rate variability (HRV) is the observed variation in the time interval between heartbeats, calculated from the RR interval derived from ECG. Studies have shown that HRV increases as sleep enters the REM phase and decreases when entering the NREM phase [4, 5]. As sleep enters the REM phase, a frequency domain indicator of HRV, the low frequency high frequency component ratio (LF/HF), increases significantly [6].

Cardiopulmonary Coupling (CPC) is a measurement of Respiratory Sinus Arrhythmia (RSA). It is a reflection of the interaction between the cardiovascular system and the respiratory system. As CPC measures, the energies of the high frequency, low frequency and ultra low frequency regions can be used to reflect different sleep depth and detect respiratory disorders. It has been applied in assessing sleep quality and determining sleep apnea events [7, 8]. Yet few papers have tried to apply CPC into sleep stage classification.

In recent years, many scholars have tried to use physiological signals other than EEG for sleep stage classification. Wilhelm Daniel Scherz et al. used three transform domain HRV features to determine sleep stage [9]. Agnes Klein et al. proposed an algorithm that combines ECG and body motion signals. The average rates for sleep stages REM, Wake, NREM-1, NREM-2, NREM-3 were 38.1%, 14%, 16%, 75%, 54.3% respectively [10]. Mourad Adnane et al. attempted to extract heart rate variability features in time domain, frequency domain, detrended fluctuation analysis and window detrended fluctuation analysis, and used support vector machine to distinguish between sleep and wake stages only with achieved average accuracy of 79.31% [11]. Martin Oswaldo Mendez et al. used a time-varying autoregressive model to extract HRV and used Hidden Markov Model to classify REM and NREM with an accuracy of 79.3% [12].

This paper explores the combination of HRV and CPC features in sleep stage classification. In the rest of this paper, Sect. 2 describes the method, including data processing, features and classifier used in the research. Classification results and discussions are presented in Sect. 3. Section 4 is the conclusion and future work.

2   Method

2.1   Data Processing

This paper uses polysomnographic data from the MIT-BIH database for training and testing. MIT-BIH Polysomnographic Database [13] from PhysioNet [14] stores recordings of multiple physiologic signals during sleep studies. Each recording contains the subject's ECG, respiratory signal and EEG with sleep stage annotations.

To obtain the NN interval for model training and testing, the following processing steps were performed:

(a) In the database, the sleep stage annotations are given in terms of 30-second data interval. HRV features, especially Cardiopulmonary Coupling (CPC) features should be extracted in a period which contains enough number of respiration cycles. Here we take the length of data segment to be 300 s. Figure 1 shows the relationship of the data segments with the original data internals in the MIT-BIH database.

(b) For each data segment, perform standard ECG processing [15, 16] to detect R points and to form sequence of RR intervals. Outliers and RR interval values that are too large or too small compared to average value are removed and replaced with cubic spline interpolation values. The processed RR intervals are called NN intervals.

(c) For each data segment, perform re-sampling to NN intervals so as to synchronize with respiration data.

(d) For each data segment, extract HRV and CPC features.

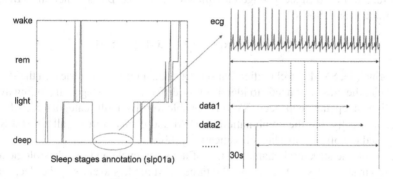

Fig. 1. Data preprocessing diagram

2.2 Feature Extraction

The classification algorithm in this paper uses a total of 15 features, consisting of 12 HRV features and 3 CPC features.

HRV Features

Following existing research [17, 18], HRV features used in this paper can be split into three categories:

three time-domain features: SDNN, RMSSD, pNN50;

four frequency-domain features: VLF power, LF power, HF power, LF/HF ratio;

five nonlinear features: SD1, SD2 and SD1/SD2 from Poincare plot, as well as sample entropy and approximate entropy.

CPC Features

CPC analysis is dependent on NN intervals and breathing signals, as well as their individual cross spectral densities. The formula for CPC [19] is given by:

$$CPC(f_n) \equiv <\Gamma_n(R, E)>^2 \Lambda_n \qquad (1)$$

$$\Lambda_n = \frac{<\Gamma_n(R, E)>^2}{<\widehat{R}_n>^2 <\widehat{E}_n>^2} \tag{2}$$

It is important to note that NN intervals used in the formula above needs to be resampled by cubic spline interpolation to the same length as the respiratory signal in the time window. CPC features used in this paper are: LCPC (0.01–0.1 Hz), HCPC (0.1–0.4 Hz) and LCPC/HCPC.

2.3 Classification Method

Support Vector Machine (SVM) is chosen for its statistical learning capabilities. The kernel function used in sleep stage classification is Radial Basis Function (RBF), and the formula is as follows:

$$K(x_i, x_j) = \exp\left(-\gamma |x_i - x_j|^2\right), \gamma > 0 \tag{3}$$

In principle, SVM is a classifier that separates data into 2 categories. In the study of this article, the classifier needs to identify four different sleep stages, including awake, REM, light sleep and deep sleep. SVM is not able to solve multi-category classification problems directly, but the combination of SVM and decision tree (called DTB-SVM) can be used to solve multi-class classification problems.

Based on the structural characteristics of the sleep cycle and the physiological features used in sleep classification, we used three SVM models to classify the sleep stages. First, a classifier is used to separate the awake phase and the sleep phase, and then within the sleep phase, the REM phase and the NREM phase are separated, and finally the light sleep and deep sleep are separated by the last classifier.

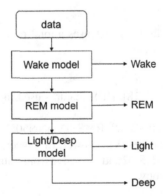

Fig. 2. Three stages DTB-SVM

Before training the model, each feature is normalized in its own dimension. Since the algorithm uses the RBF kernel of the SVM, the parameters that need to be manually set are the width parameter g of the kernel function and the error penalty factor C.

g affects the complexity of the distribution of sample data in high-dimensional space, and C presents the ratio of the confidence range and empirical risk of the learning machine in a particular space. In order to prevent over-fitting, this algorithm selects the optimal parameters in the way of K-fold cross-validation and grid search. The contour map of the classification model for determining the Wake Model is as follows:

3 Experimental Results

The MIT-BIH Polysomnographic database stores recordings of multiple physiologic signals during sleep and provides a total of 18 records. Among which, records slp01a and slp01b are data of a single subject and slp02a and slp02b are data of another subject. The remaining 14 records are all from different subjects. In the provided records, all 16 subjects are male, aged 32 to 56 (mean age 43), with weights ranging from 89 to 152 kg (mean weight 119 kg). Each recording contains an ECG signal annotated beat-by-beat, and EEG and respiration signals annotated with respect to sleep stages and apnea. All physiological signals were digitized at a sampling rate of 250 Hz with a resolution of 12 bits/sample.

Sleep stages are annotated per 30-s window according to the database. Every ten consecutive windows are grouped into 300-s epochs and the stage of each epoch is defined as the majority stage among ten annotations. If the amount of majority stage is less than seven, the epoch is defined as a transitional stage and will be removed. After that, the dataset, consisting of a total of 8069 epochs (with 2042 in wake, 564 in REM, 4968 in light and 495 in deep) was obtained. Half of the dataset is randomly selected as the training set and the other half is used as the test set. Each epoch contains 15 features.

To show the effect of adding CPC features, classification done using HRV and CPC features are compared with classification done using HRV features only.

Table 1. Confusion matrix of HRV-only classification

Predicted	Actual			
	Wake	REM	Light	Deep
Wake	959	8	41	2
REM	9	251	6	0
Light	48	19	2419	26
Deep	0	0	12	216
Total	1016	278	2478	244
Accuracy	94.39%	90.29%	97.62%	88.52%

Table 1 lists classification results for using HRV features only. Classification accuracy of wake, REM, light, deep stages are 94.39%, 90.29%, 97.62%, 88.52%, respectively, with an overall accuracy of 95.74% (k = 0.9257).

Table 2. Confusion matrix of HRV and CPC classification

Predicted	Actual			
	Wake	REM	Light	Deep
Wake	976	5	25	1
REM	5	249	2	0
Light	34	24	2441	18
Deep	1	0	10	225
Total	1016	278	2478	244
Accuracy	96.06%	89.57%	98.51%	92.21%

Table 2 contains classification results for combination of HRV and CPC features. Classification accuracy of wake, REM, light, deep stages are 96.06%, 89.57%, 98.51%, 92.21%, respectively, with an overall accuracy of 96.89% ($k = 0.9449$). There is a 1.15% increase compared to using HRV features only.

From Tables 1 and 2, there is a notable increase in deep sleep classification accuracy when CPC features are incorporated, while the increase in wake and light stages are not as prominent. That means CPC may be helpful in distinguishing deep sleep stage (with 3.69% accuracy increase) from other stages, but not as good in distinguishing REM stage from other stages. The possible physiological principles behind can be that CPC measures the synchronization between respiration and heart rate which is distinct in cases of NREM and other situations.

Further studies shall be on exploring and developing novel features for distinguishing REM stage from NREM stage, which is essential to quantify the sleep quality.

Limitation of the data sets in DTB-SVM, mainly the size and quality of the data, has posed challenging issue for our further study. There is plan to clinical trial the sleep stage classification method, and to collect more data for further investigation.

4   Conclusion and Future Work

This paper proposes a new sleep stage classification method that makes use of combination of HRV and CPC features to classify sleep data into wake, REM, light and deep stages, using DTB-SVM as classifier. When testing using data from MIT-BIH polysomnographic database, the method using combination of HRV and CPC features achieved an overall accuracy of 96.89% (k = 0.9449), an accuracy improvement of 1.15% than using HRV features only. The accuracy improvement is most prominent in distinguishing light stage from other stages. Future work has been planned to seek for more prominent features for deep sleep stage classification and to test the sleep stage classification method in clinical trials.

References

1. Surantha, N., Kusuma, G.P., Isa, S.M.: Internet of things for sleep quality monitoring system: a survey. In: 2016 11th International Conference on Knowledge, Information and Creativity Support Systems (KICSS), pp. 1–6. IEEE (2016)
2. Sztajzel, J.: Heart rate variability: a noninvasive electrocardiographic method to measure the autonomic nervous system. Swiss Med. Wkly. **134**, 514–522 (2004)
3. Widdicombe, J.G., Sterling, G.M.: The autonomic nervous system and breathing. Arch. Intern. Med. **126**, 311–329 (1970)
4. Cajochen, C., Pischke, J., Aeschbach, D., Borbély, A.A.: Heart rate dynamics during human sleep. Physiol. Behav. **55**, 769–774 (1994)
5. ŽEmaitytė, D., Varoneckas, G., Sokolov, E.: Heart rhythm control during sleep. Psychophysiology **21**, 279–289 (1984)
6. Scholz, U.J., Bianchi, A.M., Cerutti, S., Kubicki, S.: Vegetative background of sleep: spectral analysis of the heart rate variability. Physiol. Behav. **62**, 1037–1043 (1997)
7. Ibrahim, L.H., et al.: Heritability of abnormalities in cardiopulmonary coupling in sleep apnea: use of an electrocardiogram-based technique. Sleep **33**, 643–646 (2010)
8. Yang, A.C., et al.: Sleep state instabilities in major depressive disorder: detection and quantification with electrocardiogram-based cardiopulmonary coupling analysis. Psychophysiology **48**, 285–291 (2011)
9. Scherz, W.D., Fritz, D., Velicu, O.R., Seepold, R., Madrid, N.M.: Heart rate spectrum analysis for sleep quality detection. EURASIP J. Embed. Syst. **2017**, 26 (2017)
10. Klein, A., Velicu, O.R., Madrid, N.M., Seepold, R.: Sleep stages classification using vital signals recordings. In: 2015 12th International Workshop on Intelligent Solutions in Embedded Systems (WISES), pp. 47–50. IEEE (2015)
11. Adnane, M., Jiang, Z., Yan, Z.: Sleep–wake stages classification and sleep efficiency estimation using single-lead electrocardiogram. Expert Syst. Appl. **39**, 1401–1413 (2012)
12. Mendez, M.O., et al.: Sleep staging from heart rate variability: time-varying spectral features and hidden Markov models. Int. J. Biomed. Eng. Technol. **3**, 246–263 (2010)
13. Ichimaru, Y., Moody, G.: Development of the polysomnographic database on CD-ROM. Psychiatry Clin. Neurosci. **53**, 175–177 (1999)
14. Goldberger, A.L., et al.: PhysioBank, PhysioToolkit, and PhysioNet: components of a new research resource for complex physiologic signals. Circulation **101**, e215–e220 (2000)
15. Behar, J., Oster, J., Clifford, G.D.: Combining and benchmarking methods of foetal ECG extraction without maternal or scalp electrode data. Physiol. Meas. **35**, 1569 (2014)
16. Zong, W., Moody, G., Jiang, D.: A robust open-source algorithm to detect onset and duration of QRS complexes. In: Computers in Cardiology, 2003, pp. 737–740. IEEE (2003)
17. Malik, M., et al.: Heart rate variability: standards of measurement, physiological interpretation and clinical use. Task Force of the European Society of Cardiology and the North American Society of Pacing and Electrophysiology. Circulation **93**, 1043–1065 (1996)
18. Shaffer, F., Ginsberg, J.: An overview of heart rate variability metrics and norms. Front. Public Health **5**, 258 (2017)
19. Thomas, R.J., Mietus, J.E., Peng, C.-K., Goldberger, A.L.: An electrocardiogram-based technique to assess cardiopulmonary coupling during sleep. Sleep **28**, 1151–1161 (2005)

sEMG-Based Fatigue Detection for Mobile Phone Users

Li Nie[1], Xiaozhen Ye[1], Shunkun Yang[2], and Huansheng Ning[1(✉)]

[1] Department of Computer and Communication Engineering, University of Science and Technology Beijing, Xueyuan Road, Beijing, China
ninghuansheng@ustb.edu.cn
[2] School of Reliability and Systems Engineering, Beihang University, Beijing, China

Abstract. With the increasing widespread and popularity of internet connected smartphones, more and more people are becoming addicted to their mobile phones which has caused many health problems. Previous studies have proved that surface electromyographic (sEMG) signal can be used to monitor muscle fatigue in different situation such as driving environment or detect some cervical diseases such as muscle chronic pain. It inspired us an objective way to detect the fatigue status of phone users during a prolonged use of mobile phone. In this paper, an experiment was organized to collect phone users' sEMG data and four classifiers were used with multiple sets of features for fatigue detection. Results show that the sEMG signal is an effective measure for detecting users' neck fatigue, while the best classifier that achieved the highest accuracy compared to the other tested classifiers is the support vector machine (SVM).

Keywords: Mobile phone users · sEMG · Fatigue detection

1 Introduction

According to the Internet Report 2019 [1], the number of Internet users worldwide is now more than 50%. Although the development of high-speed mobile Internet brings a lot of convenience, it also brings numerous problems that are becoming alarmingly unignorable. More and more people are getting addicted to their mobile phones. The research of Zheng et al. [2] showed that the excessive use of the Internet has seriously affected individual health in potential and subtle ways. Most of the phone users are unconscious and they still spend a lot of time on their phones which will cause a series of health problems in the long term. Muscle fatigue, especially the fatigue in the neck area, is no doubt a typical health problem for phone users.

Surface electromyography (sEMG) signal is a kind of bioelectrical phenomenon that leads to the change of muscle motor unit potential due to the conduction and diffusion of muscle fiber action potential caused by skeletal muscle excitation. The synthesis of each fiber potential constitutes the action potential of the motor unit. Action potential sequences are superimposed by action potentials, and the sum of these sequences eventually forms sEMG [3]. sEMG signals are characterized by unstable random signals. The generation of sEMG signals is essentially a superposition of a set of signals.

© Springer Nature Singapore Pte Ltd. 2019
H. Ning (Ed.): CyberDI 2019/CyberLife 2019, CCIS 1138, pp. 528–541, 2019.
https://doi.org/10.1007/978-981-15-1925-3_39

A large number of nerve impulses are transmitted to the corresponding nerve center of the cerebral cortex when working or moving. They make it continuously excited and causing a large amount of increase in energy consumption. To avoid overconsumption, the brain produces protective inhibition when it reaches a certain point [4]. As fatigue happens, the central nervous system's firing frequency decreases in order to reduce the energy consumption, resulting in a decrease in the firing frequency of motor neurons [5]. So the mean frequency of muscle electrical signals decrease with the onset of fatigue, which is consistent with most studies.

Many studies [6–10] have shown that monitoring muscle fatigue by sEMG is feasible. Chowdhury et al. [6] recorded that the sEMG signals of superior trapezius muscle and sternocleidomastoid muscle under dynamic repetitive movements. Discrete wavelet transform (DWT) was used to quantitatively evaluate the fatigue of neck and shoulder muscles, and it was concluded that DWT could better predict the fatigue of neck muscles under dynamic exercise. Chen et al. [7] studied the changes of cervical muscle activity under fatigue by surface electromyography. It was concluded that the sternocleidomastoid muscle and the antagonistic muscle were more prone to fatigue during fatigue neck flexion exercise, and the periodicity of sEMG signal increased with the accumulation of fatigue. In Andersen et al. [8] research, using sEMG signals from the neck muscles under various kinds of movements have discovered a number of rehabilitation actions such as side stretch, upright and shrug training on trapezius muscle which have a significant curative effect. Hostent et al. [9] Studied the process of muscle fatigue of long-distance drivers by using sEMG signals. The results showed that when muscle fatigue occurred, the sEMG value increased and the average frequency of sEMG decreased.

Although a lot of researches have been done in this area, most of them focused on dynamic fatigue which makes the subjects of the experiment perform different degrees of motion, with different MVCs, under different loads. Obviously, such experimental requirements are not real-world situations. There are few researches based on static fatigue analysis, especially for mobile phone users. Therefore, this paper mainly focuses on the fatigue problem caused by long use of mobile phone. In this work, we collect EMG neck muscle signals of mobile phone users to develop a real-time fatigue monitoring system. we tried to find the most relevant features and compare different classification models to find the best model with the highest accuracy and performance.

The following of this paper is organized as:

Section 2 is about the experiment for data acquisition including the questionnaire, experiment design and data analysis. Section 3 is the analysis and result of the experiment data including data preprocessing, feature extraction and classification. Then we discuss the method and result of the experiment and make some conclusions in Sects. 4 and 5.

2 Experiment and Data Acquisition

2.1 Questionnaire

We designed a questionnaire to collect phone users' information such as what users often use their phones for, how long they spend on their mobile phones, how they feel when they are fatigue while using mobile phones.

We surveyed 80 people between the ages of 20 and 28. The main target of our survey is young people whose mean age is 24. 36.71% of respondents and 63.29% are female, and 74.03% of them are students. This shows that our main target group is mainly young people on campus. 55.69% respondents use mobile phones from three to seven hours a day during the working day. Whilst the average daily using time on weekends is commonly longer than that on working days. According to the survey, about 89.87% people use more than 3 h a day on weekend. Social networking and video-consumption are two main activities that respondents spent time on their phones. 74.68% respondents sit when they use their phone. The average time that they kept using their mobile was about 117.06 min. Our respondents started to feel uncomfortable after using their phones for about 105.06 min. Long-time use of mobile phones can cause discomfort in the neck, eyes and waist and most phone users suffered from neck discomfort.

2.2 Experiment

8 volunteers between 22 and 28 years old were recruited for experiment. The subjects had no strenuous exercise or physical discomfort before the experiment. Two ZTemg sensors of ZT technology were placed on the left and right sternocleidomastoid muscles [11, 12] of the subject's neck and fixed respectively as shown in Fig. 1. The subjects sat and played with the mobile phone (e.g. using social network apps, watching videos or playing games). The experiment would be stopped when subjects reported they were feeling obvious discomfort for a few seconds.

After the sensors were fixed and the signal became stable, the Borg Scale of Perceived Exertion [13] of the subject was recorded before and after the experiment. Borg Scale of Perceived Exertion also known as Rating of Perceived Exertion Scale (RPE), was invented by Gunnar Borg, a Swedish psychologist. RPE is a widely accepted scale for evaluating and quantifying subjective fatigue. Its numerical range is 0 to 10, which corresponds to different fatigue degree. The higher the score, the fatigue sensation is. The subject was required to rotate his neck at the fastest acceptable speed for relaxation after data collection. The total number of 90 degrees rotations was counted for a period of 30 s and recorded to obtain the level of fatigue. If the rotation frequency became higher

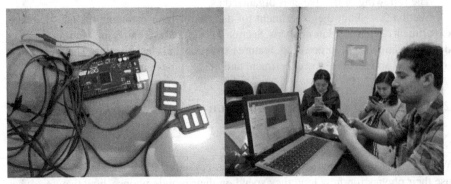

Fig. 1. Arduino and ZTemg sensors (left) and experimental collection process (right)

after the experiment it meant the subject did not experience any significant fatigue. On the contrary, if it became lower it meant the subject had experienced high level of fatigue.

2.3 Dataset

Through the experiment, we finally got dataset described as following:

1. sEMG sensor data of subjects while playing mobile games (see Table 1(a));
2. parameters related to the subjects' specific phone using scenario: subject's posture, apps being used and the experiment's duration (see Table 1(b));
3. The basic information of the subjects such as gender, age, height, weight, waist circumference, BMI, body fat rate, average time spent using mobile phones during working days and weekends, the longest continues time spent playing mobile games one time, posture, the time it takes to start feeling sore and the uncomfortable body part (see Table 1(c)).

According to the procedure of the experiment, it is easy to find that our dataset is unbalanced. The fatigue sEMG data is significantly less than the normal sEMG data. We need to use some methods such as undersampling to deal with unbalanced dataset.

We choose the under-sampling method to build the data of fatigue group and non-fatigue group. In order to achieve the balance between positive and negative sample datasets, we selectively collect a part of the whole non-fatigue data to achieve the similar quantity as the fatigue data. We select 3 s' data from each subject's experimental process data before the end as fatigue group data, select 3 s data from each subject's experiment after the experiment started 5 s as normal data. Every 1 s data as one sample, step length is 100 sampling data points and overlap is 900 sampling data points. After sampling, each kind of dataset has 336 samples.

Table 1(a). sEMG sensor data

Sensor value	Time
557,540	2019-04-26 09:48:32.293
571,541	2019-04-26 09:48:32.293
561,535	2019-04-26 09:48:32.293
566,519	2019-04-26 09:48:32.293
555,508	2019-04-26 09:48:32.293
576,506	2019-04-26 09:48:32.293
561,516	2019-04-26 09:48:32.293
549,528	2019-04-26 09:48:32.293
546,531	2019-04-26 09:48:32.293
525,538	2019-04-26 09:48:32.293

Table 1(b). Experimental parameters

Subject	Behavior	Electrode position	Number of sensors	Experiment start time	Total time of collecting	Posture	RPE before collecting	RPE after collecting	Fatigue location	Fastest acceptable rotation frequency
1	Wechat	Left, right sternocleidomastoid muscles	2	19.20 pm	40 min	Sit	5	7	Eye	60
2	Video	Left, right sternocleidomastoid muscles	2	21.00 pm	55 min	Sit	1	7	Back	20
3	Video	Left, right sternocleidomastoid muscles	2	11.00am	50 min	Sit	1	7	Neck	29
4	Phone game	Left, right sternocleidomastoid muscles	2	20.10 pm	45 min	Sit	1	7	Neck	28
5	Phone game	Left, right sternocleidomastoid muscles	2	15.42 pm	40 min	Sit	1	7	Neck	21
6	Phone game	Left, right sternocleidomastoid muscles	2	16.42 pm	30 min	Sit	1	7	Neck	25
7	Wechat	Left, right sternocleidomastoid muscles	2	19.00 pm	40 min	Sit	1	9	Neck	10
8	Phone game	Left, right sternocleidomastoid muscles	2	20.30 pm	20 min	Sit	5	10	Neck	14

Table 1(c). Basic information of the subjects

Subject	Gender (male-0, female-1)	Age	Height (m)	Weight (kg)	BMI	Waist (cm)	Fat rate	Longest time keep using phone (h)	Common posture using phone	Uncomfortable part of the body using phone long time	Time start to feel fatigue (h)
1	1	28	1.68	59	20.90	78	0.4718983	1.5 h	Sit	Neck	0.7 h
2	0	26	1.8	70	21.60	86	0.6101429	5.5 h	Sit	Neck	1 h
3	1	23	1.66	65	23.59	77	0.7124615	2 h	Sit	Neck	2 h
4	1	26	1.53	58	24.78	79	1.0036897	1 h	Sit	Neck, eye	0.5 h
5	1	24	1.65	53	19.47	70	1.1663019	1 h	Sit	Neck	1 h
6	0	27	1.7	60	20.77	78	1.4476667	2 h	Sit	Neck	0.5 h
7	1	24	1.54	47	19.82	70	1.954383	3 h	Sit	Neck, eye	2 h
8	1	24	1.68	54	19.13	66	1.818	2 h	Sit	Neck, eye	0.7 h

3   Analysis and Results

In this section, we will introduce the process, methods and results of data analysis. The whole process framework of data analysis is shown in the Fig. 2 below.

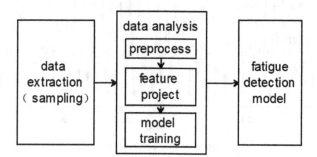

Fig. 2. Data analysis framework.

3.1   Data Preprocessing

Original data collected from the experiment include some known or unknown errors or noise usually. In order to ensure the correctness and validity of the following data analysis, we need to do some work on data preprocessing. Because of the errors of serial port transmission, data may be abnormal such as non-character or values greater than 1023. Since the normal analog output of Arduino ranges from 0 to 1023, we replaced the value that is out of 0 to 1023 by 512 during preprocessing.

According to the manual of the sEMG sensors, the sensor measured value under normal condition is between about 300–700. However, we found that sometimes the sensor values swung wildly between 0 and 1023 because of poor contact problem in the process of experiment, as shown in the Fig. 3(a). The sensor data contains time information which means each sensor data value represents a time. Therefore, we cannot delete data directly. We chose to use moving average method for the elimination of volatility. The comparison of signal before and after error elimination comparison respectively is shown in the Fig. 3(b).

The normal human sEMG signal frequency range is mainly between 5–500 Hz. According to the Nyquist sampling theorem, we used 1000 Hz sampling frequency to satisfy the requirement of the sEMG data acquisition and used the bandpass filter with 5–500 Hz to denoise. The schematic diagram of signal before and after noise removal is shown in the Fig. 3(c).

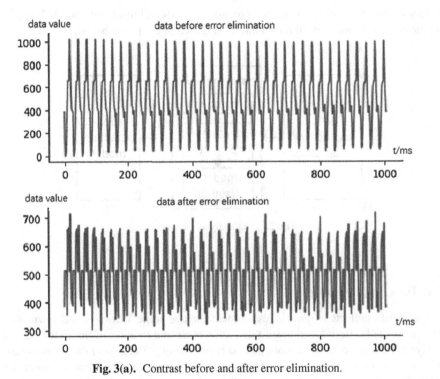

Fig. 3(a). Contrast before and after error elimination.

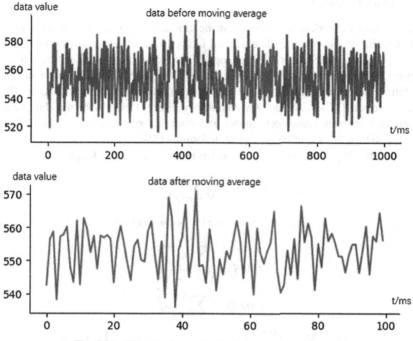

Fig. 3(b). Signal before and after using moving average.

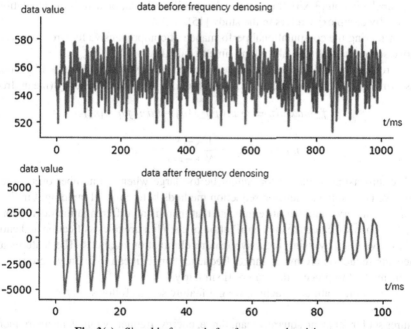

Fig. 3(c). Signal before and after frequency denoising.

3.2 Feature Extraction

In this section we introduce features used in the subsequent classification process. In most of the existing studies on muscle electrical fatigue, integral electromyogram (iEMG), root mean square (RMS), average frequency [14] are often used as indicators to analyze the characteristics of muscle fatigue. Specifically, average frequency is used in most real-time fatigue detection studies. mean (mean), range (maxmin), root mean square (RMS), energy, zero rate (ZCR), Lempel-Ziv complexity (nlz_c) Considering the classification performance, computational performance and other factors, we calculated the following characteristics in this experiment, shown in formula (1)–(7).

We record $x[n] = [x_1, x_2, ..., x_i, ..., x_n]$ as signal sequence of length n:

$$mean = \frac{1}{N} \sum_{i=1}^{N} x_i \tag{1}$$

$$maxmin = max(x[n]) - min(x[n]) \tag{2}$$

$$RMS = \sqrt{\frac{\sum_{i=1}^{N} x_i^2}{N-1}} \tag{3}$$

$$energy = \sum_{i=1}^{N} x_i^2 \tag{4}$$

$$ZCR = \sum \sum_{i=1}^{N} sgn(-x_i, x_{i+1}), \, sgn(x) = \begin{cases} 1, & x > 0 \\ 0, & x \le 0 \end{cases} \tag{5}$$

Lempel-Ziv complexity (nlz_c): For the specific calculation details of calculation of Lempel-Ziv complexity refers to the study [15].

Considering the computational performance of comprehensive features, we need to remove some features that take a long time to calculate.

We record the real part of the sequence after the signal is transformed by Fourier transform as $fq\_x[n]$, then frequency range (f_maxmin), mean frequency (mean_freq):

$$f\_maxmin = max(fq\_x[n]) - min(fq\_x[n]) \tag{6}$$

$$mean\_freq = \frac{1}{N} \sum_{i=1}^{N} i * fq\_x_i \tag{7}$$

The dimension of the vector cannot be too large when the number of samples is small. We need some dimension reduction methods before model building. Firstly, the features with very small variance indicate that there is no differentiation degree for the category. Then we calculate the contribution of each feature to classification and choose the best five ones by using recursive feature elimination methods. The features after dimension reduction are mean, range, frequency range, mean frequency, zero rate. We use this method to process the feature II similarly.

Based on the feature extraction, we got feature sets as follows:

Features set I: root mean square, mean, range, energy, frequency range, mean frequency, zero rate, complexity.

Features set I (less dimensions): mean, range, frequency range, mean frequency, zero rate

Features set II: root mean square, mean, range, zero rate, complexity

The calculation of Features I did not use moving average preprocessing, but only used 5–500 Hz bandpass filtering. The calculation of feature II used the data processed by 5–500 Hz bandpass filtering after using moving average preprocessing, and did not calculate the frequency feature. In the final classification, the best set of features is feature II, its correlation is shown in Fig. 4.

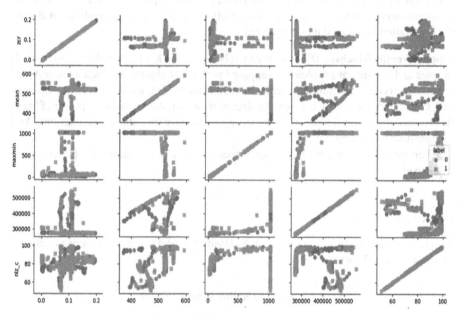

Fig. 4. Correlation of features II.

3.3 Classification

According to the feature extraction in the previous section, we got the feature group extracted from the original sensor data and the experiment dataset. The collected data was divided into training data and testing data. In the first group, we used 15% of the data as the test set to test the classifications. Then we increased the number of test sets, using 25% of the data as the test set as another group.

For classification, we choose classic supervised machine learning models not deep learning ones and unsupervised ones. There were not enough samples to train neural networks or deep learning models, so they were not used here. The situation is same for unsupervised learning. Therefore, we chose some good supervised learning models for classification.

The supervised machine learning models which are very commonly used in the field of pattern recognition are logistic regression, decision trees and SVM [16]. Their axiom involves complex mathematical derivation, we can refer to statistical learning methods. Their advantages and disadvantages are as follows.

Logistic Regression. Its computing cost is not high and easy to understand and implement. But it is easy to produce under-fitting and the classification accuracy is not high usually.

Decision Trees. It is easy to understand and interpret and extract rules. It can deal with both discrete and continuous values. When testing data sets, it runs faster. But it is difficult to process missing data. It is very easy to over-fit, resulting in poor generalization ability. The correlation between data is neglected.

Support Vector Machine (SVM). It can solve machine learning problems with small samples and non-linear problems. Compared with other algorithms such as neural networks, there is no local minimum problem. It can handle high-dimensional data sets very well. Generalization ability is relatively strong. However, the explanatory power of high-dimensional mappings for kernels is not strong, especially for radial basis functions. And it is sensitive to missing data.

Then we chose to use SVM, decision trees and logistic regression for classification respectively.

3.4   Result

The classification accuracy results of each classifier model using two groups of features are shown in the Table 2.

Table 2. Classification accuracy results

Model	SVM		Decision tree		Logistic regression	
test_size	0.15	0.25	0.15	0.25	0.15	0.25
Features I	45.54%	47.62%	50.49%	56.55%	48.51%	45.83%
Features I (less dimension)	42.57%	47.62%	46.53%	58.33%	50.49%	50.60%
Features II	59.41%	85.12%	83.17%	82.14%	48.51%	51.79%

According to the results, some conclusions can be inferred:

In the three groups of features, we can find that the performance of the first group of features is not very good in the four classifiers.

After dimensionality reduction, the classification effect of logistic regression has improved a little, and the classification effect has not improved significantly after dimensionality reduction for other classifier. But the classification accuracy of decision tree and SVM classifier has improved.

The second group of features and the SVM classifier (when we use the poly function as kernel function) have the best classification effect. When the test data is 25%, the classification accuracy is 85.12%.

Actually, the addition of frequency features did not improve the classification performance of the model. It may due to the fact that the frequency features are not processed by some methods such as Fourier transform in the case of spectrum leakage. The second group of features performs a little better than the first group of features. Some features like Lempel-Ziv complexity have poor robustness to noise. The introduction of frequency features does not necessarily improve the accuracy of model classification. These problems deserve further consideration.

4 Discussion

We organized experiment to collect data and constructed data sets. On this basis, we trained and tested four machine learning models and got the accuracy of classification. In the future work, improvements can be considered from the following aspects:

1. On the data level, due to the cost limitation, there are a few volunteers participating in the experiment which means the universality of the results still need to be verified. In order to pursue the user friendliness of the system, sensors with lower cost are selected instead of very expensive physiological instruments which has more uncertainty and instability than compared to the expensive ones.
2. On the experimental level, we used the contacting sEMG sensors in the process of collecting data and the implementation of the whole fatigue monitoring system. Wearing these sensors will bring the users negative experience and need more advanced sensor technology. We can design another experiment that do more data mining and analysis of the correlation between sEMG data and other non-contact user data. Then we can replace sEMG data with non-contact sensor data to improve the user experience.
3. On the data analysis level, more attempts can be made in data analysis methods to pursue higher accuracy and speed obviously.
4. In further work, we will convert the offline classification operation mentioned above into real-time analysis function and build a user-friendly real-time fatigue detection system. We hope that the classification model we have explored can be used to monitor user fatigue in real-time scenarios, so we need a user-friendly real-time fatigue monitoring system. If we only returned a simply fatigue determination result, the data would appear to be less responsive and less lively from the user's perspective. Therefore, our fatigue monitoring system should show to show users other information. Due to the high sampling frequency of sEMG sensors, if we want to give the user feedback about their muscle power waveform information in real time by using the traditional AJAX (Asynchronous JavaScript and XML) method, the browser will start polling data polling continuously which will cause a heavy burden on the server side. The browser will keep requesting and there will be thousands of numerical transmissions per second. It will be very slow and bring very bad experience to the user. Therefore, websocket is used to connect the server and the client to establish

a full-duplex pipeline so that the server can send data to the client at any time once it has data. We can build social network such as anti-fatigue community based on the data collected by the existing system.

5 Conclusion

Based on the sEMG sensor data of mobile phone users during a prolonged phone use, we developed a real-time fatigue detection system. By comparing different feature sets and classification methods, we found the feature II with SVM classifier has the best accuracy at 85.12%. In this case, we introduce these features with superior performance and SVM model to build the real-time monitoring system. Moreover, in order to reduce the cost of real-time system transformation, we choose the cheaper sEMG sensor data and use websocket to transmit the real-time data to improve the user experience.

References

1. 2019 "Internet queen" report essence. http://tech.ifeng.com/a/20190612/45508276_0.shtml. Accessed 12 June 2019
2. Zheng, Y., Wei, D., Li, J., et al.: Internet use and its impact on individual physical health. IEEE Access 4, 5135–5142 (2016)
3. De Luca, C.J.: The use of Surface electromyography in biomechanics. J. Appl. Biomech. 27(6), 724 (1997)
4. Ren, Y., Yang, J., Yin, S., et al.: Study on the relation between fatigue process of localized muscle and change of surface sEMG signal's fractal dimension. J. Biomed. Eng. Res. 23, 215–217 (2004)
5. Zu, X., Li, Y., Zhou, Q.: Evaluation of muscle fatigue based on surface electromyography and subjective assessment. IFMBE Proc. 39, 2003–2006 (2013)
6. Chowdhury, S.K., Nimbarte, A.D., Jaridi, M.A.: Discrete wavelet transform analysis of surface electromyography for the fatigue assessment of neck and shoulder muscles. J. Electromyogr. Kinesiol. 23(5), 995–1003 (2013)
7. Chen, Q., Ma, J., Wang, J.: The regularity of SEMG characteristics of neck muscles under fatiguing contracting condition. J. Beijing Sport. Univ. 33(9), 52–55 (2010)
8. Hostens, I., Ramon, H.: Assessment of muscle fatigue in low level monotonous task performance during car driving. J. Electromyogr. Kinesiol. 15(3), 0–274 (2005)
9. Andersen, L.L., Kjaer, M., Andersen, C.H., et al.: Muscle activation during selected strength exercises in women with chronic neck muscle pain. Phys. Ther. 88(6), 703 (2008)
10. Wang, L., Fu, R., Zhang, C., et al.: Biomechanics based investigation on the relation between index Q and cervical muscle fatigue. Chin. J. Sci. Instrum. 38, 878–885 (2017)
11. Bernhardt, P., Wilke, H.J., Wenger, K.H., et al.: Multiple muscle force simulation in axial rotation of the cervical spine. Clin. Biomech. 14(1), 32–40 (1999)
12. Jr, N.J., Sherk, H.H.: Biomechanical evaluation of the extensor musculature of the cervical spine. Spine 13(1), 9–11 (1988)
13. Hummel, A., Laubli, T., Pozzo, M., et al.: Relationship between perceived exertion and mean power frequency of the EMG signal from the upper trapezius muscle during isometric shoulder elevation. Eur. J. Appl. Physiol. 95(4), 321–326 (2005)

14. Liu, L., Zou, R., Zhang, D., et al.: Research and development trend of feature extraction methods of surface electromyographic signals. Prog. Biomed. Eng. **3**, 164–168 (2015)
15. Zhang, Y., Zou, J., Ma, J.: Damage assessment method for rolling bearings combined CEEMD with Lempel-Ziv complexity. Mech. Sci. Technol. Aerosp. Eng. **37**, 1408–1414 (2018)
16. Mitchell, T.M.: Machine Learning. McGraw-Hill Education, New York (1997)

Research on the Effect of Video Games on College Students' Concentration of Attention

Zhixin Zhu and Jiansheng Li[✉]

School of Education Science, Nanjing Normal University, Nanjing 210097,
Jiangsu, People's Republic of China
zhuzhixin@snnu.edu.cn, 42056@njnu.edu.cn

Abstract. Video games have become the main entertainment tool for people, occupying a lot of our time and attention, which is a limited resource and plays an important role in the organization and maintenance of intellectual activities. In the study, the influence of video games on college students' attention was explored from two aspects (the game content and the time of game play) through the questionnaire survey and experimental measurement. A total of 310 persons participated in the survey, 27 of whom were involved in experiment. Results are given as follows. The more video games played, the harder attention is to focus. Also, non-violent video games have a negative short-term effect on players' attention and reduced their attention level. But in the long term, non-violent video games can increase the females' attention and reduce the males' attention. Violent games affect players' attention whether in the short term or in the long term. And playing games with high frequency will bring negative effect to players' attention concentration, no matter non-violent or violent video games, no matter male or female players.

Keywords: Video games · Attention · Brain wave

1 Introduction

At present, the development of the Internet and the expansion of the information volume have become an inevitable trend. Network information occupies a large amount of attention of people, and attention is a limited resource [1]. Attention is the taking possession by the mind, in clear and vivid form, of one out of what seem several simultaneously possible objects or trains of thought. It implies withdrawal from some things in order to deal effectively with others [2]. The concentration of attention mentioned here refers to the concentration of college students' attention in the process of learning, which also means the time that attention is maintained on something.

The results of existing researches on the effect of video games on attention are more complicated. One of the research themes is the effect of video games on visual space attention. Researchers found that people who often play games perform better on visual space attention than those who have not been exposed to video games. They can track targets faster, more easily detect changes of objects stored in visual short-term memory,

© Springer Nature Singapore Pte Ltd. 2019
H. Ning (Ed.): CyberDI 2019/CyberLife 2019, CCIS 1138, pp. 542–549, 2019.
https://doi.org/10.1007/978-981-15-1925-3_40

transit tasks more flexibly and obtain more effective mental rotation [3–6]. Among them, Green and Bavelier found that playing video games has increased the player's selective attention [6].

Spence [7] found that video games, especially motion training games, have an improvement in visual spatial resolution, promote the distribution of attention and broaden the breadth of attention. The study on the performance of participants with different genders in the game shows that although mental rotation skills of women have a great difference from that of men before training, the difference is obviously narrowed after training. Towards the effect of violent content in video games on attention bias, researchers found that players who are addicted to online games have problem with attention bias [8] and so do players with violent game experience [9].

Our study divides video games into two categories: violent video games and non-violent video games. As to the former, we use screen violence and content violence as variables. The individual elements of the study include gender, the time of game play and education background. The research question is the effect of video games on college students' attention concentration.

2   Method

Two methods: questionnaires and experiments were used in our study. Specially, we investigated the players' game usage and attention via questionnaires, on basis of which participants were selected for experiments.

393 questionnaires were collected, including 323 online questionnaires (310 valid) and 70 paper questionnaires (69 valid). On the basis, 27 participants were selected for the experiment, including 12 males and 15 females, with an average age of 24 years old. Participants are all right-handed with normal color vision, and their naked vision or correct vision is above 1.0.

The experiment instruments included: (1) video games questionnaire designed by Anderson and Dill, which contains 40 items, such as names, types and violence level of games played more often, the frequency of the game play and so on; (2) Adult ADHD Self-Report Scale. Diagnostic criteria for attention in the scale are in accordance with Diagnostic and Statistical Manual of Mental Disorders-IV [10]; (3) video games ("Lianliankan" and "CS"); (4) 16-channel physiological signal recorder, collecting the data of the alpha, beta, theta and SMR; (5) the exam material including 10 questions about Two-digit addition and subtraction.

3   Data Analysis

3.1   The Effect of Non-violent Video Games on Attention Concentration is Related to Players' Gender

We first focus on non-violent games, allowing participants to play non-violent games for more than 30 min. Then 15-min and 30-min brainwave data was collected for analysis. The results are as follows.

From Figs. 1 and 2, for both males and females, the amplitude of the alpha wave measured before game is higher than that measured after 15-min game play or 30-min game play. But the amplitude of the alpha wave measured after playing for 15 min is little different from that measured after 30-min game play. Also, players' attention is more concentrated after playing games for 15 min or 30 min, but 15-min game play and 30-min game play have little different effect on attention concentration since the amplitude of the alpha wave is inversely proportional to the concentration of attention, which means the lower the amplitude of the alpha wave, the more concentrated players' attention.

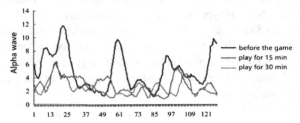

Fig. 1. The alpha wave of females in violent game group

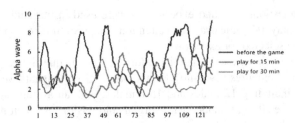

Fig. 2. The alpha wave of males in violent game group

Fig. 3. Theta/SMR of males and females in non-violent game group

As to female players, the value of theta/SMR decreases after they play games for 15 min and the value increases after they play for 30 min; but for male players, the value increases after they play for 15 min and the value drops significantly after playing for 30 min (See Fig. 3). It can be concluded that the influence of non-violent video games on players' concentration of attention differs in genders according to the negative correlation between the level of attention concentration and the value of theta/SMR, which is similar to the existing analysis result [11].

3.2 The Short-Term Effect of Violent Video Games on Attention Concentration: The Reduction of Attention Concentration

The amplitude of the alpha wave measured after 15-min game play or 30-min game play is higher than that before the game play, which indicates that the players' attention concentration decreases (see Fig. 4). Also, we found that the players' concentration of attention is more affected, comparing the males' value of theta/SMR of non-violent game group to that of violent game group.

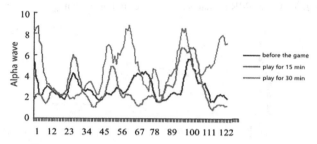

Fig. 4. The alpha wave of males in violent game group

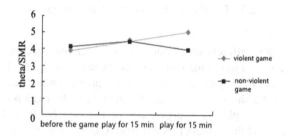

Fig. 5. Theta/SMR of males in violent game group and non-violent game group

From Fig. 5, the value of theta/SMR measured after 15-min game play or 30-min game play is higher than that before the game. Also, the longer the game time, the higher the value, which means players' concentration of attention after game play is lower than that before the game. And as the game time goes on, the influence is more obvious. It is completely different from the change of males' attention concentration in the non-violent game group.

3.3 The Long-Term Effect of Video Games on Attention Concentration: Related to Game Types and Players' Gender

According to the content of games and the time of game play, participants were divided into 6 groups: short time of non-violent game group, medium time of non-violent game group, long time of non-violent game group, short time of violent game group, medium time of violent game group and long time of violent game group. Then we compared the value of theta/SMR of all the groups. Results are shown in Figs. 6 and 7.

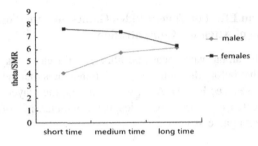

Fig. 6. Theta/SMR of males and females in non-violent game group

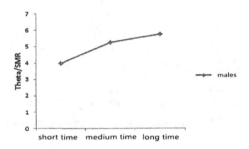

Fig. 7. Theta/SMR of males in violent game group

Video games have different effects on players' attention concentration in gender after playing games for a long time with high frequency (see Fig. 6). Besides, females who play non-violent video games for a long time have a lower theta/SMR ratio than females who play for a short time. That is, playing non-violent video games for a long time will improve female players' concentration of attention; but for males, the result is quite the opposite. In addition, theta to SMR ratio of males who play non-violent video games for a short time is much lower than that of males who play for a long time, which indicates that non-violent video games have an inhibiting effect on males' concentration of attention. However, it is contrary to the result analyzed from Fig. 4, so it can be proved that the conclusion analyzed from Fig. 4 is incorrect. Perhaps the reason is that both male and female participants are under experiment implication and focus on the experiment as much as possible when collecting EEG data, which is unavoidable in the laboratory.

We just considered the influence of violent video games on males' concentration of attention, due to no female participants that suitable for playing violent games. Figure 7 shows that males who play games for a long time have a higher ratio of theta to SMR than males who play for a short time. That is, playing violent video games for a long time will inhibit players' concentration of attention, and the longer the game time, the greater the inhibition, which may cause more problems on the attention. It is also consistent with results of some existing researches.

3.4 The More Often You Play Non-Violent Video Games, the Lower Your Concentration of Attention is

Before the analysis, it should be ensured that there is no significant difference in participants' education background, gender and the violence of game played. The result of One-Samples T Test indicates that there is non-significant difference in the violence of game ($p = 0.981 > 0.05$). Then according to the frequency of game play from low to high, participants were divided into 3 groups: low frequency group, medium frequency group and high frequency group. Results are shown in Fig. 8.

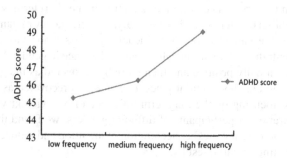

Fig. 8. ADHD score for different frequency of game play

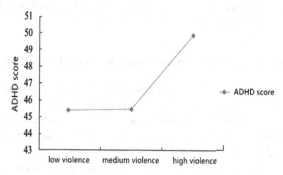

Fig. 9. ADHD score for games with different violence

It shows that the players' ADHD score increases as the frequency of game play increases, which indicates that the frequency of game play is negatively correlated with players' concentration of attention. That is, the higher the frequency of game play, the worse the players' concentration of attention.

3.5 The More Violent Video Games You Play, the Lower Your Concentration of Attention Is

The same analysis as the above, there is no significant difference in the frequency of game play of non-violent game group ($p = 1.000 > 0.05$). Then according to the violence of the game from low to high, participants were divided into 3 groups: low violence group, medium violence group and high violence group. Results are shown in Fig. 9.

4   Discussion and Conclusion

4.1   Non-violent Video Games and Players' Concentration of Attention

From above, non-violent video games have an effect on player concentration of attention with gender differences. The alpha wave showed that non-violent video games have a negative effect on players' concentration of attention in the short term with a decline in players' concentration of attention. Also, the analysis on frequency of game play support it. In addition, Fig. 3 revealed the specific influencing process of non-violent video games on players' concentration of attention, which have difference in gender. Specifically, the effect on males is initially positive and then partially recovered as the game time goes on; but the effect on females is initially negative and then recovered as the game time goes on. Besides, focusing on the long-term influence of non-violent video games on attention concentration of participants of different genders, we found that non-violent video games can increase females' concentration of attention and reduce males', which is the same as existing research results.

The results can be explained by displacement hypothesis theory. The displacement hypothesis means that video games take up the time participants spend on other activities, as a result, their ability of self-control and attention concentration decreased without being exercised. But playing video games is also an activity, which has a positive effect within a proper range or brings negative results beyond that.

4.2   Violent Video Games and Players' Concentration of Attention

Violent video games have a negative effect on players' concentration of attention. From the alpha wave, we can conclude that playing violent video games inhibits players' concentration of attention. Also, the ratios of theta and SMR also showed that the influence of violent video games on players' attention concentration in the short term is negative, and the negative effect increases as the game time increases.

However, as to the effect of non-violent video games on players' concentration of attention, it is not simple to explain that the frequency of game play and violence of games are negatively correlated with the concentration of attention. Data from EEG revealed that the effect of non-violent video games on players' attention concentration is complicated but simple straight line in the short term, but in the long term, it has positive effect on females' and negative effect on males'. Results of questionnaire analysis are the same as the long-term effect analyzed above, that is, the higher the frequency of the game play, the greater the negative effect on players' concentration of attention.

Also, many researches found that males are eager to use computers and the Internet, and people addicted to Internet are mainly male [12–14], which is similar to the use of video games. Either the non-violent game group or the violent game group, the frequency of game play of male players is much higher than that of female players, and so is the violence of games played. Violent video games always have richer graphics, powerful sound effects, complex scenes and task of appropriate difficulty, which increases male players' interest and makes them excited and play frequently. However, excessive game play will increase players' attention to the game and lead to some problems on their concentration in some things which do not seem to be attractive like learning because video games with such high appeal will change players' excitement for other things with less attraction to which players will pay little attention. But female players' interest is different from male players', which leads to different results of the effect of video games on attention concentration in learning of players of different genders.

Acknowledgements. This work was supported by the National Social Science Foundation of China (13BRK026).

References

1. Zhang, B., Tang, X.Y.: Research on the current situation of attention distribution in college students' network use. Sci. Technol. Inf. **17**, 50–51 (2013)
2. James, W.: The Principles of Psychology, vol. 1. Henry Holt & Company, New York (1890)
3. Achtman, R.L., Green, C.S., Bavelier, D.: Video games as a tool to train visual skills. Restorative Neurol. Neurosci. **26**(4–5), 435–446 (2003)
4. Boot, W.R., Kramer, A.F., Simons, D.J., et al.: The effects of video game playing on attention, memory, and executive control. Acta Psychological **129**(3), 387–398, 1–10 (2008)
5. Green, C.S., Bavelier, D.: Action video game modifies visual selective attention. Nature **423**(6939), 534–537 (2003)
6. Green, C.S., Bavelier, D.: Action-video-game experience alters the spatial resolution of vision. Psychol. Sci. **18**(1), 88–94 (2007)
7. Spence, I.: Video games and spatial cognition. Rev. Gen. Psychol. **14**(2), 92–104 (2010)
8. Zhang, Z.J., Zhao, J.B., Zhang, F., Du, K.L., Yuan, D.: Heavy online game user's attention bias and ERP patterns. Chin. J. Appl. Psychol. **14**(4), 291–296 (2009)
9. Zhen, S.J., Xie, X.D., Hu, L.P., Zhang, W.: The influence of violent video games on attention bias. J. South China Normal Univ. (Soc. Sci. Ed.) **2**, 67–73 (2013)
10. Guan, F.C.: Research on the relationship between the attention and inhibition function of ADHD and the visual cognitive processing of high school students. M.A. Thesis. Central China Normal University, Wuhan (2015)
11. Tahiroglu, A.Y., Celik, G.G., Avci, A., et al.: Short-term effects of playing computer games on attention. J. Attention Disord. **13**(6), 668–676 (2009)
12. Bayraktar, F., Gün, Z.: Incidence and correlates of Internet usage among adolescents in North Cyprus. Cyberpsychol. Behav. **10**(2), 191–197 (2007)
13. Griffiths, M.: Internet addiction. Psychologist **12**, 246–251 (1999)
14. Tahiroglu, A.Y., Celik, G.G., Uzel, M., et al.: Internet use among Turkish adolescents. Cyberpsychol. Behav. Impact Internet Multimedia Virtual Reality Behav. Soc. **11**(5), 537–543 (2008)

Study on Cardiovascular Disease Screening Model Based-Ear Fold Crease Image Recognition

Xiaowei Zhong(✉)

Changshu Health Information Center, Changshu 215500, Jiangsu, China
zhongxiaoweijkcs@163.com

Abstract. Cardiovascular disease screening is an effective means to effectively control the incidence of cardiovascular disease. The earlobe crease is an important marker for identifying cardiovascular diseases and can be used as an important sign for cardiovascular disease screening. Through the one-click uploading of human ear-based photos, the analysis of human ear crease based on image recognition is carried out, and the medical staff's initial screening of cardiovascular disease assessment for some people is transformed into intelligent primary screening for cardiovascular disease in the city.

Keywords: Ear lobe crease · Image recognition · Cardiovascular disease · Primary screening

With the development of social economy and the continuous improvement of medical technology, Changshu, an economically developed area in southern Jiangsu, which has officially entered an aging society. According to the statistics of Changshu population health, as of the end of 2018, the number of elderly people over 60 years old in the city has reached 331,300, which accounts for 31.02% of the city's registered population. The absolute value of the elderly population ranks first in all districts and cities in Suzhou, and the elderly population shows a trend of large base, fast growth and high age. According to the "China Cardiovascular Disease Report 2018" published by the National Cardiovascular Center, cardiovascular deaths accounted for the first cause of total deaths in urban and rural residents, 45.50% in rural areas and 43.16% in cities [1]. Cardiovascular diseases in Changshu City accounted for 30.32% of the total cause of death, and have become the most important health killer in Changshu residents. Because cardiovascular disease has the characteristics of high incidence, high disability rate, high mortality, high recurrence rate and many complications, prevention of cardiovascular disease is particularly important. In 2014, the Chinese Medical Association Health Management and the Editorial Board of the Chinese Journal of Health Management jointly issued the "China Health Checkup Population Health Risk Self-Measurement Form" [2], which provides the standard subjective self-test questionnaire tool for conducting cardiovascular screening of physical examination population. However, the initiative of the general population to carry out self-tests on health risks is not strong, and it is often necessary to complete self-measurement under the guidance of medical personnel. Therefore, a preliminary screening model in the paper is established for cardiovascular disease based

H. Ning (Ed.): CyberDI 2019/CyberLife 2019, CCIS 1138, pp. 550–556, 2019.
https://doi.org/10.1007/978-981-15-1925-3_41

on image recognition of earlobe creases. On the basis of the risk, the patient is guided to follow-up screening and intervention measures.

1 Medical Image Recognition Technology

1.1 Image Recognition Technology

The general development of image recognition has gone through three stages: text recognition, digital image recognition, and object recognition. The study of character recognition can be traced back to the 1950s and has been widely used. The research of digital image recognition began in the 1960s. Digital images have great advantages such as storage, transmission, and processing convenience, which provide a powerful driving force for the development of image recognition technology. Object recognition is based on digital image recognition, and image segmentation is a key technique in image processing. In recent years, research on computational techniques and VLSI technology based on image segmentation methods based on histograms and wavelet transforms has developed rapidly. Combined with the research direction of artificial intelligence, systems science and other disciplines, research on image processing has made great progress, and its research results are widely used in various industries.

1.2 Medical Image Recognition Technology

According to medical application classification, medical image recognition is mainly divided into radiology imaging, surgical guidance, pathological recognition, facial recognition, etc., involving medical image segmentation, image fusion, image registration, image reconstruction and other fields. threshold method, region growing method, pattern recognition method, artificial neural network method, variable model method, wavelet analysis are mainly applied in the field of the medical image segmentation algorithm; image fusion is a plurality of images of one or more medical imaging devices. The processing technology that comprehensively forms an image is mainly divided into single-mode fusion, multi-mode fusion, and template fusion; image registration usually puts together multiple images examined by the same patient for quantitative analysis, so that a medical image undergoes spatial transformation. Spatially consistent with corresponding points on another medical image, obtaining information about the medical image changes of the patient, thereby improving the level of medical diagnosis and treatment; image reconstruction generally refers to three-dimensional reconstruction of medical images by passing CT, MR, etc. 2D images are used to create 3D stereoscopic images through 3D visualization technology, providing doctors with a basis for pre-operative or accurate diagnosis.

2 The Relationship Between Earlobe Crease and Cardiovascular Disease

The relationship between earlobe creases and cardiovascular disease was first discovered in 1973 by American doctor Frank [3]. Therefore, the earlobe crease is medically known as the Frank sign, which starts from the tragus and ends at the posterior edge of the auricle. The diagonal at a 45-degree angle appears on one side or both sides of the earlobe, as shown in Fig. 1.

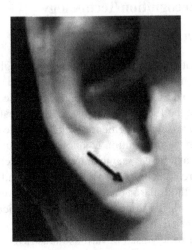

Fig. 1. Lobe crease

The earlobe is the lowest part of the human's ears, and it is also the most sensitive part. With the increase of age and high blood pressure, smoking, obesity, diabetes and other factors, the content of collagen and elastic fibers in the earlobe is reduced, resulting in earlobe creases. A study by the Heart Research Center in Copenhagen, Denmark, found that about 30% of men aged 30 or older had ear lobe creases, while men over 70 years of age increased to 50% [4]. Shmilovich et al. used computed tomography to assess the relationship between coronary artery disease and earlobe crease, and found that 71% of patients with coronary artery disease had elastic fiber tear and elastin degeneration [5]. In recent years, Japanese studies show that people who do not have coronary heart disease have a higher probability of dying from coronary heart disease each year than those without wrinkles if they have earlobe folds. By the research, Japanese men with metabolic syndrome and earlobe creases, shorter telomeres were found in these male peripheral blood leukocytes, and shorter telomeres were a sign of aging and accelerated atherosclerosis [6]. The largest study data published in 2014 confirmed that earlobe creases are a danger sign of cardiovascular disease [7]. Therefore, earlobe creases can predict the incidence of all-cause mortality and cardiovascular events [8], and can be used as an important sign of population census [9].

3 Human Ear Image Recognition Technology

Biometrics technology has developed rapidly in recent years, especially in fingerprint recognition and face recognition. Biometrics technology is widely used in various industries. As the biological representation of the human body, the human ear's structure is stable and will not change significantly with age. Therefore, the research on human ear recognition has been paid more and more attention [10]. Because of the balanced color distribution of human ears, the small size and the few calculation steps, the human ears recognition can be used as a mature biometric recognition technology [11, 12].

3.1 Human Ear Recognition Based on Convolutional Neural Network

The advantage of convolutional neural networks over traditional artificial neural networks is that they can extract more diverse high-dimensional features in the image. In the network structure, neurons in each level are not fully connected, but at the same time weight parameters are shared. The structure of the convolutional neural network mainly includes the input layer, the convolution layer, the pooling layer and the fully connected layer [13]. The purpose of the convolution operation is to extract various levels and unknown features in the image, and to extract more complete features through training. The calculation expression is as follows.

$$x_j^l = f \sum_{i \in M_j} x_i^{l-1} \times k_{ij}^l + b_j^l$$

In the formula, M_j is the set of input feature maps; k is the convolution kernel; l is the number of layers; b is the offset value. If the convolutional layer of the network has m filters $(w_1, \ldots, w_m)$, the size of the human ear sample is $n \times n$, and the image of the human ear is input, and m feature maps are generated after the filtering operation [14]. The principle of convolutional neural network mode can be used to extract the local texture features of human ear images and improve the recognition rate of human ear features.

3.2 Feature Fusion and Sparse Representation of Human Ear Recognition

The Sobel operator is applied to detect edges in four directions: horizontal, vertical, and two diagonals. Features are extracted on each edge map, and the human ear texture features in four directions are extracted using a gray level co-occurrence matrix. The human ear is classified and identified by edge features, texture features, and by sparse representation models. Assuming that the original image is A, the four operators is applied to the original image to obtain the edge intensity maps Gx, Gxy, Gy, Gyx in the four directions (0°, 45°, 90°, 135°). The calculation expression is as follows.

$$G_x = \begin{bmatrix} 1 & 0 & -1 \\ 2 & 0 & -2 \\ 1 & 0 & -1 \end{bmatrix} * A$$

$$G_{xy} = \begin{bmatrix} 0 & -1 & -2 \\ 1 & 0 & -1 \\ 2 & 1 & 0 \end{bmatrix} * A$$

$$G_y = \begin{bmatrix} -1 & -2 & -1 \\ 0 & 0 & 0 \\ 1 & 2 & 1 \end{bmatrix} * A$$

$$G_{yx} = \begin{bmatrix} 2 & 1 & 0 \\ 1 & 0 & -1 \\ 0 & -1 & -2 \end{bmatrix} * A$$

Where * denotes a two-dimensional convolution operation. Usually the area with higher edge strength and density on the edge map is the human ear feature [15]. Through experiments, the combination of features (edge features and texture features) combined with sparse representation models can improve the accuracy of human ear recognition.

3.3 Human Ear Recognition Based on Gabor Wavelet and Double Layer LLE

The two-dimensional Gabor wavelet transform actually convolves the signal to be processed with a set of Gabor filters, and the result of the convolution can be used to represent or approximate the signal [16, 17]. The human ear image is convoluted with the Gabor filter to obtain filtering results of different scales and different directions. After the above-mentioned Gabor wavelet transform, the human ear image obtains the Gabor amplitude map, and the corresponding amplitude is obtained from each pixel in the map. These amplitudes indicate the energy distribution of the pixel in different frequency domains. The Gabor amplitude map is arranged in rows as a high-dimensional row vector, which is the Gabor amplitude eigenvector.

As shown in Fig. 2, there are three types of samples, A, B, and C. When the original LLE algorithm is used to select the neighbors of the center point of one of the samples, the other two types of points may be selected as their neighbors. However, after the dimension reduction of the erroneous neighbors, the low-order spatial neighbor property is unchanged, which reduces the correctness of the classifier and ultimately affects the accuracy of the recognition.

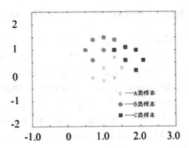

Fig. 2. Schematic diagram of three types of samples

The two-layer LLE algorithm first calculates the category center of each class, then calculates the Euclidean distance between the sample point and the center of the same category, and takes the data point with the smallest Euclidean distance in the same category as the neighbor point. In theory, the higher recognition can be obtained. rate. The fusion of Gabor wavelet and double-layer LLE makes the available information of human ear recognition feature richer, so that higher recognition rate can be obtained [18].

4   Identification of Earlobe Creases

On the basis of human ear recognition, a sub-image of the earlobe crease is obtained. Firstly, what you need to do is to locate the earlobe endpoint. Next, the crease inclination and the rotation center point of the image are obtained by the endpoint coordinates, so as to further geometrically adjust the image. Finally, the adjusted image [19] is intercepted as required to achieve recognition of the earlobe crease.

5   Conclusion

In 2016, Suzhou City issued the "Notice on Implementing the "531" Action Plan for Healthy Citizens in Suzhou City". In accordance with the concept of "emergency and acute illness must be accurate", we will explore innovative medical and health service mechanisms, enhance the professional prevention and treatment capabilities for major diseases, and actively promote the healthy Suzhou 2020 strategy. It is required to do a good job in screening and comprehensive intervention of high-risk groups of cardio-vascular diseases, and comprehensively reduce the harm of cardiovascular diseases to residents. However, traditional cardiovascular disease screening often uses the model of voluntary and doctoral screening for the general population, and the actual screen-ing population is limited. Changshu City, in conjunction with the "Changshu Wisdom Health" WeChat public account, opens the "Calcium Screening" column, provides pho-tos based on human ear photographs and one-click uploading of human ear creases, which can realize intelligent screening of cardiovascular diseases. On the basis of the preliminary screening, the people are guided to do a cardiovascular scale screening or medical advice, so as to expand the screening of cardiovascular diseases for some people to carry out intelligent screening of cardiovascular diseases for residents in the city.

References

1. Lee, J.S., Kuo, Y.M., Chung, P.C.: The adult image identification based on online sampling. In: Neural Networks, IJCNN 2006 (2006)
2. Takeda, M., Uchida, S., Hiramatsu, K., et al.: Finger image identification method for personal verification. In: Proceedings of the 10th International Conference on Pattern Recognition. IEEE (1990)
3. Frank, S.T.: Aural sign of coronary artery disease. N. Engl. J. Med. **289**(6), 327–328 (1973)
4. Kang, E.H., Kang, H.C.: Association between earlobe crease and the metabolic syndrome in a cross-sectional study. Epidemiol. Health **34**, e2012004 (2012)
5. Shmilovich, H., Cheng, V.Y., Rajani, R., et al.: Relation of diagonal ear lobe crease to the presence, extent, and severity of coronary artery disease determined by coronary computed tomography angiography. Am. J. Cardiol. **109**(9), 1283–1287 (2012)
6. Higuchi, Y., Maeda, T., Ouan, J.Z., et al.: Diagonal earlobe crease are associated with shorter telomere in male Japanese patients with metabolic syndrome. Circ. J. **73**(2), 274–279 (2009)
7. Christoffersen, M., Frikke-Schmidt, R., Schnohr, P., et al.: Visible age-related signs and risk of ischemic heart disease in the general population: a prospective cohort study. Circulation **129**(9), 990–998 (2014)

8. Korkmaz, L., Agac, M.T., Acar, Z., et al.: Earlobe crease may provide predictive information on asymptomatie peripheral arterial disease in patients clinically free of atherosclerotic vascular disease. Angiology **65**(4), 303–307 (2014)

9. Shi, W., Caballero, J., Huszár, F., et al.: Real-Time Single Image and Video Super-Resolution Using an Efficient Sub-Pixel Convolutional Neural Network (2016)

10. Burge, M., Burger, W.: Ear biometrics for machine vision. In: 21st Workshop of the Austrian Association for Pattern Recognition, pp. 275–282 (1997)

11. Yuan, L., Mu, Z.C.: Ear recognition based on local information fusion. Pattern Recogn. Lett. **33**(2), 182–190 (2012)

12. Caselles, V., Catté, F., Coll, T., et al.: A geometric model for active contours in image processing. Numer. Math. **66**(1), 1–31 (1993)

13. Yamamoto, S., Tanaka, I., Senda, M., et al.: Image processing for computer-aided diagnosis of lung cancer by CT (LSCT). Syst. Comput. Jpn. **25**(2), 67–80 (1994)

14. Hall, E.L.: Computer image processing and recognition. Proc. IEEE **69**(9), 1169–1170 (1980)

15. Giger, M., Macmahon, H.: Image processing and computer-aided diagnosis. Radiol. Clinics North Am. **34**(3), 565–596 (1996)

16. Vural, E., Frossard, P.: Learning pattern transformation manifolds for classification. In: Image Processing, Orlando, FL, 30 September–3 October 2012, pp. 1165–1168 (2012)

17. Roweis, S.T., Saul, L.K.: Nonlinear dimensionality reduction by locally linear embedding. Science(S0036-8075) **290**(5500), 2323–2326 (2000)

18. Hwang, K., Fu, K.S.: Integrated computer architectures for image processing and database management. Computer **16**(1), 51–60 (1983)

19. Sonka, M., Fitzpatrick, J.M., Masters, B.R.: Handbook of medical imaging, volume 2: medical image processing and analysis. Opt. Photonics News **13**(6), 50–51 (2002)

Author Index

Printed in the United States
By Bookmasters